Nuclear & High Energy Physics

Nuclear & High Energy Physics

Editor:

Olga Moreira, Ph.D.

ArclerPress

www.arclerpress.com

Nuclear & High Energy Physics

Editor: Olga Moreira, Ph.D.

708 3rd Avenue, 6th Floor
New York
NY 10017
United States of America
www.arclerpress.com

ISBN: 978-1-68094-434-1

Library of Congress Control Number: 2016951378

About the Editor

Olga Moreira, Ph.D.

Olga Moreira obtained her Ph.D. in Astrophysics from the University of Liege (Belgium) in 2010, her BSc. in Physics and Applied Mathematics from the University of Porto (Portugal). Her post-graduate travels and international collaborations with the European Space Agency (ESA) and European Southern Observatory (ESO) led to great personal and professional growth as a scientist. Currently, she is working as an independent researcher, technical writer, and editor in the fields of Mathematics, Physics, Astronomy and Astrophysics.

Contents

List of Contributors

Sinya Aoki
Graduate School of Pure and Applied Sciences, University of Tsukuba, Tsukuba, Ibaraki 305-8571, Japan
Center for Computational Sciences, University of Tsukuba, Tsukuba, Ibaraki 305-8577, Japan

Takumi Doi
Theoretical Research Division, Nishina Center, RIKEN, Wako, Saitama 351-0198, Japan

Tetsuo Hatsuda
Theoretical Research Division, Nishina Center, RIKEN, Wako, Saitama 351-0198, Japan
IPMU, The University of Tokyo, Kashiwa, Chiba 277-8583, Japan

Yoichi Ikeda
Department of Physics, Tokyo Institute of Technology, Meguro, Tokyo 152-8551, Japan

Takashi Inoue
Nihon University, College of Bioresource Sciences, Fujisawa, Kanagawa 252-0880, Japan

Noriyoshi Ishii
Center for Computational Sciences, University of Tsukuba, Tsukuba, Ibaraki 305-8577, Japan

Keiko Murano
Theoretical Research Division, Nishina Center, RIKEN, Wako, Saitama 351-0198, Japan

Hidekatsu Nemura
Center for Computational Sciences, University of Tsukuba, Tsukuba, Ibaraki 305-8577, Japan

Kenji Sasaki
Center for Computational Sciences, University of Tsukuba, Tsukuba, Ibaraki 305-8577, Japan

W. Nazarewicz
Department of Physics and Astronomy, University of Tennessee Knoxville, Tennessee 37996, USA
Oak Ridge National Laboratory, Oak Ridge, Tennessee 37831, USA
Faculty of Physics, University of Warsaw, ul. Ho˙za 69, 00-681 Warsaw, Poland

P.-G. Reinhard
Institut f¨ur Theoretische Physik, Staudtstr. 7 D-90158 Universit¨at Erlangen/N¨urnberg, Erlangen, Germany

W. Satula
Faculty of Physics, University of Warsaw, ul. Ho˙za 69, 00-681 Warsaw, Poland

D. Vretenar
Physics Department, Faculty of Science, University of Zagreb, Zagreb, Croatia

Zbigniew Sosin
M. Smoluchowski Institute of Physics, Jagiellonian University, Lojasiewicza 11, 30-348 Krak´ow, Poland

Jan Blocki
NCBJ, Theoretical Physics Division (BP2), Swierk, Poland ´

Jinesh Kallunkathariyil
M. Smoluchowski Institute of Physics, Jagiellonian University, Lojasiewicza 11, 30-348 Krak´ow, Poland

Jerzy Lukasik
IFJ PAN, Krak´ow, Poland

Piotr Pawlowski
M. Smoluchowski Institute of Physics, Jagiellonian University, Lojasiewicza 11, 30-348 Krak´ow, Poland
IFJ PAN, Krak´ow, Poland

David Foster
School of Physics, HH Wills Physics Laboratory, University of Bristol, Tyndall Avenue, Bristol BS8 1TL, UK

Nicholas S. Manton
Department of Applied Mathematics and Theoretical Physics, University of Cambridge, Wilberforce Road, Cambridge CB3 0WA, UK

Jun Xu
Shanghai Institute of Applied Physics, Chinese Academy of Sciences, Shanghai 201800, China
Kavli Institute for Theoretical Physics China, Chinese Academy of Sciences, Beijing 100190, China

Bao-An Li
Department of Physics and Astronomy, Texas A&M University–Commerce, Commerce, TX 75429-3011, USA

Wen-Qing Shen
Shanghai Institute of Applied Physics, Chinese Academy of Sciences, Shanghai 201800, China

Yin Xia
Shanghai Institute of Applied Physics, Chinese Academy of Sciences, Shanghai 201800, China
University of Chinese Academy of Sciences, Beijing 100049, China

A. A. Godizov
SRC Institute for High Energy Physics of NRC "Kurchatov Institute", 142281 Protvino, Russia

U. U. Abdurakhmanov
Institute for Physics and Technology, Fizika-Solntse Research and Production Association, Uzbek Academy of Sciences, Bodomzor yoli street 2, 100084 Tashkent, Uzbekistan

K. G. Gulamov
Institute for Physics and Technology, Fizika-Solntse Research and Production Association, Uzbek Academy of Sciences, Bodomzor yoli street 2, 100084 Tashkent, Uzbekistan

V. Sh. Navotny
Institute for Physics and Technology, Fizika-Solntse Research and Production Association, Uzbek Academy of Sciences, Bodomzor yoli street 2, 100084 Tashkent, Uzbekistan

A. A. Godizov
Institute for High Energy Physics, NRC "Kurchatov Institute", 142281 Protvino, Russia

Adam J. Freese
Florida International University, Miami, FL 33199, USA

Misak M. Sargsian
Florida International University, Miami, FL 33199, USA

Mark I. Strikman
Pennsylvania State University, University Park, PA 16802, USA

J. Schneitzer
Institut für Theoretische Physik, Goethe-Universität Frankfurt am Main, 60438 Frankfurt am Main, Germany

T. Wolkanowski
Institut für Theoretische Physik, Goethe-Universität Frankfurt am Main, 60438 Frankfurt am Main, Germany

F. Giacosa
Institut für Theoretische Physik, Goethe-Universität Frankfurt am Main, 60438 Frankfurt am Main, Germany
Institute of Physics, Jan Kochanowski University, 25406 Kielce, Poland

Hiroyuki Koura
Advanced Science Research Center, Japan Atomic Energy Agency, 2-4 Shiakata-shirane, Tokai-mura, Naka-gun, Ibaraki 319-1195, Japan

J.J. Kolata
Physics Department, University of Notre Dame, Notre Dame, IN, 46556-5670, USA

V. Guimar˜aes
Instituto de F`ısica, Universidade de S˜ao Paulo, Rua do Mat˜ao, 1371, 05508-090, S˜ao Paulo, SP, Brazil

E.F. Aguilera
Departamento de Aceleradores, Instituto Nacional de Investigaciones Nucleares, Apartado Postal 18-1027, C´odigo Postal 11801, M´exico, Distrito Federal, Mexico

S. Oberstedt
European Commission, Joint Research Centre IRMM, 2440 Geel, Belgium

R. Billnert
European Commission, Joint Research Centre IRMM, 2440 Geel, Belgium
Fundamental Fysik, Chalmers Tekniska H¨ogskola, 41296 G¨oteborg, Sweden

F.-J. Hambsch
European Commission, Joint Research Centre IRMM, 2440 Geel, Belgium

M. Lebois
Institut de Physique Nucleaire Orsay, 91406 Orsay, France

A. Oberstedt
Fundamental Fysik, Chalmers Tekniska H¨ogskola, 41296 G¨oteborg, Sweden
Ossolution Consulting, 70353 Orebro, Sweden

J.N. Wilson
Institut de Physique Nucleaire Orsay, 91406 Orsay, France

Boris Pritychenko
National Nuclear Data Center, Brookhaven National Laboratory, Upton, NY 11973-5000 USA

P. G. Giannaka
Division of Theoretical Physics, University of Ioannina, 45100 Ioannina, Greece

T. S. Kosmas
Division of Theoretical Physics, University of Ioannina, 45100 Ioannina, Greece

Gary A. Wegner
Department of Physics and Astronomy, Dartmouth College, 6127 Wilder Laboratory, Hanover, NH 03755, USA

Roberto Onofrio
Dipartimento di Fisica e Astronomia 'Galileo Galilei', Università di Padova, Via Marzolo 8, 35131 Padova, Italy
ITAMP, Harvard-Smithsonian Center for Astrophysics, 60 Garden Street, Cambridge, MA 02138, USA

S. Bilenky
Joint Institute for Nuclear Research, Dubna, R-141980, Russia
TRIUMF 4004 Wesbrook Mall, Vancouver BC, V6T 2A3 Canada

Michele Frigerio
Laboratoire Charles Coulomb, UMR 5221 (CNRS/Université Montpellier 2), 34095 Montpellier, France

Carlos E. Yaguna
Institut für Theoretische Physik, Universität Münster, Wilhelm-Klemm-Straße 9, 48149 Münster, Germany

G. Dufour
Laboratoire Kastler-Brossel, CNRS, ENS, Collège de France, UPMC, Campus Jussieu, 75252 Paris, France

D. B. Cassidy
Department of Physics and Astronomy, University College London, Gower Street, London WC1E 6BT, UK

P. Crivelli
ETH Zurich, 8093 Zurich, Switzerland

P. Debu
Institut de Recherche sur les lois Fondamentales de l'Univers, CEA Saclay, 91191 Gif-sur-Yvette, France

A. Lambrecht
Laboratoire Kastler-Brossel, CNRS, ENS, Collège de France, UPMC, Campus Jussieu, 75252 Paris, France

V. V. Nesvizhevsky
Institut Max von Laue-Paul Langevin, 6 rue Jules Horowitz, 38042 Grenoble, France

S. Reynaud
Laboratoire Kastler-Brossel, CNRS, ENS, Collège de France, UPMC, Campus Jussieu, 75252 Paris, France

A. Yu. Voronin
P.N. Lebedev Physical Institute, 53 Leninsky Prospekt, 117924 Moscow, Russia

T. E. Wall
Department of Physics and Astronomy, University College London, Gower Street, London WC1E 6BT, UK

Harald Fritzsch
Physik-Department, Universitat M¨ unchen, 80333 Munich, Germany ¨
Institute for Advanced Study, Nanyang Technological University, 50 Nanyang Avenue, Singapore 639798

Joan Solà
Institute for Advanced Study, Nanyang Technological University, 50 Nanyang Avenue, Singapore 639798
High Energy Physics Group, Department of ECM and Institut de Ciencies del Cosmos, Universitat de Barcelona, ` Avinguda Diagonal 647, Catalonia, 08028 Barcelona, Spain

Preface

This book is a collection of 20 contemporary research articles which offer an overview of the recent progress in the modelling of fundamental interactions, nuclear structures, stellar nucleosynthesis and non-baryonic matter. The first part of the book is focused mainly on hadron interactions and nuclear structure modelling. It starts with a review of lattice quantum chromodynamics (QCD) methods for the study of hadron interactions, then, proceeds to more specific topics such as the study of the symmetry energy in nuclear density functional theory; the alpha-cluster model of nuclei; spin dynamics and heavy-ion reactions at low and intermediate energies. The second part of the book is focused on the domain of nuclear astrophysics. It reviews the tools, methods and data available for the study of stellar nucleosynthesis, including the calculation of electron capture cross sections. It continues further by discussing the use of neutron stars for determining bounds on the Higgs-Kretschmann coupling, neutrino oscillations experiments, reviewing sterile neutrinos as a potential solution to neutrino masses, the baryon asymmetry, and the dark matter problems. Finally, the last two chapters provide a status of the ongoing experiments and future prospects of observing gravitational quantum states as well as a discussion on how, at the Planck scale, the unification of quantum haplodynamics (QHD), chromodynamics, and electrodynamics couplings could explain the dark energy of the universe.

Editor
Olga Moreira, Ph.D.

INTRODUCTION

High-energy nuclear physics studies the behavior of nuclear matter in energy regimes typical of high energy physics. The primary focus of this field is the study of heavy-ion collisions, as compared to lower atomic mass atoms in other particle accelerators. At sufficient collision energies, these types of collisions are theorized to produce the quark–gluon plasma. In peripheral nuclear collisions at high energies one expects to obtain information on the electromagnetic production of leptons and mesons which are not accessible in electron-positron colliders due to their much smaller luminosities.

Previous high-energy nuclear accelerator experiments have studied heavy-ion collisions using projectile energies of 1 GeV/nucleon up to 158 GeV/nucleon. Experiments of this type, called "fixed target" experiments, primarily accelerate a "bunch" of ions (typically around 10^6 to 10^8 ions per bunch) to speeds approaching the speed of light (0.999c) and smash them into a target of similar heavy ions. While all collision systems are interesting, great focus was applied in the late 1990s to symmetric collision systems of gold beams on gold targets at Brookhaven National Laboratory's Alternating Gradient Synchrotron (AGS) and uranium beams on uranium targets at CERN's Super Proton Synchrotron.

Currently, high-energy nuclear physics experiments are being conducted at Brookhaven National Laboratory's Relativistic Heavy Ion Collider (RHIC) and in CERN's new Large Hadron Collider. The four primary experiments at RHIC (PHENIX, STAR, PHOBOS, and BRAHMS) study collisions of highly relativistic nuclei. Unlike fixed target experiments, collider experiments steer two accelerated beams of ions toward each other at (in the case of RHIC) six interaction regions. At RHIC, ions can be accelerated (depending on the ion size) from 100 GeV/nucleon to 250GeV/nucleon. Since each colliding ion possesses this energy moving in opposite directions, the maximum energy of the collisions can achieve a center of mass collision energy of 200GeV/nucleon for gold and 500GeV/nucleon for protons.

History

The exploration of hot hadron matter and of multiparticle production has a long history initiated by theoretical work on multiparticle production by Enrico Fermi in the USA, and Lev Landau in the USSR. These efforts paved the way to the development in the early sixties of the statistical bootstrap model description of hadron production by Rolf Hagedorn.

Traditional nuclear physics has been devoted to the study of nuclei which are gently perturbed. Using high energy beams of heavy nuclei (ions), we can create states of nuclear matter that are far removed from the ground state. For example, at sufficiently high densities and temperatures, neutrons and protons should "melt" into their constituent quarks and gluons, forming the so-called quark-gluon plasma (QGP). The density or temperature for this to happen is predicted by Quantum Chromodynamics (the fundamental theory governing the strong interaction) to be about 1 GeV/fm^3 (10^{12}Kg/cm^3) or 170 MeV (10^{12} °K). The QGP state is believed to have existed for the first 10 μ sec of the universe after the Big Bang. The QGP may exist in the core of neutron stars. However, the only viable way to recreate the QGP in today's laboratory is to collide heavy nuclei at relativistic energies.

Creating and studying the QGP state will provide fundamental information about Nature. The Purdue High Energy Nuclear Physics (HENP) group is in the forefront of this exciting area of research. We create and study the QGP in experiments conducted at the Relativistic Heavy Ion Collider (RHIC) at the Brookhaven National Laboratory (BNL) and the Large Hadron Collider (LHC) at CERN. Heavy-ion beams are accelerated to nearly the speed of light. A collision of these beams produces thousands of particles. We measure those particles with the STAR (Solenoidal Tracker At RHIC) and CMS (Compact Muon Solenoid at LHC) experiments. From the measurements we investigate the state of matter formed in the collisions between the high energy heavy ion beams.

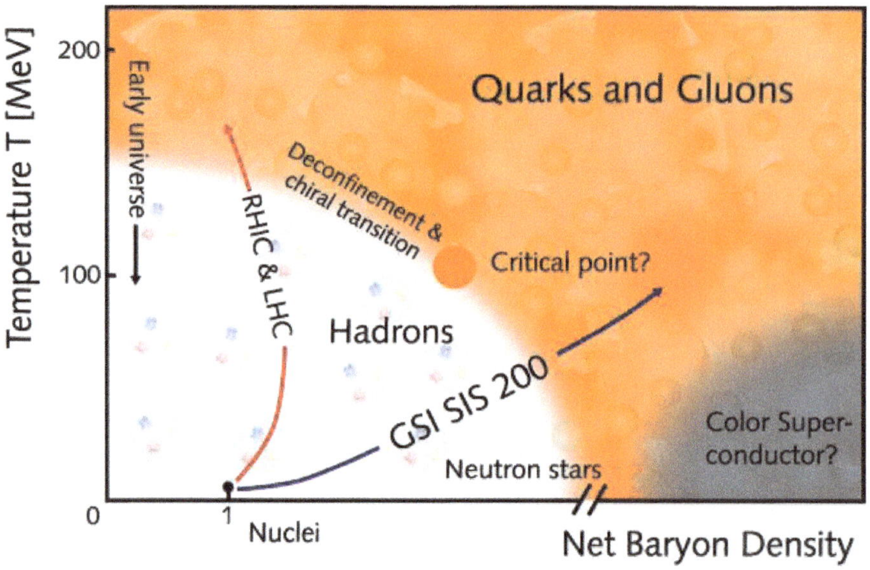

In contrast to earlier beliefs, the QGP created in relativistic heavy ion collisions is strongly interacting. It behaves like a nearly perfect liquid where its shear viscosity to entropy density ratio is close to the conjectured quantum lower limit of $1/4\pi$. The created QGP can be described by macroscopic thermodynamic quantities and hydrodynamics. The QGP is sufficiently opaque to energetic gluons and light and heavy quarks due to their strong interactions with the QGP, so that high momentum quarks and gluons are strongly attenuated. We are a world leading group in the measurements of collective hydrodynamic flow and jet correlations arising from energetic gluons and light and heavy quarks. The experimental effort is supplemented by the High Energy Nuclear Theory group. The theory group provides critical inputs to the interpretation of experimental data using Quantum Chromodynamics and hydrodynamics.

Particle Physics

Particle physics is the study of fundamental particles and the forces that guide them. Because many of the fundamental particles only appear during relativistic collisions in particle accelerators, colloquially called "atom smashers," particle physics is also known as "high-energy physics". Physicists have been slamming together particles at extreme speeds since 1929.

The best picture of particle physics we have today is called the Standard Model, developed painstakingly in the 1970s. It was a reaction to the "particle zoo", a huge proliferation of unusual fundamental particles discovered during high-energy physics experiments in the 1950s and 1960s. The final particle count ended up to be around 31, including 24 fermions (quarks, electrons, neutrinos, and their antiparticles), 6 bosons (one of which, the graviton, has yet to be observed), and one elusive particle responsible for the property of mass itself, which also has yet to observed, the Higgs boson. Basically, the fermions make up matter and the bosons mediate interactions between matters. The light coming from your computer screen is made up of photons which are bosons. They are interacting with the fermions making up your eyeball.

Most of the matter all around us is composed of only a few fundamental particles: up quarks, down quarks, and electrons. There are also 50 trillion low-mass neutrinos streaming through our bodies every second, passing through the entire Earth almost as if it wasn't even there. Neutrinos, whose name means "little neutral particle," are so elusive that it wasn't even known that they had a mass until 1998. One of the newest areas of astronomy is known as neutrino astronomy, whereby neutrino flux from the Sun and supernovae are observed using huge detectors.

Although only a few of particles in the particle zoo make up the matter we are familiar with, particle physics gives us great insight into the structure of reality by showing us the less common variants and how they fit together into a unified family. Particle physics can be said to be responsible for the existence of nuclear power, nuclear medicine, and nuclear bombs. Particle physics is considered one of the most respected areas of science because it ends up producing insights which are useful to other areas, sort of like math.

Fluence

Fluence is a flux measurement in physics for calculating the number of elementary particles in a square area, or the amount of energy in that area in values of particles per square meter or joules of energy per square meter. It is often used in the fields of nuclear medicine and space travel to determine the amount of ionizing radiation delivered to human beings or equipment, known as radiation dosimetry. When talking about fluence, the positively-charged particles that make up atomic nuclei, known as protons, and the energy of negatively-charged particles of orbiting electrons are considered most important.

To be accurate in three dimensions, fluence calculations are actually based on the number of particles or amount of energy that falls on the surface area of a predefined sphere within a specific time. Proton fluence is another term for particle fluence, and is often called particle flux density, which is measured in particles per square meter multiplied by a time value. Energy fluence from electron-based ionizing radiation is measured in joules per square meter multiplied by a time value.

In space research, the fluence rate is calculated to determine the amount of ionizing radiation that energetic heavy nuclei contain, known as HZE particles. These particles are the primary component of cosmic rays that can lead to carcinogenic effects in human tissue with significant exposure levels. While the magnetic field of the Earth shields orbiting spacecraft within it from some of this cosmic ray damage, in deep space or for prolonged missions farther from Earth, exposure levels will be much higher. Fluence rates by square area for radiation exposure are also seen as more accurate than typical radiation dose level calculations, as research using lab animals has shown that heavy ions, such as those of iron, promote tumor expression as much as lighter ones made from lanthanum or niobium.

Neutron fluence is also an important measure in science, which is used to calibrate equipment for its level of radiological protection as well as to calibrate radiation dose meters. The rate for neutron flow is an important element of nuclear particle colliders in high-energy physics research. Research facilities such as the Large Hadron Collider (LHC) near Geneva, Switzerland, can have neutron fluence energy levels of up to 1,000,000 electron volts. Neutron energy levels are also an important component of nuclear weapons research and production, nuclear fission power generation, and astronomical research into supernova stars.

Routine exposure to ionizing radiation does not merely occur in narrow research areas of physics or space travel, however. Fluence rates are also a significant factor in air travel, as passengers and crew on commercial aircraft are exposed to more radiation from space than they would be on the ground. Photon fluence is another more routine exposure method for non-ionizing radiation that poses less risks to human health, and is calculated for fields such as laser dentistry.

Nuclear Physics

Nuclear physics is a branch of the physics field which is concerned with the structure of atomic nuclei, and the understanding of potential ways in which to manipulate atomic nuclei. This branch of physics dates to the early 20th century, when scientists began to realize that the atom had a structure, and that understanding this structure could be important. The most infamous application of nuclear physics was probably the development of the atom bomb in the 1940s, but the field has many more applications, including highly beneficial ones.

Some people group particle physics in with nuclear physics, although technically particle physics should be viewed as its own field. Particle physics is an offshoot of this field, but particle physicists take their research in very different directions, and work with different kinds of natural phenomena.

One of the most diverse applications of nuclear physics is in medicine. The work of nuclear physicists is behind a number of medical imaging techniques which are used to provide non-invasive looks into the body. Radioactive isotopes utilized in the treatment of some medical conditions such as cancer are also the product of research in nuclear physics, with physicists learning about such isotopes and how they can be applied safely and effectively to address medical problems.

Certain aspects of engineering require a knowledge of nuclear physics, most notably in nuclear engineering, a field which involves the development of nuclear power plants which can do anything from generating electricity to powering submarines. Radiocarbon dating, a technique used in geology and archeology, is also a product of nuclear physics. People in this profession may also work with astronomers, using their knowledge to help date the universe, explain physical phenomena, and design experiments.

Nuclear Engineering

Nuclear engineering is a field in which nuclear engineers work with radiation and radioactive materials to design processes and develop their use in power generation, industry, and nuclear medicine. Those who work in the nuclear engineering industry solve real-world problems like the management of nuclear waste, response protocols for nuclear power plant accidents, and how radiation can effectively treat cancer. People who work in the nuclear engineering industry have undergone extensive training and have a high aptitude for advanced mathematics, chemistry and physics. There are many different specialized areas within nuclear engineering, including the generation of nuclear power, medical use of using nuclear power, and the environmental impact of nuclear by-products.

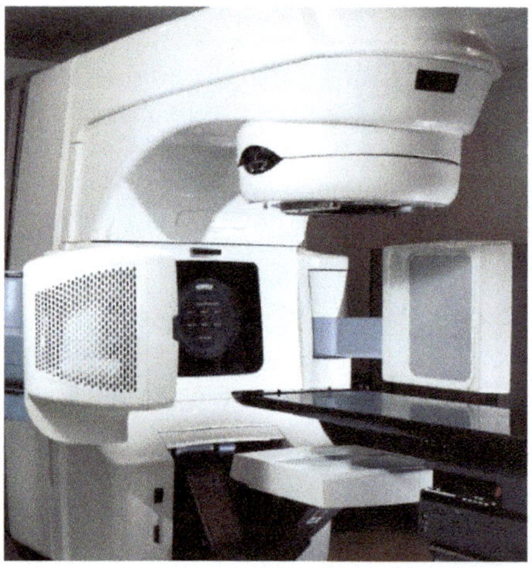

Since the late 19th century, engineers have imagined all sorts of uses for the power stored in atoms, including the use of steam, produced by nuclear power, to turn turbines and produce electricity. In 1957, the first full-scale nuclear power plant began operation. In 1984, the amount of electricity generated by nuclear power surpassed that generated by hydro-electric power, and in 2011, there are 109 nuclear power plants in the US alone which generate 610 billion kilowatt hours of electricity.

In the world of medicine, nuclear power is used in applications ranging from magnetic resonance imaging (MRI), the treatment of cancer, and surgery using a nuclear gamma knife. While early types of radiation therapy caused unnecessary damage to surrounding tissues, doctors can now accurately place the precise amount of radiation needed to kill tumors without harming other vital tissues and organs. The focused beams of radiation used in the gamma knife allow surgeons to perform a non-invasive medical procedure, treating tumors deep within the body. Other nuclear medical devices include stereotactic devices, laser beams, and linear accelerators which target and treat patients without surgery. One-third of all medical procedures done in the US use some type of technique developed by nuclear engineering.

Preserving the environment also an important part of nuclear engineering. There are many national and international agencies like the Nuclear Regulatory Commission (NRC) and the International Atomic Energy Agency (IAEA) that are tasked with keeping a close eye on the use and disposal of nuclear materials. Organizations like the Nuclear Energy Institute protect the environment by working with other organizations to develop methods and procedures for the safe handling of the hazardous nuclear waste produced by nuclear power generation. Other nuclear engineers are working to design safer nuclear power plant equipment, developing new safety protocols to prevent and handle accidental contamination, and finding new methods to dispose of nuclear waste.

Nuclear Reaction

A nuclear reaction is a process that occurs when the nucleus of an atom loses subatomic particles to the degree that its properties are altered. The original atom of an element that undergoes a nuclear reaction can either become a different isotope, or variety, of the same element or it can become a different element entirely. Nuclear reactions are closely related to radiation in general, which can occur spontaneously outside of a reaction. Radiation simply describes the process of energy or articles being emitted from an atom or other particle. The term nuclear reaction, however, usually specifically refers to a situation in which the nuclei of two atoms collide and alter the properties of at least one of the nuclei.

A nuclear reaction can occur in many different forms, each giving considerably different results. In a fission reaction, a large and often unstable particle, either spontaneously or as the result of a collision, splits apart into two different particles. The opposite occurs in a fusion nuclear reaction: two smaller particles collide and their nuclei combine to form a larger particle. Fusion reactions occur naturally in stars, but most human attempts to effectively and efficiently control them have failed. In a spallation reaction, a nucleus is struck with enough momentum to dislodge several neutrons or protons, thereby reducing the atomic weight of the particle.

Fission nuclear reactions are used in nuclear reactors to produce usable energy. Unstable particles collide and split apart, generating a significant amount of kinetic and thermal energy. This energy can be harvested by the nuclear reactor and put to use for human concerns. There is significant interest in the use of fusion reactions to generate power, as they tend to release a considerable amount of energy. Unfortunately, fusion reactions are extraordinarily difficult to control--they occur naturally in the high-pressure, high-energy conditions present on stars, and such conditions are very difficult to replicate.

There are several different kinds of particles that are commonly emitted from nuclei during a nuclear reaction. Alpha particles are essentially the same as the nuclei of helium atoms and are composed of two neutrons and two protons bound together. Beta particles are simply electrons; they have a much smaller mass and a negative charge. Neutrons are also released in nuclear reactions; they are very penetrating because they have a neutral charge so there are few forces that prevent them from passing through various substances, including human skin. Gamma rays are rays that leave the nucleus in the form of pure energy; they are also highly penetrating and can pass through almost anything because of their nonexistent mass and neutral charge.

Nuclear Fusion

Nuclear fusion is the process by which multiple atoms having the same charge join together in order to form a heavier nucleus. In some cases, depending on the mass, energy can be released or absorbed during this process. It is a very important energy source. Although it's often confused with nuclear fission, likely because of the similar name and — sometimes — similar results, nuclear fusion is a very different process. Nuclear fission involves splitting the nucleus of an atom, which normally has the effect of generating a large release of energy, such as is seen in nuclear bombs. This is also how most nuclear power plants generate energy. Nuclear fusion as a source of manmade energy is still largely in the developmental stage, though some fusion power plants are online. Most of the energy produced this way that benefits humans and other life forms comes from the sun. Fusion is the process by which all stars generate energy.

The problem with generating nuclear fusion lies in getting two atoms having the same charge close to each other because these atoms generally repel each other rather than moving together. Once brought together, however, nuclear force begins to take over. This force will attract the nuclei of two or more atoms toward each other and start the fusion process, but this happens only if they are in close enough proximity.

To get the nuclei close enough together to fuse requires finding some way to get the atoms together. In most cases, this is accomplished with very high levels of heat. Heat causes the atoms to speed up, allowing them to overcome their electromagnetic tendencies to repel each other. Though this can require a great deal of energy in the beginning, the energy that results is often much greater than what is initially put in. In some cases, the heat from the explosion can create ongoing nuclear fusion reactions, called self-sustaining reactions.

SECTION I: MANY-BODY INTERACTIONS AND NUCLEI MODELS

CHAPTER 1

LATTICE QUANTUM CHROMODYNAMICAL APPROACH TO NUCLEAR PHYSICS

Sinya Aoki, Takumi Doi, Tetsuo Hatsuda, Yoichi Ikeda, Takashi Inoue, Noriyoshi Ishii, Keiko Murano, Hidekatsu Nemura, and Kenji Sasaki

INTRODUCTION

One of the ultimate goals in nuclear physics is to describe hadronic many-body problems on the basis of hadronic S-matrices calculated from first-principles quantum chromodynamics (QCD). In particular, baryon-baryon interactions are the most fundamental quantities: once they are obtained from QCD, one can solve finite nuclei, hypernuclei, nuclear matter, and hyperon matter by employing various many-body techniques developed in nuclear physics. Phenomenological nucleon–nucleon (N N) potentials, which are designed to reproduce a large amount of proton–proton and neutron–proton scattering data as well as deuteron properties, were constructed in the 1990s and are called high-precision N N potentials. Some examples are shown in Fig. 1; these examples reflect the characteristic features of the N N interaction for different values of the relative distance r as reviewed in [1–5]: the long range part of the N N force ($r > 2$ fm) is dominated by the one-pion exchange originally introduced by Yukawa [10]. Since the pion is the Nambu–Goldstone boson associated with spontaneous breaking of chiral symmetry, it couples to the nucleon's spin–isospin density and leads to not only the central force but also the tensor force. The medium range part (1 fm $< r < 2$ fm) of the N N force receives significant contributions from twopion ($\pi\pi$) exchange [11] and/or heavy meson (ρ, ω, and σ) exchanges. In particular, the spin–isospin independent attraction of about 50–100 MeV in this region plays an essential role in binding the atomic nuclei and nuclear matter. The short range part ($r < 1$ fm) of the N N force is best described by the phenomenological repulsive core introduced by Jastrow [12]. Nuclear saturation, nuclear shell structure, nuclear superfluidity, and the structure of neutron stars are all related to the properties of the nuclear force [13–15]. Furthermore, hyperon–nucleon (Y N) and

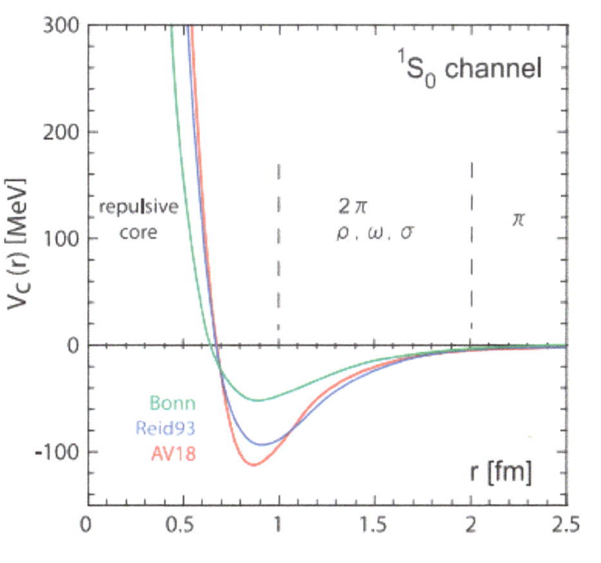

Figure. 1: Three examples of the modern N N potential in the $^1S_0$ (spin-singlet and S-wave) channel: Bonn [6], Reid93 [7], and Argonne v_{18} [8]. Taken from Ref. [9].

hyperon–hyperon (Y Y) forces, information on which is still quite limited experimentally, are crucial to understand the structure of hypernuclei and the cores of neutron stars. Three-nucleon forces (and three-baryon forces in general) are also

Citation: Sinya Aoki, Takumi Doi, Tetsuo Hatsuda, Yoichi Ikeda, Takashi Inoue, Noriyoshi Ishii, Keiko Murano, Hidekatsu Nemura, and Kenji Sasaki, Lattice quantum chromodynamical approach to nuclear physics, DOI: 10.1093/ptep/pts010.

important to understand the binding energies of finite nuclei and the equation of state of dense hadronic matter. It has been a long-standing challenge in theoretical particle and nuclear physics to extract hadron–hadron interactions from first principles. A framework suitable for such a purpose in lattice QCD was first proposed by Lüscher [16]: For two hadrons in a finite box with a size L × L × L under periodic boundary conditions, an exact relation between the energy spectra in the box and the elastic scattering phase shift at these energies has been derived. If the range of hadron interaction R is sufficiently smaller than the size of the box R < L/2, the behavior of the two-particle Nambu–Bethe–Salpeter (NBS) wave function ϕ(r) in the interval R < |r| < L/2 is sufficient to relate the phase shift and the two-particle spectrum. This method, Lüscher's finite volume method, bypasses the difficulty in treating the real-time scattering process on the Euclidean lattice. Furthermore, it utilizes the finiteness of the lattice box effectively to extract information on the on-shell scattering matrix and the phase shift.

A closely related but new approach to hadron interactions from lattice QCD has been proposed recently by three of the present authors [9,17,18] and has been developed extensively by the HAL QCD Collaboration. (The approach is therefore now called the HAL QCD method.) Its starting point is the same NBS wave function ϕ(r) as discussed in Ref. [16]. Instead of looking at the wave function outside the range of the interaction, the authors consider the internal region |r| < R and define an integral kernel (or the non-local "potential" in short) U(r,r) from ϕ(r) so that it obeys the Schrödinger type equation in a finite box. This potential can be shown to be energy-independent by construction. Since U(r,r) in QCD is localized in its spatial coordinates due to confinement of quarks and gluons, it receives only a weak finite volume effect in a large box. Therefore, once U is determined and is appropriately extrapolated to L → ∞, one may simply use the Schrödinger type equation in infinite space to calculate the scattering phase shifts and bound state spectra to compare the results with experimental data. Since U is a smooth function of the quark masses, it is relatively easy to handle on the lattice. This is in sharp contrast to the scattering length, which shows singular behavior in the quark mass corresponding to the formation of the hadronic bound state. A further advantage of the HAL QCD method is that it can be generalized directly to many-body forces and also to the case of inelastic scattering.

Studying the structure of S(strangeness) = −1 and S = −2 hypernuclei is one of the key challenges in modern nuclear physics. Also, central cores of neutron stars will have hyperonic matter if neutron beta-decays to hyperons become possible at high density. The hyperon–nucleon (Y N) and hyperon–hyperon (Y Y) interactions are crucial to determine the level structures of hypernuclei as well as the onset-density of hyperonic matter in neutron stars [19]. By generalizing the N N scattering in the flavor SU(2) space to baryon–baryon (B B) scatterings in the flavor SU(3) space, the HAL QCD method can give Y N and Y Y potentials as natural extensions of N N potentials. Such extensions are also useful for identifying the origin of the short-range repulsive core of the N N potential and for studying possible S = −2 six-quark states, such as the H-dibaryon. In this article, we review the basic ideas and recent progress of the HAL QCD method to hadron interactions. (As for Lüscher's finite volume method, see [20] for a recent review.) In Sect. 2, the basic strategy for defining the N N potential in QCD is explained. In Sect. 3, we introduce lattice formulations of the time-independent HAL QCD method originally proposed in Refs. [9,17,18], as well as its time-dependent generalization. In Sect. 4, some recent results of lattice QCD calculations for the N N potential are given in both quenched and full QCD. The magnitude of the non-locality in U is also discussed in this section. In Sect. 5, the method is applied to hyperon–nucleon interactions such as Nand N systems. In Sect. 6, interactions between octet baryons are investigated in the flavor SU(3) limit, where up, down, and strange quark masses are all equal. In Sect. 7, a generalization of the HAL QCD method to the case of inelastic scattering is given. In Sect. 8, we show the results of the three-nucleon potential, especially its short distant structure. In Sect. 9, an application to kaon–nucleon scattering is considered. Section 10 is devoted to a summary and concluding remarks.

DEFINING THE POTENTIAL IN QCD

Nambu–Bethe–Salpeter (NBS) Wave Function

A key quantity to define the baryon–baryon (BB) "potential" in QCD is the equal-time Nambu–Bethe–Salpeter wave function,

$$\varphi^W(\boldsymbol{x})e^{-Wt} = \langle 0|T\{B(\boldsymbol{r}+\boldsymbol{x},t)B(\boldsymbol{r},t)\}|2B, W, s_1 s_2\rangle, \tag{2.1}$$

where $|2B, W, s_1 s_2\rangle$ is a QCD eigenstate for two baryons with equal mass m_B, helicity s_1, and s_2, total energy $W = 2\sqrt{k^2 + m_B^2}$, relative momentum k, and total momentum p (we take p = 0 in this paper). Generalization to the unequal mass can be formulated in a similar manner. In the case of two nucleons, the local interpolating operator B(x) is taken as

$$B_\alpha(x) \equiv \begin{pmatrix} p_\alpha(x) \\ n_\alpha(x) \end{pmatrix} = \varepsilon_{abc}(u_a^T(x)C\gamma_5 d_b(x))q_{c,\alpha}(x), \quad q(x) = \begin{pmatrix} u(x) \\ d(x) \end{pmatrix}, \tag{2.2}$$

where x = (x, t), a, b, c are the color indices, and α is the spinor index. The charge conjugation matrix is given by $C = \gamma_2\gamma_4$, p, n are the proton and neutron operators, and u, d denote the up and down quark operators. Here ϕ^W implicitly has two pairs of spinor-flavor indices from $B_\alpha(r + x, t)B_\beta(r, t)$ as well as two helicity indices s_1 and s_2

The most important property of the above NBS wave function is as follows. If the total energy W lies below the threshold of meson production (i.e. $W < 2m_B + m_M$ with the meson mass m_M), it satisfies the Helmholtz equation with k = |k| at r = |r|→∞,

$$[k^2 + \nabla^2]\varphi^W(\boldsymbol{r}) \simeq 0. \tag{2.3}$$

Furthermore, the asymptotic behavior of the radial part of the NBS wave function for a given orbital angular momentum L and total spin S reads [18,21]

$$\varphi^W(r; LS) \propto \frac{\sin(kr - L\pi/2 + \delta_{LS}(k))}{kr} e^{i\delta_{LS}(k)}. \tag{2.4}$$

Here $\delta_{LS}(k)$ is nothing but the phase shift obtained from the baryon–baryon S-matrix in QCD below the inelastic threshold. It should be remarked here that only the upper components of the spinor indices for the NBS wave function ($\alpha = 1, 2$ and $\beta = 1, 2$) are enough to reproduce all B B scattering phase shifts δL S(k) with L = 0, 1, 2, 3,... and S = 0, 1. (See Appendix A of Ref. [18] for the precise expression of Eq. (2.4) and its relation to the S-matrix in QCD.)

Non-Local Potential from the NBS Wave Function

From the NBS wave function, we define a non-local potential through the relation [9,17,18]

$$(E_k - H_0)\varphi^W_{\alpha\beta}(\boldsymbol{x}) = \int U_{\alpha\beta;\gamma\delta}(\boldsymbol{x}, \boldsymbol{y})\varphi^W_{\gamma\delta}(\boldsymbol{y})d^3y, \quad \left(E_k = \frac{k^2}{2\mu},\ H_0 = \frac{-\nabla^2}{2\mu}\right), \tag{2.5}$$

where U(x, y) is expected to be short-ranged because of the absence of massless particle exchanges between two baryons. As mentioned in the previous subsection, it is enough to consider the upper spinor indices of α, β, γ , δ: Then 16 components of $U_{\alpha\beta;\gamma\delta}$ can be determined from 4 components of $\varphi^W_{\alpha\beta}$ for 4 different combinations of (s_1,s_2). Since the NBS wave function ϕ^W is multiplicatively renormalized, the potential U(x, y) is finite and does not depend on the particular renormalization scheme. Note that, while Lorentz covariance is lost by using the equal-time NBS wave function and Eq. (2.5) is written as a Schrödinger type equation, no non-relativistic approximation is employed here to define U(x, y). The non-local potential U(x, y) has been shown to be energy-independent [17,18]. To see this, let V_{th} be the space spanned by the wave function at $W \le W_{\rm th} \equiv 2m_B + m_M$: $V_{\rm th} = \{\varphi^W_c | W \le W_{\rm th}\}$, where c represents the quantum numbers of the NBS wave function other than energy W. Then the projection operator to V_{th} is given by

$$P^{W_{\rm th}}(\boldsymbol{x}, \boldsymbol{y}) = \sum_{W_1, W_2 \le W_{\rm th}} \sum_{c_1, c_2} \varphi^{W_1}_{c_1}(\boldsymbol{x}) N^{-1}_{c_1,c_2}(W_1, W_2)\varphi^{W_2}_{c_2}(\boldsymbol{y})^\dagger$$
$$\equiv \sum_{W \le W_{\rm th}} \sum_c P^{W_{\rm th}}_c(W; \boldsymbol{x}, \boldsymbol{y}), \tag{2.6}$$

where $N^{-1}_{c_1,c_2}(W_1, W_2)$ is defined as the inverse of the Hermitian operator

$$N_{c_1,c_2}(W_1, W_2) = \int \varphi^{W_1}_{c_1}(\boldsymbol{r})^\dagger \varphi^{W_2}_{c_2}(\boldsymbol{r})\, d^3r, \quad W_{1,2} \le W_{\rm th}, \tag{2.7}$$

which satisfies

$$\sum_{W \le W_{\rm th}} \sum_c N_{c_1,c}(W_1, W) N^{-1}_{c,c_2}(W, W_2) = \delta_{c_1,c_2}\delta_{W_1,W_2} \tag{2.8}$$

in the restricted indices that $W_{1,2} \le$ Wth. (We assume here that $N(W_1, W_2)$ does not have zero eigenvalues in this restricted space.)

Using these, the non-local potential is defined by

$$U^{W_{\rm th}}(\boldsymbol{x}, \boldsymbol{y}) = \sum_{W_{1,2} \le W_{\rm th}} \sum_{c_1, c_2} \rho(W_1)[E_k - H_0]\varphi^{W_1}_{c_1}(\boldsymbol{x}) N^{-1}_{c_1,c_2}(W_1, W_2)\varphi^{W_2}_{c_2}(\boldsymbol{y})^\dagger$$
$$= \sum_{W \le W_{\rm th}} \sum_c [E_k - H_0] P^{W_{\rm th}}_c(W; \boldsymbol{x}, \boldsymbol{y}). \tag{2.9}$$

Then, it is easy to observe that the above non-local potential satisfies Eq. (2.5) at $W \le W_{th}$:

$$\int U^{W_{\rm th}}(\boldsymbol{x}, \boldsymbol{y})\varphi^W_c(\boldsymbol{y})\, d^3y = \sum_{W_1 \le W_{\rm th}} \sum_{c_1} [E_k - H_0]\varphi^{W_1}_{c_1}(\boldsymbol{x})\delta_{c,c_1}\delta_{W_1,W}$$
$$= \theta(W_{\rm th} - W)[E_k - H_0]\varphi^W_c(\boldsymbol{x}). \tag{2.10}$$

This non-local potential U(x, y) is energy independent by construction. It is also easy to see that we can make the potential local but energy-dependent. A similar trade-off between non-locality and energy dependence was also discussed a long time ago in Ref. [22] in a different context. Note however that the non-local potential U(x, y) which satisfied Eq. (2.5) at $W \le W_{th}$ is not unique. For example, one may add a term such as f (x) $[1 - P^{Wth}(x, y)]$ with arbitrary functions f (x) to the non-local potential U(x, y) without affecting Eq. (2.5) at $W \le W_{th}$. We remark that one may define a non-local potential different from Eq. (2.9) as

$$U^\infty(\boldsymbol{x}, \boldsymbol{y}) = \sum_{W \le \infty} \sum_c [E_k - H_0] P^\infty_c(W; \boldsymbol{x}, \boldsymbol{y}), \tag{2.11}$$

which satisfies Eq. (2.5) for all W. This potential, however, becomes long-ranged, due to the presence of inelastic contributions above Wth. An extension of the HAL QCD method, which keeps the shortrange nature of the potential while inelastic channels are open, will be discussed in Sect. 7. The most general form of the Schrödinger type equation for the NBS wave function has an energydependent and non-local potential as shown in Ref. [16]. However, one can always remove its energy dependence, as demonstrated in the above derivation.

Velocity Expansion of the Non-Local Potential

If one knows the NBS wave functions ϕ^W for all $W \le W_{th}$, the non-local potential U can be constructed according to Eq. (2.9). In lattice QCD simulations in a finite box, however, only a limited number of wave functions at low energies (the ground state and possibly a few low-lying excited states) can be obtained. In such a situation, it is useful to expand the non-local potential in terms of the velocity (derivative) with local coefficient functions [23]:

$$U(\boldsymbol{x}, \boldsymbol{y}) = V(\boldsymbol{x}, \nabla)\delta^3(\boldsymbol{x} - \boldsymbol{y}). \tag{2.12}$$

In the lowest few orders we have

$$V(\boldsymbol{r}, \nabla) = \underbrace{V_0(r) + V_\sigma(r)\boldsymbol{\sigma}_1 \cdot \boldsymbol{\sigma}_2 + V_T(r)S_{12}}_{\rm LO} + \underbrace{V_{\rm LS}(r)\mathbf{L} \cdot \mathbf{S}}_{\rm NLO} + O(\nabla^2), \tag{2.13}$$

where $r = |r|$, σ_i is the Pauli matrix acting on the spin index of the ith baryon, $S = (\sigma_1 + \sigma_2)/2$ is the total spin, $L = r \times p$ is the angular momentum, and

$$S_{12} = 3\frac{(\boldsymbol{r}\cdot\boldsymbol{\sigma}_1)(\boldsymbol{r}\cdot\boldsymbol{\sigma}_2)}{r^2} - \boldsymbol{\sigma}_1\cdot\boldsymbol{\sigma}_2 \tag{2.14}$$

is the tensor operator. Each coefficient function is further decomposed into its flavor components. In the case of nucleons (i.e. $N_f = 2$), we have

$$V_X(r) = V_X^0(r) + V_X^\tau(r)\boldsymbol{\tau}_1\cdot\boldsymbol{\tau}_2, \quad X = 0, \sigma, \mathrm{T}, \mathrm{LS}, \ldots, \tag{2.15}$$

where τ_i is the Pauli matrix acting on the flavor index of the ith nucleon. The form of the velocity expansion (2.13) agrees with the form determined by symmetries [24]. At the leading order of the velocity expansion, the local potential is given by

$$V^{\mathrm{LO}}(\boldsymbol{r}) = V_0(r) + V_\sigma(r)\boldsymbol{\sigma}_1\cdot\boldsymbol{\sigma}_2 + V_T(r)S_{12}, \tag{2.16}$$

which is obtained from the NBS wave function at one value of W. Since $S_{12} = 0$ for the spin-singlet state, for example, one has

$$V_C(r, S=0) \equiv V_0(r) - 3V_\sigma(r) = \frac{(E_k - H_0)\varphi^W(\boldsymbol{r})}{\varphi^W(\boldsymbol{r})}. \tag{2.17}$$

Remarks on the "Scheme" Dependence of the Potential

We emphasize that the potential itself is not a physical observable, and is therefore not a unique quantity in quantum mechanics and in field theory. In fact, the baryon–baryon potential in QCD depends on the choice of the interpolating baryon operator to define the NBS wave function. Among others, the local baryon operator used in the HAL QCD method is a most convenient choice, since the reduction formula for composite particles can be derived in the simplest way for this choice [25–27]. Nevertheless, one may adopt other interpolating operators (such as higher dimensional operators and non-local operators). The particular choice of baryon operator and associated potential may be considered a "scheme" to describe physical observables such as the scattering phase shift and binding energies. The potential, although being "scheme"-dependent, is still useful to understand physical phenomena, as we well know in quantum mechanics. The repulsive core of the nucleon–nucleon potential in the coordinate space, which is known to be the best way to summarize the N N scattering phase shift at high energies, is one such example.1 Good convergence of the velocity expansion is an important check of the choice of our present scheme. Such a check can be carried out by examining the W dependence of the lower order potentials. For example, if we have φ^{W_n} for $n = 1, 2, \ldots N$, we can determine the $N - 1$ unknown local functions of the velocity expansion in N different ways. The variation among N different determinations gives an estimate of the size of the higher order terms. Furthermore, one of these higher order terms can be determined from ϕ^{W_n} for n = 1, 2,... N. The convergence of the velocity expansion will be investigated explicitly in Sect. 4.

The analysis in this section shows that the use of a Schrödinger type equation with a non-local potential is justified for describing B B scattering in QCD. The key quantity is the NBS wave function, whose asymptotic behavior encodes phases of the S-matrix for B B scattering. If the velocity expansion of the non-local potential is reasonably good at low energies, one can use the LO and NLO potentials to investigate various nuclear many-body problems

LATTICE FORMULATION

We now discuss procedures to extract the NBS wave function from lattice QCD simulations. For this purpose, we consider the correlation function on the lattice defined by

$$F(\boldsymbol{r}, t - t_0) = \langle 0|T\{B(\boldsymbol{x}+\boldsymbol{r}, t)B(\boldsymbol{x}, t)\}\overline{\mathcal{J}}(t_0)|0\rangle, \tag{3.1}$$

Where $\mathcal{J}(t_0)$ is a source operator which creates two-baryon states. Inserting a complete set and considering baryon number conservation, we have

$$\begin{aligned} F(\boldsymbol{r}, t - t_0) &= \langle 0|T\{B(\boldsymbol{x}+\boldsymbol{r}, t)B(\boldsymbol{x}, t)\}\sum_{n,s_1,s_2}|2B, W_n, s_1, s_2\rangle\langle 2B, W_n, s_1, s_2|\overline{\mathcal{J}}(t_0)|0\rangle + \cdots \\ &= \sum_{n,s_1,s_2} A_{n,s_1,s_2}\varphi^{W_n}(\boldsymbol{r})e^{-W_n(t-t_0)} + \cdots, \end{aligned} \tag{3.2}$$

Where $A_{n,s_1,s_2} = \langle 2B, W_n, s_1, s_2|\overline{\mathcal{J}}(t_0)|0\rangle$ and ellipses represent contributions from inelastic states such as $NN\pi$, $NNN\bar{N}$, etc. At large time separation $(t - t_0) \to \infty$, we obtain

$$\lim_{(t-t_0)\to\infty} F(\boldsymbol{r}, t - t_0) = A_0\varphi^{W_0}(\boldsymbol{r})e^{-W_0(t-t_0)} + O(e^{-W_{n\neq 0}(t-t_0)}), \tag{3.3}$$

where W_0 is the lowest B B state energy. Since the source-dependent term A_0 is just a multiplicative constant to the NBS wave function ϕ^{W0} (r), the potential defined from ϕ^{W0} (r) is manifestly sourceindependent. For this extraction of the wave function to work, the ground state saturation for F in Eq. (3.3) must be satisfied by taking large $t - t_0$. In practice, however, F becomes very noisy at large $t - t_0$. In Sect. 3.4, we will discuss more on this point.

Choice of Source Operators

We choose the source operator $\bar{\mathcal{J}}$ to fix the quantum numbers of $|2B, W, s_1, s_2\rangle$. Since lattice QCD simulations are usually performed on a hyper-cubic lattice, the cubic transformation group SO(3, Z) instead of SO(3, R) is considered as the symmetry of 3D space. Therefore, the quantum number is classified in terms of the irreducible representation of SO(3, Z), which is denoted by A_1, A_2, E, T_1, and T_2 whose dimensions are 1, 1, 2, 3, and 3, respectively. The relationship of irreducible representations between SO(3, Z) and SO(3, R) is given in Table 1 for $L \leq 6$, where L denotes the angular momentum for the irreducible representation of SO(3, R). For example, the source operator $\bar{\mathcal{J}}(t_0)$ in the A_1 representation with positive parity generates states with L = 0, 4, 6, ··· at t = t0, while the operator in the T_1 representation with negative parity produces states with L = 1, 3, 5, ··· . For two octet-baryons, the total spin S becomes 1/2 ⊗ 1/2 = 1

$\oplus$ 0, which corresponds to T_1(S = 1) and A_1(S = 0) of SO(3, Z). The total representation J for a two-baryon system is thus determined by the product $R_1 \otimes R_2$, where $R_1 = A_1, A_2, E, T_1, T_2$ for the orbital "angular momentum" while $R_2 = A_1, T_1$ for the total spin. In Table 2, the product $R_1 \otimes R_2$ is decomposed into the direct sum of irreducible representations.

We often use the wall source at $t = t_0$ defined by

$$\mathcal{J}^{\rm wall}(t_0)_{\alpha\beta, fg} = B^{\rm wall}_{\alpha, f}(t_0) B^{\rm wall}_{\beta, g}(t_0), \tag{3.4}$$

where α, β = 1, 2 are the upper component of the spinor indices while f, g are the flavor indices. Here $B^{\rm wall}$(t0) is obtained by replacing the local quark field q(x) of B(x) by the wall source,

$$q^{\rm wall}(t_0) \equiv \sum_{x} q(\mathbf{x}, t_0), \tag{3.5}$$

Table 1: The number of each representation of SO(3,Z) that appears in the angular momentum L representation of SO(3, R). P = $(-1)^L$ denotes the eigenvalue under parity transformation

L	P	A_1	A_2	E	T_1	T_2
0 (S)	+	1	0	0	0	0
1 (P)	−	0	0	0	1	0
2 (D)	+	0	0	1	0	1
3 (F)	−	0	1	0	1	1
4 (G)	+	1	0	1	1	1
5 (H)	−	0	0	1	2	1
6 (I)	+	1	1	1	1	2

Table 2: The decomposition of a product of two irreducible representations, $R_1 \otimes R_2$, into irreducible representations in SO(3,Z). Note that $R_1 \otimes R_2 = R_2 \otimes R_1$ by definition

	A_1	A_2	E	T_1	T_2
A_1	A_1	A_2	E	T_1	T_2
A_2	A_2	A_1	E	T_2	T_1
E	E	E	$A_1 \oplus A_2 \oplus E$	$T_1 \oplus T_2$	$T_1 \oplus T_2$
T_1	T_1	T_2	$T_1 \oplus T_2$	$A_1 \oplus E \oplus T_1 \oplus T_2$	$A_2 \oplus E \oplus T_1 \oplus T_2$
T_2	T_2	T_1	$T_1 \oplus T_2$	$A_2 \oplus E \oplus T_1 \oplus T_2$	$A_1 \oplus E \oplus T_1 \oplus T_2$

with the Coulomb gauge fixed at $t = t_0$. Note that this gauge dependence of the source operator disappears for the potential. All states created by the wall source have zero total momentum. Among them, the state with zero relative momentum has the largest magnitude. One reason for employing the wall source here is that the ground state saturation for the potential at long distances is better achieved for the wall source than for other sources.

Let us consider the case of two nucleons. The source operator $\bar{\mathcal{J}}^{\rm wall}(t_0)$ has zero orbital angular momentum at $t = t_0$, which corresponds to the A_1 representation with positive parity. There fore, the total angular momentum can be fixed by using the spin recoupling matrix $M^{(S,S_z)}$, e.g.,

$M^{(S=0,S_z=0)} \equiv \sigma_2/\sqrt{2}$ and $M^{(S=1,S_z=m)} \equiv (\sigma_2\sigma_m)/\sqrt{2}$ for $m = 0, \pm 1$ as

$$\mathcal{J}(t_0; J^{P=+}, J_z = m, I) = M^{(S)}_{\beta\alpha} \mathcal{J}^{\rm wall}(t_0)_{\alpha\beta, fg}. \tag{3.6}$$

Here P = ± is the parity and I = 1, 0 is the total isospin of the system. Since the nucleon is a fermion, exchange of the nucleon operators in the source should give a minus sign. This fact fixes the total isospin given the total spin: (S, I) = (0, 1) or (1, 0). (Note that S, I = 0 are antisymmetric while S, I = 1 are symmetric under the exchange.) Since $A_1^+ \otimes A_1(S = 0) = A_1^+$ and $A_1^+ \otimes T_1(S = 1) = T_1^+$, the state with either $(J^P, I) = (A_1^+, 1)$ for the spin-singlet or $(J^P, I) = (T_1^+, 0)$ for the spin-triplet is created at $t = t_0$ by the corresponding source operator. The NBS wave function extracted at $t > t_0$ has the same quantum numbers(J^P, I) as they are conserved under QCD interactions. In addition, the total spin S is conserved at t > t0 for a two-nucleon system with equal up and down quark masses: under the exchange of the two particles, the constraint $(-1)^{S+1+I+1}P = -1$ must be satisfied due to the fermionic nature of the nucleon. Also, parity P and isospin I are conserved in this system. Therefore S is conserved. However, L is not conserved in general. While the state with $(J^P, I) = (A_1^+, 1)$ always has $L = A_1^+$ even at $t > t_0$, the one with $(J^P, I) = (T_1^+, 0)$ has both $L = A_1^+$ and $L = E^+, T_2^+$ components[2] at $t > t_0$, which correspond to $L = 0$ and $L = 2$ in SO(3, R), respectively. Note that J and L are used to represent the total and orbital quantum numbers respectively for SO(3, Z) as well as for SO(3, R).

The orbital angular momentum L of the NBS wave function for N N can be fixed to a particular value by the projection operator $P^{(L)}$ as

$$\varphi^W(\mathbf{r}; J^P, I, L, S) = P^{(L)} P^{(S)} \varphi^W(\mathbf{r}; J^P, I), \tag{3.7}$$

Where $\varphi^W(\mathbf{r}; J^P, I)$ is extracted from

$$F(\mathbf{r}, t - t_0; J^P, I) \simeq A(J^P, I)\varphi^W(\mathbf{r}; J^P, I)e^{-W(t-t_0)},$$
$$A(J^P, I) = \langle 2B, W|\bar{\mathcal{J}}(t_0; J^P, I)|0\rangle \quad (3.8)$$

for large t − t_0. The total spin projection operator is $(P^{(S=0)})_{\alpha\beta;\alpha'\beta'} \equiv \frac{1}{2}(\sigma_2)_{\alpha\beta}(\sigma_2)_{\beta'\alpha'}$ for spinsinglet and $P^{(S=1)} \equiv \mathbb{I} - P^{(S=0)}$ for spin-triplet, but this is redundant since the total spin S, already fixed by the source, is conserved as mentioned before. The projection operator $P^{(L)}$ of the orbital angular momentum for an arbitrary function ϕ(r) is defined in general by

$$P^{(L)}\varphi^W(\mathbf{r}) \equiv \frac{d_L}{24}\sum_{g\in SO(3,\mathbf{Z})} \chi^L(g)^*\varphi^W(g^{-1}\cdot\mathbf{r}) \quad (3.9)$$

for L = A_1, A_2, E, T_1, T_2, where χL denotes the character of the representation L in SO(3, Z), ∗ is its complex conjugate, g is one of 24 elements in SO(3, Z), and d_L is the dimension of L.

Leading Order N N Potential: Spin-Singlet Case

We present the procedure to determine potentials at the leading order (LO):

$$V^{\mathrm{LO}}(\mathbf{r}) = V_0(r) + V_\sigma(r)(\boldsymbol{\sigma}_1\cdot\boldsymbol{\sigma}_2) + V_T(r)S_{12}. \quad (3.10)$$

Since $S_{12} = 0$ and $\sigma_1 \cdot \sigma_2 = -3$ for the spin-singlet case, the LO central potential for the spin-singlet case is extracted from the $(J^P, I) = (A_1^+, 1)$ state as

$$V_C(r)^{(S,I)=(0,1)} \equiv V_0^{I=1}(r) - 3V_\sigma^{I=1}(r) = \frac{[E_k - H_0]\varphi^W(\mathbf{r}; A_1^+, I=1, L=A_1, S=0)}{\varphi^W(\mathbf{r}; A_1^+, I=1, L=A_1, S=0)}, \quad (3.11)$$

Where $V_X^{I=1} = V_X^0 + V_X^\tau$ in isospin space. The potential $V_C(\mathbf{r})^{(S,I)=(0,1)}$ in the above is often referred to as the central potential for the $^1S_0$ state, where the notation $^{2S+1}L_J$ represents the orbital angular momentum L (see Table 1), the total spin S and the total angular momentum J of J = L + S. It is noted, however, that in the leading order of the velocity expansion, the potential does not depend on the quantum number of the state J = L = A_1. Moreover, the A_1 state may contain L = 4, 6,... components other than L = 0, though the L = 0 component may dominate. Therefore it is more precise to refer to $V_C(\mathbf{r})^{(S,I)=(0,1)}$ as the spin-singlet (isospin-triplet) central potential determined from the state with J = L = A_1. A possible difference of spin-singlet central potentials between this determination and others such as the one determined from J = L = E gives an estimate for contributions from higher order terms in the velocity expansion.

Leading Order Potential: Spin-Triplet Case

Both the tensor potential V_T and central potential V_C appear in the LO for the spin-triplet case. Let us consider the determination from the $(J^P, I) = (T_1^+, 0)$ state. The Schrödinger equation for this state becomes

$$[H_0 + V_C(r)^{(S,I)} + V_T(r)S_{12}]\varphi^W(\mathbf{r}; J^P = T_1^+, I) = E_k\varphi^W(\mathbf{r}; J^P = T_1^+, I) \quad (3.12)$$

We separate the Schrödinger Eq. (3.12) into A_1 and non-A_1 components by using projection operators $\mathcal{P} \equiv P^{(A_1)}$ and $\mathcal{Q} \equiv \mathbb{I} - \mathcal{P}$ as

$$(V_C(r)^{(1,0)} - E_k)\mathcal{P}\varphi^W(\mathbf{r}) + V_T(r)\mathcal{P}S_{12}\varphi^W(\mathbf{r}) = -H_0\mathcal{P}\varphi^W(\mathbf{r})$$
$$(V_C(r)^{(1,0)} - E_k)\mathcal{Q}\varphi^W(\mathbf{r}) + V_T(r)\mathcal{Q}S_{12}\varphi^W(\mathbf{r}) = -H_0\mathcal{Q}\varphi^W(\mathbf{r}). \quad (3.14)$$

Note that P and Q commute with H_0, $V_C(r)$ and V_T (r), whereas they do not commute with S_{12}. The non-A_1 component receives contributions from E, T_1, and T_2, among which only E and T_2 contribute to the D-wave. Since the contribution from the T_1 component turns out to be negligible in the numerical simulation, the non-A_1 component is dominated by D-wave contributions.

Using these projections, V_C and V_T can be extracted as

$$V_C(r)^{(1,0)} = E_k - \frac{1}{\Delta(\mathbf{r})}([\mathcal{Q}S_{12}\varphi^W]_{\alpha\beta}(\mathbf{r})H_0[\mathcal{P}\varphi^W]_{\alpha\beta}(\mathbf{r}) - [\mathcal{P}S_{12}\varphi^W]_{\alpha\beta}(\mathbf{r})H_0[\mathcal{Q}\varphi^W]_{\alpha\beta}(\mathbf{r})) \quad (3.15)$$

$$V_T(r) = \frac{1}{\Delta(\mathbf{r})}([\mathcal{Q}\varphi^W]_{\alpha\beta}(\mathbf{r})H_0[\mathcal{P}\varphi^W]_{\alpha\beta}(\mathbf{r}) - [\mathcal{P}\varphi^W]_{\alpha\beta}(\mathbf{r})H_0[\mathcal{Q}\varphi^W]_{\alpha\beta}(\mathbf{r})) \quad (3.16)$$

$$\Delta(\mathbf{r}) \equiv [\mathcal{Q}S_{12}\varphi^W]_{\alpha\beta}(\mathbf{r})[\mathcal{P}\varphi^W]_{\alpha\beta}(\mathbf{r}) - [\mathcal{P}S_{12}\varphi^W]_{\alpha\beta}(\mathbf{r})[\mathcal{Q}\varphi^W]_{\alpha\beta}(\mathbf{r}). \quad (3.17)$$

In numerical simulations, (α, β) = (2, 1) in the $J_z = 0$ state is mainly employed.

One may focus only on the A_1 component of the wave function and define the so-called effective central potential for the spin-triplet (isospin-singlet), often used in nuclear physics:

$$V_C^{\text{eff}}(r)^{(1,0)} = \frac{[E_k - H_0]\mathcal{P}\varphi^W_{\alpha\beta}(\boldsymbol{r})}{\mathcal{P}\varphi^W_{\alpha\beta}(\boldsymbol{r})}. \tag{3.18}$$

The effect of V_T, which leads to a transition from the A_1 component to the non-A_1 component of the wave function, is implicitly included in this effective central potential: For small V_T, the difference between V_C and V_C^{eff} is $O(V_T^2)$, as the second order perturbation tells us.

Time-Dependent Hal QCD Method

One of the practical difficulties of extracting the NBS wave function and the potential from the correlation function Eq. (3.1) is to achieve ground state saturation in numerical simulations at large but finite $t - t_0$ with reasonably small statistical errors. While the stability of the potential against $t - t_0$ has been confirmed within statistical errors in numerical simulations [9,18], the determination of W for the ground state suffers from systematic errors due to the contamination of possible excited states. There exist three different methods to determine W. The most well-known method is to determine W from the $t - t_0$ dependence of the correlation function Eq. (3.1) summed over r to pick up the zero momentum state. On the other hand, one may determine k^2 of W by fitting the r dependence of the NBS wave function with its expected asymptotic behavior at large r or by reading off the constant shift of the Laplacian part of the potential from zero at large r. Although the latter two methods usually give consistent results within statistical errors, the first method sometimes leads to a result different from those determined by the latter two at the value of $t - t_0$ employed in numerical simulations. Although, in principle, the increase of $t - t_0$ is needed in order to see an agreement between the three methods, it is difficult in practice due to the larger statistical errors at large $t - t_0$. The problem above is common in various applications of lattice QCD. Fortunately, the original HAL QCD method can be improved to overcome this difficulty as follows. Let us consider the normalized correlation function defined from Eq. (3.1) as

$$R(\boldsymbol{r}, t) \equiv \frac{F(\boldsymbol{r}, t)}{e^{-2m_B t}} = \sum_n A_n \varphi^{W_n}(\boldsymbol{r}) \exp(-t\Delta W_n) + O(e^{-\Delta W_{\text{th}} t}), \tag{3.19}$$

Where

$\Delta W_n = W_n - 2m_B$ and $\Delta W_{\text{th}} = W_{\text{th}} - 2m_B = m_M$. By neglecting the inelastic contributions above the meson production threshold, represented by $O(e^{-\Delta W_{\text{th}} t})$, for large enough t^3, the nonrelativistic approximation $W_n - 2m_B \simeq k_n^2/m_B$ leads us to

$$\begin{aligned} R(\boldsymbol{r}, t) &\simeq \sum_n A_n \varphi^{W_n}(\boldsymbol{r}) \exp\left(-t\frac{k_n^2}{m_B}\right) = e^{-t(H_0+U)} \sum_n A_n \varphi^{W_n}(\boldsymbol{r}) \\ &= e^{-t(H_0+U)} R(\boldsymbol{r}, t=0), \end{aligned} \tag{3.20}$$

where the Schrödinger Eq. (2.5), the defining relation of the non-local potential U, is used to replace $e^{-tk_n^2/m_N}$ by $e^{-t(H_0+U)}$ By applying a time derivative on both side, we have the time-dependent Schödinger equation in imaginary time

$$\left(-\frac{\partial}{\partial t} - H_0\right) R(\boldsymbol{r}, t) \simeq \int d^3\boldsymbol{r}' U(\boldsymbol{r}, \boldsymbol{r}') R(\boldsymbol{r}', t). \tag{3.21}$$

Now the velocity expansion of the non-local potential leads us to the formula of the leading order potential

$$V^{\text{LO}}(\boldsymbol{r}) = -\frac{(\partial/\partial t)R(\boldsymbol{r}, t)}{R(\boldsymbol{r}, t)} - \frac{H_0 R(\boldsymbol{r}, t)}{R(\boldsymbol{r}, t)}. \tag{3.22}$$

Once the ground state saturation is achieved in R(r, t), Eq. (3.22) reduces to Eq. (2.17), for example for the spin-singlet case. Indeed, in this case, $-\partial/\partial t$ is safely replaced by the non-relativistic energy E_k of the ground state under the non-relativistic approximation.

The non-relativistic formula for $V^{\text{LO}}(r)$ above can be generalized to the case that the masses of the two particles are different by the replacement, $R(\boldsymbol{r}, t) = F(\boldsymbol{r}, t)/e^{-(m_A+m_B)t}$. Note also that the potential extracted in this method automatically satisfies $V^{\text{LO}}(r \to) \to 0$ without constant shift. This property can be used to check whether this extraction works correctly or not.

The non-relativistic approximation used to derive Eq. (3.19) can be removed by using the second order derivative in t:

$$\left(\frac{1}{4m_B}\frac{\partial^2}{\partial t^2} - \frac{\partial}{\partial t} - H_0\right) R(\boldsymbol{r}, t) = \int d^3\boldsymbol{r}' U(\boldsymbol{r}, \boldsymbol{r}') R(\boldsymbol{r}', t), \tag{3.23}$$

which leads to

$$V^{\text{LO}}(\boldsymbol{r}) = \frac{1}{4m_B}\frac{(\partial/\partial t)^2 R(\boldsymbol{r}, t)}{R(\boldsymbol{r}, t)} - \frac{(\partial/\partial t)R(\boldsymbol{r}, t)}{R(\boldsymbol{r}, t)} - \frac{H_0 R(\boldsymbol{r}, t)}{R(\boldsymbol{r}, t)}. \tag{3.24}$$

Here we have assumed that the inelastic contributions are negligibly small and that the two particles have the same mass. Since t is discrete on the lattice, the t derivatives have to be carefully performed. One has to employ the numerical derivative scheme which reduces the statistical as well as the systematic errors of $V^{LO}(r)$.

One can generalize Eq. (3.19) to the correlation function with two relative coordinates x and y:

$$R(x,y,t) = \frac{1}{e^{-2m_B t}} \int d^3x_1 d^3y_1 \langle 0|T\{B(x_1+x,t)B(x_1,t)\overline{B}(y_1+y,0)\overline{B}(y_1,0)\}|0\rangle, \tag{3.25}$$

which leads to

$$\left(\frac{1}{4m_B}\frac{\partial^2}{\partial t^2} - \frac{\partial}{\partial t} - H_0\right) R(x,y,t) = \int d^3z\, U(x,z) R(z,y,t). \tag{3.26}$$

Then, we obtain the non-local potential as

$$U(x,y) = \int d^3z \left(\frac{1}{4m_B}\frac{\partial^2}{\partial t^2} - \frac{\partial}{\partial t} - H_0\right) R(x,z,t) \cdot \tilde{R}^{-1}(z,y,t), \tag{3.27}$$

Where $\tilde{R}^{-1}(x,y,t)$ is a "truncated" inverse of the Hermitian operator R(x, y, t),

$$\tilde{R}^{-1}(x,y,t) = \sum_{\lambda_n \neq 0} \frac{1}{\lambda_n(t)} v_n(x,t) v_n^\dagger(y,t), \tag{3.28}$$

with $\lambda_n(t)$ and $v_n(x, t)$ being the eigenvalues and corresponding eigenvectors of R(x, y, t), respectively. Note that zero eigenvalues are removed in the above summation. Suppose we introduce a modified potential as

$$\hat{U}(x,y) = U(x,y) + \sum_{\lambda_n = 0} c_n v_n(x,t) v_n^\dagger(y,t). \tag{3.29}$$

Then it satisfies the same Schrödinger equation for all possible values of c_n; the non-local potential is not unique, as discussed before.

NN POTENTIAL FROM LATTICE QCD

Central Potential in Quenched QCD

Let us first show the results in quenched QCD, where creations and annihilations of virtual quark– antiquark pairs are neglected: the standard plaquette gauge action is employed on a 32^4 lattice at the bare gauge coupling constant $\beta = 6/g^2 = 5.7$. This corresponds to the lattice spacing a $a \simeq 0.137$ fm (1/a = 1.44(2) GeV), determined from the ρ meson mass in the chiral limit, and the physical size of the lattice $L \simeq 4.4$ fm [9]. As for the quark action, the standard Wilson fermion action is used at three different values of the quark mass corresponding to the pion mass $m_\pi \simeq 731, 529, 380$ MeV and the nucleon mass $m_N \simeq 1560, 1330, 1200$ MeV, respectively.

Figure 2 (left) shows the NBS wave functions for the spin-singlet and spin-triplet channels in the orbital A_1 representation a $m_\pi \simeq 529$ MeV. These wave functions are normalized to be 1 at the largest spatial point $r \simeq 2.2$ fm. The central potential in the spin-singlet channel and the effective central potential in the spin-triplet channel extracted from the wave functions at $m_\pi \simeq 529$ MeV are shown in Fig. 2 (right). These potentials reproduce the qualitative features of the phenomenological N N potentials, namely the repulsive core at short distances surrounded by the attractive well at medium and long distances. From this figure one observes that the interaction range of the potential is smaller than 1.5 fm, showing that the box size $L \simeq 4.4$ fm is large enough for the potential. Labels

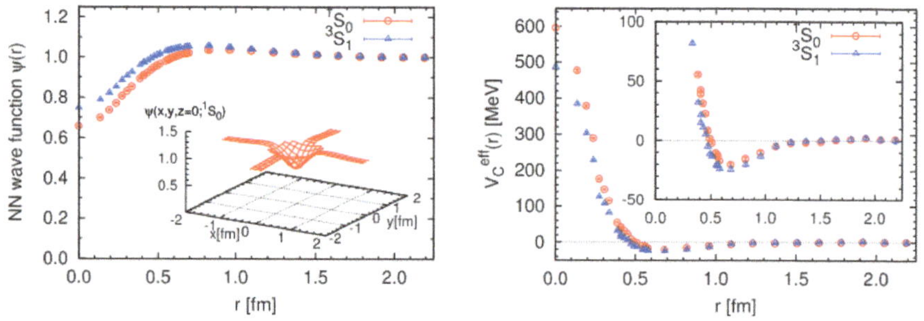

Figure. 2: (Left) The N N wave function for the spin-singlet and spin-triplet channels in the orbital A_1^+ representation at $m_\pi \simeq 529$ MeV and $a \simeq 0.137$ fm m in quenched QCD. The inset is a 3D plot of the spin-singlet wave function $\varphi^W(x,y,z=0)$. (Right) (Right) The N N (effective) central potential for the spin-singlet (spin-triplet) channel determined from the orbital A_1^+ wave function. Both figures are taken from Ref. [18].

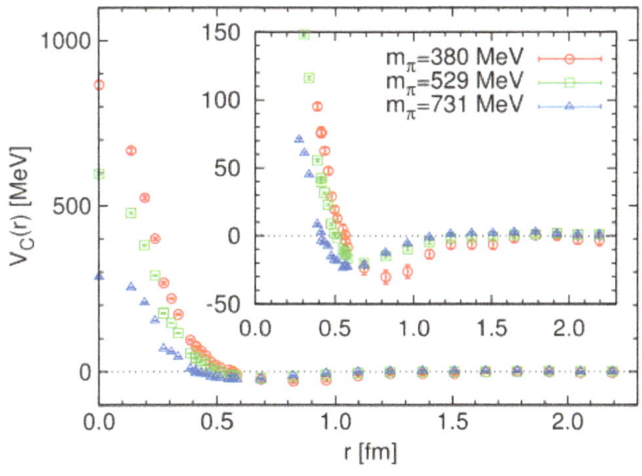

Figure. 3: The central potentials for the spin-singlet channel from the orbital A+ 1 representation at three different pion masses in quenched QCD. Taken from Ref. [18]

[1] S0 and [3] S1 of the potentials in the figure represent the fact that potentials are determined from A_1 wave functions, which are dominated by the S-wave component. Note here that the lattice artifacts are expected to be large for potentials (as well as wave functions) at short distances such that $r \simeq O(a)$.. Therefore, our results at short distances should be considered to be qualitative, not quantitative, and this caution should be applied to all of our results in this paper otherwise stated. The continuum extrapolation is necessary to predict the short distance behavior of potentials quantitatively. Indeed, B B potentials in the continuum limit are shown to diverge as r → 0 [28,29]. In Fig. 3, N N central potentials in the spin-singlet channel are shown for three different pion masses. The repulsion at short distances and the attraction at medium distances are simultaneously enhanced as the pion mass decreases.

In Fig. 3, N N central potentials in the spin-singlet channel are shown for three different pion masses. The repulsion at short distances and the attraction at medium distances are simultaneously enhanced as the pion mass decreases.

TENSOR POTENTIAL IN QUENCHED QCD

In Fig. 4 (left), we show the A_1 and non-A_1 components of the NBS wave function obtained from the $J^P = T_1^+$ (and $J_z = S_z = 0$) states at $m_\pi \simeq 529$ MeV. The non-A_1 wave function is multivalued as a function ofr due to its angular dependence. For example, the (α, β) = (2, 1) component of the L = 2 part of the non-A_1 wave function is proportional to the spherical harmonics $Y_{20}(\theta, \phi) \propto 3\cos^2\theta - 1$. Figure 4 (right) shows the non-A_1 component divided by $Y_{20}(\theta, \varphi)$. The non-A1 wave function seems

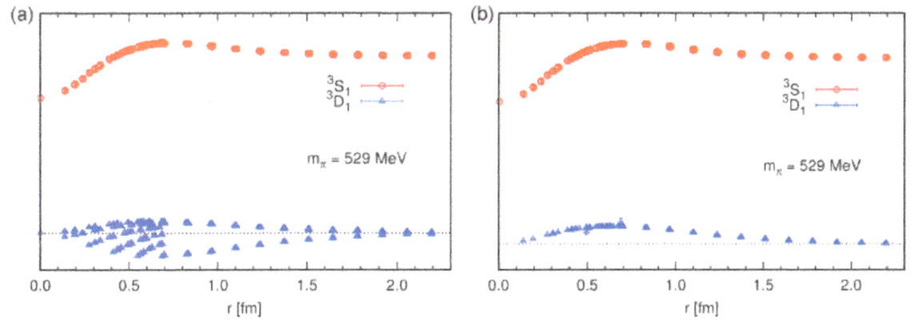

Figure. 4: (Left) (α, β) = (2, 1) components of the orbital A_1^+ and non-A_1^+ wave functions from $J^P = T_1^+$ (and $J_z = S_z = 0$) states at $m_\pi \simeq 529$ MeV. (Right) The same wave functions but the spherical harmonics components are removed from the non- A_1^+ part. Taken from Ref. [18].

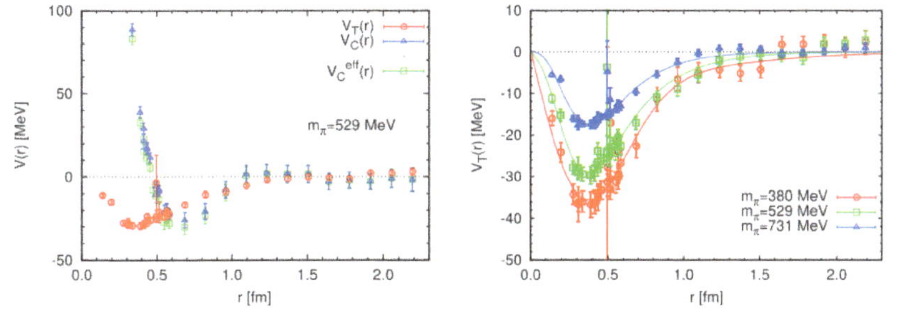

Figure. 5: (Left) The central potential $V_C(r)^{(1,0)}$ and the tensor potential $V_T(r)$ obtained from the $J^P = T_1^+$ NBS wave function, together with the effective central potential $V_C^{\rm eff}(r)^{(1,0)}$, at $m_\pi \simeq 529$ MeV. (Right) Pion mass dependence of the tensor potential. The lines show the four-parameter fit using one-pion-exchange + one-rho-exchange with a Gaussian form factor. Taken from Ref. [18]. to be dominated by

the D (L = 2) state, since its multivaluedness is mostly absorbed to $Y_{20}(\theta, \varphi)$. Figure 5 (left) shows the central potential $V_C(r)^{(1,0)}$ and tensor potential $V_T(r)$, together with the effective central potential $V_C^{\text{eff}}(r)^{(1,0)}$, at the leading order of the velocity expansion as given in Eqs. (3.15), (3.16), and (3.18), respectively

Note that $V_C^{\text{eff}}(r)$ contains the effect of V_T (r) implicitly as higher order effects through processes such as ${}^3S_1 \to {}^3D_1 \to {}^3S_1$. At the physical pion mass, $V_C^{\text{eff}}(r)$ is expected to gain sufficient attraction from the tensor potential, which leads to the appearance of a bound deuteron in the spin-triplet (and flavor-singlet) channel and an absence of the bound dineutron in the spin-singlet (and flavor-triplet) channel. The difference between $V_C(r)^{(1,0)}$ and $V_C^{\text{eff}}(r)$ in Fig. 5 (left) is still small in this quenched simulation due to the relatively large pion mass

The tensor potential in Fig. 5 (left) is negative for the whole range of r within statistical errors and has a minimum around 0.4 fm. If the tensor potential receives a significant contribution from onepion exchange, as expected from meson theory, V_T (r) is rather sensitive to changes in the pion mass. As shown in Fig. 5 (right), this is indeed the case: the attraction of V_T (r) is substantially enhanced as the pion mass decreases.

The central and tensor potentials in lattice QCD are given at discrete data points. For practical applications to nuclear physics, however, it is more convenient to parameterize the lattice results by

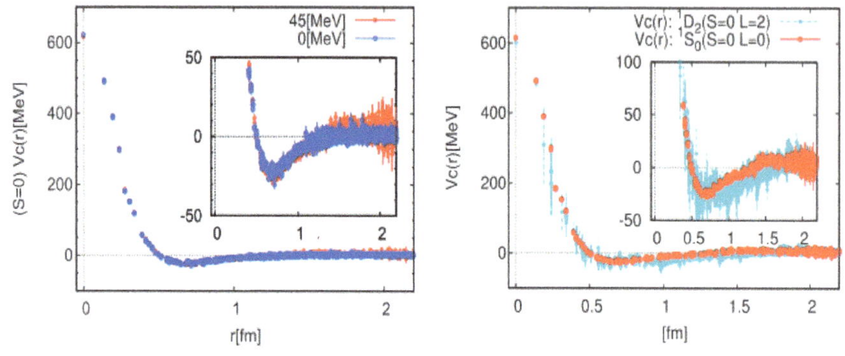

Figure. 6: (Left) The spin-singlet central potential $V_C(r)^{(0,1)}$ obtained from the orbita A_1^+ channel at $E \simeq 45$ MeV (red solid circles) and at $E \simeq 0$ MeV (blue open circles) in quenched QCD at $m_\pi \simeq 529$ MeV. (Right) The same potentials at $E \simeq 45$ MeV, obtained from the orbital A_1^+ representation (red open circles) and from the T_2^+ representation (gray solid circles). Taken from Ref. [33].

known functions. Such a fit for V_T (r)is given by the form of one-pion-exchange + one-rho-exchange with Gaussian form factors as

$$V_T(r) = b_1(1 - e^{-b_2 r^2})^2 \left(1 + \frac{3}{m_\rho r} + \frac{3}{(m_\rho r)^2}\right) \frac{e^{-m_\rho r}}{r}$$
$$+ b_3(1 - e^{-b_4 r^2})^2 \left(1 + \frac{3}{m_\pi r} + \frac{3}{(m_\pi r)^2}\right) \frac{e^{-m_\pi r}}{r}, \quad (4.1)$$

where $b_{1,2,3,4}$ are the fitting parameters while m_π (m_ρ) is taken to be the pion mass (the rho meson mass) calculated at each pion mass. The fit line for each pion mass is shown in Fig. 5 (right). It may be worth mentioning that the pion–nucleon coupling constant extracted from the parameter b_3 in the case of the lightest pion mass $(m_\pi = 380\,\text{MeV})$ gives $g_{\pi N}^2/(4\pi) = 12.1(2.7)$, which is encouragingly close to the empirical value.

Validity of the Velocity Expansion in Quenched QCD

The potentials so far are derived in the leading order of the velocity expansion. It is therefore important to investigate the convergence of the velocity expansion. In the absence of non-locality for the N N potentials, the leading order approximation for the potentials would give exact results at all energies below the inelastic threshold. The non-locality of the potentials therefore becomes manifest in the energy dependence of the potentials.

To study the energy dependence, the leading order local potentials at $E \simeq 45$ MeV, realized by anti-periodic boundary conditions in spatial directions, are calculated in quenched QCD at $m_\pi \simeq 529$ MeV and $L \simeq 4.4$ fm [30–33]. In this calculation, four types of momentum-wall sources, defined by

$$q(t_0; f) \equiv \sum_{\mathbf{x}} q(\mathbf{x}, t_0) f(\mathbf{x}), \quad (4.2)$$

are employed, where $f(\mathbf{x}) = \cos((\pm x \pm y + z)\pi/L)$. Note that $f(\mathbf{x}) = 1$ corresponds to the wall source used in the periodic boundary condition. These momentum-wall sources induce $L = T_2^+$ as well as $L = A_1^+$ states.

In Fig. 6 (left), the spin-singlet potential $V_C(r)^{(S,I)=(0,1)}$ obtained from the $L = A_1^+$ state at $E \simeq$ 45 MeV (red circles) is compared with that at $E \simeq 0$ MeV (blue circles), while a comparison is made in Fig. 7 for the spin-triplet potentials, $V_C(r)^{(S,I)=(1,0)}$ (left) and $V_T(r)$) (right). Good agreements between the results at two energies indicate that higher order contributions are rather small in this

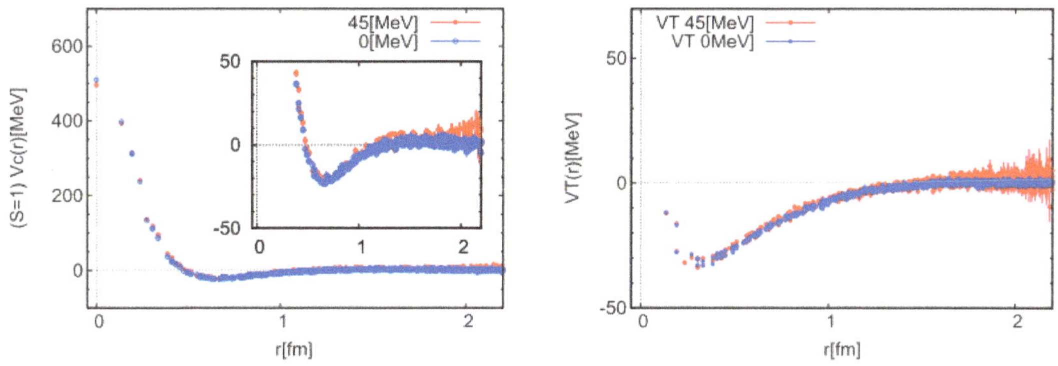

Figure. 7: (Left) The spin-triplet central potential $V_C(r)^{(1,0)}$ obtained from the orbital $A_1^+ - T_2^+$ coupled channel in quenched QCD at $m_\pi \simeq 529$ MeV. (Right) The tensor potential $V_T(r)$ from the orbital $A_1^+ - T_2^+ + 2$ coupled channel. For these two figures, the symbols are same as in Fig. 6 (left). Taken from Ref. [33].

energy interval. In other words, these local potentials obtained at $E \simeq 0$ MeV can be safely used to describe the N N scattering phase shift in the range between E = 0 MeV and E = 45 MeV at this pion mass in quenched QCD.

The non-locality of the potential may also become manifest in its angular momentum dependence, since the orbital angular momentum L = r × p contains a derivative. In Fig. 6 (right), the spin-singlet potential $V_C(r)^{(S,I)=(0,1)}$ obtained from the $L = T_2^+$ state, whose main component has L = 2, is compared to the one from the $L = A_1^+$ state, whose main component has L = 0. In this comparison, local potentials are determined at the same energy $E \simeq 45$ MeV, but different orbital angular momenta. Although the statistical errors are rather large in the case of $L = T_2^+$ a good agreement between the two is again observed, suggesting that the L dependence of the potential is small at least for the spin-singlet case. By these comparisons, it is observed that both energy and orbital angular momentum dependences for local potentials are very weak within statistical errors. We therefore conclude that contributions from higher order terms in the velocity expansion are small and that the LO local potentials in the expansion obtained at $E \simeq 0$ MeV and $L = 0$ are good approximations for non-local potentials, at least up to the energy $E \simeq 45$ MeV and orbital angular momentum L = 2.

Central Potential in Full QCD

Needless to say, it is important to carry out calculations of N N potentials in full QCD on larger volumes at lighter pion masses. The PACS-CS Collaboration is performing 2 + 1 flavor QCD simulations, which cover the physical pion mass [34,35]. Gauge configurations are generated with the Iwasaki gauge action and O(a)-improved Wilson quark action on a $32^3 \times 64$ lattice. The lattice spacing a is determined from m_π, m_K, and m_Ω as $a \simeq 0.089$ fm, leading to $L \simeq 2.9$ fm. Three ensembles of gauge configurations are used to calculate N N potentials at $(m_\pi, m_N) \simeq$(701 MeV, 1583 MeV), (570 MeV, 1412 MeV), and (411 MeV, 1215 MeV) [36]. To overcome the difficulty of achieving ground state saturations in full QCD simulations, the time-dependent HAL QCD method in Sect. 3.4 is employed [37].

Figure 8 (left) shows the spin-singlet N N central potential $V_C(r)$ obtained at $E \simeq 0$ from the PACS-CS configurations with $m_\pi \simeq 701$ MeV and $m_N \simeq 1583$ MeV. This central potential $V_C(r)$ is fitted with a multi-Gaussian function that $g(r) = \sum_{n=1}^{N_{\mathrm{Gauss}}} V_n \exp(-\nu_n r^2)$ with fit parameters V_n and $\nu_n (> 0)$. The solid line in the figure represents the fit result with $N_{\mathrm{Gauss}} = 5$.

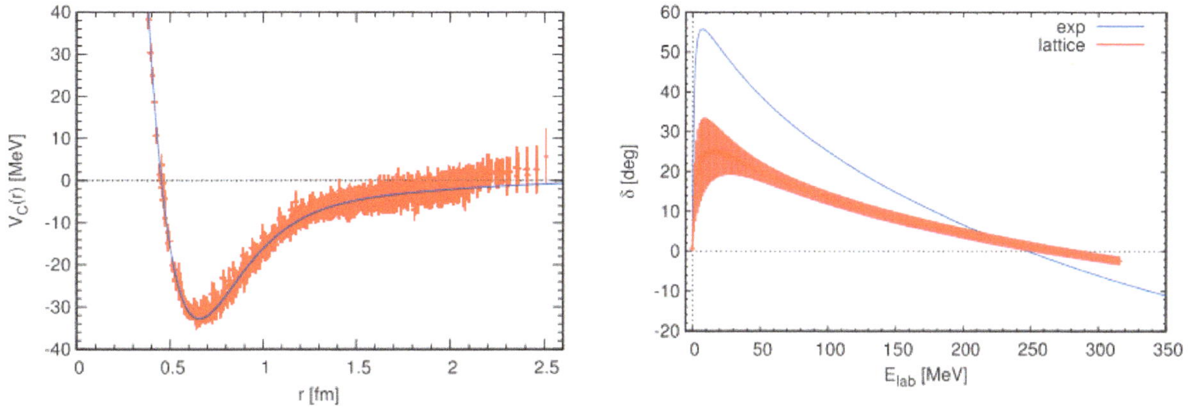

Figure. 8: (Left) The multi-Gaussian fit of the central potential $V_C(r)$ with $N_{\mathrm{Gauss}} = 5$. (Right) The scattering phase in $^1S_0$ channel in the laboratory frame obtained from the lattice N N potential, together with experimental data [38].

We solve the Schrödinger equation in the $^1S_0$ channel with this fitted potential $V_C(r)$, in order to calculate the scattering phase shift. Figure 8 (right) shows the scattering phase $\delta(k)$ in the laboratory frame, together with the experimental data [38] for a comparison. A qualitative feature of the experimental data is well reproduced by the lattice potential, though the strength is weaker, most likely due to the heavier pion mass, $m_\pi \simeq 701\,\text{MeV}$. The scattering length obtained from the derivative of the phase shift at k = 0 becomes $a(^1S_0) = \lim_{k\to 0} \tan\delta(k)/k = 1.6(1.1)$ fm, which is compared to the experimental value $a^{\exp}(^1S_0) \simeq 20\,\text{fm}$.

Nuclear Force in the Odd Parity Sector and the Spin-Orbit Force in Full QCD

In this subsection, we consider the potentials in odd parity sectors. Together with nuclear forces in even parity sectors, information on odd parity sectors is necessary for studying many-nucleon systems with Schrödinger equations. In particular, we are interested in the spin-orbit (LS) force, which gives rise to part of the spin-orbit coupling in the average single-particle potential of nuclei. It is also expected to induce superfluidity in neutron stars by providing an attraction between two neutrons in the $^3P_2$ channel [13].

The LS force appear at the NLO of the derivative expansion as

$$[H_0 + V_C(r)^{(S,I)} + V_T(r)S_{12} + V_{LS}(r)\boldsymbol{L}\cdot\boldsymbol{S}]\varphi^W(\boldsymbol{r}; J^-, I) = E_k\varphi^W(\boldsymbol{r}; J^-, I) \quad (4.3)$$

To obtain the three unknown potentials, V_C, V_T, and V_{LS}, we need three independent NBS wave functions. We therefore generalize the two-nucleon source for odd parity sectors, by imposing a momentum on the composite nucleon fields as

$$\mathcal{J}_{\alpha\beta}(t_0; f^{(i)}) \equiv N_\alpha(t_0; f^{(i)})N_\beta(t_0; f^{(i)*}) \quad \text{for} \quad i = \pm 1, \pm 2, \pm 3, \quad (4.4)$$

where N denotes a composite nucleon source field carrying a momentum,

$$N_\alpha(t_0; f^{(i)}) \equiv \sum_{\mathbf{x}_1,\mathbf{x}_2,\mathbf{x}_3} \epsilon_{abc}(u_a^T(\mathbf{x}_1)C\gamma_5 d_b(\mathbf{x}_2))q_{c,\alpha}(\mathbf{x}_3)f^{(i)}(\mathbf{x}_3), \quad (4.5)$$

With $f^{(\pm j)}(\mathbf{x}) = \exp[\pm 2\pi i x_j/L]$. The star "*" in the r.h.s. of Eq. (4.4) represents the complex conjugation, which is used to invert the direction of the plane wave. A cubic group analysis shows that the two-nucleon source Eq. (4.4) contains the orbital contribution $A_1^+ \oplus E^+ \oplus T_1^-$, whose main components are S-, D-, and P-waves, respectively. Thus the two-nucleon source Eq. (4.4) covers all the two-nucleon channels with $J \leq 2$.

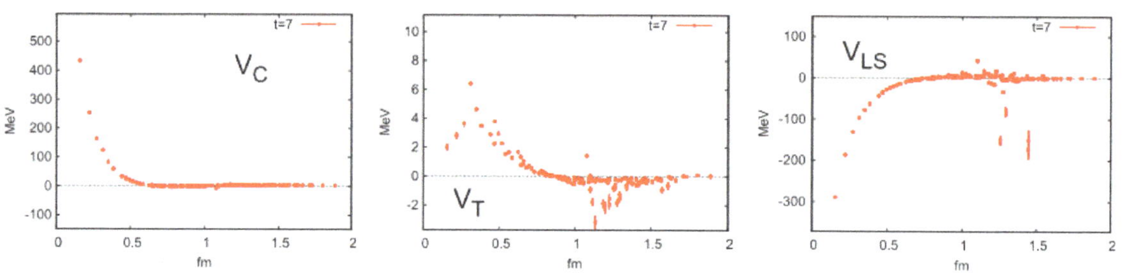

Figure. 9: Potentials in the odd parity sector obtained from $^3P_0$, $^3P_1$, and $^3P_2 + ^3F_2$ NBS wave functions calculated at m_π = 1136 MeV. The left, middle, and right figures show central, tensor, and spin-orbit force in the odd parity sector, respectively.

For the spin-triplet odd-parity sector, Eq. (4.4) generates the lowest-lying NBS wave functions for $(J^P, I) = (A_1^-, 1), (T_1^-, 1), (E^-, 1)$, and $(T_2^-, 1)$, which roughly correspond to $J^P = 0^-, 1^-, 2^-$, and 2^-, respectively. Among these, we consider Schrödinger equations for three NBS wave functions in $J^P = A_1^-, T_1^-, T_2^-$ as

$$[H_0 + V_C(r) + V_T(r)S_{12} + V_{LS}(r)\boldsymbol{L}\cdot\boldsymbol{S}]\varphi^W(\boldsymbol{r}; J^P) = E_0(J^P)\varphi^W(\boldsymbol{r}; J^P), \quad (4.6)$$

where $E_0(J^P) = k^2/m_N$ from the lowest-lying energy $W = 2\sqrt{m_N^2 + k^2}$ for the J^P sector. In order to obtain $V_C(r)$, $V_T(r)$, and $V_{LS}(r)$ in odd parity sectors, Eq. (4.6) for $J^P = A_1^-, T_1^-, T_2^-$, which correspond to $^3P_0$, $^3P_1$, and $^3P_2 + ^3F_2$, are solved.

Numerical calculations are performed by using 2 flavor QCD gauge configurations on a $16^3 \times 32$ lattice generated by the CP-PACS Collaboration [39? ,40], with Iwasaki gauge action at $\beta = 1.95$ and O(a) improved Wilson (clover) quark action at $\kappa = 0.1375$. This setup leads to the lattice spacing $a^{-1} = 1.27\,\text{GeV}$ ($a \simeq 1.555$ fm), the pion mass $m_\pi \simeq 1136\,\text{MeV}$, and the nucleon mass $m_N \simeq 2165\,\text{MeV}$. The spatial extension amounts to $L = 16a \simeq 2.5$ fm.

Figure 9 shows the preliminary results for the central potential $V_C(r)$, tensor potential $V_T(r)$, and the spin-orbit force $V_{LS}(r)$ in the spin-triplet odd parity sector. They have the following qualitative features. (1) $_{VC}(r)$ has a repulsive core at short distances. (2) $V_T(r)$ is positive and very small. (3) $V_{LS}(r)$ is large and negative at short distances. These features qualitatively agree with those of phenomenological potentials [8].

HYPERON INTERACTIONS

A study of hyperon–nucleon (Y N) and hyperon–hyperon (Y Y) interactions is one of the challenges in contemporary nuclear physics. These potentials give a key to understanding nuclear many-body systems with strangeness. Also they are essential for exploring the structure of the neutron star core, where a strangeness degree of freedom is expected to appear. At present, experimental data on Y N and Y Y scatterings are not sufficient to make precise constraints on the hyperon potentials, while spectroscopic studies of hypernuclei, performed by employing various reactions such as (π^+, K^+), (K^-, π^-), and $(e, e'K^+)$ [41,42], give some information on the N interactions. Under these circumstances, studies on the basis of lattice QCD are quite important as an alternative method to access Y N and Y Y interactions. In this section we mainly consider potentials in the strangeness S = −1 sector, obtained from 2 + 1 flavor lattice QCD simulations with PACS-CS gauge configurations. A study of potentials between octet barons in the flavor SU(3) limit and a coupled channel analysis on potentials in the strangeness S = −2 sector beyond the SU(3) limit will be discussed in the next two sections.

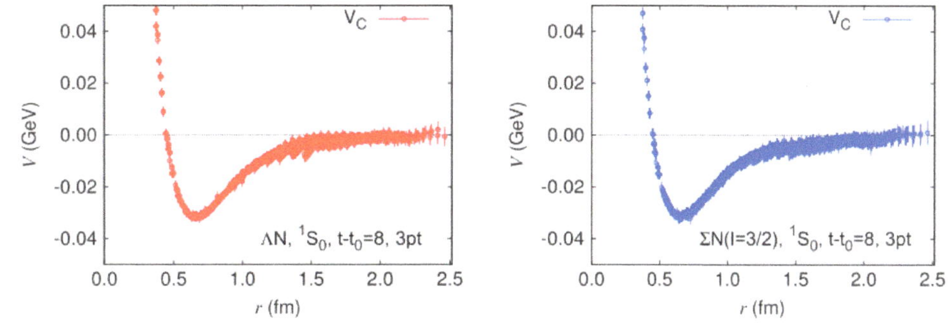

Figure. 10: (Left) The central potential in the $^1S_0$ channel of the ΛN system in 2 + 1 flavor QCD as a function of r. (Right) The central potential in the $^1S_0$ channel of the $\Sigma N(I = 3/2)$ system as a function of r.

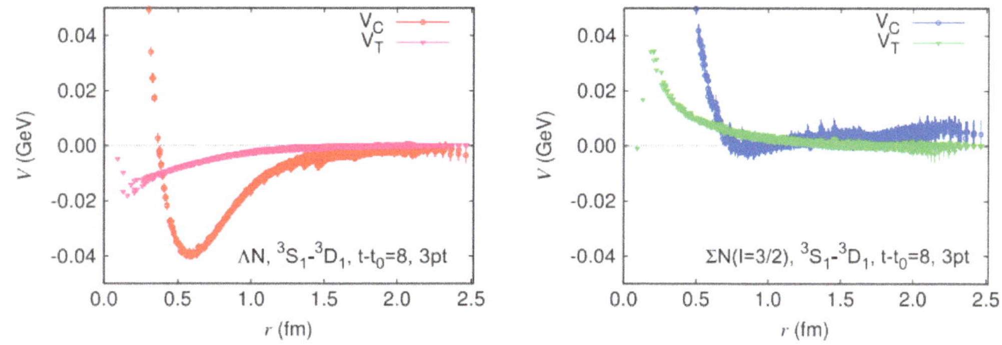

Figure. 11: (Left) The central potential (circle) and the tensor potential (triangle) in the $^3S_1 -^3 D_1$ 1 channel of the ΛN system as a function of r. (Right) The central potential (circle) and the tensor potential (triangle) in the $^3S_1 -^3 D_1$ channel of the $\Sigma N(I = 3/2)$ system as a function of r.

ΛN and ΛN Potentials in Full QCD

ΛN and ΣN $(I = 3/2)$ are the lowest states in the strangeness S = −1 systems with I = 1/2 and I = 3/2, respectively. Therefore potentials for these states can be calculated as in the case of the N N potential. In Ref. [43], the ΛN potential and the ΛN potential with I = 3/2 are calculated by using 2 + 1 flavor full QCD gauge configurations with the original time-independent HAL QCD method. In the following, we show improved results on a $32^3 \times 64$ lattice at a = 0.091(1)fm with the time-dependent HAL QCD method discussed in Sect. 3.4.

ΛN (left panel) and the $\Sigma N(I = 3/2)$ (right panel) potentials in the $^1S_0$ channel are shown in Fig. 10. In 2 + 1 flavor QCD, while the ΛN (I = 3/2) potential still belongs directly to the 27(I = 3/2) representation thanks to the isospin symmetry, an energy eigenstate of a ΣN system in the $^1S_0$ channel is a mixture of **27**(I=1/2) and $\mathbf{8}_s$ in the flavor representation, so that these two potentials are not necessarily equal. In the present 2 + 1 flavor QCD calculation shown in Fig. 10, these potentials look similar due to small flavor-SU(3) breaking: For example, our hadron masses are $(m_\pi, m_K, m_N, m_\Lambda, m_\Sigma) = (0.7006(4), 0.7879(4), 1.574(3), 1.635(3), 1.650(3))$ GeV TheleftpanelofFig.11showsthecentralpotential (circle) and the tensor potential (triangle) of the ΛN system in the $^3S_1 -^3 D_1$ channel, whose eigenstate is a mixture of $\overline{\mathbf{10}}$ and $\mathbf{8}_a$. The attractive well at distance $^3S_1 -^3 D_1$ channel, whose eigenstate is a mixture of $\overline{\mathbf{10}}$ and $\mathbf{8}_a$. channel, while the tensor potential itself (triangle) is weaker than the tensor potential in the N N system [44].

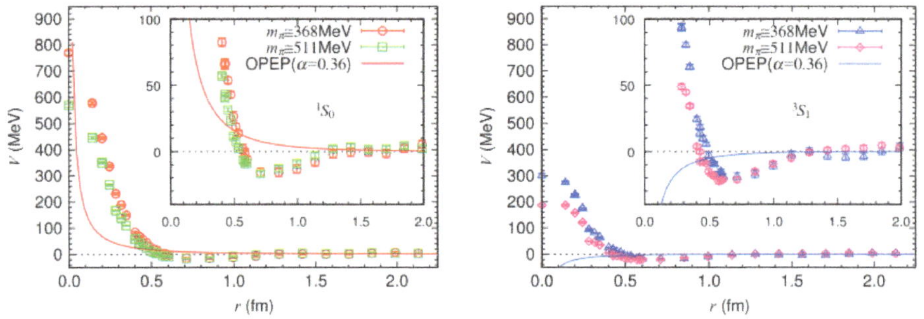

Figure. 12: (Left) The spin-singlet central potential for $p\Xi^0$ obtained from the orbital A_1^+ channel at $m_\pi \simeq 368$ MeV (circle) and $m_\pi \simeq 511$ MeV (box). The central part of the OPEP $(F/(F+D) = 0.36)$ in Eq.(5.2) is also given by a solid line. (Right) The spin-triplet effective central potential from the orbital A_1^+ channel at at $m_\pi \simeq 368$ MeV (triangle) and $m_\pi \simeq 511$ MeV V (diamond), together with the OPEP (solid line). Taken from Ref. [52].

The right panel of Fig. 11 shows the central potential (circle) and the tensor potential (triangle) of the $\Sigma N(I = 3/2)$ system in the $^3S_1 - ^3D_1$ channel. Due to the isospin symmetry, this channel belongs solely to the flavor 10 representation without a mixture of $\overline{10}$ or 8_a. As seen from the figure, there is no clear attractive well in the central potential (circle). This repulsive nature of the $\Sigma N(I = 3/2, ^3S_1 - ^3D_1)$ central potential is consistent with the prediction from the naive quark model [45,46]. The tensor force is a little stronger that that of the ΛN system but is still weaker in magnitude than that of the N N system.

ΞN Potential in Quenched QCD

Experimentally, not much information is available on the NΞ interaction except for a few studies: a recent report gives the upper limit of elastic and inelastic cross sections [47] while earlier publications suggest weak attractions of Ξ− nuclear interactions [48–50]. The Ξ−nucleus interactions will soon be studied as one of the day-one experiments at J-PARC [51] via the (K^-, K^+) reaction with a nuclear target. Ref. [52] gives the first result of the potential in $I = 1$ $N\Xi$ system, which does not show strong decay into other channels. The lattice parameters are the same as for the quenched N N potential in Sect. 4.1, but the method to determine the lattice spacing in Ref. [52] is a little different from the one in Sect. 4.1. The potential is calculated at $(m_\pi, m_N, m_\Xi) = (511(1)\,\text{MeV}, 1300(4)\,\text{MeV}, 1419(4)\,\text{MeV})$, and $(368(1)\,\text{MeV}, 1167(7)\,\text{MeV}, 1383(6)\,\text{MeV})$ with the interpolation operators

$$p_\alpha(x) = \varepsilon_{abc}(u_a^T(x)C\gamma_5 d_b(x))u_{c,\alpha}(x), \quad \Xi_\alpha^0(x) = \varepsilon_{abc}(u_a^T(x)C\gamma_5 s_b(x))s_{c,\alpha}(x). \tag{5.1}$$

Since both p and Ξ^0 have $(I, I_z) = (1/2, 1/2)$, the $p\Xi^0$ system has $I = 1$ with the strangeness S = −2.

The left (right) of Fig. 12 gives the (effective) central potential of the $p\Xi^0$ system obtained from the $L = A_1^+$ representation for the spin-singlet (triplet) at $m_\pi = 511$ MeV and 368 MeV. Potentials in the $I = 1$ $N\Xi$ system for both channels show a repulsive core at $r \leq 0.5$ fm surrounded by an attractive well, similar to the N N systems. In contrast to the N N case, however, the repulsive core of the $p\Xi^0$ potential in the spin-singlet channel is substantially stronger than in the triplet channel. The attraction in the medium to long distance region (0.6 fm ≤ r ≤ 1.2 fm) is similar in both channels. The height of the repulsive core increases as the light quark mass decreases, while a significant difference is not seen for the attraction in the medium to long distance within statistical errors. The potentials in Fig. 12 are weakly attractive on the whole in both spin channels at both pion masses, in spite of the repulsive core at short distances, and the attraction in the triplet is a little stronger than that in the singlet.

The solid lines in Fig. 12 are the one-pion exchange potential (OPEP), given by

$$V_C^\pi = -(1-2\alpha)\frac{g_{\pi NN}^2}{4\pi}\frac{(\boldsymbol{\tau}_N \cdot \boldsymbol{\tau}_\Xi)(\boldsymbol{\sigma}_N \cdot \boldsymbol{\sigma}_\Xi)}{3}\left(\frac{m_\pi}{2m_N}\right)^2\frac{e^{-m_\pi r}}{r} \tag{5.2}$$

with $(m_\pi, m_N) = (368\,\text{MeV}, 1167\,\text{MeV})$, where the pseudo-vector $\pi\Xi\Xi$ coupling $f_{\pi\Xi\Xi}$ is related to the πNN coupling as $f_{\pi\Xi\Xi} = -f_{\pi NN}(1-2\alpha)$ with the parameter $\alpha = F/(F+D)$, and $g_{\pi NN} = f_{\pi NN}\frac{m_\pi}{2m_N}$. The empirical values, $\alpha \simeq 0.36$ and $g_{\pi NN}/(4\pi) \simeq 14.0$, are used for the plot. Unlike the NN potential, the OPEP in the present case has opposite signs for the spin-singlet channel and spin-triplet channel. In addition, the absolute magnitude is smaller due to the factor 1 − 2α. No clear signature of the OPEP at long distances (r ≥ 1.2 fm) is yet observed in Fig. 12 within statistical errors.

BARYON INTERACTION IN THE FLAVOR SU(3) LIMIT

Potentials in the Flavor Su(3) Limit

In order to reveal the nature of the hyperon interactions in various channels, it is more convenient to consider an idealized flavor SU(3) symmetric world, where u, d, and s quarks are all degenerate with a common finite mass. In this limit, one can capture the essential features of the interaction, in particular, the short range force, without contamination from the quark mass difference. In the flavor SU(3) limit, the ground state baryon belongs to the flavor-octet with spin 1/2, and two-baryon states with a given angular momentum can be labeled by the irreducible representation of SU(3) as

$$\mathbf{8} \otimes \mathbf{8} = \underbrace{\mathbf{27} \oplus \mathbf{8} \oplus \mathbf{1}}_{\text{symmetric}} \oplus \underbrace{\overline{\mathbf{10}} \oplus \mathbf{10} \oplus \mathbf{8}}_{\text{anti-symmetric}}, \quad (6.1)$$

where "symmetric" and "anti-symmetric" stand for the symmetry under the flavor exchange for two baryons. For a system with an orbital S-wave, the Pauli principle for baryons requires 27, 8, and 1 to be spin-singlet $({}^1S_0)$, and $\overline{\mathbf{10}}$, $\mathbf{10}$, and $\mathbf{8}$ to be spin-triplet $({}^3S_1 - {}^3D_1)$.

Calculations in the SU(3) limit allow us to extract potentials for these six flavor irreducible multiplets as follows.

A two-baryon operator B $B^{(X)}$, which belongs to one definite flavor representation X, can be given in terms of the baryon base operator with the corresponding Clebsch–Gordan (CG) coefficients C_{ij}^X as $BB^{(X)} = \sum_{ij} C_{ij}^X B_i B_j$. By using this operator at source and/or sink, the NBS wave function for the two-baryon system in the flavor representation can be obtained. Potentials in the flavor base, $V^{(X)}(r)$, are extracted from such wave functions in the same manner as nuclear forces, explained in Sect. 3.

In Ref. [53], the (effective) central potentials are calculated in the original time-independent HAL QCD method by using the 3 flavor full QCD gauge configuration [54] on a $16^3 \times 32$ lattice at $a \simeq 0.12$ fm, and at two values of quark hopping parameter corresponding to $(M_{\rm ps}, M_B) =$ (1014(1) MeV, 2026(3) MeV) and (835(1) MeV, 1752(3) MeV), where M_{ps} and M_B denote the octet pseudo-scalar (PS) meson mass and the octet baryon mass, respectively. In Refs. [55–57], on the other hand, the central and tensor potentials are calculated in the time-dependent HAL QCD method discussed in Sect. 3.4 by using the 3 flavor full QCD gauge configuration generated by the HAL QCD Collaboration on a $32^3 \times 32$ lattice at $a \simeq 0.12$ 2 fm, and at five values of quark hopping parameter that correspond to (M_{ps}, M_B) = (1170.9(7) MeV, 2274(2) MeV), (1015(1) MeV, 2030(2) MeV), (837(1) MeV, 1748(1) MeV), (673(1) MeV, 1485(2) MeV), and (468.6(7) MeV, 1161(2) MeV).

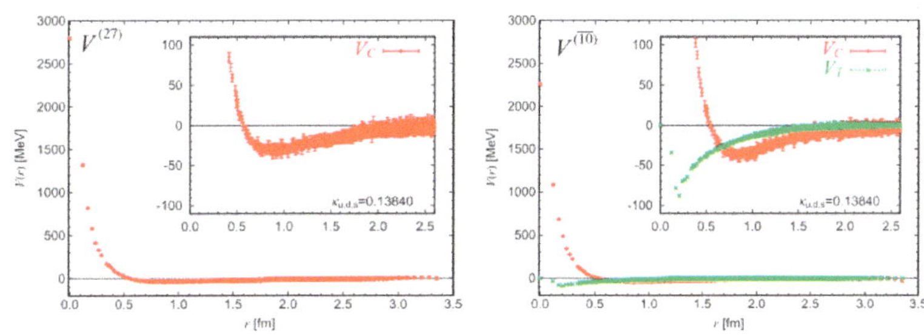

Figure. 13: The B B potentials in 27 (left) and $\overline{\mathbf{10}}$ (right) representations extracted from the lattice QCD simulation at M_{ps} = 469 MeV. Taken from Ref. [57].

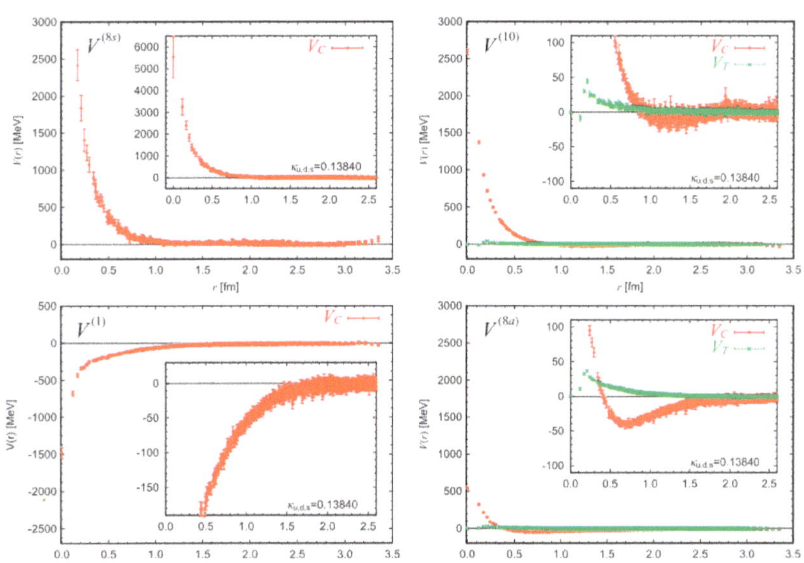

Figure. 14: The B B potentials in 8_s (upper left), 10 (upper right), 1 (lower left), and 8_a (lower right) extracted from the lattice QCD simulation at M_{ps} = 469 MeV. Taken from Ref. [57].

Figures 13 and 14 show the flavor basis potentials for M_{ps} = 469 MeV [57]. The left panels show central potentials for the spin-singlet channel from the $J^P = A_1^+$ state, while the right panels give central (V_C) and tensor (V_T) potentials for the spin-triplet channel from the $J^P = T_1^+$ state. As listed in Table 3, some octet-baryon pairs belong solely to an irreducible representation of flavor SU(3). For example, symmetric N N belongs to the 27 representation. Therefore, $V^{(27)}(r)$ can be considered as the flavor SU(3) symmetric limit of the N N spin-singlet (^{1} S0) potential. Similarly, $V^{(\overline{10})}$, $V^{(10)}$, and $V^{(8_a)}$ can be considered as the flavor SU(3) symmetric limit of some B B potentials of the particle basis, while $V^{(1)}$ and $V^{(8s)}$ are always mixtures of different B B potentials in the particle basis.

Table 3: Baryon pairs that belong to an irreducible flavor SU(3) representation, where {B B } and [B B] denote B B + B B and B B − B B, respectively. Flav

Flavor multiplet	Baryon pair (isospin)
27	$\{NN\}(I=1)$, $\{N\Sigma\}(I=3/2)$, $\{\Sigma\Sigma\}(I=2)$, $\{\Sigma\Xi\}(I=3/2)$, $\{\Xi\Xi\}(I=1)$
8$_s$	none
1	none
10*	$[NN](I=0)$, $[\Sigma\Xi](I=3/2)$
10	$[N\Sigma](I=3/2)$, $[\Xi\Xi](I=0)$
8$_a$	$[N\Xi](I=0)$

Figure 13 shows $V_C^{(27)}(r)$ and $V_{C,T}^{(\overline{10})}(r)$, which correspond to spin-singlet and spin-triplet N N potentials, respectively. Both central potentials have a repulsive core at short distances with an attractive pocket around 0.8 fm. These qualitative features are consistent with the results found for the N N potential in the previous section. The upper right panel of Fig. 14 shows that $V_C^{(10)}(r)$) has a stronger repulsive core and a weaker attractive pocket compared to $V_C^{(27)}(r)$ and $V^{(\overline{10})}{}_C(r)$. Furthermore, $V_C^{(8_s)}(r)$ in the upper left panel of Fig. 14 has a very strong repulsive core among all 6 channels, while $V_C^{(8_a)}(r)$ in the lower right panel has a very weak repulsive core. In contrast to all other cases, $V_C^{(1)}(r)$ shows attraction at short distances instead of repulsion, as shown in the lower left panel. The above features are consistent with what has been observed in an SU(6) quark model [45,46]. In particular, the potential in the 8_s channel in this quark model becomes strongly repulsive at short distances, since the six quarks cannot occupy the same orbital state due to the Pauli exclusion for quarks. On the other hand, the potential in the 1 channel does not suffer from quark Pauli exclusion at all, and can become attractive due to the short-range gluon exchange. Such agreements between the lattice data and the quark model suggest that quark Pauli exclusion plays an essential role for the repulsive core in B B systems. The potential for the flavor singlet is entirely attractive even at very short distances. This might produce a bound state, the H-dibaryon, in this channel. We will discuss this possibility in the next subsection. In the flavor SU(3) limit, the B B potentials in the particle basis can be obtained from those in the flavor basis by a unitary rotation as

$$V_{ij}(r) = \sum_X U_{iX} V^{(X)}(r) U_{Xj}^*, \quad (6.2)$$

where U_{iX} is a unitary matrix that rotates the flavor basis $|X\rangle$ to the particle basis $|i\rangle$ as $|i\rangle = U_{iX}|X\rangle$, and is given in terms of the CG coefficients. The explicit forms of the unitary matrix U are found in Ref. [53]. In Fig. 15, we show B B potentials for the $S=-2$, $I=0$, $^1S_0$ sector at $M_{PS} = 469$ MeV, as a characteristic example. The flavor base potentials are fitted by the analytic function composed of an attractive Gaussian core plus a long range (Yukawa)^2 attraction,

$$V(r) = b_1 e^{-b_2 r^2} + b_3(1 - e^{-b_4 r^2})^2 \left(\frac{e^{-b_5 r}}{r}\right)^2, \quad (6.3)$$

with five parameters $b_{1,2,3,4,5}$. The left panel of Fig. 15 shows the diagonal potentials. One observes that all three diagonal potentials have a repulsive core. The repulsion is most strong in the $\Sigma\Sigma$ (I=0) channel, reflecting its largest CG coefficient of the 8s state among the three channels, while the

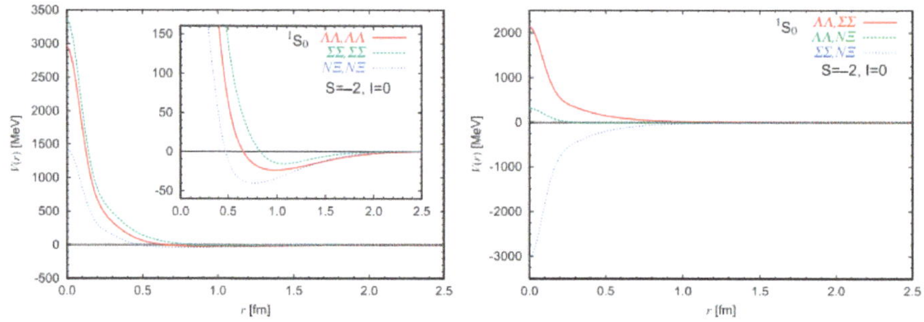

Figure. 15: B B potentials in the particle basis for the S = −2, I = 0, ^{1} S_0 sector. Three diagonal (off-diagonal) potentials are shown in the left (right) panel.

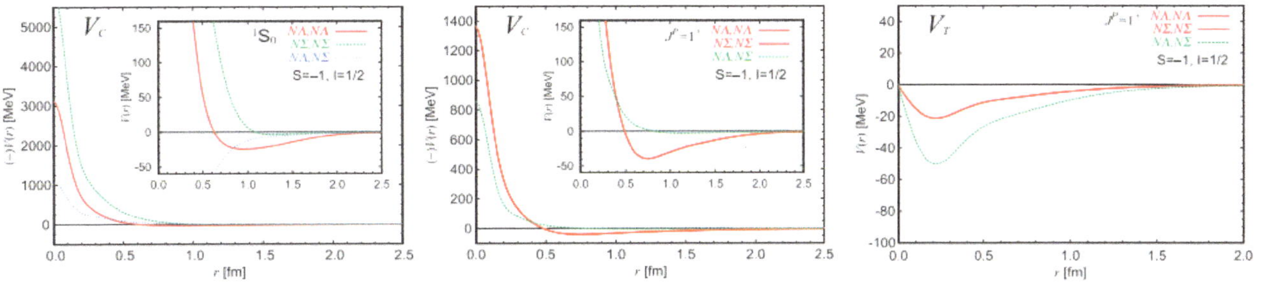

Figure. 16: B B potentials in the particle basis for the S = −1, I = 1/2 sector, $^1S_0$ (left), spin-triplet V_C (center), and spin-triplet V_T (right), extracted from the lattice QCD simulation at M_{ps} = 469 MeV

attraction in the 1 state is reflected most in the $N\Xi$(I=0) potential due to its largest CG coefficient. The right panel of Fig. 15 shows the off-diagonal potentials, which are comparable in magnitude to the diagonal ones, except for the $\Lambda\Lambda$-$N\Xi$ transition potential. Since the off-diagonal parts are not negligible in the particle basis, a fully coupled channel analysis is necessary to study observables in this system. This is important when we study the real world with flavor SU(3) breaking, where only the particle base is meaningful.

As another example, we show B B potentials for the S = −1, I = 1/2 sector at M_{PS} = 469 MeV in Fig. 16. The left panel of Fig. 16 shows the potential in the $^1S_0$ channel, while the center (right) panel of Fig. 16 shows the central (tensor) potential for the $J^P = T_1^+$ spin-triplet channel. We observe that the off-diagonal $N\Lambda$-$N\Sigma$ potentials are particularly large, especially in the spin-triplet channel, so that a full $N\Lambda$-$N\Sigma$ coupled channel analysis is also necessary to study observables. In addition, the repulsive cores in the spin-single channel are much stronger than those in the spin-triplet channel, due to the strong repulsion of the 8_s state, which couples only to the spin-singlet channel.

Although all quark masses of 3 flavors are degenerate and rather heavy in the present simulations, these particle base potentials, shown in Fig. 15 and Fig. 16, may provide useful information for the behavior of hyperons in hyper-nuclei and in neutron stars [42,58]. B B potentials in all other sectors can be found in Ref. [53].

Bound H Dibaryon in the Flavor Su(3) Limit

In this subsection, we investigate the potential of the flavor singlet channel in order to see whether or not the bound H dibaryon exists in the flavor SU(3) limit case.

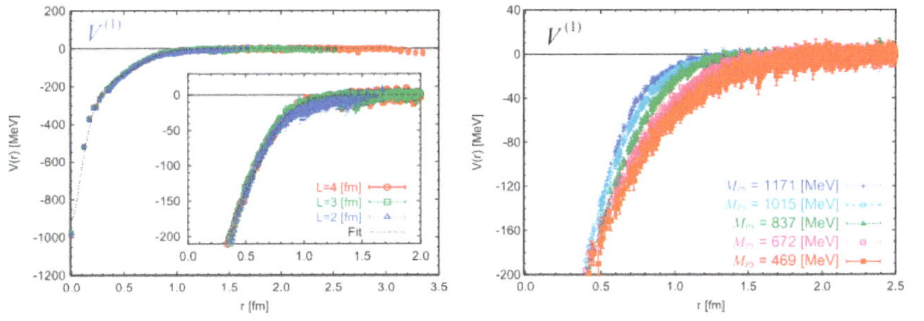

Figure. 17: The flavor-singlet potential $V_C^{(1)}(r)$ at $(t - t_0)/a = 10$. (Left) Results for $L = 1.94, 2.90, 3.87$ fm at $M_{ps} = 1015$ MeV. (Right) Results for $L = 3.87$ fm at $M_{ps} = 1171, 1015, 837, 673, 469$ MeV.

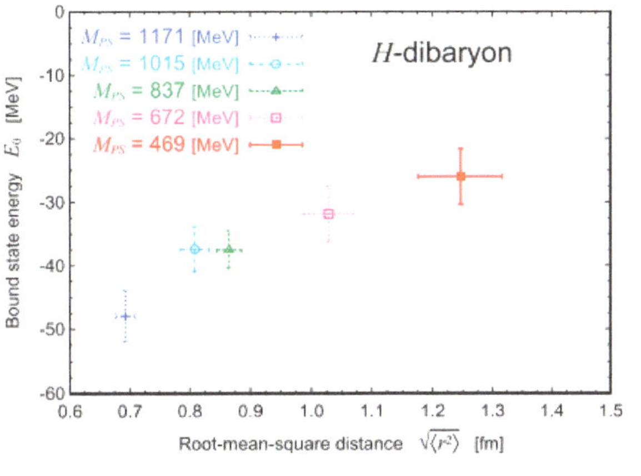

Figure. 18: The energy E_0 and the root-mean-square distance $\sqrt{\langle r^2 \rangle}$ of the bound state in the flavor singlet channel at each quark mass. Bars represent statistical errors only

Potentials for the flavor irreducible channels in the SU(3) limit have been calculated in Refs. [55–57] on $16^3 \times 32$, $24^3 \times 32$, and $32^3 \times 32$ lattices at a = 0.121(2)fm and five values of the quark mass, as mentioned before.

Figure 17 shows the volume and the quark mass dependences of the central potential in the flavorsinglet channel $V_C^{(1)}(r)$ at $(t - t_0)/a = 10$, where the potentials do not show an appreciable change with respect to the choice of t. The flavor-singlet potential is shown to have an "attractive core" and to be well localized in space. Because of the latter property, no significant volume dependence of the potential is observed within the statistical errors, as seen in Fig. 17 (left). As the quark mass decreases in Fig. 17 (right), the long range part of the attraction tends to increase. The resultant potential is fitted by the form in Eq. (6.3). With the five parameters, $b_{1,2,3,4,5}$, the lattice results can be fitted reasonably well with $\chi^2/\mathrm{dof} \simeq 1$. The fitted result for $L = 3.87$ fm is shown by the dashed line in Fig. 17 (left).

Solving the Schrödinger equation with the fitted potential in infinite volume, the energies and the wave functions are obtained at the present quark masses in the flavor SU(3) limit. It turns out that, at each quark mass, there is only one bound state with a binding energy of 20–50 MeV. Figure 18 shows the energy and the root-mean-squared (rms) distance of the bound state at each quark mass obtained from the potential at L = 3.87 fm and $(t - t_0)/a = 10$, where errors are estimated by the jackknife method. Despite the fact that the potential becomes more attractive as the quark mass decreases, the resultant binding energies of the H-dibaryon decrease in the present range of the quark masses, since the increase of the attraction toward the lighter quark mass is compensated by the increase of

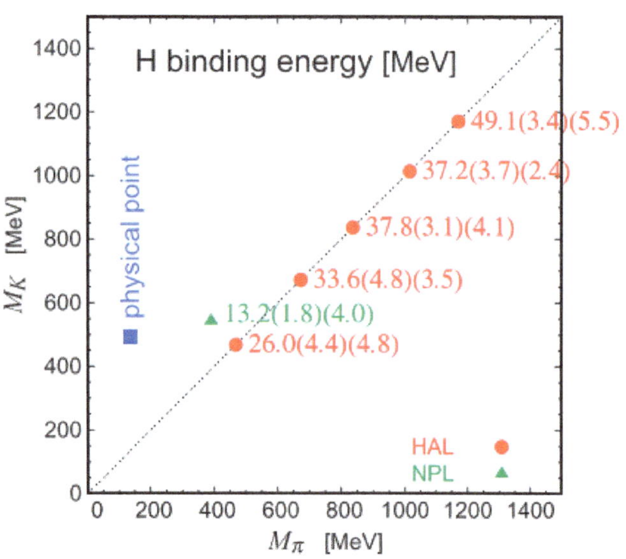

Figure. 19: Summary of the H-dibaryon binding energy in recent full QCD simulations. HAL stands for the present results and NPL stands for the result in Ref. [60].

the kinetic energy for the lighter baryon mass. It is noted that there appears no bound state for the potential of the 27-plet channel or the $\overline{10}$-plet channel ("deuteron") in the present range of quark masses. By including systematic errors caused by the choice of sink-time t in $R(r, t - t_0)$, the final results of the binding energy B_H and the rms distance $\sqrt{\langle r^2 \rangle}$ are summarized below, where the 1st and 2nd parentheses correspond to statistical and systematic errors, respectively

$$M_{ps} = 1171\,\mathrm{MeV}: \quad B_H = 49.1(3.4)(5.5)\,\mathrm{MeV} \quad \sqrt{\langle r^2 \rangle} = 0.685(13)(25)\,\mathrm{fm}$$

$$M_{ps} = 1015\,\mathrm{MeV}: \quad B_H = 37.2(3.7)(2.4)\,\mathrm{MeV} \quad \sqrt{\langle r^2 \rangle} = 0.809(23)(10)\,\mathrm{fm}$$

$$M_{ps} = 837\,\mathrm{MeV}: \quad B_H = 37.8(3.1)(4.2)\,\mathrm{MeV} \quad \sqrt{\langle r^2 \rangle} = 0.865(20)(25)\,\mathrm{fm}$$

$$M_{ps} = 672\,\mathrm{MeV}: \quad B_H = 33.6(4.8)(3.5)\,\mathrm{MeV} \quad \sqrt{\langle r^2 \rangle} = 1.029(41)(23)\,\mathrm{fm}$$

$$M_{ps} = 469\,\mathrm{MeV}: \quad B_H = 26.0(4.4)(4.8)\,\mathrm{MeV} \quad \sqrt{\langle r^2 \rangle} = 1.247(70)(59)\,\mathrm{fm}.$$

Recently, the existence of the H-dibaryon has also been investigated by a direct calculation of its binding energy in 2 + 1 full QCD simulations [59,60], where B_H = 13.2(1.8)(4.0) MeV is reported in the L → ∞ extrapolation at $m_\pi \simeq 389$ MeV, $m_K \simeq 544$ MeV. Figure 19 gives a summary of the H-dibaryon binding energy from full QCD simulations recently reported. Since the binding energy is comparable to the splitting between physical hyperon masses and is not so sensitive to the quark mass, there may be a possibility of a weakly bound or resonant H-dibaryon, even in the real world with lighter quark masses and flavor SU(3) breaking. Our phenomenological trial analysis using 3-flavor lattice QCD results suggests a resonant H-dibaryon above the $\Lambda\Lambda$ but below the $N\Xi$ thresholds [57]. To make a definite conclusion on this point, however, a $\Lambda\Lambda - N\Xi - \Sigma\Sigma$ coupled channel analysis is necessary for H in the (2 + 1)-flavor lattice QCD simulations, as will be discussed in Sect. 7.

HADRONIC INTERACTIONS ABOVE THE INELASTIC THRESHOLD

In this section, we discuss a method to investigate hadron interactions above the inelastic threshold by generalizing the HAL QCD method. Then we apply the method to coupled channel potentials in the S = −2 and I = 0 sectors.

Coupled-Channel Approach to Inelastic Scattering

For simplicity, let us discuss a case of A + B → C + D scattering where A, B,C, D represent some 1-particle states. This is a simplified version of the octet baryon scattering in the strangeness S = −2 and isospin I = 0 channel, where $\Lambda\Lambda$, $N\Xi$, and $\Sigma\Sigma$ appear as asymptotic states of the strong interaction if the total energy is larger than $2m_\Sigma$. We here assume $m_A + m_B < m_C + m_D < W$, where $W = E_k^A + E_k^B$ is the total energy of the system, and $E_k^X = (m_X^2 + \boldsymbol{k}^2)^{1/2}$.

In this case, the QCD eigenstate with the quantum numbers of the AB state and center of mass energy W is expressed as

$$|W\rangle = c_{AB}|AB, W\rangle + c_{CD}|CD, W\rangle + \cdots \quad (7.1)$$

$$|AB, W\rangle = |A, \boldsymbol{k}\rangle_{\mathrm{in}} \otimes |B, -\boldsymbol{k}\rangle_{\mathrm{in}}, \quad |CD, W\rangle = |C, \boldsymbol{q}\rangle_{\mathrm{in}} \otimes |D, -\boldsymbol{q}\rangle_{\mathrm{in}}, \quad (7.2)$$

where $W = E_k^A + E_k^B = E_q^C + E_q^D$. Then we define the following NBS wave functions,

$$\varphi_{AB}(\boldsymbol{r}, \boldsymbol{k})e^{-Wt} = \langle 0|T\{\varphi_A(\boldsymbol{x} + \boldsymbol{r}, t)\varphi_B(\boldsymbol{x}, t)\}|W\rangle, \quad (7.3)$$

$$\varphi_{CD}(\boldsymbol{r}, \boldsymbol{q})e^{-Wt} = \langle 0|T\{\varphi_C(\boldsymbol{x}+\boldsymbol{r}, t)\varphi_D(\boldsymbol{x}, t)\}|W\rangle. \tag{7.4}$$

Using the partial wave decomposition such that[4]

$$\varphi_X(\boldsymbol{r}, \boldsymbol{k}) = 4\pi \sum_{l,m} i^\ell \varphi_X^\ell(r, k) Y_{\ell m}(\Omega_{\boldsymbol{r}})\overline{Y_{\ell m}(\Omega_{\boldsymbol{k}})} \tag{7.5}$$

for X = AB or C D, it can be shown [62] that these wave functions satisfy

$$(\nabla^2 + \boldsymbol{k}^2)\varphi_{AB}(\boldsymbol{r}, \boldsymbol{k}) = 0, \quad (\nabla^2 + \boldsymbol{q}^2)\varphi_{CD}(\boldsymbol{r}, \boldsymbol{q}) = 0 \tag{7.6}$$

for $r \to \infty$

Let us now consider QCD in the finite volume V where $|AB, W\rangle$ and $|CD, W\rangle$ are no longer eigenstates of the Hamiltonian. True eigenvalues are shifted from W to $W_i = W + O(V^{-1})$ $(i = 1, 2)$. By the diagonalization method in lattice QCD, it is relatively easy to determine W_1 and W_2. With these values, Lüscher's finite volume formula gives two conditions, which, however, are insufficient to determine three observables: two scattering phase shifts $\delta_\ell^1, \delta_\ell^2$ and one mixing angle θ. We here explain a new approach proposed in Refs. [61,62] to overcome this difficulty. Let us consider the NBS wave functions at two different values of energy, W_1 and W_2, in the finite volume:

$$\varphi_X(\boldsymbol{x}, \boldsymbol{p}_i^X)e^{-W_i t} = \langle 0|T\{\varphi_{X_1}(\boldsymbol{x}+\boldsymbol{r}, t)\varphi_{X_2}(\boldsymbol{r}, t)\}|W_i\rangle, \quad i = 1, 2. \tag{7.7}$$

where $X(= X_1X_2) = AB$ or CD with $\boldsymbol{p}^{AB} = \boldsymbol{k}$ or $\boldsymbol{p}^{CD} = \boldsymbol{q}$. We then define the coupled channel non-local potentials from the coupled channel Schrödinger equation as

$$\left[\frac{(p_i^X)^2}{2\mu_X} - H_0\right]\varphi_X(\boldsymbol{x}, \boldsymbol{p}_i^X) = \sum_Y \int d^3y\, U_{X,Y}(\boldsymbol{x}, \boldsymbol{y})\, \varphi_Y(\boldsymbol{y}, \boldsymbol{p}_i^Y) \tag{7.8}$$

for i = 1, 2 where the reduced mass is defined by $1/\mu_X = 1/m_{X_1} + 1/m_{X_2}$. In the leading order of the velocity expansion, we have

$$K_X(\boldsymbol{x}, \boldsymbol{p}_i^X) \equiv \left[\frac{(p_i^X)^2}{2\mu_X} - H_0\right]\varphi_X(\boldsymbol{x}, \boldsymbol{p}_i^X) = \sum_Y V_{X,Y}(\boldsymbol{x})\, \varphi_Y(\boldsymbol{x}, \boldsymbol{p}_i^Y) \tag{7.9}$$

These equations for i = 1, 2 can be solved as

$$\begin{pmatrix} V_{AB,AB}(\boldsymbol{x}) & V_{AB,CD}(\boldsymbol{x}) \\ V_{CD,AB}(\boldsymbol{x}) & V_{CD,CD}(\boldsymbol{x}) \end{pmatrix} = \begin{pmatrix} K_{AB}(\boldsymbol{x}, \boldsymbol{k}_1) & K_{AB}(\boldsymbol{x}, \boldsymbol{k}_2) \\ K_{CD}(\boldsymbol{x}, \boldsymbol{q}_1) & K_{CD}(\boldsymbol{x}, \boldsymbol{q}_2) \end{pmatrix} \begin{pmatrix} \varphi_{AB}(\boldsymbol{x}, \boldsymbol{k}_1) & \varphi_{AB}(\boldsymbol{x}, \boldsymbol{k}_2) \\ \varphi_{CD}(\boldsymbol{x}, \boldsymbol{q}_1) & \varphi_{CD}(\boldsymbol{x}, \boldsymbol{q}_2) \end{pmatrix}^{-1} \tag{7.10}$$

Once we obtain the coupled channel local potentials $V_{X,Y}$ (x), we solve the coupled channel Schrödinger equation in infinite volume with some appropriate boundary condition such that the incoming wave has a definite and consists of the AB state only, in order to extract three observables for each ℓ ($\delta_\ell^1(W)$, $\delta_\ell^2(W)$, and $\theta_\ell(W)$) at all values of W. Of course, since $V_{X,Y}$ is the leading order approximation in the velocity expansion of $U_{X,Y}$ (x, y), the results for the three observables $\delta_\ell^1(W)$, $\delta_\ell^2(W)$, and $\theta(W)$ at $W \neq W_1, W_2$ are also approximate ones and might be different from the exact values. By performing an additional extraction of $V_{X,Y}(\boldsymbol{x})$ at $(W_3, W_4) \neq (W_1, W_2)$, we can test how good the leading order approximation is. The method considered above can be generalized to inelastic scattering where a number of particles is not conserved such that A + B → A + B and A + B → A + B + C. See Ref. [62] for more details.

Coupled-Channel Potentials in the (S, I) = (−2, 0) Channel

As an application of the method in the previous subsection, let us consider B B potentials for the S = −2 and I = 0 system, i.e., the coupled $\Lambda\Lambda$–$N\Xi$–$\Sigma\Sigma$ system. It is interesting to investigate this system since it involves the flavor single state, which is free from the Pauli blocking of quark degrees of freedom at short distances. Since mass differences in these B B systems are quite small, all 3 states appear in NBS wave functions. The coupled channel method in the previous subsection can be applied to treat such complicated systems. At the leading order of the velocity expansion, the coupled channel 3 × 3 potential matrix in this case is given by

where i is a label for energy W_i and $X, Y = \Lambda\Lambda, N\Xi$ or $\Sigma\Sigma$. Here the last factor is the inverse of the 3 × 3 matrix $\varphi_A^{W_i}(\boldsymbol{r}, \mathbf{k}_i^A)$ with indices i and A.

Gauge configurations generated by CP-PACS/JLQCD Collaborations on a $16^3 \times 32$ lattice at $a \simeq 0.12$ fm (therefore $L \simeq 1.9$ fm) in 2 + 1-flavor full QCD simulations are employed to calculate the coupled channel potentials at three different values of the light quark mass with a fixed bare strange quark mass [63]. Quark propagators are calculated with the spatial wall source at t_0 with the Dirichlet boundary condition in time at $t = t_0 + 16$. The corresponding hadron masses are given in Table 4.

Table 4: Hadron masses in units of MeV and number of configurations for each set adopted in Sect. 7.2.

	m_π	m_K	m_N	m_Λ	m_Σ	m_Ξ
Set 1	875(1)	916(1)	1806(3)	1835(3)	1841(3)	1867(2)
Set 2	749(1)	828(1)	1616(3)	1671(2)	1685(2)	1734(2)
Set 3	661(1)	768(1)	1482(3)	1557(3)	1576(3)	1640(3)

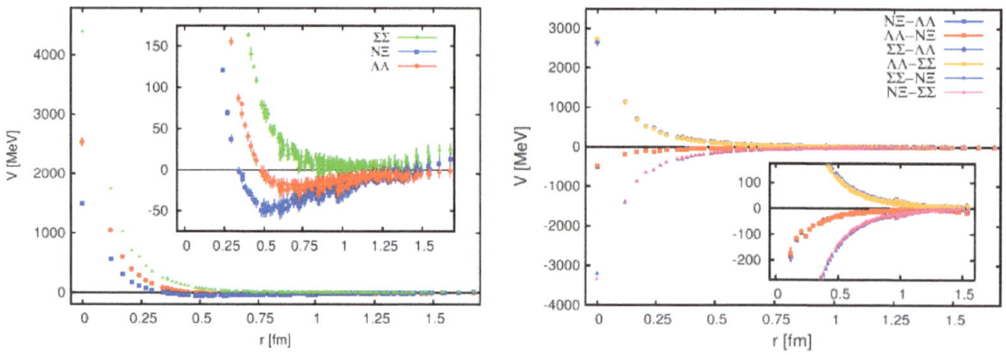

Figure. 20: The coupled channel potential matrix from the NBS wave function for Set 1. The vertical axis is the potential strength in units of MeV, while the horizontal axis is the relative distance between two baryons in units of fm.

The coupled channel potential matrix $V_{AB,C D}$ from the NBS wave function for Set 1 is shown in Fig. 20. All diagonal components of the potential matrix have a repulsion at short distance. The strength of the repulsion in each channel, however, varies, reflecting the properties of its main component in the irreducible representation of the flavor SU(3). In particular, the $\Sigma\Sigma$ potential has the strongest repulsive core of these three components. It is important to note that off-diagonal parts of the potential matrix satisfy the hermiticity relation $V_{AB,C D} = V_{C D,AB}$ within statistical errors. In addition, the off-diagonal parts of the $\Lambda\Lambda$ to $N\Xi$ transition $V_{\Lambda\Lambda,N\Xi}$ are much smaller than the other two off-diagonal potentials, $V_{\Lambda\Lambda,\Sigma\Sigma}$ and $V_{N\Xi,\Sigma\Sigma}$.

In order to compare the results of the potential matrix calculated in three configuration sets, the potentials from the particle basis are transformed by the unitary rotation U to those in the flavor SU(3) irreducible representation (IR) basis as

$$V^{IR} = U^{\dagger} V U = \begin{pmatrix} V_{1,1} & V_{1,8} & V_{1,27} \\ V_{8,1} & V_{8,8} & V_{8,27} \\ V_{27,1} & V_{27,8} & V_{27,27} \end{pmatrix} \quad (7.12)$$

The potential matrix in the IR basis give a good measure of flavor SU(3) breaking effects since it is diagonal in the SU(3) symmetric limit. In Fig. 21, the results of the potential matrix in the IR basis are compared between different configuration sets, together with that in the flavor SU(3) symmetric limit. As the pion mass decreases, the attraction in the $V_{1,1}$ potential increases over the whole range. While the V8,8 potential in the flavor SU(3) limit deviates from others, we do not observe a clear pion mass dependence of the potentials among these flavor breaking cases. In the $V_{27,27}$ potential, we observe a growth in the attraction range of the potential and an enhancement of the repulsive core. The $V_{1,27}$ and $V_{8,27}$ transition potentials

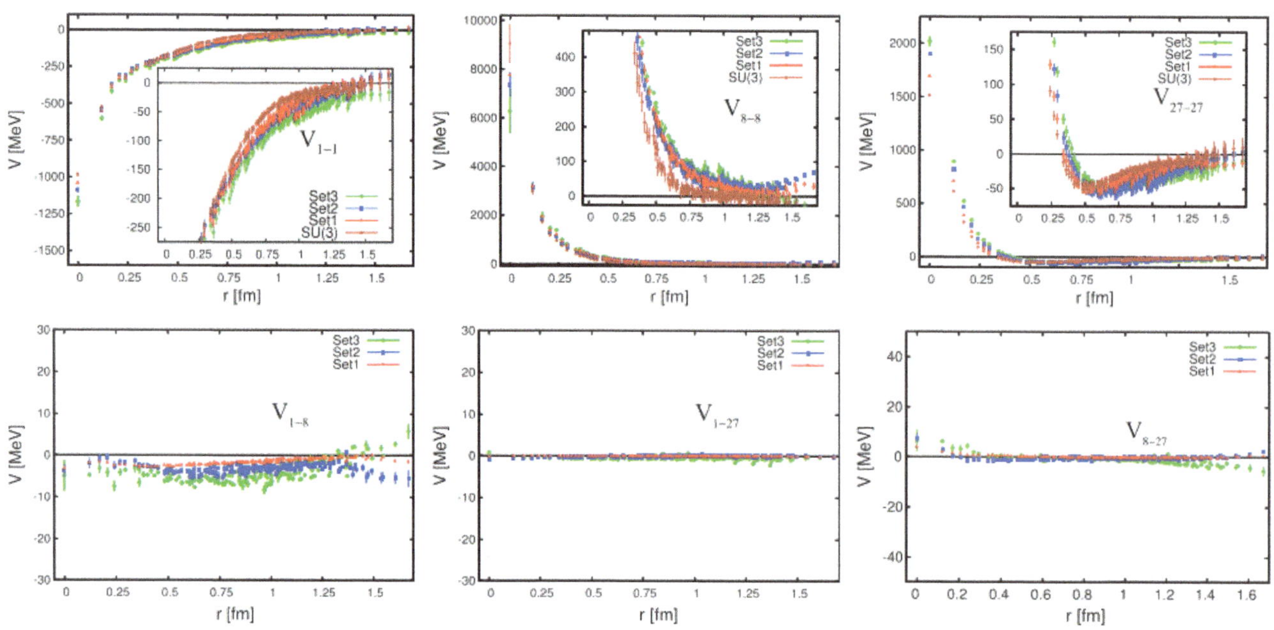

Figure. 21: Transition potentials in the flavor SU(3) IR basis. Red, blue, and green symbols correspond to the results of Set 1, Set 2, and Set 3, respectively. The result of the flavor SU(3) symmetric limit at the same strange quark mass is also plotted with brown symbols [53].

are consistent with zero within statistical errors. On the other hand, it is noteworthy that the flavor SU(3) symmetry breaking effect becomes manifest in the $V_{1,8}$ transition potential.

THREE-NUCLEON FORCES

In this section, we expand the scope of our studies from two-nucleon (2N) systems to A-body nucleon systems. Generally speaking, there could exist not only 2-body forces but also 3, 4,... A-body forces in such systems. In particular, the determination of three-nucleon forces (3NF) attracts a great deal of interest, since it has been revealed that 3NF play an important and nontrivial role in various phenomena. Some examples include (a) binding energies of light nuclei [64], (b) deuteron–proton elastic scattering experiments [65], (c) the anomaly in the oxygen isotopes near the neutron drip-line [66], and (d) the nuclear equation of state (EoS) at high density relevant to the physics of neutron stars [67]. Universal short-range repulsion for three baryons (nucleons and hyperons) has also been suggested in relation to the maximum mass of neutron stars with hyperon cores [68,69]. Despite its phenomenological importance, microscopic understanding of 3NF is still limited. Pioneered by Fujita and Miyazawa [70], the long range part of 3NF has been modeled by the two-pion exchange (2πE) [71], particularly with the Δ-resonance excitation. This 2πE-3NF component is known to have an attractive nature at long distances. An additional repulsive component of 3NF at short distances is often introduced in a purely phenomenological way [72]. An approach based on the chiral effective field theory (EFT) is quite useful to classify two-, three-, and more-nucleon forces and has been studied intensively [73,74]. A completely different approach based on holographic QCD was recently proposed, which obtains repulsive 3NF at short distances [75]. To go beyond phenomenology, it is most desirable to determine 3NF directly from the fundamental degrees of freedom (DoF), the quarks and the gluons, on the basis of quantum chromodynamics (QCD). In this section, we present an exploratory study of a first-principles lattice QCD calculation of 3NF in the quantum numbers of $(I, J^P) = (1/2, 1/2^+)$ (the triton channel). For details of this study, see Ref. [76].

Formalism

We consider the NBS wave function ψ3N (r, ρ) extracted from the six-point correlator as

$$G_{3N}(\boldsymbol{r}, \boldsymbol{\rho}, t - t_0) \equiv \frac{1}{L^3} \sum_{\boldsymbol{R}} \langle 0 | (N(\boldsymbol{x}_1) N(\boldsymbol{x}_2) N(\boldsymbol{x}_3))(t) \, \overline{(N'N'N')}(t_0) | 0 \rangle,$$
$$\xrightarrow[t \gg t_0]{} A_{3N} \psi_{3N}(\boldsymbol{r}, \boldsymbol{\rho}) e^{-E_{3N}(t - t_0)}, \quad A_{3N} = \langle E_{3N} | \overline{(N'N'N')}(0) | 0 \rangle, \tag{8.1}$$

$$\psi_{3N}(\boldsymbol{r}, \boldsymbol{\rho}) \equiv \langle 0 | N(\boldsymbol{x}_1) N(\boldsymbol{x}_2) N(\boldsymbol{x}_3)(0) | E_{3N} \rangle, \tag{8.2}$$

Where E_{3N} and $|E_{3N}\rangle$ denote the energy and the state vector of the 3N ground state, respectively, N (N) the nucleon operator in the sink (source), and $\boldsymbol{R} \equiv (\boldsymbol{x}_1 + \boldsymbol{x}_2 + \boldsymbol{x}_3)/3$, $\boldsymbol{r} \equiv \boldsymbol{x}_1 - \boldsymbol{x}_2$, $\boldsymbol{\rho} \equiv \boldsymbol{x}_3 - (\boldsymbol{x}_1 + \boldsymbol{x}_2)/2$ the Jacobi coordinates. We consider the following Schrödinger equation of the 3N system with the derivative expansion of the potentials,

$$\left[-\frac{1}{2\mu_r} \nabla_r^2 - \frac{1}{2\mu_\rho} \nabla_\rho^2 + \sum_{i<j} V_{2N}(\boldsymbol{r}_{ij}) + V_{3NF}(\boldsymbol{r}, \boldsymbol{\rho}) \right] \psi_{3N}(\boldsymbol{r}, \boldsymbol{\rho}) = E_{3N} \psi_{3N}(\boldsymbol{r}, \boldsymbol{\rho}), \tag{8.3}$$

Where

$V_{2N}(\boldsymbol{r}_{ij})$ with $\boldsymbol{r}_{ij} \equiv \boldsymbol{x}_i - \boldsymbol{x}_j$ denotes the 2NF between the (i, j) pair, $V_{3NF}(\boldsymbol{r}, \boldsymbol{\rho})$ the 3NF, and $\mu_r = m_N/2$, $\mu_\rho = 2m_N/3$ the reduced masses. 3NF can be determined as follows. We first calculate ψ3N (r, ρ) and obtain the total potential of the 3N system through Eq. (8.3). Once we obtain all necessary $V_{2N}(\boldsymbol{r}_{ij})$ by performing (separate) lattice simulations for genuine 2N systems, we can extract $V_{3NF}(\boldsymbol{r}, \boldsymbol{\rho})$ by subtracting $\sum_{i<j} V_{2N}(\boldsymbol{r}_{ij})$ from the total potential. The extension to four- and more-nucleon forces can be immediately understood. Note that potentials determined in this way reproduce the energy of the system by construction. An important remark is that 3NF are always determined in combination with 2NF, and 3NF alone do not make too much sense. Therefore a comparison between lattice 3NF and phenomenological 3NF can be made only at a qualitative level. Rather, our purpose is to determine two-, three-, and (more-) nucleon forces systematically, and to provide them as a consistent set. One of the difficulties in the 3NF study from lattice QCD is that computational costs become enormous. Since there are 9 valence quarks, the DoF of color and spinor are significantly enlarged. In addition, the number of diagrams in the Wick contraction tends to diverge with a factor of $N_u! \times N_d!$,

where N_u (N_d) are numbers of up (down) quarks in the system. We here develop several techniques to reduce these computational costs. We first take advantage of symmetries (such as isospin symmetry) to reduce the number of Wick contractions. Second, we utilize a freedom for the choice of a nucleon interpolating operator. In particular, a potential is independent of the choice of a nucleon operator at the source, N , and we employ the non-relativistic operator as $N' = N_{nr} \equiv \epsilon_{abc}(q_a^T C\gamma_5 P_{nr} q_b) P_{nr} q_c$ with $P_{nr} = (1 + \gamma_4)/2$, which reduces the spinor DoF. Similar techniques are (independently) developed in Ref. [77]. On the other hand, a potential depends on the choice of a nucleon operator at the sink, N. As discussed in Sect. 2.4, choosing N corresponds to choosing the "scheme" to calculate nuclear forces. Note that physical observables calculated from these different potentials, such as phase shifts and binding energies, are unique. In order to determine 3NF and 2NF in the same "scheme", we employ the same sink operator in the 3NF study as that employed in 2NF: $N = N_{std} \equiv \epsilon_{abc}(q_a^T C\gamma_5 q_b) q_c$. Recall that the choice of $N = N_{std}$ is shown to have good convergence in the derivative expansion in Sect. 4.3, and can be considered to be a good "scheme" for lattice nuclear forces. We next consider the geometry of the 3N. Since the spatial DoF for general (r, ρ) is too large, it is necessary to find an efficient way to restrict the geometry. In this exploratory study, we propose to use a "linear setup" with $\rho = 0$, in which 3N are aligned linearly with equal spacings of $r_2 \equiv |r|/2$. In this setup, the third nucleon is attached to (1, 2)-nucleon pair with only an S-wave. Considering the total 3N quantum numbers of $(I, J^P) = (1/2, 1/2+)$, the triton channel, the wave function can be completely spanned by only three bases, which can be labeled by the quantum numbers of the (1, 2) pair as ${}^1S_0, {}^3S_1, {}^3D_1$ Therefore, the Schrödinger equation leads to 3 × 3 coupled channel equations with bases of $\psi_{^1S_0}, \psi_{^3S_1}, \psi_{^3D_1}$.

The reduction of the dimension of bases is expected to improve the S/N as well. It is worth mentioning that considering the linear setup is not an approximation: Among the various geometric components of the wave function of the ground state, we calculate the (exact) linear setup component as a convenient choice to study 3NF. While we can access only a part of 3NF from it, we plan to extend the calculation to more general geometries step by step, toward the complete determination of the full 3NF. Finally, we emphasize that 3NF study requires the precise determination of 2NF. This is not surprising, because the interactions in 3N systems are mostly dominated by 2NF, and thus small uncertainties in 2NF could easily obscure the signal of 3NF. Note that we generally need precise 2NF in both parity-even and parity-odd channels, since a 2N pair inside the 3N system could be of either positive or negative parity. On this point, we have shown that 2NF in the parity-even channel can be determined with good precision in lattice QCD. On the other hand, the determination of 2NF in the parity-odd channel is much more difficult. While the formulation for parity-odd 2NF is developed in Sect. 4.5, the results are found to suffer from larger statistical errors than parity-even 2NF. This is considered to be a general tendency, since one has to inject a non-zero momentum in the parity-odd 2NF study. Therefore, in the 3NF study, it is essential to suppress the uncertainties originating from parity-odd 2NF. In order to address this issue, we propose that we consider the following channel [76]:

$$\psi_S \equiv \frac{1}{\sqrt{6}}[-p_\uparrow n_\uparrow n_\downarrow + p_\uparrow n_\downarrow n_\uparrow - n_\uparrow n_\downarrow p_\uparrow + n_\downarrow n_\uparrow p_\uparrow + n_\uparrow p_\uparrow n_\downarrow - n_\downarrow p_\uparrow n_\uparrow], \tag{8.4}$$

which is anti-symmetric in spin/isospin spaces for any 2N pair. Combined with the Pauli principle, it is automatically guaranteed that any 2N pair couples with even parity only. Therefore, parity-odd 2NF vanish in $\langle \psi_S | H | \psi_{3N} \rangle$, where H is the Hamiltonian of the 3N system, and we can extract 3NF unambiguously without referring to parity-odd 2NF. Note that no assumption on the choice of 3D geometry of r, ρ is imposed in this argument, and we can take advantage of this feature for future 3NF calculations with various 3N geometries.

Numerical Results

We employ $N_f = 2$ dynamical configurations with a mean field improved clover fermion and RGimproved gauge action generated by the CP-PACS Collaboration [39]. We use 598 configurations at $\beta = 1.95$ and a lattice spacing of $a^{-1} = 1.269(14)$ GeV; a lattice size of $V = L^3 \times T = 16^3 \times 32$ corresponds to a (2.5 fm)^3 box in physical spatial size. For u, d quark masses, we take the hopping parameter at the unitary point as $\kappa_{ud} = 0.13750$, which corresponds to $m_\pi = 1.13$ GeV, $m_N = 2.15$ GeV, and $m_\Delta = 2.31$ GeV. We use a wall quark source with Coulomb gauge fixing, and the periodic (Dirichlet) boundary condition is imposed in the spatial (temporal) direction. In order to enhance

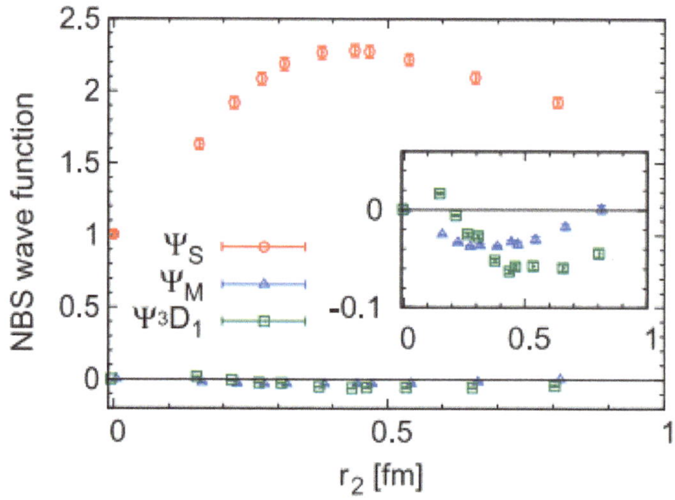

Figure. 22: 3N wave functions at $(t - t_0)/a = 8$. Circle (red), triangle (blue), and square (green) points denote ψ_S, ψ_M, and $\psi_{^3D_1}$, respectively.

the statistics, we perform the measurement at 32 source time slices for each configuration, and the forward and backward propagations are averaged. The results from both of total angular momentum $J_z = \pm 1/2$ are averaged as well. Due to the enormous computational cost, we can only perform the simulations at a few sink time slices. Looking for the range of sink times where ground state saturation is achieved, we carry out preparatory simulations for effective 2NF in the 3N system [76] in the triton channel at $2 \leq (t - t_0)/a \leq 11$, and find that the results are consistent with each other as long as $(t - t_0)/a \geq 7$ [76]. Being on the safer side, we perform linear setup calculations of 3NF at $(t - t_0)/a = 8$ and 9. We perform the simulation at eleven values of the spatial distance r_2.

In Fig. 22, we plot the radial part of each wave function of $\psi_S = (-\psi_{^1S_0} + \psi_{^3S_1})/\sqrt{2}$, $\psi_M \equiv (\psi_{^1S_0} + \psi_{^3S_1})/\sqrt{2}$, and $\psi_{^3D_1}$ obtained at $(t - t_0)/a = 8$. Here, we normalize the wave functions by the central value of $\psi_S(r_2 = 0)$. What is noteworthy is that the wave functions are obtained with good precision, which is quite nontrivial for the 3N system. We observe that ψ_S overwhelms the wave function, and ψ_M, ψ_{3D1} are much smaller by one to two orders of magnitude. This indicates that the higher partial wave components in ψ_S are also strongly suppressed, and the wave function is completely dominated by the component with which all three nucleons are in the S-wave in this lattice setup.

We determine 3NF by subtracting 2NF from total potentials in the 3N system. As discussed before, we have only one channel, $\langle\psi_S|H|\psi_{3N}\rangle = \langle\psi_S|H|\psi_S\rangle + \langle\psi_S|H|\psi_M\rangle + \langle\psi_S|H|\psi_{^3D_1}\rangle$, which is free from parity-odd 2NF. Correspondingly, we can determine one type of spin/isospin functional form for 3NF. In this study, 3NF are effectively represented in a scalar–isoscalar functional form. This form is an efficient representation, since ψS overwhelms the wave function and thus $|\langle\psi_S|V_{3NF}|\psi_S\rangle| \gg |\langle\psi_S|V_{3NF}|\psi_M\rangle|, |\langle\psi_S|V_{3NF}|\psi_{^3D_1}\rangle|$ is expected. Note also that the scalar– isoscalar functional form is often employed for the short-range part of 3NF in phenomenological models [72].

In Fig. 23, we plot the results for the effective scalar–isoscalar 3NF at $(t-t_0)/a = 8$. We here include the r_2-independent shift by energy, $\delta_E \simeq 5$ MeV, which is determined by the long-range behavior of potentials [76]. While δ_E suffers from $\lesssim 10$ MeV systematic error, this does not affect the following discussions much, since δ_E merely serves as an overall offset. In order to check the dependence on the sink time slice, we compare 3NF from $(t - t_0)/a = 8$ and 9. While the results with $(t - t_0)/a = 9$ suffer from quite large errors, they are consistent with each other within statistical fluctuations.

Figure 23 shows that 3NF are small at the long distance region of r_2. This is in accordance with the suppression of 2πE-3NF for the heavy pion. At short distances, however, an indication of repulsive

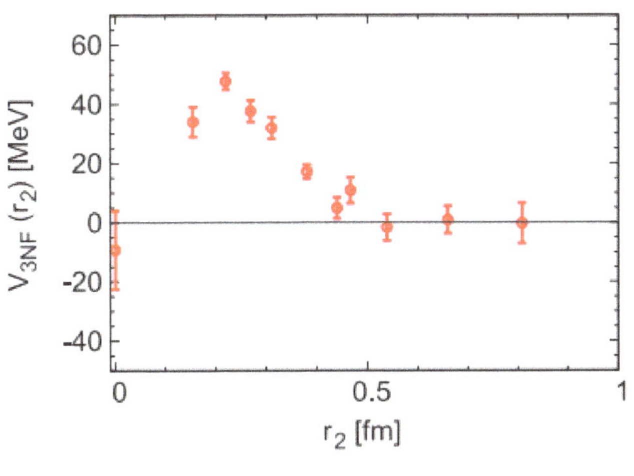

Figure. 23: The effective scalar–isoscalar 3NF in the triton channel with the linear setup obtained at $(t - t_0)/a = 8$. r_2 is the distance between the center and edge in the linear setup.

3NF is observed. Note that a repulsive short-range 3NF is phenomenologically required to explain the properties of high density matter. Since multi-meson exchanges are strongly suppressed for the large quark mass, the origin of this short-range 3NF may be directly attributed to the quark and gluon dynamics. In fact, we recall that the short-range repulsive (or attractive) cores in the generalized two-baryon potentials in the flavor SU(3) limit discussed in Sect. 6 are consistent with the Pauli exclusion principle in the quark level [45,46,53]. In this context, it is intuitive to expect that the 3N system is subject to an extra Pauli repulsion effect, which could be the origin of the observed shortrange repulsive 3NF. Further investigation along this line would certainly be an interesting subject for future research.

We remark here that the quark mass dependence of 3NF is certainly an important issue, since the lattice simulations are carried out only at a single large quark mass. In the case of 2NF, shortrange cores have enhanced strength and broaden their range by decreasing the quark mass [18]. We would therefore expect a significant quark mass dependence to exist in short-range 3NF as well. In addition, long-range 2πE-3NF will emerge at lighter quark masses, in particular, at the physical point. Quantitative investigations through lattice simulations with lighter quark masses are currently underway

MESON–BARYON INTERACTIONS

Since our potential method can be naturally extended to meson–baryon systems as well, we consider meson–baryon interactions in this section. The first application is a study on kaon–nucleon (K N) interactions in the $I(J^P) = 0(1/2-)$ and $1(1/2-)$ channels. The elastic K N scattering allows us to study the origin of "non-resonant" nuclear forces, since kaon contains $u\bar{s}$ quarks, and these quarks do not annihilate in non-strange nucleons. Therefore, direct production of conventional baryon resonances is ruled out. Also, the K N systems in the $I(J^P) = 0(1/2-)$ and $1(1/2-)$ channels may be relevant for a possible exotic state Θ^+ [78], the existence of which is still controversial. It is important to emphasize that one-pion exchange is absent in K N systems, so that short- and mid-range interactions dominate elastic K N scattering. Theoretical studies of K N interactions have so far been carried out by constituent quark models and meson-exchange models. In both models, it was found that genuine quark–gluon dynamics become important to describe the empirical scattering phase shifts [79].

To investigate the K N potentials in 2 + 1 flavor full QCD, we have utilized the gauge configurations of the JLDG (Japan Lattice Data Grid)/ILDG (International Lattice Data Grid) generated by the PACS-CS Collaboration on a $32^3 \times 64$ lattice [80]. The renormalization group improved Iwasaki

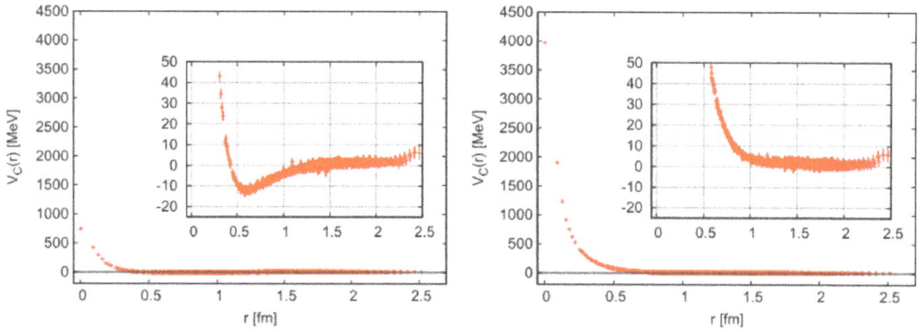

Figure. 24: The LO S-wave central potentials for K N states in the I = 0 (left) and I = 1 channels (right).

gauge action and non-perturbatively O(a)improved Wilson quark action are used at β = 1.90, which corresponds to a lattice spacing a = 0.09 fm, determined from π, K, and Ω masses. The physical size of the lattice is about (2.9 fm)^3 and the the hopping parameters are taken to be $\kappa_u = \kappa_d = 0.1370$ and $\kappa_s = 0.1364$. In the present simulation, we adopt the wall source located at t_0 with the Dirichlet boundary condition at time slice $t = t_0 + 32$ in the temporal direction and the periodic boundary condition in each spatial direction. Coulomb gauge fixing is employed at $t = t_0$. The number of gauge configurations used in the simulation is 399. With this setup, we obtain $m_\pi = 705(2), m_K = 793(2)$, and $M_N = 1590(8)$ MeV [81].

Figure 24 shows the LO S-wave potential V(r) for the K N state in I = 0 (left) and I = 1 (right). Repulsive interactions are observed at short distances in both channels, while the attractive well appears at mid-distances (0.4 < r < 0.8 fm) in the I = 0 channel; this is not found in the constituent quark model of hadrons [82]. These results indicate that there are no bound states in the $I(J^\pi) = 0(1/2^-)$ and $1(1/2^-)$ states at $m_\pi \simeq 705$ MeV.

The strong repulsions near the origin in the I = 1 channel can be expected from quark Pauli blocking effects. This was first pointed out by Machida and Namiki [83] for meson–baryon systems. In the $I = 1$ KN (K^+p) state whose configuration is $K^+p \sim (u\bar{s})(uud)$, one of the u-quarks cannot be in the S-state. The strong repulsion at short distances in this channel found in our simulations suggests a manifestation of quark Pauli blocking. In addition, the repulsive interactions in the S-wave I = 1 K N state can be expected to be much stronger than those of the I = 0 K N state in the constituent quark model. Our simulation shows that the repulsion at short distances for the K N potential becomes significantly smaller in the I = 0 channel compared to the I = 1 channel. This again confirms the expectation from quark Pauli blocking effects.

Kawanai and Sasaki investigate charmonium–nucleon $(c\bar{c}{-}N)$ interactions using the potential method. Since charmonia do not share the same quark flavor as the nucleon, the $c\bar{c}{-}N$ interaction is mainly induced by the genuine QCD effect of multi-gluon exchanges and does not manifest the repulsive core near the origin. In Ref. [84], the charmonium–nucleon potentials are calculated in quenched QCD on $16^3 \times 48$ and $32^3 \times 48$ lattices at $a \simeq 0.94$ fm at three different values of the light quark mass corresponding to $(m_\pi, m_N) \simeq (640, 1430), (720, 1520), (870, 1700)$ in units of MeV and one fixed value of the charm quark mass corresponding to $m_{\eta_c} \simeq 2920$ MeV and $m_{J/\Psi} \simeq 3000$ MeV.

They found that the effective central $c\bar{c}{-}N$ potentials clearly exhibit entirely attractive interactions without any repulsion at all distances. The absence of the short range repulsion (the repulsive core) is related to the absence of the Pauli exclusion between the heavy quarkonium and the light hadron.

CONCLUSION

In this report, we reviewed the basic notion of the HAL QCD method for the baryon–baryon (B B) potential and its field-theoretical derivation from the equal-time Nambu–Bethe–Salpeter (NBS) wave function in QCD. The potential U(x, y) (or the integral kernel of the Schrödinger type equation) in the HAL QCD method has three characteristic features: (i) non-locality in relative coordinates, (ii) energy independence, and (iii) scheme dependence. Each of these features has been discussed in detail in this report. By construction, this potential correctly reproduces the scattering phase shift defined from the S-matrix in QCD below the inelastic threshold. One can construct U(x, y) once all the NBS wave functions for scattering energies below the inelastic threshold are obtained. In lattice QCD simulations in a finite box, however, it is more practical to adopt the velocity (derivative) expansion of U(x, y) by its non-locality and determine the local potentials V(x) order by order. This is also in conformity with phenomenological potentials widely used in nuclear physics: An advantage of the HAL QCD method is that one can check the accuracy of this velocity expansion by changing the scattering energies on the lattice. To avoid the well-known problem of exponential error-growth in the temporal correlation of multi-hadrons, we have introduced a time-dependent HAL QCD method on the basis of the (imaginary) time Schrödinger type equation. Due to this improved method, we could achieve better construction of the potential as demonstrated in this report. The leading order (LO) terms of the velocity expansion correspond to the central and tensor potentials: Those for the nucleon–nucleon (N N), hyperon–nucleon (Y N) and hyperon–hyperon (Y Y) interactions have been investigated in full QCD simulations, some of which are recapitulated in this report. The next-to-leading order (NLO) term is the spin–orbit potential: By introducing finite momentum to the nucleons, we could extract the N N spin–orbit force for the first time. The origin of the repulsive core of the N N interaction has been also investigated by extending the SU(2)-flavor to the degenerate SU(3)-flavor. The role of the Pauli principle in the quark level to describe the short range part of the interaction becomes clear. In particular, there arises a short range "attractive" core in the flavor singlet channel; we found that it is strong enough to form a bound state, the H-dibaryon, in the SU(3) limit. The HAL QCD method can be extended to the case beyond the inelastic threshold. This is necessary to treat Y N and Y Y interactions with SU(3)-flavor symmetry breaking. We have presented its application to the (S, I) = (−2, 0) system and derived the coupled channel potentials among $\Lambda\Lambda$, $N\Xi$, and $\Sigma\Sigma$. The HAL QCD method is also applied to the three-nucleon force relevant for the extra binding of finite nuclei and also for the maximum mass of neutron stars, and to meson–baryon interactions relevant for meson–baryon resonances and the pentaquark. So far, our full QCD simulations of the B B interactions have been performed at non-zero lattice spacing on a finite volume with relatively large quark masses. We therefore need careful studies of systematic errors on the finite volume effect, quark mass dependence, and the lattice spacing effect. Among others, the most important direction is to carry out (2 + 1)-flavor simulations on a large volume (e.g. L = 6 − 9 fm) at the physical quark mass (m_π = 135 MeV) to extract realistic B B and BBB potentials. Such simulations are planned at the 10 PFlop "K computer" at the Advanced Institute for Computational Science (AICS), RIKEN.

If it turns out that the program described in this paper indeed works in lattice QCD at the physical quark mass, it would be a major step toward the understanding of atomic nuclei and neutron stars from the fundamental law of strong interaction, quantum chromodynamics.

ACKNOWLEDGMENTS

We thank the CP-PACS, JLQCD, and PACS-CS Collaborations and ILDG/JLDG for providing us with the gauge configurations [34,39,54,85,86]. We are grateful for the authors and maintainers of CPS++ [87], a modified version of which is used for the measurement done in this work. The numerical simulations were carried out on Blue Gene/L at KEK, T2K at University of Tsukuba and at the University of Tokyo, SR16000 at YITP at Kyoto University, and SX9 and SX8R at RCNP at Osaka University. This research is supported in part by Grants-in-Aid for Scientific Research on Innovative Areas (No. 2004:20105001,20105003) and for Scientific Research (C) 23540321, 24740146, JSPS 21 · 5985, and SPIRE (Strategic Program for Innovative REsearch).

REFERENCES

1. M. Taketani et al., Prog. Theor. Phys. Suppl. 39, 1 (1967).
2. N. Hoshizaki et al., Prog. Theor. Phys. Suppl. 42, 1 (1968).
3. G. E. Brown and A. D. Jackson, Nucleon–Nucleon Interaction (North-Holland, Amsterdam, 1976).
4. R. Machleidt, Adv. Nucl. Phys. 19, 189 (1989).
5. R. Machleidt and I. Slaus, J. Phys. G 27, R69 (2001).
6. R. Machleidt, Phys. Rev. C 63, 024001 (2001).
7. V. G. J. Stoks, R. A. M. Klomp, C. P. F. Terheggen, and J. J. de Swart, Phys. Rev. C 49, 2950 (1994).
8. R. B. Wiringa, V. G. J. Stoks, and R. Schiavilla, Phys. Rev. C 51, 38 (1995).
9. N. Ishii, S. Aoki, and T. Hatsuda, Phys. Rev. Lett. 99, 022001 (2007).
10. H. Yukawa, Proc. Math. Phys. Soc. Jpn. 17, 48 (1935).
11. M. Taketani, S. Machida, and S. Onuma, Prog. Theor. Phys. 7, 45 (1952).
12. R. Jastrow, Phys. Rev. 81, 165 (1951).
13. R. Tamagaki et al., Prog. Theor. Phys. Suppl. 112, 1 (1993).
14. H. Heiselberg and V. Pandharipande, Annu. Rev. Nucl. Part. Sci. 50, 481 (2000).
15. J. M. Lattimer and M. Prakash, Phys. Rep. 333, 121 (2000).
16. M. Lüscher, Nucl. Phys. B 354, 531 (1991).
17. S. Aoki, T. Hatsuda, and N. Ishii, Comput. Sci. Discovery 1, 015009 (2008).
18. S. Aoki, T. Hatsuda, and N. Ishii, Prog. Theor. Phys. 123, 89 (2010).
19. A. Gal, O. Hashimoto, and J. Pochodzalla, (eds.) Special Issue on Progress in Strangeness Nuclear Physics, Nucl. Phys. A 881 (2012).
20. S. R. Beane, W. Detmold, K. Orginos, and M. J. Savage, Prog. Part. Nucl. Phys. 66, 1 (2011).
21. N. Ishizuka, PoS LAT2009, 119 (2009).
22. W. Królikowski and J. Rzewuski, Nuovo Cimento 4, 1212 (1956).
23. R. Tamagaki and W. Watari, Prog. Theor. Phys. Suppl. 39, 23 (1967).
24. S. Okubo and R. E. Marshak, Ann. Phys. (N.Y.) 4, 166 (1958).
25. K. Nishijima, Phys. Rev. 111, 153 (1958).
26. W. Zimmermann, Nuovo Cimento 10, 597 (1958) and MPI-PAE/PTh-61/87 (1987) (unpublished).
27. R. Haag, Phys. Rev. 112, 669 (1958).
28. S. Aoki, J. Balog, and P. Weisz, J. High Energy Phys. 1005, 008 (2010).
29. S. Aoki, J. Balog, and P. Weisz, J. High Energy Phys. 1009, 083 (2010).
30. S. Aoki, J. Balog, T. Hatsuda, N. Ishii, K. Murano, H. Nemura, and P. Weisz, PoS LAT2008, 162 (2008).
31. K. Murano, N. Ishii, S. Aoki, and T. Hatsuda, PoS LAT2009, 126 (2009).
32. K. Murano, N. Ishii, S. Aoki, and T. Hatsuda, PoS LAT2010, 150 (2010).
33. K. Murano, N. Ishii, S. Aoki, and T. Hatsuda, Prog. Theor. Phys. 125, 1225 (2011).
34. S. Aoki et al.
35. PACS-CS Collaboration., Phys. Rev. D 79, 034503 (2009).
36. S. Aoki et al.
37. PACS-CS Collaboration., Phys. Rev. D 81, 074503 (2010).
38. N. Ishii, S. Aoki, and T. Hatsuda, PoS LATTICE2008, 155 (2008).
39. N. Ishii et al.

40. HAL QCD Collaboration., Phys. Lett. B 712, 437 (2012).
41. http://www.nn-online.org/
42. A. Ali Khan et al.
43. CP-PACS Collaboration., Phys. Rev. D 65, 054505 (2002); 67, 059901 (2003)
44. erratum..
45. CP-PACS Collaboration, http://www.jldg.org/ildg-data/CPPACSconfig.html
46. B. F. Gibson and E. V. Hungerford, Phys. Rep. 257, 349 (1995).
47. Reviewed in O. Hashimoto and H. Tamura, Prog. Part. Nucl. Phys. 57, 564 (2006).
48. H. Nemura, for the HAL QCD Collaboration, arXiv:1203.3320
49. hep-lat..
50. N. Ishii, for PACS-CS and HAL QCD Collaborations, PoS LAT2009, 019 (2009).
51. Reviewed in M. Oka, K. Shimizu, and K. Yazaki, Prog. Theor. Phys. Suppl. 137, 1 (2000).
52. Y. Fujiwara, Y. Suzuki, and C. Nakamoto, Prog. Part. Nucl. Phys. 58, 439 (2007).
53. J.K. Ahn et al., Phys. Lett. B 633, 214 (2006).
54. K. Nakazawa, Nucl. Phys. A 639, C345 (1998).
55. T. Fukuda et al., Phys. Rev. C 58, 1306 (1998).
56. P. Khaustov et al., Phys. Rev. C 61, 054603 (2000).
57. See, for example, http://j-parc.jp/NuclPart/index_e.html
58. H. Nemura, N. Ishii, S. Aoki, and T. Hatsuda, Phys. Lett. B 673, 136 (2009).
59. T. Inoue et al.
60. HAL QCD Collaboration., Prog. Theor. Phys. 124, 591 (2010).
61. CP-PACS and JLQCD Collaborations, http://www.jldg.org/ildg-data/CPPACS+JLQCDconfig.html
62. T. Inoue et al.
63. HAL QCD Collaboration., Phys. Rev. Lett. 106, 162002 (2011).
64. T. Inoue, for HAL QCD Collaboration, arXiv:1111.5098
65. hep-lat..
66. T. Inoue et al.
67. HAL QCD Collaboration., Nucl. Phys. A 881, 28 (2012).
68. Reviewed in J. Schaffner-Bielich, Nucl. Phys. A 835, 279 (2010).
69. S. R. Beane et al.
70. NPLQCD Collaboration., Phys. Rev. Lett. 106, 162001 (2011).
71. S. R. Beane et al.
72. NPLQCD Collaboration., arXiv:1109.2889
73. hep-lat..
74. N. Ishii, for HAL QCD Collaboration, PoS LATTICE2010, 145 (2010).
75. S. Aoki et al.
76. HAL QCD Collaboration., Proc. Japan Acad. B 87, 509 (2011).
77. K. Sasaki, for HAL QCD Collaboration, PoS LATTICE2010, 157 (2010).
78. S. C. Pieper, Riv. Nuovo Cimento 31, 709 (2008).
79. K. Sekiguchi et al., Phys. Rev. C 83, 061001 (2011).
80. T. Otsuka, T. Suzuki, J. D. Holt, A. Schwenk, and Y. Akaishi, Phys. Rev. Lett. 105, 032501 (2010).
81. A. Akmal, V. R. Pandharipande, and D. G. Ravenhall, Phys. Rev. C 58, 1804 (1998).
82. S. Nishizaki, T. Takatsuka, and Y. Yamamoto, Prog. Theor. Phys. 108, 703 (2002).
83. T. Takatsuka, S. Nishizaki, and R. Tamagaki, Prog. Theor. Phys. Suppl. 174, 80 (2008).
84. J. Fujita and H. Miyazawa, Prog. Theor. Phys. 17, 360 (1957).
85. S. A. Coon and H. K. Han, Few Body Syst. 30, 131 (2001).
86. S. C. Pieper, V. R. Pandharipande, R. B. Wiringa, and J. Carlson, Phys. Rev. C 64, 014001 (2001).
87. S. Weinberg, Phys. Lett. B 295, 114 (1992).
88. See the review E. Epelbaum, H.-W. Hammer, and U.-G. Meissner, Rev. Mod. Phys. 81, 1773 (2009).
89. K. Hashimoto and N. Iizuka, J. High Energy Phys. 1011, 058 (2010).
90. T. Doi et al.
91. HAL QCD Collaboration., Prog. Theor. Phys. 127, 723 (2012).
92. T. Yamazaki, Y. Kuramashi, and A. Ukawa
93. PACS-CS Collaboration., Phys. Rev. D 81, 111504 (2010).
94. T. Nakano et al.
95. LEPS Collaboration., Phys. Rev. Lett. 91, 012002 (2003).
96. K. Hashimoto, Phys. Rev. C 29, 1377 (1984).
97. PACS-CS Collaboration, http://www.jldg.org/ildg-data/PACSCSconfig.html
98. Y. Ikeda et al.
99. HAL QCD Collaboration., arXiv:1111.2663
100. hep-lat..
101. T. Barnes and E. S. Swanson, Phys. Rev. C 49, 1166 (1994).
102. S. Machida and M. Namiki, Prog. Theor. Phys. 33, 125 (1965).
103. T. Kawanai and S. Sasaki, Phys. Rev. D 82, 091501 (2010).
104. M. G. Beckett, B. Joo, C. M. Maynard, D. Pleiter, O. Tatebe, and T. Yoshie, Comput. Phys. Commun. 182, 1208 (2011).
105. http://www.lqcd.org/ildg, http://www.jldg.org
106. CPS++. http://qcdoc.phys.columbia.edu/cps.html (maintainer: Chulwoo Jung).

CHAPTER 2

SYMMETRY ENERGY IN NUCLEAR DENSITY FUNCTIONAL THEORY

W. Nazarewicz, P.-G. Reinhard, W. Satula, and D. Vretenar

INTRODUCTION

Density Functional Theory (DFT) is a universal approach used to describe properties of complex, strongly correlated many-body systems. Originally developed in the context of many-electron systems in condensed matter physics and quantum chemistry [1,2] (also known under the name of Kohn-Sham DFT), it is also a tool of choice in microscopic studies of complex heavy nuclei. The basic implementation of this framework is in terms of self-consistent mean-field (SCMF) models [3–5]. Extending the DFT to atomic nuclei, the nuclear DFT, is not straightforward as nuclei are self-bound, small, superfluid aggregations of two kinds of fermions, governed by strong surface effects. Their smallness leads to appreciable quantal fluctuations (finite-size effects) which are difficult to incorporate into the energy density functional (EDF). The lack of external binding potential implies that the nuclear DFT must be necessarily formulated in terms of intrinsic normal and anomalous (pairing) densities [6]. A density matrix expansion of the effective interaction suggests that, in addition to the standard local nucleon density, superior EDFs should also include more involved nucleon aggregates such as the kinetic-energy density and spin-orbit density [3–5] The commonly used single-reference SCMF methods include the local (Skyrme), non-local (Gogny) and covari-ant (relativistic) approaches [5,7,8]. All these approaches are thought to be different realizations of an underlying effective field theory [9] with the ultraviolet physics hidden in free parameters adjusted to observations. For that reason, predictions for low-energy (infrared) physics should be fairly independent of the particular variant used in calculations [10–13]. The underlying EDFs are constructed in a phenomenological way, with coupling constants optimized to selected nuclear data and expected properties of homogeneous nuclear matter. In practice, nuclear EDFs differ in their functional form and are subject to different optimization strategies causing that their predictions vary even within a single family of EDFs. In particular, large uncertainties remain in the isovector channel, which is poorly constrained by experiment. A key quantity characterizing the interaction in the isovector channel is the nuclear symmetry energy (NSE) describing the static response of the nucleus to the neutron-proton asymmetry. As discussed in this Topical Issue, the NSE influences a broad spectrum of phenomena, ranging from subtle isospin mixing effects in N ~ Z nuclei to particle stability of neutron-rich nuclei, to nuclear collective modes, and to radii and masses of neutron stars. Various nuclear observables are sensitive probes of NSE, and numerous phenomenological indicators can be constructed to probe its various aspects. It is the aim of this contribution to analyze the relations between NSE and measurable observables in fi- nite nuclei. The most promising observables for isovector properties that have stimulated vigorous experimental and theoretical activity include neutron radii, neutron skins, dipole polarizability, and neutron star radii. The ongoing efforts are focused on better constraining the uncertainties concerning the equation of state (EOS) of the symmetric and asymmetric nucleonic matter (NM) and, in particular, the symmetry energy and its density dependence. Parameters that characterize the NSE are not entirely independent. They are affected by key nuclear observables in different ways. Thus it is the sine qua non of a further progress in this area to understand the correlation

Citation: W. Nazarewicz, P.-G. Reinhard, W. Satula, and D. Vretena, Symmetry energy in nuclear density functional theory, DOI 10.1140/epja/i2014-14020-3.

pattern between NSE parameters and finite-nuclei observables, and to provide uncertainty quantification on theoretical predictions using the powerful methods of statistical analysis [14]. A second aim is to understand the dependences from a formal perspective and to explore the impact of con- figuration mixing. Within the independent particle picture the isovector response can be described in terms of a charge-dependent symmetry potential that shifts the neutron well with respect to the proton average potential. The effect can be estimated quantitatively within the Fermi-gas model (FGM) augmented by a schematic isospin-isospin interaction [15],

$$V_{TT} = \frac{1}{2}\kappa \hat{\boldsymbol{T}} \cdot \hat{\boldsymbol{T}}. \tag{1}$$

In the Hartree approximation this model gives rise to a quadratic dependence of the NSE on the neutron excess $I = (N - Z)/A,$

$$E_{\mathrm{sym}}/A = a_{\mathrm{sym}} I^2 = (a_{\mathrm{sym,kin}} + a_{\mathrm{sym,int}}) I^2, \tag{2}$$

as $T = |T_z| = |N - Z|/2$ in the ground states of almost all nuclei. The FGM, in spite of its simplicity, has played an important role in our understanding of the NSE. In particular, it separates the NSE strength into kinetic and interaction (potential) contributions, and predicts a nearequality $a_{\mathrm{sym'kin}} \approx as_{\mathrm{ym,int}}$ of these contributions. It also provides an estimate $a_{\mathrm{sym}} \approx 25\mathrm{MeV}$ for the NSE coeffi- cient (see ref. [16] for a recent discussion).

Furthermore, we note that the SCMF approach can lead to spontaneous breaking of symmetries. This apparent drawback can be turned into an advantage, as the symmetry breaking mechanism allows to incorporate many inter-nucleon correlations within a single product state or, alternatively, within a single-reference DFT sacrificing good quantum numbers; broken symmetries have to be restored a posteriori. We will address this topic using the example of isospin mixing which naturally has an impact on isovector properties. This survey is organized as follows. Section 2 outlines the SCMF approaches and details various theoretical ingredients of the models employed in this work. Observables pertaining to bulk NM and finite nuclei that are essential for NSE are discussed in sect. 3. Constraints on NSE and correlations between observables and NSE parameters, using the statistical covariance technique, are presented in sect. 4. Section 5 summarizes the current status of NSE parameters. The planned extensions of the current DFT work are laid out in sect. 6. Finally, sect. 7 contains the conclusions of this survey.

NUCLEAR DFT

The nuclear EDF constitutes a crucial ingredient for a set of DFT-based theoretical tools that enable an accurate description of ground-state properties, collective excitations, and large-amplitude dynamics over the entire chart of nuclides, from relatively light systems to superheavy nuclei, and from the valley of β-stability to the nucleon drip-lines. In general EDFs are not directly related to any specific microscopic inter-nucleon interaction, but rather represent universal functionals of nucleon densities and currents. With a small set of global parameters adjusted to empirical properties of nucleonic matter and to selected data on finite nuclei [17,18], models based on EDFs enable a consistent description of a variety of nuclear structure phenomena. The unknown exact and universal nuclear EDF is approximated by simple, mostly analytical, functionals built from powers and gradients of nucleonic densities and currents, representing distributions of matter, spins, momentum and kinetic energy. When pairing correlations are included, they are represented by pair (anomalous) densities. In the field of nuclear structure this method is analogous to Kohn-Sham DFT. SCMF models effectively map the nuclear many-body problem onto a one-body problem using auxilliary Kohn-Sham single-particle orbitals. By including many-body correlations in EDF, the Kohn-Sham method in principle goes beyond the Hartree-Fock (HF) or Hartree-Fock-Bogolyubov (HFB) approximations and, in addition, it has the advantage of using local potentials. A broad range of nuclear properties have been very successfully described using SCMF models based on Skyrme EDFs, relativistic EDFs, and the Gogny interaction [5,7, 8,19–22]. (Note that the Gogny model is not strictly local as the other EDFs.) In the remainder of this section we briefly outline the Skyrme-Hartree-Fock (SHF) method and the relativistic mean-field (RMF) approach. As both methods are widely used and extensively described in the literature, we keep the presentation short and concentrate on a side-by-side comparison of the models. The basis of any mean-field approach is a set of singlenucleon canonical (Kohn-Sham) orbitals $\psi_\alpha(r)$, with occupations amplitudes v_α. The ψ_α denote Dirac fourspinor wave functions in the RMF framework, and twocomponent-spinor wave functions in the SHF which is a classical mean-field model. The canonical occupation amplitudes v_α are determined by the pairing interaction. The starting point of a particular model is an EDF expressed in terms of ψ_{α}, v_α and the local densities derived therefrom. The energy functional for the SHF method reads

$$\begin{aligned} E &= \int \mathrm{d}^3 r \, (\mathcal{E}_{\mathrm{kin}} + \mathcal{E}_{\mathrm{pot}}) + E_{\mathrm{Coul}} + E_{\mathrm{pair}} + E_{\mathrm{cm}}, \\ \mathcal{E}_{\mathrm{kin}} &= \frac{\hbar^2}{2m_{\mathrm{p}}}\tau_{\mathrm{p}} + \frac{\hbar^2}{2m_{\mathrm{n}}}\tau_{\mathrm{n}} \\ E_{\mathrm{cm}} &= -\frac{1}{2mA}\langle (\hat{P}_{\mathrm{cm}})^2 \rangle. \end{aligned} \tag{3}$$

The kinetic energy $\mathcal{E}_{\mathrm{kin}}$ is expressed in terms of single-nucleon wave functions. The Skyrme functional is

Table 1: Upper: The basic isoscalar (T = 0) and isovector (T = 1) local densities of SHF (left) and RMF (right). Lower: The potential-energy densities in the three considered SCMF models. Model parameters (third row) defining the coupling constants are indicated by lowercase latin letters. For further explanation see text.

Densities			
SHF		RMF	
$T=0$	$T=1$	$T=0$	$T=1$
$\rho_0(\mathbf{r})=\sum_\alpha v_\alpha^2\,\psi_\alpha^\dagger\psi_\alpha$	$\rho_1(\mathbf{r})=\sum_\alpha v_\alpha^2\,\psi_\alpha^\dagger\hat{\tau}_3\psi_\alpha$	$\rho_0(\mathbf{r})=\sum_\alpha v_\alpha^2\,\psi_\alpha^\dagger\psi_\alpha$	$\rho_1(\mathbf{r})=\sum_\alpha v_\alpha^2\,\psi_\alpha^\dagger\hat{\tau}_3\psi_\alpha$
$\tau_0(\mathbf{r})=\sum_\alpha v_\alpha^2\,\nabla\psi_\alpha^\dagger\nabla\psi_\alpha$	$\tau_1(\mathbf{r})=\sum_\alpha v_\alpha^2\,\nabla\psi_\alpha^\dagger\hat{\tau}_3\nabla\psi_\alpha$	$\rho_S(\mathbf{r})=\sum_\alpha v_\alpha^2\,\psi_\alpha^\dagger\hat{\gamma}_0\psi_\alpha$	
$\mathbf{J}_0(\mathbf{r})=-\mathrm{i}\sum_\alpha v_\alpha^2\,\psi_\alpha^\dagger\nabla\times\hat{\boldsymbol{\sigma}}\psi_\alpha$	$\mathbf{J}_1(\mathbf{r})=-\mathrm{i}\sum_\alpha v_\alpha^2\,\psi_\alpha^\dagger\hat{\tau}_3\nabla\times\hat{\boldsymbol{\sigma}}\psi_\alpha$		

Potential-energy density						
	SHF		RMF-PC		RMF-ME	
	$T=0$	$T=1$	$T=0$	$T=1$	$T=0$	$T=1$
$\rho\rho$	$C_0^\rho\rho_0^2$	$C_1^\rho\rho_1^2$	$G_\omega\rho_0^2$	$G_\rho\rho_1^2$	$G_\omega\rho_0\frac{1}{-\Delta+m_\omega^2}\rho_0$	$G_\rho\rho_1\frac{1}{-\Delta+m_\rho^2}\rho_1$
mass	$C_0^\tau\rho_0\tau_0$	$C_1^\tau\rho_1\tau_1$	$G_\sigma\rho_S^2$		$G_\sigma\rho_S\frac{1}{-\Delta+m_\sigma^2}\rho_S$	
$\boldsymbol{\ell}\cdot\boldsymbol{s}$	$C_0^{\nabla\mathbf{J}}\rho_0\nabla\cdot\mathbf{J}_0$	$C_1^{\nabla\mathbf{J}}\rho_1\nabla\cdot\mathbf{J}_1$	"		"	
gradient	$C_0^{\Delta\rho}(\nabla\rho_0)^2$	$C_1^{\Delta\rho}(\nabla\rho_1)^2$	$f_S(\nabla\rho_S)^2$			
dens. dep.	$C_0^\rho=c_0^\rho+d_0^\rho\rho_0^a$	$C_1^\rho=c_1^\rho+d_1^\rho\rho_0^a$	$G_i=a_i+(b_i+c_ix)e^{-d_ix}$		$G_i=a_i\frac{1+b_i(x+d_i)^2}{1+c_i(x+d_i)^2}$	$G_\rho=g_\rho e^{-a_\rho(x-1)}$
	$C_T^\tau=c_T^\tau$, $C_T^{\Delta\rho}=c_T^{\Delta\rho}$, $C_T^{\nabla\mathbf{J}}=c_T^{\nabla\mathbf{J}}$		$i\in\{\sigma,\omega,\rho\}$, $x=\frac{\rho_0}{\rho_{sat}}$		$i\in\{\sigma,\omega\}$, $x=\frac{\rho_0}{\rho_{sat}}$	

contained in the interaction part with the potential-energy density E_{pot}. The Coulomb energy E_{Coul} consists of the direct Coulomb term, and the Coulomb exchange that is usually taken into account at the level of the Slater approximation. In most applications the center-of-mass correction E_{cm} is applied a posteriori because its variation would considerably complicate the mean-field equations. The pairing functional Epair will be detailed later. The RMF approach is usually formulated in terms of a Lagrangian,

$$L=\int \mathrm{d}^3r\,(\mathcal{L}_{\mathrm{kin}}-\mathcal{E}_{\mathrm{pot}})-E_{\mathrm{Coul}}-E_{\mathrm{pair}}-E_{\mathrm{cm}}, \tag{4}$$

$$\mathcal{L}_{kin}=\sum_\alpha v_\alpha^2\psi_\alpha^\dagger\hat{\gamma}_0(i\hat{\gamma}\cdot\boldsymbol{\partial}-m)\psi_\alpha, \tag{5}$$

where $\hat{\gamma}$ is the Dirac matrix. Again, the kinetic part is expressed explicitly in terms of Dirac spinor wave functions, whereas interaction terms are included in the potential energy density E_{pot}. Further contributions from Coulomb, pairing and center-of-mass motion are treated similarly as in the SHF approach.

The basic building blocks of an EDF are local densities and currents built from single-nucleon wave functions [5, 23]. These are summarized in the upper part of table 1. All densities appear in two flavors [24,25]: isoscalar (T = 0), or total density (sum of proton and neutron densities), and isovector (T = 1) density (difference between neutron and proton densities). Both can be conveniently expressed using the isospin operator $\hat{\tau}_3$. The basic ingredients of an EDF are the local densities ρ_0 and ρ_1. In RMF these can be associated with the zero-component of the four-vector current, where ρ_0 is often called the vector density and ρ_1 the isovector-vector density. RMF uses one more ingredient, the isoscalar-scalar density denoted here as ρ_S. SHF instead employs the kinetic-energy densities $\tau_0/_1$ and the spin-orbit densities $J_0/1$. One can show that τ_0 and J_0 emerge in the non-relativistic limit of ρ_S [26]. The principal difference between SHF and RMF is that the quantities τ_0 and J_0 are independent in SHF, whereas they are tightly related through ρ_S in RMF. Moreover, the RMF does not invoke an isovector counterpart of ρ_S thus being more restricted in the isovector channel.

The lower part of table 1 displays the main components of the potential-energy density. The underlying is to take all bi-linear isoscalar combinations of the local densities and to associate a coupling constant with each term [25]. The SHF confines the combinations to have at most second order of derivatives (the term J^2 is also dropped). In the RMF approach one keeps only terms that form a Lorentz scalar. Moreover, two bi-linear realizations of RMF will be considered. First there is the straightforward point-coupling (RMF-PC) realization that corresponds to contact interactions between nucleons and, second, the meson-exchange folding (RMF-ME). The folding is motivated by the traditional route to RMF as a model of nucleons coupled to classical meson fields. Of course, at energies characteristic for nuclear binding meson exchange represents just a convenient representation of the effective nuclear interaction. In practice RMF-PC and RMF-ME present equivalent realizations of the relativistic SCMF, differing in the range of effective interactions (zero range vs. finite range) and the choice of density dependence for the couplings. In practical applications one restricts the density dependence of coupling (vertex) functions to keep the number of free parameters to a minimum. In SHF, only the leading terms $\propto\rho_0^2$ and ρ_1^2 are given a (simple) density dependence as shown in table 1. In RMF-PC and RMF-ME, each term has some density dependence, but not all of these parameters are actually used. In RMF-PC, in particular, the parameters c_1 and a_1 (isovector channel) and c_v (isoscalar vector channel) are set to zero [27]. In RMF-ME, the parameters are correlated by additional boundary conditions on G_1 [28,29]. In total, there are 11 adjustable parameters for SHF, 10 for RMF-PC, and 8 for RMF-ME. From a formal perspective, SHF and RMFPC are rather similar, differing mainly in the relativistic kinematics, while RMF-ME includes a significantly different density dependence of the couplings, in addition to the finite range. These three models thus allow to display separately effects of kinematics, density dependence, and range of the effective nuclear interaction.

As far as particle-particle interaction is concerned, in the SHF we use the pairing functional derived from a density-dependent zero-range force,

$$E_{\mathrm{pair}}=\frac{1}{4}\sum_{q\in\{p,n\}}v_{0,q}\int \mathrm{d}^3r\tilde{\rho}_q^2\left[1-\frac{\rho(\mathbf{r})}{\rho_{\mathrm{pair}}}\right], \tag{6a}$$

$$\tilde{\rho}_q(\mathbf{r}) = \sum_{\alpha \in q} u_\alpha v_\alpha |\psi_\alpha(\mathbf{r})|^2, \tag{6b}$$

where q runs over protons and neutrons. It involves the pair-density $\tilde{\rho}_q$ and is usually augmented by some density dependence. We consider here $v_{0,p}$, $v_{0,n}$, and ρ_{pair} as free parameters of the pairing functional in SHF. Note that we do not recouple to isoscalar and isovector terms because pairing is considered independently for protons and neutrons. Actually, the zero-range pairing force works only together with a limited phase space for pairing. We use here a soft cut-off in the space of single-nucleon energies [30] according to ref. [31].

In RMF calculations we use the recently developed separable pairing force [32,33]. It is separable in momentum space, and is completely determined by two parameters that are adjusted to reproduce in symmetric nuclear matter the pairing gap of the Gogny force. We have verified that both pairing prescriptions yield comparable results for the pairing gaps.

OBSERVABLES

In this section, we discuss observables pertaining to nuclear matter (NM) and finite nuclei that are essential for discussion of NSE. Those observables can be roughly divided [14] into good isovector indicators that correlate very well with NSE (such as weak-charge form factor, neutron skins, dipole polarizability, slope of the symmetry energy, and neutron pressure) and poor isovector indicators (such as nuclear and neutron matter binding energy, giant resonance energies, isoscalar and isovector effective mass, incompressibility, and saturation density).

Table 2: Definitions of NMP used in this work. All derivatives are to be taken at the equilibrium point corresponding to the saturation density ρ_{eq}

Incompressibility:	K_∞	=	$9\rho_0^2 \frac{d^2}{d\rho_0^2} \frac{E}{A}\|_{eq}$
Symmetry energy:	a_{sym}	=	$\frac{1}{2}\rho_0^2 \frac{d^2}{d\rho_1^2} \frac{E}{A}\|_{eq}$
Slope of a_{sym}:	L	=	$3\rho_0 \frac{da_{sym}}{d\rho_0}\|_{eq}$
Effective mass – SHF:	$\frac{\hbar^2}{2m^*}$	=	$\frac{\hbar^2}{2m} + \frac{\partial}{\partial\tau_0} \frac{E}{A}\|_{eq}$
Effective mass – RMF:	m^*	=	$m + G_\sigma \rho_S$
TRK enhancement:	κ_{TRK}	=	$\frac{m}{m^*} - 1 + \frac{1}{2}\rho_0 \frac{d}{d\rho_1} \frac{m}{m_n^*}$

Nuclear Matter Properties

Bulk properties of symmetric nuclear matter, called nuclear matter properties (NMP), are often used to characterize the properties of a model, or functional respectively. Starting point for the definition of NMP is the binding energy per nucleon in the symmetric nuclear matter $E/A = E/A(\rho_0,\rho_1,\tau_0,\tau_1)$. Variation with respect to Kohn-Sham wave functions establishes a relation $\tau_i = \tau_i(\rho_0,\rho_1)$ between kinetic densities τ_i and densities ρ_i. This yields the commonly used binding energy at equilibrium, $E/A[\rho_0,\rho_1,\tau_0(\rho_0,\rho_1),\tau_1(\rho_0,\rho_1)]$, as a function of the densities ρ_i alone. Table 2 lists the NMP discussed in this work. We consider τ_i as independent variables for the purpose of a formally compact definition of the effective mass in SHF. This is indicated by the partial derivatives with respect to τ_i for m^*/m in SHF and κ_{TRK}. Static properties are deduced from the binding energy at equilibrium, which depends on ρ_i only. This is indicated by using the total derivatives for K_∞, asym, and L. The slope of the symmetry energy L parametrizes the density dependence of a_{sym}. This quantity is essential for the characterization of the EOS of neutron matter and the mass-radius relation in neutron stars [34–39]. The notion of effective mass is ambiguous in RMF. In this survey, we have decided to use what is called the Dirac mass. (Alternatively, one can talk about the effective mass at the Fermi surface, which is typically 10% larger than the Dirac mass.) The Thomas-Reiche-Kuhn (TRK) enhancement factor κ_{TRK} [40] is merely used to characterize the isovector effective mass. (The alternative value deduced from the mass at the Fermi momentum is typically 20% smaller.) The quantity m_n^* is the neutron effective mass. It is defined in the same way as the isoscalar effective mass m^* but using the derivative with respect to the neutron kinetic density τ_n.

Next to NMP come the corresponding bulk surface parameter, the (isoscalar) surface energy a_{surf} and the (isovector) surface symmetry energy a_{ssym}. These surface parameters can be determined from the leptodermous expansion of the liquid drop model (LDM) energy per nucleon, $E_{LDM} = E_{LDM}/A$, in terms of inverse radius

$(\propto A^{-1/3})$ and neutron excess I [41],

$$\mathcal{E}_{\mathrm{LDM}}(A,I) = a_{\mathrm{vol}} + a_{\mathrm{surf}}A^{-1/3} + a_{\mathrm{curv}}A^{-2/3} + a_{\mathrm{sym}}I^2 + a_{\mathrm{ssym}}A^{-1/3}I^2 + a_{\mathrm{sym}}^{(2)}I^4. \tag{7}$$

The LDM energy $E_{LDM}(A,I)$ is obtained from the DFT calculation by subtracting the fluctuating shell correction energy. The general strategy behind this correction and leptodermous expansion is detailed in refs. [41,42]. In essence, we combine NM calculations ($A=\infty$) with (shellcorrected) DFT calculations for a huge set of spherical nuclei and extract the surface parameters by a fit to the expansion (7). Alternatively and simpler, one can compute the surface energy and surface symmetry energy through a semi-classical approximation (extended Thomas-Fermi) for the semi-infinite nuclear matter [43]. In this paper, we shall apply both strategies, the semi-classical approach whenever RMF is involved.

An important parameter characterizing the pure neutron matter is the neutron pressure

$$P(\rho_n) = \rho_n^2 \frac{d}{d\rho_n}\left(\frac{E}{A}\right)_n, \tag{8}$$

a quantity that is proportional to the slope of the binding energy of neutron matter at a given neutron density (derivative of neutron EOS). As discussed below, P is excellent isovector indicator.

Observables from Finite Nuclei

The total energy of a nucleus E(Z,N) is the most basic observable described by SCMF. It is also the most important ingredient for calibrating the functional, see sect. 4.1. We often consider binding energy differences. Of great importance for stability analysis are separation energies and Q_α values. Another energy observable, potentially useful in the context of NSE, is the indicator

$$\delta V_{pn} = -\frac{1}{4}[E(N,Z) - E(N-2,Z) \\ -E(N,Z-2) + E(N-2,Z-2)], \quad (9)$$

involving the double difference of binding energies [44]. Since δV_{pn} approximates the mixed partial derivative of binding energy with respect to N and Z, for nuclei with an appreciable neutron excess, the average value of δV_{pn} probes the symmetry energy term of LDM [45]: $\delta V_{pn}^{\rm LDM} \approx 2(a_{\rm sym} + a_{\rm ssym}A^{-1/3})/A$. That is, the shell-averaged trend of δV_{pn} is determined by the symmetry and surface symmetry energy coefficients.

It has been shown in [46] that effective SCMF provide a pertinent description of the form factors in the momentum regime $q < 2_{qF}$ where q_F is the Fermi momentum. The key features of the nuclear density are related to this low-q range. The basic parameters characterizing nuclear density distributions are: r.m.s. charge radius r_C, diffraction radius R_C, and surface thickness σ_C [47]. The diffraction radius R_C, also called the box-equivalent radius, parametrizes the gross diffraction pattern which resemble those of a hard sphere of radius R_C [47]. The actual charge form factor $F_C(q)$ falls off faster than the box-equivalent form factor F_{box}. This is due to the finite surface thickness σ which, in turn, can be determined by comparing the height of the first maximum of F_{box} with FC from the realistic charge distribution. The charge halo parameter h_C is composed from the three basic charge form parameters and serves as a nuclear halo parameter found to be a relevant measure of the outer surface diffuseness [48].

The charge distribution is basically a measure of the proton distribution. It is only recently that the parityviolating electron scattering experiment PREX has provided some information on the weak-charge form factor F_W (q) of ^{208}Pb [49,50]. These unique data gives access to neutron properties, such as the neutron r.m.s. radius r_n. Closely related and particularly sensitive to the asymmetry energy is the neutron skin $r_{\rm skin} = r_{\rm n} - r_{\rm p}$, which is the difference of neutron and proton r.m.s. radii. (As discussed in ref. [48], it is better to define the neutron skin through neutron and proton diffraction radii and surface thickness. However, for well-bound nuclei, which do not exhibit halo features, the above definition of r_{skin} is practically equivalent.) Neutron radii and skins are excellent isovector indicators [14,36,51–57] that help to check and improve isovector properties of the nuclear EDF [14].

Nuclear excitations are characterized by the strength distributions S_{JT} (E) where J is the angular momentum of the excitation, T its isospin, and E the excitation energy. For example, the cross section for photo-absorption is proportional to $S_{11}(E)$. The strengths functions can be obtained from the excitation spectrum,

$$S_{JT}(E) = \sum_n E_n B_n(EJT)\delta_\Delta(E - E_n), \quad (10)$$

where E_n is the excitation energy of state n, $B_n(EJT)$ the corresponding transition matrix element of multipolarity J and isospin T, and δ_Δ as finite width fold ing function —if S_{JT} (E) is calculated theoretically using, e.g., the random phase approximation (RPA). In our RPA estimates, we use an energy-dependent width $\Delta = \max(\Delta_{min},(E_n - E_{th}r)/E_{slope})$ which simulates the broadening mechanisms beyond RPA. The parameters for ^{208}Pb are Δ_{min} = 0.2MeV, E_{thr} = 10MeV, and E_{slope} = 5MeV.

The resulting spectral distributions for heavy nuclei, as ^{208}Pb, show one clear giant resonance peak at E_{GR}(JT) for (J,T) = (0, 0),(1, 1),(2, 0). We will consider these resonance energies as characteristic observables of dynamical response in heavy nuclei. The strength functions S_{JT} (E) in light nuclei are much more fragmented and cannot be reduced to one single characteristic number.

There are other key observables that can be extracted from the strength distributions. In particular, for the dipole case $S_{11}(E)$, a key observable is the electric dipole polarizability,

$$\alpha_{\rm D} = \sum_n E_n^{-1} B_n(E11). \quad (11)$$

In the following, we will consider α_D for ^{208}Pb. It has been demonstrated [14,57] that α_D strongly correlates with NSE; hence, it can serve as excellent isovector indicator that can be precisely extracted from measured E1 strength [58]. On the other hand, the low-energy E1 strength, sometimes referred to as the pygmy dipole strength, exhibits weak collectivity. The correlation between the accumulated low-energy strength and the symmetry energy is weak, and depends on the energy cutoff assumed [14,59,60]. Giant resonances are small amplitude excitations and belong to the regime of linear response. The low-energy branch of isoscalar quadrupole excitations is often associated with large amplitude collective motion along nuclear shapes with substantial quadrupole deformation. Of particular importance is nuclear fission, which determines existence of heavy and superheavy nuclei. As a simple and robust measure of fission, we shall consider the axial fission barrier height in ^{266}Hs. Unlike actinides, most superheavy nuclei have one single fission barrier [61–63], which simpli- fies the analysis for our purpose. It has to be kept in mind that the inner barrier is often lowered by triaxial shapes, but this is not important for the study of large-amplitude nuclear deformability.

SYMMETRY ENERGY: CONSTRAINTS AND CORRELATIONS

Brief Review of Χ² Technique and Correlation Analysis

As discussed in sect. 2, the nuclear EDF is characterized by about a dozen of coupling constants $p = (p_1,...,p_F)$ that are determined by confronting DFT predictions with experiment. The standard procedure is to adjust the parameters p to a large set of nuclear observables in carefully selected nuclei [5,17,18,64,65]. This is usually done by the standard least-squares optimization technique. The starting point is the χ^2 objective function

$$\chi^2(\mathbf{p}) = \sum_{\mathcal{O}} \left(\frac{\mathcal{O}^{(\rm th)}(\mathbf{p}) - \mathcal{O}^{(\rm exp)}}{\Delta\mathcal{O}} \right)^2, \quad (12)$$

where "th" stands for the calculated values, "exp" for experimental data, and ΔO for adopted errors. The optimum parametrization p_0 is the one which minimizes χ^2 with the minimum value $\chi_0^2 = \chi^2(\mathbf{p}_0)$. Around the minimum p_0, there is a range of "reasonable" parametrizations p that can be considered as delivering a good fit, *i.e.*, $\chi^2(\mathbf{p}) \le \chi_0^2 + 1$. As this range is usually rather small, we can expand χ^2 as

$$\chi^2(\mathbf{p}) - \chi_0^2 \approx \sum_{i,j=1}^{F} (p_i - p_{i,0}) \mathcal{M}_{ij} (p_j - p_{j,0}), \tag{13}$$

$$\mathcal{M}_{ij} = \tfrac{1}{2} \partial_{p_i} \partial_{p_j} \chi^2 |_{\mathbf{p}_0}. \tag{14}$$

The reasonable parametrizations thus fill the confidence ellipsoid given by

$$(\mathbf{p} - \mathbf{p}_0) \hat{\mathcal{M}} (\mathbf{p} - \mathbf{p}_0) \le 1, \tag{15}$$

see sect. 9.8 of [66]. Given a set of parameters p, any observable $A = \langle \hat{A} \rangle$ can be uniquely computed. In this way, $A = A_{(p)}$. The value A thus varies within the confidence ellipsoid, and this results in some uncertainty ΔA. Let us assume for simplicity that the observable varies weakly with p such that one can linearize in the relevant range $A_{(p)} = A_0 + (p - p_0) \cdot \partial_p A$. Let us, furthermore, associate a weight $\propto \exp(-\chi^2(\mathbf{p}))$ with each parameter set. A weighted average over the parameter space yields the covariance between two observables $\hat{A}$ and $\hat{B}$, which represents their combined uncertainty,

$$\overline{\Delta A \, \Delta B} = \sum_{ij} \partial_{p_i} A (\hat{\mathcal{M}}^{-1})_{ij} \partial_{p_j} B. \tag{16}$$

For A = B, eq. (16) gives the variance $\overline{\Delta^2 A}$ that defines a statistical uncertainty of an observable. Variance and covariance are useful quantities that allow to estimate the impact of an observable on the model and its parametrization. We shall explore the covariance analysis in three different ways,

1) We perform a constrained fit during which the observable of interest is kept fixed at a desired value. In the present paper, we consider the symmetry energy a_{sym} as constraining observable. Comparing uncertainties from a constrained fit with those from an unconstrained fit provides a first indicator on the impact of the constrained observable on other observables.
2) The next step is a trend analysis, in which one performs a series of constrained fits with systematically varied values of the constraining observable. One then studies other observables as a function of the constrained quantity. This provides valuable information on possible inter-dependences.
3) Finally, we compute correlation (16) between a_{sym} and other observables. Here, a useful dimensionless measure is given by the Pearson product-moment correlation coefficient [66],

$$c_{AB} = \frac{|\overline{\Delta A \, \Delta B}|}{\sqrt{\overline{\Delta A^2} \; \overline{\Delta B^2}}}. \tag{18}$$

A value $c_{AB} = 1$ means fully correlated and $c_{AB} = 0$ – uncorrelated.

In the following, we will apply these three ways of studying correlations with a_{sym} to different groups of observables. To this end, we have produced a series of parametrizations with systematically varied a_{sym} for the SV Skyrme family and for the RMF-ME and RMF-PC models.

The optimization and covariance analysis carried out in this survey for all three EDFs (SHF-SV, RMF-PC, and RMF-ME) is based on the same standard set of data on spherical nuclei (masses, diffraction radii, surface thickness, charge radii, separation energies, isotope shifts, and odd-even mass differences) that has originally been proposed in ref. [17] and recently employed in refs. [67,38]. We wish to emphasize that this is the first time that one consistent phenomenological input has been used to constrain SHF and RMF EDFs. A slightly modified variant of the fitting protocol has been used for RMF-ME. This EDF did not lead to stable results in the fits which were unconstrained by NMP. Consequently, we included the nuclear matter information on $(E/A)_{eq}$ into the dataset. This is still much less than in the previously published optimization protocols of RMF-ME, in which all NMP were constrained [28,29,68].

Correlations with Nuclear Matter Properties

The NMP corresponding to unconstrained optimization of SHF-SV, RMF-PC, and RMF-ME EDFs —using the same standard dataset— are shown in table 3. They are compared with NMP of SHF-RD [67] (employing a modified density dependence and the standard dataset) and SHFTOV [38] (using neutron star data in addition the standard dataset in the optimization process). As expected, isoscalar effective mass is significantly lowered in RMF as compared to SHF, and the opposite holds for κ_{TRK}. The slope parameter L is predicted to be very different in all five models. In particular, RMF-ME has very low value of L, and —at the same time— the uncertainty on a_{sym} in this model is very small.

Figure 1 shows the trends for selected properties of symmetric nuclear matter with a_{sym}. The purpose of this analysis is to relate systematic variations with a_{sym} to statistical uncertainties. The isoscalar properties K_∞, m^*/m as well as the isovector dynamical response κ_{TRK} are fairly insensitive to a_{sym}. Their variation with a_{sym} are much smaller than the typical statistical uncertainties. This independency is also indicated by the fact that the uncertainty obtained in the unconstrained fit is not visibly larger than those from the constrained optimizations. The trend is markedly different for the density dependence of the symmetry energy L: variations with a_{sym} well exceed the statistical error bars and the uncertainties from unconstrained fits are larger than those from constrained calculations. It is to be noted that the dedicated variations of a_{sym} stay within the uncertainty of a_{sym} in the unconstrained optimization. The uncertainty of L in the free fit thus covers nicely the uncertainty of the constrained calculations plus the variation of L with a_{sym}. Anyway, the results shows that L cannot be used as independent NMP although the formal structure of the EDF would allow that. There seems to be a strong link established by the data which yet has to be worked out.

Correlations with Properties of Finite Nuclei

Figure 2 illustrates the trends with a_{sym} and extrapolation uncertainties for three observables in ^{208}Pb: weakcharge form factor at $q = 0.475 fm^{-1}$ (q-value of PREX), neutron skin, and dipole polarizability. These observables are all known to be sensitive to isovector properties of EDF [14,57]. This is confirmed by the trends in the present results. The comparison of uncertainties shows a large growth when going from constrained to unconstrained optimizations. This corroborates the close relation between the symmetry energy and the three isovectors indicators shown in fig. 2. It is, furthermore, interesting to note that SHF and RMF-PC stay safely within the bands given by experimental data and RMF-ME is not far away. A better discrimination between models requires more precise data, a task on which presently many experimental groups are heavily engaged.

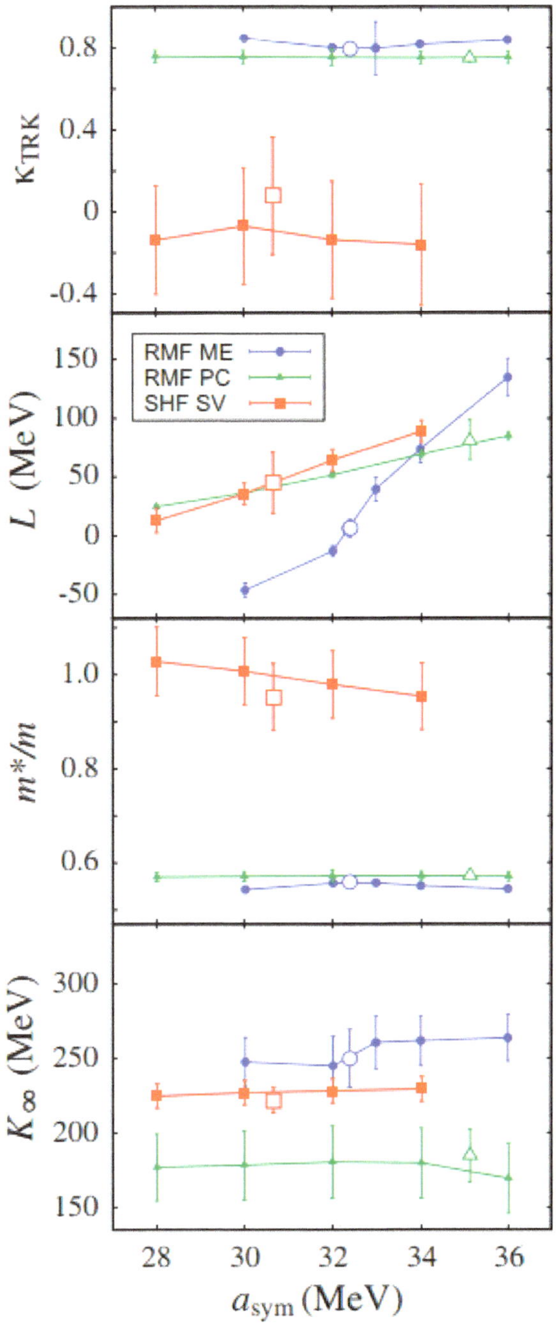

Figure. 1: Behavior of selected nuclear matter properties with symmetry energy a_{sym} for the SV Skyrme family and for the ME and PC RMF model families. The statistical uncertainties are indicated by error bars. The result of the unconstrained fits are shown by large open symbols with corresponding error bars.

Table 3: Nuclear matter parameters of SHF-SV, RMF-PC, and RMF-ME EDFs used in this paper (with error bars) obtained by means of unconstrained optimization. Also shown are the values of NMP of SHF-RD [67] and SHF-TOV [38]

Model	ρ_{eq} (fm^{-3})	E/A (MeV)	K_∞ (MeV)	m^*/m	a_{sym} (MeV)	L (MeV)	κ_{TRK}
SHF-SV	0.161(1)	−15.91(4)	222(9)	0.95(7)	31(2)	45(26)	0.08(29)
RMF-PC	0.159(1)	−16.14(3)	185(18)	0.57(1)	35(2)	82(17)	0.75(2)
RMF-ME	0.159(3)	−16.2(2)	250(19)	0.56(1)	32.4(1)	6(7)	0.79(2)
SHF-RD	0.161	−15.93	231	0.90	32(2)	60(32)	0.04(32)
SHF-TOV	0.161	−15.93	222	0.94	32(1)	76(15)	0.21(26)

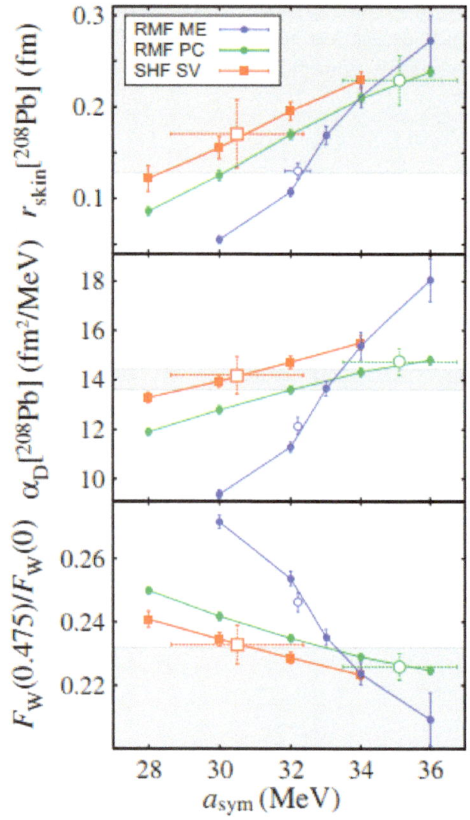

Figure. 2: Similar to fig. 1, but for selected properties of ^{208}Pb: neutron skin (top), dipole polarizability (middle), and weak-charge form factor (bottom). The current experimental ranges are shaded grey $r_{skin} = 0.33^{+0.16}_{-0.18}$ fm [49], $\alpha_D = 14.0 \pm 0.4\,\mathrm{fm}^2/\mathrm{MeV}$ [58], and $F_W(0.475)/F_W(0) = 0.204 \pm 0.028$ [50].

To explore the usefulness of δV_{pn} as an isovector indicator, we choose the heavy deformed nucleus ^{168}Er, as its even-even neighbors have similar structure and the calculated values of δV_{pn} for even-even Er isotopes show little variations around N = 100. The results displayed in fig. 3 show a gradual decrease of this quantity with a_{sym}, but the magnitude of the variation is very small and cannot account for the deviation from experiment (around 50keV). It is apparent that this quantity is too strongly influenced by shell effects (given by the deviation from the LDM estimate; also around 50keV) to probe NSE, see refs. [45, 69] and sect. 4.4 below.

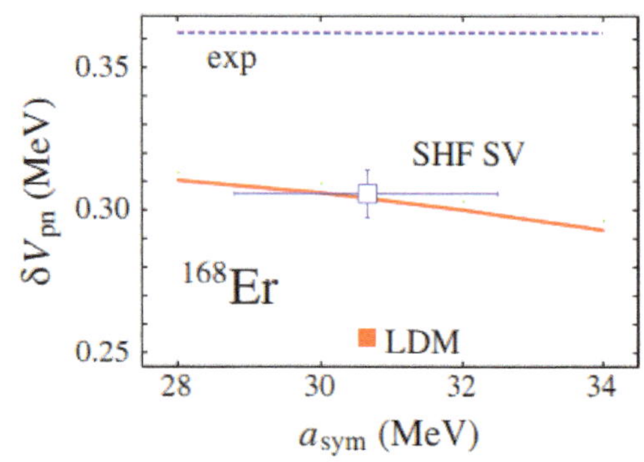

Figure. 3: Behavior of δV_{pn} in 168Er with symmetry energy a_{sym} for SHF-SV (solid line) as compared to experiment (dashed line) and the LDM value (filled square). The result of the unconstrained fit is marked by a large open square with corresponding error bars.

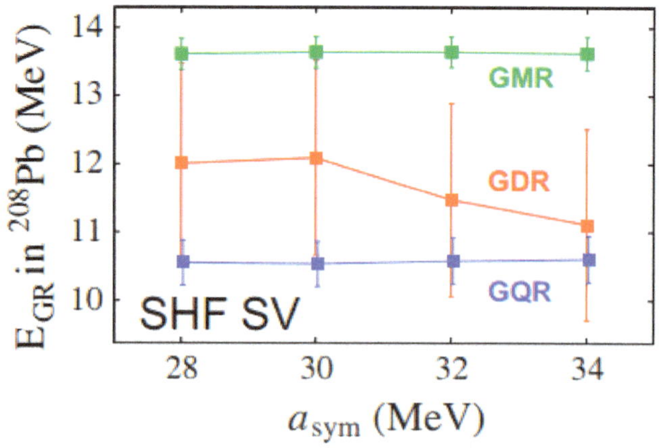

Figure. 4: Behavior of giant resonance energies in ^{208}Pb with symmetry energy a_{sym} for the SV Skyrme family [17]. In order not to make the graph too busy the uncertainties from the unconstrained fit are not shown; they have the same size as those from the constrained fits.

Figure 4 shows the trends of the three major giant resonances in ^{208}Pb: isoscalar monopole resonance (GMR), isovector dipole resonance (GDR), and isoscalar quadrupole resonance (GQR). For technical reasons, we only show results obtained with the SV Skyrme family. The isoscalar resonances show no dependence on a_{sym} at all; this is understandable for the symmetry energy belongs to the isovector sector. Somewhat surprisingly, the GDR exhibits very little dependence on a_{sym} as well, with

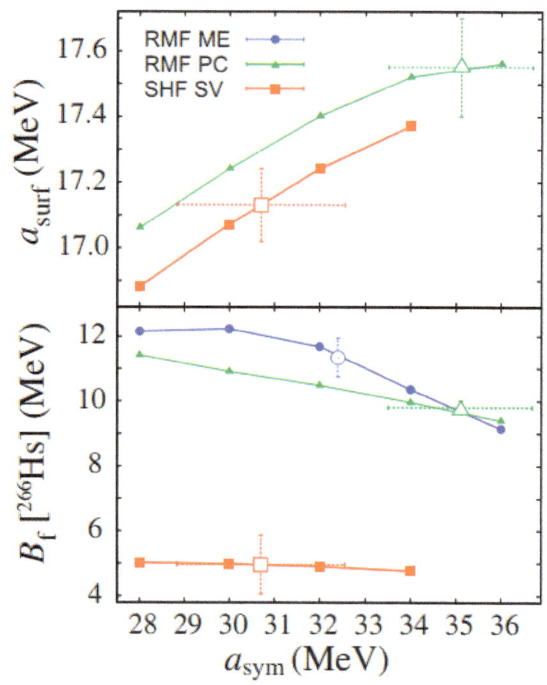

Figure. 5: Similar to fig. 1, but for surface energy (top) and fission barrier (bottom) in ^{266}Hs. The surface energy from RMF-ME is not shown.

the magnitude of variations well below the statistical uncertainties. As demonstrated earlier [17,57], it is the sumrule enhancement factor κ_{TRK} that has the dominant impact on the GDR peak frequency rather than a_{sym}. The covariance analysis of fig. 4 confirms that the energies of GMR, GDR, and GQR do not obviously relate to a_{sym}.

Figure 5 shows behavior of surface energy a_{surf} and the inner fission barrier B_f in 266Hs with a_{sym}. The surface energy was computed by means of the extended ThomasFermi method. (The surface energy from RMF-ME is not shown: we do not have at our disposal a code for semiinfinite matter which could handle this particular RMF functional.) The trends of a_{surf} predicted by SHF and RMF are similar. An offset of about 2MeV is most likely due to very different effective masses in both models. Much larger differences are seen for the fission barriers. The basic difference between SHF and RMF can again be explained predominantly in terms of effective masses. Barriers are produced by shell effects and shell effects are larger for lower effective masses. There is also a difference between the two RMF models. This could be due to a different handling of gradient terms (only RMF-PC contains such) and a much different parametrization of density dependence. All three models show not only different values as such, but also different trends.

The statistical errors differ substantially between the models. RMF-ME shows a small uncertainty in B_f . This may be due to the missing gradient term in this model which would also restrict the uncertainty in the surface energy. We note, however, that the gradient term in RMFPC is to a certain extent equivalent the mass term of the sigma meson in RMF-ME, which is considered a free parameter. The plot of the B_f demonstrates nicely the relative role of statistical and systematic errors, with the statistical errors being much smaller than inter-model differences. As discussed in refs. [70,42], fission barriers are strongly affected by a_{surf} and a_{ssym} of EDF. In particular, the recently developed EDF UNEDF1, suitable for studies of strongly elongated nuclei, has relatively low values of a_{surf} and a_{ssym} (see fig. 7 below) that reflect the constraints on the fission isomer data. The reduced surface energy coefficients result in a reduced effective surface coefficient $a^{(\mathrm{eff})}_{\mathrm{surf}} = a_{\mathrm{surf}} + a_{\mathrm{ssym}} I^2$, which has profound consequences for the description of fission barriers, especially in the neutron-rich nuclei that are expected to play a role at the final stages of the r-process through the recycling mechanism [71].

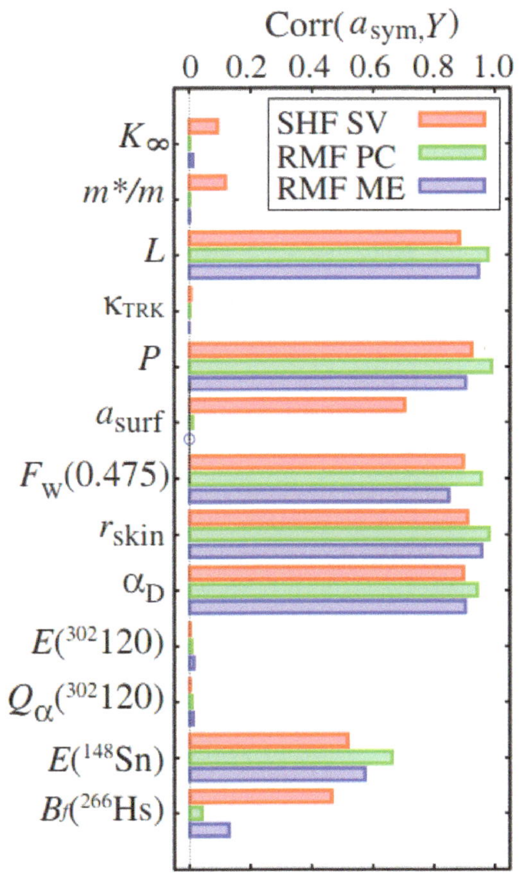

Figure. 6: The correlation (17) between symmetry energy and selected observables (Y) for three models: SHF-SV, RMF-PC and RMF-ME. Results correspond to unconstrained optimization employing the same strategy in all three cases. For RMFME no reliable numbers could be obtained for a_{surf}; this is indicated by an open circle.

Correlations Summary

The summary of our correlation analysis for a_{sym} is given in fig. 6. The first four entries concern the same nuclear matter properties as in fig. 1. It is only for L, the slope of symmetry energy, that a strong correlation with a_{sym} is seen. This complies nicely with the findings of the trend analysis in fig. 1. The next entry concerns the neutron pressure (8) at ρ_n = 0.08neutrons/fm3. It is also strongly correlated with a_{sym}, which is no surprise because it is an excellent isovector indicator [14,36,53,54,72,73]. The diagram shows, furthermore, the (isoscalar) surface energy a_{surf} computed in semi-classical approximation. This quantity is well correlated with a_{sym} for SHF and practically uncorrelated for RMF.

The next three entries are observables in ^{208}Pb: weakcharge form factor, neutron skin, and dipole polarizability. All three are known to be strong isovector indicators [14, 57,36]. This is confirmed here for all three models.

The remaining four entries deal with exotic nuclei. These are: binding energy and α-decay energy in yet-tobe-measured superheavy nucleus Z = 120,N = 182, binding energy in an extremely neutron rich ^{148}Sn, and the fission barrier in 266Hs (for which trends had been shown already in fig. 5). The data on Z = 120, N = 182 consistently do not correlate with a_{sym}. The binding energy of ^{148}Sn shows some correlation with a_{sym}, about equally strong in the three models. This is expected as a large neutron excess surely explores the static isovector sector. Finally, the correlation with fission barrier in ^{266}Hs exhibits an appreciable model dependence with some correlation in SHF and practically none in RMF.

We also studied correlations between δV_{pn} in ^{168}Er and other observables for finite nuclei and NM. We did not find a single observable that would correlate well with this binding-energy indicator. In particular, the correlation coefficient (17) with a_{sym} is 0.41, with α_D in ^{208}Pb is 0.6, and with r_{skin} in ^{208}Pb is 0.54. This results demonstrates that δV_{pn} in one single nucleus is too strongly influenced by shell effects to be used as an isovector indicator.

SYMMETRY ENERGY PARAMETERS OF EDFS

The actual values of symmetry energy parameters depend on i) the form of EDF and ii) the optimization strategy used. The first point is nicely illustrated in table 3, which compares NMP for different functional forms (SHFSV, SHF-RD, RMF-PC, and RMF-ME) using the same dataset and the same optimization technique. As far as the second point, it is instructive to compare SHF-SV and SHF-TOV NMP; namely, the inclusion of additional data on neutron stars in SHF-TOV has significantly impacted L and κTRK. Many other examples can be found in refs. [19,74] that demonstrate divergent predictions of Skyrme EDFs for neutron and nuclear matter.

The actual values of symmetry energy parameters depend on i) the form of EDF and ii) the optimization strategy used. The first point is nicely illustrated in table 3, which compares NMP for different functional forms (SHFSV, SHF-RD, RMF-PC, and RMF-ME) using the same dataset and the same optimization technique. As far as the second point, it is instructive to compare SHF-SV and SHF-TOV NMP; namely, the inclusion of additional data on neutron stars in SHF-TOV has significantly impacted L and κ_{TRK}. Many other examples can be found in refs. [19,74] that demonstrate divergent predictions of Skyrme EDFs for neutron and nuclear matter.

The range of a_{sym} is fairly narrowly constrained by various data and ab initio theory [34]; it is 28MeV < a_{sym} < 34MeV. The recent Finite-Range Droplet Model (FRDM) result [75] is a_{sym} = 32.5 ± 0.5MeV. All EDFs listed in table 3 are consistent with these expectations.

The values of L are less precisely determined [34–39, 76–79]; there is more dependence on specific observables or methodology used. Recent surveys [34,39] suggest that a reasonable range of L is 40MeV <L< 80MeV, and FRDM gives L = 70 ± 15MeV [75]. Except for RMF-ME, all models shown in table 3 are consistent with these estimates. The low value of L in RMF-ME is troublesome; here we note that while SHF-SV and RMF-PC EDFs fall within the error bars of the current experimental data in fig. 2, RMF-ME (as defined by the present optimization protocol) does not. In this context, it is worth mentioning that in the case of RMF-ME the objective function is very steep in the direction of a_{sym}. This is seen in table 3, which shows the uncertainties of a_{sym} for the three models

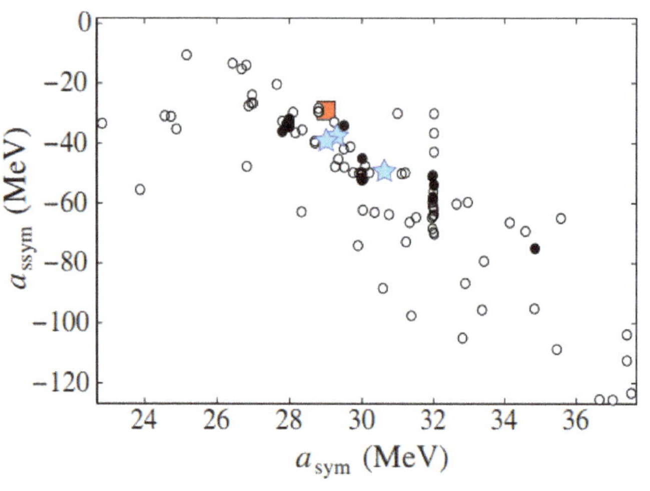

Figure. 7: Correlation between the symmetry and surface symmetry coefficients taken from ref. [70] (Skyrme EDFs, dots; LDM values, stars) and ref. [81] (Skyrme EDFs, circles). The UNEDF1 values [42] are marked by a square. (Adopted from [70].)

employed. Consequently, the RMF-ME results are mostly in a sub-optimal range of the model in a broad range of a_{sym}. For instance, at a_{sym} = 30MeV, the average deviation for binding energies with RMF-ME is about twice as large as for the optimal value a_{sym} = 32.4MeV.

As discussed in ref. [41], the leading surface and symmetry terms appear relatively similar within each family of EDFs, with a clear difference for asym between SHF and RMF. By averaging over Skyrme-EDF results of refs. [41,80], one obtains: $a_{sym} \approx 30.9 \pm 1.7$MeV, $a_{ssym} \approx -48 \pm 10$MeV. Older relativistic models provide systematically larger values [41]: $a_{sym} \approx 40.4 \pm 2.7$MeV and $a_{ssym} \approx -103 \pm 18$MeV. (Codes for a leptodermous expansion of the recent RMF-PC and RMF-ME models have yet to be developed.)

The coefficient a_{ssym} is poorly constrained in the current EDF parameterizations and there are large differences between models, see fig. 7. In addition, the values of a_{sym} and a_{ssym} have been shown to be systematically (anti)correlated [51,70,81,82]. Figure 7, displays the pairs (a_{sym},a_{ssym}) for various Skyrme EDFs and LDM parametrizations. While a correlation between a_{sym} and a_{ssym} is apparent, a very large spread of values is seen that demonstrates that the data on g.s. nuclear properties are not able to constrain a_{ssym}. It is interesting to note that the LDM values and phenomenological estimates cluster around a_{sym} = 30MeV and a_{ssym} = −45MeV. The values for UNEDF1 functional, additionally constrained by the data on very deformed fission isomers (thus probing the surface-isospin sector of EDF) are a_{sym} = 29MeV and a_{ssym} = −29MeV.

ISOSPIN PHYSICS AND SYMMETRY ENERGY

The emergence of NSE is rooted in the isobaric symmetry and its breaking as a function of neutron excess and mass. Single-reference DFT is essentially the only framework allowing for understanding global behavior of isospin effects throughout the entire nuclear landscape. While the nuclear interaction part of the nuclear EDF is constructed to be an isoscalar [24,25], the Coulomb interaction breaks isospin manifestly. There are, therefore, two different sources of isospin symmetry breaking in the nuclear DFT: spontaneous isospin breaking associated with the selfconsistent response to the neutron excess, and the explicit breaking due to the electric charge of the protons [83]. Effects related to isospin breaking and restoration are difficult to treat theoretically within the nuclear DFT. Below, we discuss two ways of dealing with this problem: isocranking and isospin projection.

1d and 3d Isocranking

The isocranking model [84,80] attributes the kinetic coefficient $a_{sym,kin}$ contribution to the mean level spacing at the Fermi energy ε(A) rather than to the total kinetic energy itself. The SHF calculations also revealed that the isovector mean potential of the Skyrme EDF can be quite well characterized by an effective V_{TT} interaction (1) characterized by a strength parameter κ(A). The actual isovector part of the Skyrme mean-field potential is composed of several terms [24,25]. As can be seen from table 1, in the uniform NM limit, two terms contribute in SHF, $C_1^\rho \rho_1^2$ and $C_1^\tau \rho_1 \tau_1$, and the NSE strength reads

$$a_{sym} = \frac{1}{8}\frac{m}{m^*}\varepsilon_{FG} + \left[\left(\frac{3\pi^2}{2}\right)^{2/3} C_1^\tau \rho_0^{5/3} + C_1^\rho \rho_0\right], \quad (18)$$

where ε_{FG} is the average level splitting in FGM. Therefore, within this scenario, a_{sym} is non-trivially modified by momentum-dependent effects introducing, in the leading order, the dependence of $a_{sym,kin}$ and $a_{sym,int}$ on the isoscalar and isovector effective mass, respectively.

Within the nuclear shell model, NSE appears through a contribution to the binding energy proportional to $T(T + 1)$ [85]. However, the local enhancement of binding around $N = Z$ (the Wigner energy) suggest an enhancement of the linear term to $T(T + \lambda)$ with $\lambda \approx 1.26$ [86–88]. Since the Wigner energy is neither fully understood nor included properly within the SCMF models [89], the microscopic origin of λ is still a matter of debate. Within the isocranking model, the Fock exchange (isovector) potential gives rise to $\lambda \approx 0.5$, at variance with enhancement seen in experimental data. The Wigner energy can be explained by shell-model calculations [89] in terms of configuration mixing. The Wigner term is usually associated with the isoscalar neutron-proton (np) pairing [90,25], but its understanding is poor as realistic calculations involving simultaneous np mixing in both the particle-hole (p-h) and particle-particle (p-p) channels have not been carried out. It is only very recently that 3D isocranking calculations including np mixing in the p-h channel have been reported [91]. This is the first step towards developing the nuclear superfluid DFT including np mixing in both p-h and p-p channels. An improved treatment of isospin within the 3D isocranking will open new opportunities for quantitative studies of isobaric analogue states and, in turn, the NSE.

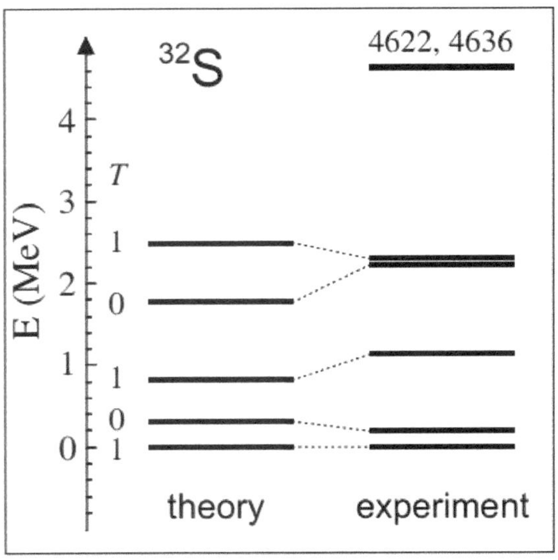

Figure. 8: Energies of $I^{\pi} = 1^{+}$ states in ^{32}S normalized to the isobaric analogue state $I^{\pi} = 1^{+}$, $T = 1$. The results of projected SHF-SkV calculations involving configuration mixing [94] (left) are compared to experiment (right). The calculations are based on 24 $I = 1^{+}$ states projected from 6 HF determinants representing low-lying 1p-1h configurations.

Isospin Projected DFT

The isospin and isospin-plus-angular-momentum projected DFT models have been developed recently to describe isospin mixing effects. These new tools open new avenues to probe NSE. To gain insight on this line of models, it is instructive to to consider the spontaneous isospin symmetry breaking effect in the so-called anti-aligned ph configurations in $N = Z$ nuclei, which are mixtures of $T = 0$ and $T = 1$ states [92]. Restoration of the isospin symmetry results in the energy splitting, ΔE_T, between the actual $T = 0$ and $T = 1$ configurations. Since these states are projected from a single mean-field determinant, the splitting is believed to be insensitive to kinematics, and the method can be used to probe dynamical effects giving rise to the interaction term $a_{sym,int}$. The results of SHF calculations [92] performed in finite nuclei confirm that $a_{sym,int}$ is indeed correlated with the isoscalar effective mass in agreement with the NM relation (18).

The isospin and isospin plus angular momentum projected DFT were designed and applied to study the isospin impurities [83] and isospin symmetry breaking corrections to the superallowed $0^{+} \rightarrow 0^{+}$ β-decay rates [93]. Unfortunately, the calculations show that these two observables are not directly correlated with the symmetry energy. Ambiguities associated with these calculations stimulated further development of the formalism in the direction of the Resonating-group method. The scheme proceeds in three steps. i) First, a set of low-lying (multi)p-(multi)h SHF states $\{\Phi_i\}$ is calculated. These states form a basis for a subsequent projection. ii) Next, the projection techniques are applied to calculate a family $\{\Psi_I^{(\alpha)}\}$ of good angular momentum states with properly treated K-mixing and isospin mixing. iii) Finally, a configuration mixing of $\{\Psi_I^{(\alpha)}\}$ states is performed using techniques suitable for non-orthogonal ensambles.

Although at present the calculations can be realized only for the SkV EDF, the preliminary results [94] are encouraging, as shown in fig. 8. Since the projected approach treats rigorously the angular momentum conservation and the long-range polarization due to the Coulomb force, it opens up a possibility of detailed studies of the isovector terms of the nuclear EDF that are sources of the NSE.

CONCLUSIONS

This paper surveys various aspects of NSE within the nuclear DFT represented by non-relativistic and relativistic self-consistent mean-field frameworks. After defining the models and statistical tools, we reviewed key observables pertaining to bulk nucleonic matter and finite nuclei. Using the statistical covariance technique, constraints on the symmetry energy were studied, together with correlations between observables and symmetry energy parameters. Through the systematic correlation analysis, we scrutinized various observables from finite nuclei that are accessible by current and future experiments. We confirm that by far the most sensitive isovector indicators are observables related to the neutron skin (neutron radius, diffraction radius, weak charge form factor) and the dipole polarizability [14,57]. In this context, PREX-II measurement of the neutron skin in ^{208}Pb [95] (a follow-up measurement to PREX [49] designed to improve the experimental precision), CREX measurement of the neutron skin in 48Ca [96], and on-going measurements of α_D in neutronrich nuclei [97] are indispensable.

The masses of heavy neutron-rich nuclei also seem to correlate well with NSE parameters. Other observables, such as Qα-values, δV_{pn}, barrier heights, and low-energy dipole strength [14,59,60] are too strongly impacted by shell effects to be useful as global isovector indicators. A major challenge is to develop the universal nuclear EDF with improved isovector properties. Various improvements are anticipated in the near future. Those include constraining the EDF at sub-saturation densities using ab initio models [98,99] and using the density matrix expansion to develop an EDF based on microscopic nuclear interactions [100]. This work will be carried out under the Nuclear Low Energy Computational Initiative (NUCLEI) [101]. Other exciting avenues are related to multireference isospin projected DFT, which will enable us to make reliable predictions for isobaric analogues, isospin mixing, and mirror energy differences.

This work was supported by the U.S. Department of Energy under Contract No. DE-FG02-96ER40963 (University of Tennessee), No. DE-SC0008499 (NUCLEI SciDAC Collaboration); by BMBF under Contract No. 06 ER 142D; and by NCN under Contract No. 2012/07/B/ST2/03907

REFERENCES

1. 1. P. Hohenberg, W. Kohn, Phys. Rev. 136, B864 (1964).
2. 2. W. Kohn, L. Sham, Phys. Rev. 140, A1133 (1965).
3. 3. D. Vautherin, D.M. Brink, Phys. Rev. C 5, 626 (1972).
4. 4. J.W. Negele, D. Vautherin, Phys. Rev. C
5. 5, 1472 (1972). 5. M. Bender, P.H. Heenen, P.G. Reinhard, Rev. Mod. Phys. 75, 121 (2003).
6. 6. J. Messud, M. Bender, E. Suraud, Phys. Rev. C 80, 054314 (2009).
7. 7. G.A. Lalazissis, P. Ring, D. Vretenar, Extended Density Functionals in Nuclear Structure Physics, Lecture Notes in Physics, Vol. 641 (Springer, 2004).
8. 8. D. Vretenar, A.V. Afanasjev, G. Lalazissis, P. Ring, Phys. Rep. 409, 101 (2005).
9. 9. S. Weinberg, The Quantum Theory of Fields, Vols. I-III (Cambridge University Press, Cambridge, 1996-2000).
10. 10. S. Puglia, A. Bhattacharyya, R. Furnstahl, Nucl. Phys. A 723, 145 (2003).
11. 11. B.G. Carlsson, J. Dobaczewski, M. Kortelainen, Phys. Rev. C 78, 044326 (2008).
12. 12. J. Drut, R. Furnstahl, L. Platter, Prog. Part. Nucl. Phys. 64, 120 (2010).
13. 13. J. Dobaczewski, K. Bennaceur, F. Raimondi, J. Phys. G 39, 125103 (2012).
14. 14. P.G. Reinhard, W. Nazarewicz, Phys. Rev. C 81, 051303 (2010).
15. 15. A. Bohr, B. Mottelson, Nuclear Structure, Vol. I (W. A. Benjamin, New York, 1969).
16. 16. A.Z. Mekjian, L. Zamick, Phys. Rev. C 85, 057303 (2012).
17. 17. P. Kl¨upfel, P.G. Reinhard, T.J. B¨urvenich, J.A. Maruhn, Phys. Rev. C 79, 034310 (2009).
18. 18. M. Kortelainen, T. Lesinski, J. Mor´e, W. Nazarewicz, J. Sarich, N. Schunck, M.V. Stoitsov, S. Wild, Phys. Rev. C 82, 024313 (2010).
19. 19. J. Stone, P.G. Reinhard, Prog. Part. Nucl. Phys. 58, 587 (2007).
20. 20. J. Erler, P. Kl¨upfel, P.G. Reinhard, J. Phys. G 38, 033101 (2011).
21. 21. J. Meng, H. Toki, S. Zhou, S. Zhang, W. Long, Prog. Part. Nucl. Phys. 57, 470 (2006).
22. 22. T. Nikˇsi´c, D. Vretenar, P. Ring, Prog. Part. Nucl. Phys. 66, 519 (2011).
23. 23. Y.M. Engel, D.M. Brink, K. Goeke, S.J. Krieger, D. Vautherin, Nucl. Phys. A 249, 215 (1975).
24. 24. E. Perli´nska, S.G. Rohozi´nski, J. Dobaczewski, W. Nazarewicz, Phys. Rev. C 69, 014316 (2004).
25. 25. S.G. Rohozi´nski, J. Dobaczewski, W. Nazarewicz, Phys. Rev. C 81, 014313 (2010).
26. 26. P.G. Reinhard, Rep. Prog. Phys. 52, 439 (1989).
27. 27. T. Nikˇsi´c, D. Vretenar, P. Ring, Phys. Rev. C 78, 034318 (2008).
28. 28. T. Nikˇsi´c, D. Vretenar, P. Finelli, P. Ring, Phys. Rev. C 66, 024306 (2002).
29. 29. G.A. Lalazissis, T. Nikˇsi´c, D. Vretenar, P. Ring, Phys. Rev. C 71, 024312 (2005).
30. 30. P. Bonche, H. Flocard, P.H. Heenen, S.J. Krieger, M.S. Weiss, Nucl. Phys. A 443, 39 (1985).
31. 31. M. Bender, K. Rutz, P.G. Reinhard, J. Maruhn, Eur. Phys. J. A 8, 59 (2000).
32. 32. Y. Tian, Z. Ma, P. Ring, Phys. Lett. B 676, 44 (2009).
33. 33. Y. Tian, Z. Ma, P. Ring, Phys. Rev. C 80, 024313 (2009).
34. 34. J.M. Lattimer, Annu. Rev. Nucl. Part. Sci. 62, 485 (2012).
35. 35. M.B. Tsang et al., Phys. Rev. C 86, 015803 (2012).
36. 36. F.J. Fattoyev, J. Piekarewicz, Phys. Rev. C 86, 015802 (2012).
37. 37. F.J. Fattoyev, W.G. Newton, J. Xu, B.A. Li, Phys. Rev. C 86, 025804 (2012).
38. 38. J. Erler, C.J. Horowitz, W. Nazarewicz, M. Rafalski, P.G. Reinhard, Phys. Rev. C 87, 044320 (2013).
39. 39. A.W. Steiner, J.M. Lattimer, E.F. Brown, Astrophys. J. 765, L5 (2013).
40. 40. P. Ring, P. Schuck, The Nuclear Many-Body Problem (Springer, 2000).
41. 41. P.G. Reinhard, M. Bender, W. Nazarewicz, T. Vertse, Phys. Rev. C 73, 014309 (2006).
42. 42. M. Kortelainen, J. McDonnell, W. Nazarewicz, P.G. Reinhard, J. Sarich, N. Schunck, M.V. Stoitsov, S.M. Wild, Phys. Rev. C 85, 024304 (2012).
43. 43. D. Von-Eiff, J.M. Pearson, W. Stocker, M.K. Weigel, Phys. Lett. B 324, 279 (1994).
44. 44. J.Y. Zhang, R. Casten, D. Brenner, Phys. Lett. B 227, 1 (1989).
45. 45. M. Stoitsov, R.B. Cakirli, R.F. Casten, W. Nazarewicz, W. Satutla, Phys. Rev. Lett. 98, 132502 (2007).
46. 46. P.G. Reinhard, Phys. Lett. A 169, 281 (1992). 47. J. Friedrich, N. V¨ogler, Nucl. Phys. A 373, 192 (1982).
47. 48. S. Mizutori, J. Dobaczewski, G. Lalazissis, W. Nazarewicz, P.G. Reinhard, Phys. Rev. C 61, 044326 (2000).
48. 49. PREX Collaboration (S. Abrahamyan et al.), Phys. Rev. Lett. 108, 112502 (2012).
49. 50. C.J. Horowitz et al., Phys. Rev. C 85, (2012).

50. 51. F. Tondeur, M. Brack, M. Farine, J. Pearson, Nucl. Phys. A 420, 297 (1984).
51. 52. P.G. Reinhard, Nucl. Phys. A 649, 305c (1999).
52. 53. R. Furnstahl, Nucl. Phys. A 706, 85 (2002).
53. 54. S. Yoshida, H. Sagawa, Phys. Rev. C 69, 024318 (2004).
54. 55. B. Li, L. Chen, C. Ko, Phys. Rep. 464, 113 (2008).
55. 56. M. Warda, X. Vi˜nas, X. Roca-Maza, M. Centelles, Phys. Rev. C 80, 024316 (2009).
56. 57. J. Piekarewicz, B.K. Agrawal, G. Col`o, W. Nazarewicz, N. Paar, P.G. Reinhard, X. Roca-Maza, D. Vretenar, Phys. Rev. C 85, 041302 (2012).
57. 58. A. Tamii et al., Phys. Rev. Lett. 107, 062502 (2011).
58. 59. I. Daoutidis, S. Goriely, Phys. Rev. C 86, 034328 (2012).
59. 60. P.G. Reinhard, W. Nazarewicz, Phys. Rev. C 87, 014324 (2013).
60. 61. J. Erler, K. Langanke, H.P. Loens, G. Martinez-Pinedo, P.G. Reinhard, Phys. Rev. C 85, 025802 (2012).
61. 62. A. Staszczak, A. Baran, W. Nazarewicz, Phys. Rev. C 87, 024320 (2013).
62. 63. M. Warda, J.L. Egido, Phys. Rev. C 86, 014322 (2012).
63. 64. F.J. Fattoyev, J. Piekarewicz, Phys. Rev. C 84, 064302 (2011).
64. 65. Y. Gao, J. Dobaczewski, M. Kortelainen, J. Toivanen, D. Tarpanov, Phys. Rev. C 87, 034324 (2013).
65. 66. S. Brandt, Statistical and computational methods in data analysis (Springer, New York, 1997) third English edition.
66. 67. J. Erler, P. Kl¨upfel, P.G. Reinhard, Phys. Rev. C 82, 044307 (2010).
67. 68. S. Typel, H. Wolter, Nucl. Phys. A 656, 331 (1999).
68. 69. M. Bender, P.H. Heenen, Phys. Rev. C 83, 064319 (2011).
69. 70. N. Nikolov, N. Schunck, W. Nazarewicz, M. Bender, J. Pei, Phys. Rev. C 83, 034305 (2011).
70. 71. I. Panov, I.Yu.Korneev, F.K. Thielemann, Astron. Lett. 34, 189 (2008).
71. 72. B. Alex Brown, Phys. Rev. Lett. 85, 5296 (2000).
72. 73. S. Typel, B.A. Brown, Phys. Rev. C 64, 027302 (2001).
73. 74. M. Dutra, O. Louren¸co, J.S. S´a Martins, A. Delfino, J.R. Stone, P.D. Stevenson, Phys. Rev. C 85, 035201 (2012).
74. 75. P. M¨oller, W.D. Myers, H. Sagawa, S. Yoshida, Phys. Rev. Lett. 108, 052501 (2012).
75. 76. B. Li, L. Chen, C. Ko, Phys. Rep. 464, 113 (2008).
76. 77. L. Trippa, G. Col`o, E. Vigezzi, Phys. Rev. C 77, 061304 (2008).
77. 78. A. Steiner, S. Gandolfi, Phys. Rev. Lett. 108, 081102 (2012).
78. 79. F.J. Fattoyev, J. Piekarewicz, arXiv:nucl-th/1306.6034 (2013).
79. 80. W. Satutla, R.A. Wyss, M. Rafalski, Phys. Rev. C 74, 011301 (2006).
80. 81. P. Danielewicz, J. Lee, Nucl. Phys. A 818, 36 (2009).
81. 82. M. Farine, J. Pearson, B. Rouben, Nucl. Phys. A 304, 317 (1978).
82. 83. W. Satutla, J. Dobaczewski, W. Nazarewicz, M. Rafalski, Phys. Rev. Lett. 103, 012502 (2009).
83. 84. W. Satutla, R.A. Wyss, Phys. Lett. B 572, 152 (2003).
84. 85. I. Talmi, Rev. Mod. Phys. 34, 704 (1962).
85. 86. J. J¨anecke, Nucl. Phys. A 73, 97 (1965).
86. 87. J. J¨anecke, T. O'Donnell, V. Goldanskii, Nucl. Phys. A 728, 23 (2003).
87. 88. S. Gtlowacz, W. Satutla, R. Wyss, Eur. Phys. J. A 19, 33 (2004).
88. 89. W. Satutla, D. Dean, J. Gary, S. Mizutori, W. Nazarewicz, Phys. Lett. B 407, 103 (1997)
89. 90. W. Satutla, R. Wyss, Phys. Lett. B 393, 1 (1997).
90. 91. K. Sato, J. Dobaczewski, T. Nakatsukasa, W. Satutla, RIKEN Accel. Prog. Rep., 46 (2013) arXiv:1308.1997.
91. 92. W. Satutla, J. Dobaczewski, W. Nazarewicz, M. Borucki, M. Rafalski, Int. J. Mod. Phys. E 20, 244 (2011).
92. 93. W. Satutla, J. Dobaczewski, W. Nazarewicz, M. Rafalski, Phys. Rev. Lett. 106, 132502 (2011).
93. 94. W. Satutla, J. Dobaczewski, M. Konieczka, W. Nazarewicz, arXiv:1307.1550 (2013).
94. 95. PREX-II Proposal to Jefferson Lab, http://hallaweb.jlab.org/parity/prex/prexII.pdf.
95. 96. CREX Proposal to Jefferson Lab, http://hallaweb.jlab. org/parity/prex/c-rex2013 v7.pdf.
96. 97. A. Tamii, Acta Phys. Pol. B 44, 571 (2013).
97. 98. S.K. Bogner, R.J. Furnstahl, H. Hergert, M. Kortelainen, P. Maris, M. Stoitsov, J.P. Vary, Phys. Rev. C 84, 044306 (2011).
98. 99. P. Maris, J.P. Vary, S. Gandolfi, J. Carlson, S.C. Pieper, Phys. Rev. C 87, 054318 (2013).
99. 100. M. Stoitsov, M. Kortelainen, S.K. Bogner, T. Duguet, R.J. Furnstahl, B. Gebremariam, N. Schunck, Phys. Rev. C 82, 054307 (2010).
100. 101. Nuclear Computational Low-Energy Initiative, http:// computingnuclei.org.

CHAPTER 3 ALPHA-CLUSTER MODEL OF ATOMIC NUCLEI

Zbigniew Sosin, Jan Blocki, Jinesh Kallunkathariyil, Jerzy Lukasik, and Piotr Pawlowski

INTRODUCTION

In 1936 Bethe and Bacher [1] and in 1938 Hafstad and Teller [2] predicted that alpha-particle structures [3, 4] could be present in atomic nuclei. Convincing arguments for the existence of such structures were provided by systematics of the binding energies of the even-even nuclei with equal number of protons and neutrons (see also [5]), as well as by systematics of the binding energy of the additional neutron in nuclei like ^{9}B, ^{13}C and ^{17}O. The latter systematics could be explained by assuming that the valence neutron moves in a multi-center potential with centers identified with alpha particles. Today, from a commonly accepted point of view one considers that alphaparticle clusters can appear in the nuclear matter at subsaturation densities. Such conditions are met, e.g., at the nuclear surface where, in not too heavy nuclei, almost the whole nuclear mass is concentrated. Subtle changes of the ground-state energy of atomic nuclei as a function of the nucleon number can be described by the shell model, which neglects correlations between nucleons. The main idea of the present approach is to verify whether these subtle changes, in the case of n-alpha nuclei, could be attributed to the appearance of alpha-like clusters which modify the structure of the wave function and introduce strong correlations between nucleons. In such an approach the system in its ground state behaves like a crystal, with stationary configuration and shape and with definite distances between the wave function centers for a given nucleus. The subtle changes of the binding energy due to appearance of alpha-like clusters in the wave functions of n-alpha nuclei will be determined within an extended version of the model proposed in [6]. It will be shown in the following that the hypothesis of alpha structures in the n-alpha nuclei can indeed describe the binding energy systematics of Bethe [1, 2]. It will be also shown that the hypothesis of the alpha structures helps to understand the evolution of the root-mean-square (rms) radii and of the density profiles as a function of the number of alpha particles in the n-alpha nuclei. The paper is organized as follows: sect. 2 describes the adopted interactions in the nuclear matter. Section 3 presents constraints on the introduced parametrization resulting from the properties of light charged particles. Section 4 is devoted to the ground-state properties of n-alpha nuclei. First, it investigates the effects of clusterization on the ground-state configurations. Then, some corrections to the Hamiltonian, related to the many-body character of the nuclear interaction and to the bosonic nature of the wave function, are introduced. It will be shown that the introduced corrections make the structural corrections stronger. Finally, the model predictions for binding energies, rms radii and density profiles are presented and compared with the experimental values. Section 5 presents the conclusions.

INTERACTIONS IN NUCLEAR MATTER AROUND SATURATION

In order to predict correctly a state and evolution of a nuclear system, a reliable description of the interaction between nucleons is needed. In the case of a strong interaction it is not an easy task. As we know, the fundamental theory governing the nucleon-nucleon interactions is the quantum chromodynamics (QCD), where interactions are calculated from the physics of quarks and gluons. Unfortunately, very limited progress has

Citation: Zbigniew Sosin, Jan Blocki, Jinesh Kallunkathariyil, Jerzy Lukasik, and Piotr Pawlowski, Alpha-cluster model of atomic nuclei, DOI 10.1140/epja/i2016-16120-4.

been achieved in the application of QCD for the description of the interaction between quark-gluon clusters. Therefore, a phenomenological approach, associated to some specific conditions has been applied. Since we are interested in the low-energy region in which nucleons are not excited internally, one can treat nucleons as quasi-elementary particles. Additionally, if we apply a non-relativistic approximation, the mutual nucleon interaction can be described by a respective potential. As can be found in the text-books, details of the form of such a potential are deduced from nuclear interaction symmetries and the mechanism of exchange of bosons. For a final test of this type of potential and the determination of its parameters, one uses a comparison with the experimental data. Here the nucleon-nucleon scattering data or light-nuclei properties are usually used. A general conclusion from this kind of research is that a simple two-body potential is not sufficient to describe the behavior of the system of more than two nucleons. One should add at least the three-body force. In particular, in the nuclear system, the three-body force is acceptable because the strong interaction is generated by the exchange process of gluons. The n-body interaction can be interpreted as coming from the multi-pion exchange processes. These considerations show the difficulties in the description of even the simplest nuclear system. The extension of the description of the interaction by the three-body forces shows that the exchanged particles (bosons) play a significant role in the interaction process. In order to describe the dynamics of the system, one has to express the change of energy associated with its evolution. In the description of energy through the potential one has to define its form. Hence the wave function of a system of identical particles must be either symmetric or antisymmetric and it must be taken into account when defining the energy of the ground state. In the case of the fermionic systems one should use an antisymmetric wave function which is a serious difficulty in the description of the collision dynamics. Some proposals to solve this problem are shown for instance in [7–9]. Other approaches, introducing some kinds of momentum-dependent Pauli potentials and Pauli-blocking schemes, which are meant to mimic the effects of antisymmetrization and are much less time consuming, are also proposed, see, e.g., [10–14]. Therefore, in the present approach, we propose a description in which particles are distinguishable and the energy associated with the fermionic motion affects the resulting energy density which is parametrized and used instead of the potential. Such a concept is present in the description of the liquid drop model (LDM) and in some density functional models, e.g., in [15]. It is obvious that such an approach requires the knowledge of the local density of matter and the knowledge of the wave function describing the system. In the present description we use a standard one, commonly applied in such cases, i.e. the wave function is defined as a simple product of minimal wave packets (eq. (1)), as, e.g., in [16]. The main argument for disregarding the antisymmetrization in the present approach is the form of the proposed EoS which includes the spin and isospin terms. These terms force the unlike nucleons to form "bosonic" alpha-like clusters for which, presumably, the antisymmetrization may no longer play a decisive role. Each minimal wave packet in the adopted wave function has a defined spin and a specific charge for a selected quantization axis:

$$\Phi = \prod_{k=1}^{A} {}^{k}\phi_{I_k S_k}, \tag{1}$$

$$ {}^{k}\phi_{I_k S_k} = \frac{1}{(2\pi\sigma_k^2(r))^{3/4}} \exp\left(\frac{-(\mathbf{r}_k - \langle \mathbf{r}_k \rangle)^2}{4\sigma_k^2(r)} + \frac{i}{\hbar}\mathbf{r}_k \langle \mathbf{p}_k \rangle \right), \tag{2}$$

Where $\sigma_k^2(r)$, $\langle \mathbf{r}_k \rangle$, $\langle \mathbf{p}_k \rangle$ are the width (the position variance of the k-th nucleon) of a Gaussian wave packet and the mean position and mean momentum of each of the A nucleons, respectively. Every partial wave function is labeled by $I_k = n$ or $I_k = p$ and $S_k = \uparrow$ or $S_k = \downarrow$, denoting the isospin and spin (or precisely, their projections on a specified quantization axis) associated with a given nucleon. Variables $\langle \mathbf{r}_k \rangle$, $\langle \mathbf{p}_k \rangle$ and $\sigma_k^2(r)$ are, in general, time-dependent parameters describing the wave functions. In the present approach the nuclear matter is treated as a four-component fluid characterized by the respective densities:

- $\rho_{p\uparrow}$ for protons with spin up;
- $\rho_{p\downarrow}$ for protons with spin down;
- $\rho_{n\uparrow}$ for neutrons with spin up;
- $\rho_{n\downarrow}$ for neutrons with spin down.

Density distributions in the system are uniquely determined by parameters of the wave function (mean position, mean momentum and variance of the position of each nucleon). The description of the system evolution due to the time evolution of these parameters is, in general, defined by the Dirac-Frenkel time-dependent variational principle [17]. This approach describes the interaction in a selfconsistent way

Assuming that nucleons are moving inside the nuclear system, the average energy of the system is a sum of the average potential energy and of the kinetic energy associated with the fermionic motion. This is the case in which one disregards the energy of the ordered motion, which, for the low excitations of the system, is considered later on as a correction (see formula (14)). According to the above idea we now present a method for determining the average energy associated with each nucleon. For the description of this average energy we assume the existence of a scalar field $\varepsilon(\rho_{p\uparrow}, \rho_{p\downarrow}, \rho_{n\uparrow}, \rho_{n\downarrow})$, determined by the local densities $\rho_{p\uparrow}(\mathbf{r})$, $\rho_{p\downarrow}(\mathbf{r})$, $\rho_{n\uparrow}(\mathbf{r})$, $\rho_{n\downarrow}(\mathbf{r})$. If, for every k-th nucleon, the corresponding wave packet ${}^{k}\phi_{I_k S_k}$ determines the probability $P_k(r) = |{}^{k}\phi_{I_k S_k}|^2$ of finding a nucleon at a given point r, then it is assumed that the average energy e_k associated with this nucleon is defined by the mean value and variance of the field $\varepsilon(\rho_{p\uparrow}, \rho_{p\downarrow}, \rho_{n\uparrow}, \rho_{n\downarrow})$ and can be expressed as

$$e_k = \langle \varepsilon \rangle_k + \lambda \sigma_k(\varepsilon), \tag{3}$$

where

$$\langle \varepsilon \rangle_k = \int P_k(\mathbf{r}) \varepsilon (\rho_{p\uparrow}, \rho_{p\downarrow}, \rho_{n\uparrow}, \rho_{n\downarrow}) \, \mathrm{d}^3\mathbf{r}, \tag{4}$$

$$\sigma_k^2(\varepsilon) = \int P_k(\mathbf{r}) \left(\varepsilon (\rho_{p\uparrow}, \rho_{p\downarrow}, \rho_{n\uparrow}, \rho_{n\downarrow}) - \langle \varepsilon \rangle_k \right)^2 \mathrm{d}^3\mathbf{r} \tag{5}$$

and λ is a parameter related to the surface energy, similarly as in [18].

The EoS of the nuclear matter is defined for an in- finite system, assuming its homogeneity and isotropy. For such a matter, the variance of the associated field $\varepsilon(\rho_{p\uparrow}, \rho_{p\downarrow}, \rho_{n\uparrow}, \rho_{n\downarrow})$ vanishes and the average value of the energy per nucleon is determined by the mean value of the field alone. The field $\varepsilon(\rho_{p\uparrow}, \rho_{p\downarrow}, \rho_{n\uparrow}, \rho_{n\downarrow})$ is in this case equivalent to the EoS of the nuclear matter. Thus, the functional form of the field $\varepsilon(\rho_{p\uparrow}, \rho_{p\downarrow}, \rho_{n\uparrow}, \rho_{n\downarrow})$ can be obtained from the EoS by applying the local density approximation. In the present work we use a 12-parameter cubic form of the EoS proposed in [6]. In the vicinity of the saturation density it can be approximated using a second-order Taylor expansion:

$$\begin{aligned} e = e_{00} &+ \frac{K_0}{18}\xi^2 \\ &+ \delta^2 \left(e_{I0} + \frac{L_I}{3}\xi + \frac{K_I}{18}\xi^2 \right) \\ &+ \left(\eta_n^2 + \eta_p^2\right) \left(e_{ii0} + \frac{L_{ii}}{3}\xi + \frac{K_{ii}}{18}\xi^2 \right) \\ &+ 2\eta_n\eta_p \left(e_{ij0} + \frac{L_{ij}}{3}\xi + \frac{K_{ij}}{18}\xi^2 \right), \end{aligned} \tag{6}$$

where

$$\xi = \frac{\rho - \rho_0}{\rho_0}, \tag{7}$$

$$\delta = \frac{\rho_n - \rho_p}{\rho}, \tag{8}$$

$$\eta_n = \frac{\rho_{n\uparrow} - \rho_{n\downarrow}}{\rho}, \tag{9}$$

$$\eta_p = \frac{\rho_{p\uparrow} - \rho_{p\downarrow}}{\rho} \tag{10}$$

in which ρ and ρ_0 are the total nuclear matter density and the density of isospin and spin-balanced matter at saturation, respectively.

The first two terms in (6) are just a standard, 6-parameter form of the EoS approximation commonly used in the present-day experimental and theoretical studies, see, e.g., [19]. The first term describes the symmetric matter in a balanced system (zero isospin and spin), with K_0 being the compressibility of the symmetric nuclear matter. The second one is a standard approximation of the symmetry energy:

$$e_I = e_{I0} + \frac{L_I}{3}\xi + \frac{K_I}{18}\xi^2 \tag{11}$$

with e_{I0} being the Wigner constant and the coefficients L_I and K_I being the slope and curvature, respectively. The last two terms of (6) are the main novelty of the present approach (see also [6]). In analogy to the isospin symmetry energy, they describe the spin symmetry energies for neutrons and protons separately:

$$e_{ii} = e_{ii0} + \frac{L_{ii}}{3}\xi + \frac{K_{ii}}{18}\xi^2 \tag{12}$$

and the energy of the mutual spin interaction of protons and neutrons:

$$e_{ij} = e_{ij0} + \frac{L_{ij}}{3}\xi + \frac{K_{ij}}{18}\xi^2. \tag{13}$$

The terms with indices ii, *i.e.* the e_{ii0}, L_{ii} and K_{ii} (constant, slope and curvature), describe the energy of protons or neutrons. Indices ij indicate that the symmetry energy refers to the mutual interaction of protons and neutrons. The determination of the expansion parameters in (6) will be discussed in the next sections.

Using the field $\varepsilon(\rho_{p\uparrow}, \rho_{p\downarrow}, \rho_{n\uparrow}, \rho_{n\downarrow})$ one can express the average value of the Hamiltonian for a given system of A nucleons as

$$\begin{aligned} \langle \Phi | H | \Phi \rangle = \sum_{k=1}^{k=A} \frac{\langle \mathbf{p}_k \rangle^2}{2m_N} &+ \sum_{k=1}^{k=A} \langle \varepsilon \rangle_k \\ &+ \lambda \sum_{k=1}^{k=A} \sigma_k(\varepsilon) + \langle \Phi | V_C | \Phi \rangle, \end{aligned} \tag{14}$$

where m_N is the nucleon mass and the last term describes the Coulomb energy. As later on we will be searching for the ground-state configurations of the even-even nuclei with N = Z, it can be assumed that the average momenta of the wave packets $\langle \mathbf{p}_k \rangle$ are equal to zero and, thus, the first term in the expression (14) can be neglected. Following a standard approach (see, e.g., [19]) it is assumed that the parametrization of the EoS includes implicitly the kinetic contribution coming from the Fermi motion. As has been demonstrated in [18], the adopted cubic approximation of the EoS is flexible enough to reproduce typical predictions of sophisticated models of the nuclear matter which explicitly take into account the kinetic contribution resulting from the Fermi motion. Since the local density can be written as

$$\rho(\mathbf{r}) = \sum_{k=1}^{k=A} P_k(\mathbf{r}), \tag{15}$$

the average value of the Hamiltonian can be expressed as:

$$\begin{aligned} \langle \Phi | H | \Phi \rangle = \int \varepsilon &\left(\rho_{p\uparrow}, \rho_{p\downarrow}, \rho_{n\uparrow}, \rho_{n\downarrow}\right) \rho(\mathbf{r}) \mathrm{d}^3\mathbf{r} \\ &+ \lambda \sum_{k=1}^{k=A} \sigma_k(\varepsilon) + \langle \Phi | V_C | \Phi \rangle. \end{aligned} \tag{16}$$

As one can see the Hamiltonian is a sum of three components, which can be interpreted as volume, surface and Coulomb energies. Therefore, the present description can be interpreted as a microscopic realization of the LDM. Finding the ground-state configuration of a group of A nucleons within the adopted parametrization of the wave function and for given parameters of the EoS, consists in finding the values of parameters $\sigma_k^2(r)$, $\langle \mathbf{r}_k \rangle$, and $\langle \mathbf{p}_k \rangle$, which minimize the value of the Hamiltonian. The procedure of finding the ground-state configurations has been described in sect. 4 of [18]. We would like to emphasize that the proposed parametrization of the EoS, with two more terms in (6), is responsible for driving the nucleons of different types, $p\uparrow$, $p\downarrow$, $n\uparrow$, $n\downarrow$, to form quadruplets, i.e. to group into α-like clusters during the minimization procedure. These two terms play a similar role as the standard symmetry term which attains a minimum when the neutron-proton asymmetry δ of (7) tends to zero, i.e. when the neutron-proton couples group together. This α-like clustering property of (6) is shown in fig. 1. The histogram in fig. 1 presents the evolution of the binding energy of 12 nucleons during the process of searching for the ground-state configuration which minimizes the model Hamiltonian (16). The 6 panels of the inset show "snapshots" of the projections of the mean positions of the nucleon wave packets onto the x-y plane, at selected stages of the minimization procedure. Panel 1 presents the initial random configuration, panels 2–5 some intermediate ones and panel 6 shows the final ground-state configuration of 12C. As can be seen, the minimum of the Hamiltonian is achieved for a configuration with three α-like clusters. These clusters appear in the corners of an equilateral triangle. Note that in the figure the triangle is not perfectly equilateral due to the 2-dimensional projection. It has to be emphasized, that if clusters are allowed to appear in the considered system the description has to be modified. Modifications of the Hamiltonian due to clustering will be discussed in sect. 4.2.

CONSTRAINTS ON THE EOS FROM THE PROPERTIES OF LIGHT CHARGED PARTICLES

Ground-state properties of light charged particles, LCP, such as d, t, ^{3}He or α, can provide important constraints on parameters which describe interactions in the nuclear matter for at least four reasons. First, one can assume that in their structure clusters of matter (obviously beyond nucleons) can be formed. Therefore, one may expect that clustering of matter has no effect on their binding energies and sizes. Second, these LCPs do not contain more than one α-particle and, thus, one can neglect the α-α interactions (see sect. 4.2) and use the simpler liquid-drop–like Hamiltonian (16). Third, the measured density pro- files of these light particles are relatively well described by Gaussian distributions, which is in line with the adopted parametrization of the wave function. Fourth, the simplicity of these particles assures the vanishing of some types of interactions in (6) and the selective increase of sensitivity to some other ones. For instance, parameters related to the isospin symmetry energy (second term in (6)) in principle do not affect the binding energy of a deuteron and alpha particle. Similarly, due to the saturation of the proton (neutron) spin interactions in 3He (t), the mutual spin interaction (fourth term in (6)) vanishes in their case. In the case of LCPs we can assume that the average value of the Hamiltonian is completely determined by the field $\varepsilon(\rho_{p\uparrow}, \rho_{p\downarrow}, \rho_{n\uparrow}, \rho_{n\downarrow})$, by the Coulomb interaction and by the surface energy. This assumption may not be valid for more complex systems in which mutually interacting clusters can be formed, e.g., $^8\mathrm{Be} \to \alpha\text{-}\alpha$ or $^6\mathrm{Li} \to \alpha\text{-}d$. For such systems one can expect additional interactions between clusters due to the specific many-body forces. The mean value of the Hamiltonian (16) depends on 13 parameters: 12 parameters of the field ε(r) plus the surface tension parameter λ. Binding energies and rms radii of the ground states for d, t, 3He and α provide 8 equations, thus an insufficient number of constraints for a unique choice of parameters describing the interaction. In the present section we will try to deduce a subspace of this 13-parameter space for which the above-mentioned properties of LCPs are reproduced.

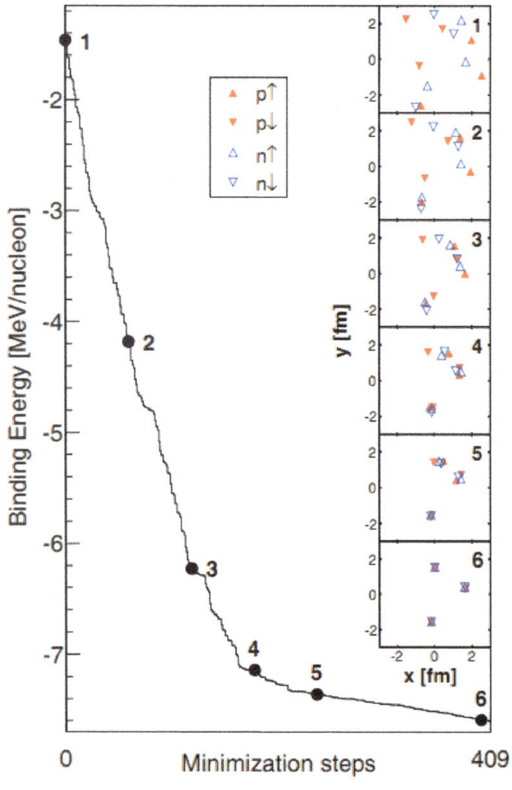

Figure. 1: Evolution of the binding energy of 12 nucleons (3×p↑, 3 × p↓, 3 × n↑ and 3 × n↓) during the minimization procedure. The insets 1–6 show the projections of the configurations of nucleons on the x-y plane at selected stages of the minimization. Panel 1 represents the random initial configuration. Panel 6 shows the projection of the final ground-state configuration of ^{12}C.

Constraints on P_0, E_{00}, K_0 and Λ from the Properties of an Alpha Particle

It is clear that the greatest reduction of the number of parameters can be expected in the case of an alpha particle. Here, the spin- and isospin-dependent interactions are practically absent. Thus, properties of alpha particles should be described by four parameters only: ρ_0, e_{00}, K_0 and λ. Constraints on these parameters are imposed by the rms radius and the binding energy of an alpha particle. These parameters cannot be determined unambiguously. However, there is a flexibility in determining a twodimensional hypersurface, in the space ρ_0, e_{00}, K_0 and λ, in which the searched values of these parameters exist. In order to determine this hypersurface the following χ^2 variable is defined:

$$\chi^2 = \left(\frac{B_{\alpha e} - B_{\alpha m}}{B_{\alpha e}}\right)^2 + \left(\frac{r_{\alpha e} - r_{\alpha m}}{r_{\alpha e}}\right)^2, \qquad (17)$$

where $B_{\alpha e}$, $B_{\alpha m}$, $r_{\alpha e}$, $r_{\alpha m}$ are the experimental and model values of the binding energy and rms radius, respectively. Note that a variable defined in this way should be zero in the searched subspace. In order to determine the correlations among the parameters in a searched subspace, the variables ρ_0 and K_0 are predefined on a grid, and the remaining two parameters are found by requiring that $\chi^2 < 10^{-7}$ for the model ground-state configurations of an alpha particle. The results are shown in fig. 2.

The figure presents the constant saturation density curves as a function of the other three parameters specified on the axis, which ensure a correct binding energy and rms radius of an alpha particle for the model groundstate configurations.

The dotted and dashed lines show how one can determine the parameters λ and e00 for a given compressibility K_0 and saturation density ρ_0. In the illustrated example the parameters ρ_0 = 0.185fm−3 and K_0 = 270MeV determine the values of λ = 0.4 and e_{00} = −16.79MeV. These values of the parameters have been suggested by additional studies of the competition among Coulomb, surface and volume energies in the binding energy calculated for the ground-state configurations of the heavier nuclei (see table 1). This problem will be considered later on.

Constraints on e_{I0}, L_I, K_I, e_{ii0}, L_{ii} and K_{ii} from the Properties of t and ³He

After fixing the parameters ρ_0, e_{00}, K_0 and λ let us try to examine the other ones. For t and 3He the ground-state energy does not depend on the choice of parameters e_{ij0},

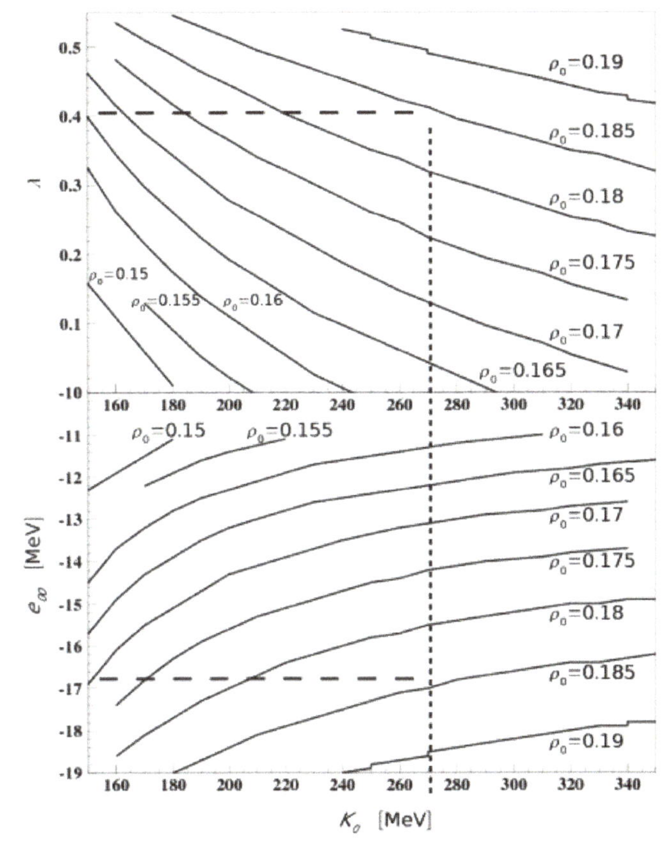

Figure. 2: Constant saturation density curves, ρ_0, as a function of e_{00}, K_0 and λ, constrained by the ground-state properties of an α-particle.

Table 1: Parameters of the standard EoS which, together with a parameter of the surface energy λ, are reproducing the ground-state energy and rms radius of the alpha particle, t and ³He

ρ_0	e_{00}	K_0	e_{I0}	L_I	K_I	λ
$(\frac{1}{fm^3})$	(MeV)	(MeV)	(MeV)	(MeV)	(MeV)	1
0.185	−16.79	270	32	59.405	−1250	0.4

L_{ij} and K_{ij}. This results from the vanishing of the $\eta_n\eta_p$ product in the last term of (6). Thus, the binding energies and rms radii for t and 3He can provide four equations for six unknown EoS parameters: e_{I0}, L_I, K_I, e_{ii0}, L_{ii} and K_{ii}, and so, two equations are still missing. The two free parameters can be constrained from the properties of heavier nuclei as was done in the case of the fully symmetric nuclear matter. At this stage it is only possible to examine the subspace defining the limits for the considered parameters from the properties of the t and ³He nuclei. We notice the fact that the parameters associated with isospin, e_{I0}, LI and K_I, differentiate between the binding energy and the rms radii of t and ³He, while the parameters e_{ii0}, L_{ii} and K_{ii} affect equally the energy and the size of these particles.

Using a similar method as in the case of an alpha particle one can find relationships among the parameters e_{I0}, K_I and L_I, by requiring a precise description of the binding energies and radii of the t and ³He particles. Here, the values of e_{I0} and K_I were varied independently on a grid and the value of L_I was found from the minimization of (17). The results of such calculations are shown in fig. 3.

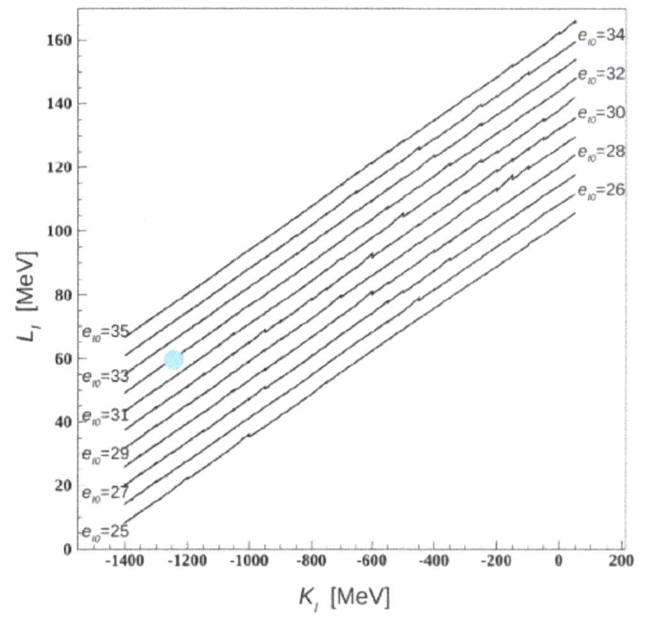

Figure. 3: Correlation between the slope, L_I, and curvature, K_I, of the isospin symmetry energy for different Wigner energies, e_{I0}, obtained from fitting the binding energies and rms radii of the t and ³He.

Table 2: Parameters of the spin interaction chosen in such a way that together with parameters from table 1 they reproduce binding energies and sizes of d, t and ³He.

e_{ii0}	L_{ii}	K_{ii}	e_{ij0}	L_{ij}	K_{ij}
(MeV)	$(\frac{1}{fm^3})$	(MeV)	(MeV)	(MeV)	(MeV)
82.37	242.48	19.909	−1.5	182.0	1210.0

The figure presents a relationship between the parameters e_{I0}, L_I and K_I. The selection of two of them allows to determine the third one. It should be stressed that the above correlations have been obtained as a result of the procedure reproducing the t and ³He ground-state proper ties with the parameters ρ_0 = 0.185fm^{-3}, K_0 = 270MeV, $\lambda = 0.4$ and $e_{00} = -16.79$ MeV fixed from the groundstate properties of an alpha particle (fig. 2). Three other parameters, e_{ii0}, L_{ii} and K_{ii}, were found to be almost independent of eI0 and K_I.

Discussion of the Parameter Values

The final values of the EoS parameters adopted in this work have been obtained by taking into account also the binding energy and size of a deuteron. The values have been summarized in tables 1 and 2. For this example set of values the model reproduces exactly the experimental values of the rms radii and of the binding energies of the d, t, ^{3}He and α particles (see also figs. 5 and 7).

From the point of view of the present work, which focuses on the nuclear systems with even and equal numbers of protons and neutrons (i.e. the spin-balanced matter), the parameters e_{ii0}, L_{ii}, K_{ii} and e_{ij0}, L_{ij}, K_{ij} j from the expansion (6) are irrelevant. However, they do play an important role when constructing the ground-state configurations and determining the properties of nuclei composed of asymmetric matter.

As far as the interactions between neutrons and protons are concerned, the Coulomb repulsion between protons causes that the proton matter tends to locate itself in the outer regions of the nucleus. This leads to differences between the neutron and proton density distributions, even for symmetric nuclei, and triggers additional interactions associated with the isospin symmetry energy (proton or neutron skins can be created). This effect can be negligible in the case of light nuclei, while for heavier ones it may be more significant. Therefore, for heavier nuclei, one has to use a reliable set of isospin symmetry energy parameters e_{I0}, L_I and K_I , even for symmetric nuclear matter.

As has been shown above, the binding energies and the rms radii of t, $_3$He and α impose certain constraints on the parameters of the EoS. In the case of the e_{00}, ρ_0, K_0 and λ parameters, once two of them are fixed by some other constraints or means, or from some other evidences, the remaining ones can be determined using the obtained correlations (fig. 2). Here, the compressibility parameter of symmetric matter K_0 should fall within the range of about 250 ± 50MeV according to the recent experimental constraints (flow interpretation [20], monopole vibrations [21, 22], subthreshold kaon production [23, 24]). It has been fixed to 270MeV in the present approximation. The volume energy parameter e_{00} should take the value of about −16MeV according to the LDM. In the present approximation it has been assumed to have the value of −16.79MeV. The values of the remaining two parameters, ρ_0 and λ have been then uniquely determined from the correlation plot of fig. 2.

In the case of the symmetry energy, the choice of any two parameters out of eI0, L_I , K_I , allows to determine the third one (fig. 3). The compilation [25] of the latest results of the nuclear-physics experiments and astrophysical observations suggests that the value of the symmetry energy constant, e_{I0}, and of the slope parameter, L_I , should fall within a range of about 32 ± 1MeV and of about 59 ± 17MeV, respectively. In the present approximation the values of these parameters have been assumed to be 32 and 59.405MeV, respectively. These values yield the value of K_I to be about −1250MeV, according to the correlation plot of fig. 3. The experimental constraint of [26] on the symmetry energy compressibility parameter, K_I , yields its value to be about −50 ± 200MeV. Another compilation [19] locates the theoretical values of K_I between −400 and +466MeV and its experimental values in a range from −566±1350MeV to 34±159MeV. Thus, this parameter is least constrained experimentally and theoretically so far. Its value obtained within the model, preferring very soft symmetry energy, is consistent with the negative experimental constraint, thanks to its large error bar. However, we should stress that the adopted set of values is not at all unique. As seen from the correlation plots, some other choice of parameters is equally well possible. It should be also emphasized that the assumed value of ρ0 equal to 0.185fm^{-3}, which fits well the properties of the LCPs, and is higher than the commonly used values of 0.16–0.17fm^{-3}, may influence the values of the remaining parameters.

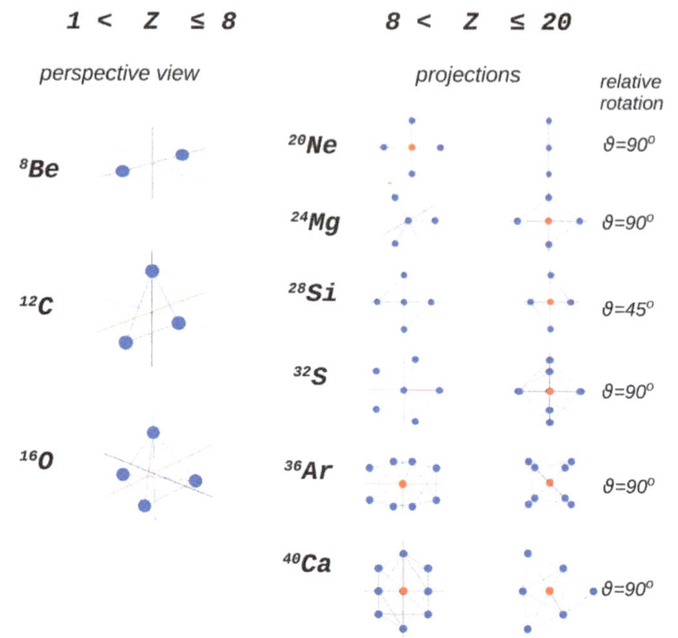

Figure. 4: Proposed structures of n-alpha nuclei. These structures ensure a high level of symmetry for a given number of alpha particles (see [3]). The density distribution of the central cluster (the red one) has quite a big spread and, therefore, it cannot be identified as an alpha particle.

In order to describe the exotic nuclei and provide more precise parameters of the symmetry energy, defi- nitely some more advanced parameter searches will be needed, involving, e.g., fitting the properties of the neutron skin nuclei and also of those approaching the proton drip line across the chart of the nuclides, but this is beyond the scope of the present work.

ENERGY AND SIZE OF NUCLEI WITH EQUAL AND EVEN NUMBER OF NEUTRONS AND PROTONS

A method of finding the ground-state configuration based on minimization of the average value of the liquid-drop– like Hamiltonian (16) with respect to the wave function parameters $\langle \mathbf{r}_k \rangle$ and $\sigma_k(r)$ with the $\langle \mathbf{p}_k \rangle = 0$ constraint, is described in [6,18]. The ground-state configuration parameters determined in this way indicate that the corresponding wave function has some numbers of centers, which can be interpreted as the centers of alpha-like clusters in the nuclear matter. Some features of these ground-state con- figurations have been discussed in [6] without taking into account interactions and correlations among clusters. In the present approach the modification of the ground-state configurations due to the presence of the alpha-alpha interactions will be investigated. The main aim is to identify such symmetries of the ground-state configuration, and thus of the corresponding wave function, which could be responsible for the experimentally observed changes of the binding energies and rms radii of the n-alpha nuclei as a function of n. We would like to check whether it is possible to identify specific con- figurations related to the magic numbers. In other words, we would like to verify whether the shell corrections necessary to describe deviations from the smooth predictions of the LDM around the magic numbers, could be interpreted in terms of a change of the ground-state configuration resulting from the corrections to the Hamiltonian related to the cluster-cluster interactions. As will be justified in the following, the ground-state configurations corresponding to the local minima of the Hamiltonian can be classified according to the nuclear charge, Z, in the following way:

- $Z \leq 8$: The structures are identical with the structures proposed in [2] and also in [6] and are characterized by no alpha cluster in the center of the system.
- $8 < Z \leq 20$: A core represented by one alpha-like cluster appears in the center of the system.
- $20 < Z \leq 28$: A core with a mass corresponding to two alpha particles appears in the center.

The structures for cases i) and ii) are visualized in fig. 4 (see also [3]). The structures presented in fig. 4 result from the analysis of a large number of configurations obtained within a procedure of minimizing the Hamiltonian (16) when starting from a random configuration (see the ^{12}C example in fig. 1). The majority of the thus-obtained configurations could be classified as those presented in the previous paper [6], with alpha-like clusters located mostly at the surface and with a moderate level of symmetry. Binding energies for these structures followed the smooth LDM trend and could not reproduce the "fine" structure of binding energies around the magic numbers. In some cases, however, the minimization procedure did not yield the global minimum, but ended up in some local one, just about 0.3–0.5MeV above the ground-state energy of majority of configurations. The configurations corresponding to these local minima were characterized by a higher level of symmetry and possibly contained alpha-like cluster(s) in the middle. Surprisingly, the binding energy distribution for these structures did show characteristic "kinks" around magic nuclei, despite some energy offset. These specific structures are those presented in fig. 4. The question that arose from the above observation was whether it might be possible to convert the local minima into global ones, while keeping the highly symmetric structures, by introducing the corrections to the Hamiltonian which take into account the cluster-cluster interactions. The following sections try to verify this hypothesis.

Matter without Alpha Clusters

The structures from fig. 4 have been used as structural constraints in the minimization procedure by providing the starting points for the preformed alpha-like clusters. In other words, the random initial configurations, as in fig. 1, have been replaced by those from fig. 4. The actual ground-state configurations were determined through minimization of the liquid-drop–like Hamiltonian (16) by varying the mean positions and the widths of the individual wave packets. The following analysis is restricted to nuclei smaller or equal to ^{40}Ca which is the heaviest stable n-alpha nucleus. Beyond Z = 20 the nuclei become radioactive and there are no data about their sizes available. The liquid-drop–like Hamiltonian (16) with the parameters from tables 1 and 2 has been used together with structural constraints to obtain the ground-state configurations for each of the considered nuclei. The minimization procedure provided the final parameters determining the ground-state wave function: $\langle \mathbf{r}_i \rangle$ and σ_i. Figure 5 presents the obtained ground-state binding energies and rms radii as a function of the atomic number Z. The experimental data and the model results are represented by the squares and the circles, respectively. As one can see the calculated binding energies change their slope at the same place (Z = 8) as the experimental ones and the same concerns the radii. The overall calculated binding energies are somewhat bigger by about 0.3–0.5MeV/nucleon than the

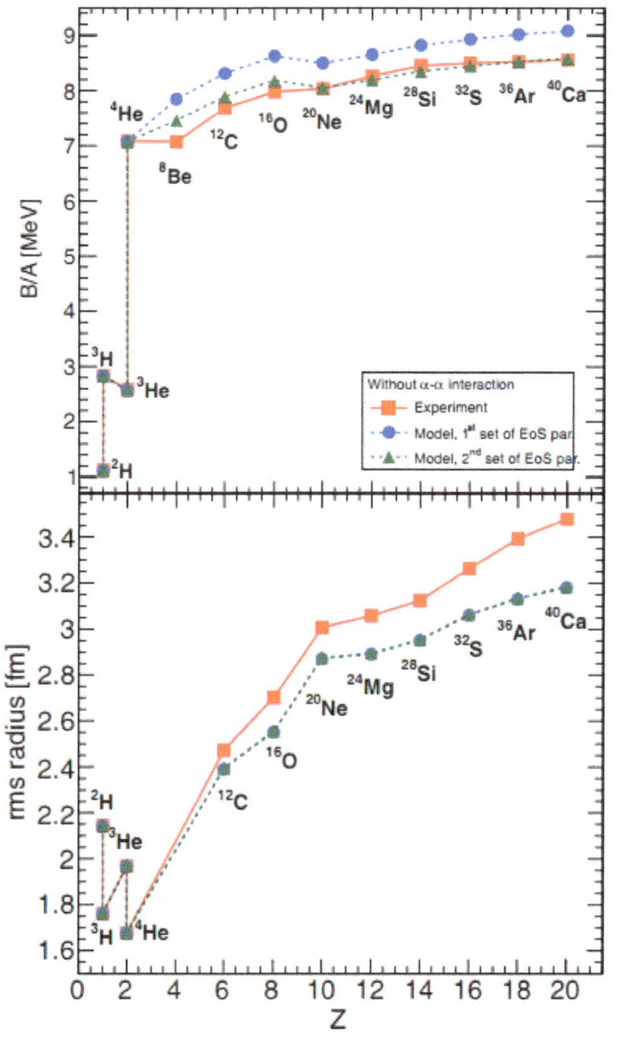

Figure. 5: The binding energies and rms radii of even-even nuclei (N = Z) obtained for the ground-state wave functions and the Hamiltonian (16). The solid line connects the experimental points. Circles and triangles represent model results for two sets of parameters of the EoS.

experimental ones and the calculated radii are somewhat smaller; however both follow the experimental trends. The properties of particles up to ^{4}He are reproduced perfectly well, because they served as the constraints in the parameter fitting procedure. In order to check the sensitivity of the results on the volume energy parameter e_{00}, similar calculations for e_{00} equal to −16.1MeV instead of −16.79MeV have been performed. The results are presented with a dotted green line in fig. 5. In the case of the rms radii there is almost no difference observed. The binding energies still deviate from the experimental values, but what is important, similar changes of the slopes in theoretical and experimental trends are observed. Results of other trial calculations in which all the remaining parameters from e_0 to λ were varied, did not yield a better reproduction of the experimental values with the use of the liquid drop form of the Hamiltonian together with the structural constraints i)–iii). A better reproduction could only be obtained with the corrections to the Hamiltonian discussed in the next section. The structural constraints assure configurations for which global minima of energy are obtained, provided that the alpha-alpha interactions are taken into account. Otherwise, they correspond to some local minima.

Modification of the Hamiltonian due to Alpha Clusters

Let us consider now the formation of the nucleus by randomly located nucleons remaining within the range of their mutual interaction. Due to the Hamiltonian (16), nucleons are forming groups in such a way that the energy of the system is minimized. At a certain density of the matter alpha-like clusters can be formed as a result of the spinisospin interactions of nucleon groups. As alpha clusters are formed, additional extra interactions connected with multi-body effects can appear having a strong influence on the final structure of the system. Figure 6 illustrates a ground state of such a system. One can see that inside a matter drop alpha clusters appear and interactions between them are changing the energy of the system. Below, corrections to the Hamiltonian arising from these extra interactions will be defined. Assuming that alpha-like clusters appear in the structure of the nucleus, one can expect two kinds of minor adjustments of the Hamiltonian, which should describe the essential part of the α-α interaction. The first correction corresponds to the possible interaction caused by the appearance of specific many-body forces. Such many-body effect can be interpreted as a modification of the type and numbers of exchange bosons in the interaction process. It appears possible that, in such an exchange, the charge currents are strongly limited because they change the energy of the considered clusters. In this case, we can expect an additional term in the interaction potential which gives the repulsive forces between groups of nucleons forming a cluster of alpha-like particles (see fig. 4). Here we test this hotfix in a form given by the product of the harmonic-oscillator potential $V_{\alpha\alpha}(dij)$ and probabilities $P_\alpha(i)$, $P_\alpha(j)$ that, given four nucleons (groups named i or j), form the alpha clusters:

$$\langle \Phi | \Delta H_1 | \Phi \rangle = \sum_{i \neq j} P_\alpha(i) P_\alpha(j) V_{\alpha\alpha}(d_{ij}), \tag{18}$$

where the potential $V_{\alpha\alpha}(d_{ij})$ is taken in the form

$$V_{\alpha\alpha}(d_{ij}) = \kappa (d_{ij} - d_0)^2 \tag{19}$$

with d_{ij} being the average distance between clusters i and j, and $P_\alpha(i)$ describing the probability that a group of four nucleons $4_n(i)$ forms an alpha cluster i. We assume that $P_\alpha(i)$ can be expressed as

$$P_\alpha(i) = \exp\left(-\nu \frac{\sigma_{4n(i)} - \sigma_{\alpha 0}}{\sigma_{\alpha 0}}\right)^2, \tag{20}$$

where $\sigma_{4n}(i)$ is the variance of the distribution describing the probability of finding a nucleon in a group $4_n(i)$ and $\sigma_{\alpha 0}$ describes such a variance (experimentally measured) for alpha particles. In the above formulas κ, d_0 and ν are free parameters. It is assumed that the potential is equal to 0 when $d_{ij} > 4$fm with a shape close to the van der Waals one.

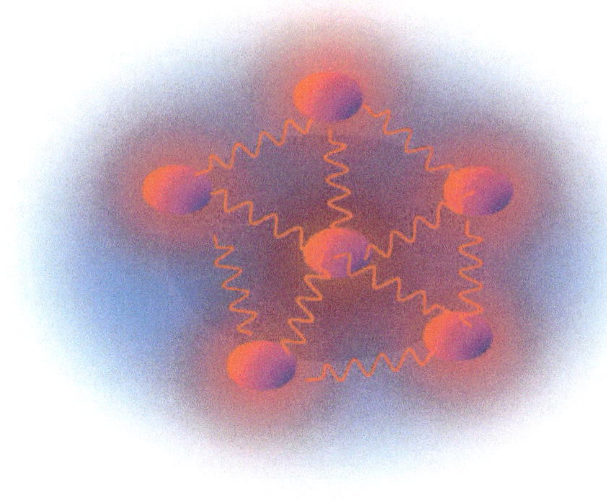

Figure. 6: Visualization of the additional interactions between alpha clusters due to the many-body interactions.

The second correction is related to the additional symmetry of the wave function, which can be associated with the bosonic character of an alpha cluster. In the present approach nucleons and their clusters are distinguishable objects (1). For nucleons their energy connected with the fermionic motion is taken into account by applying the energy density functional. For alpha particles a relevant correction to energy should also appear and it is an empirical one. When calculating the minimum of the Hamiltonian by varying the parameters of the wave function, the position of the alpha-like cluster and its size and spread are established. These quantities can be different for different clusters depending on the adopted structural constraint for a given nucleus. The exchange of clusters must cause an increase of the energy of the system as a minimum of the Hamiltonian defines definite spreads of clusters for their different positions. For the indistinguishable alpha particles due to the symmetrization with respect to the exchange of cluster positions there is no chance to relate cluster sizes to their positions. The second correction is an empirical one:

$$\langle \Phi | \Delta H_2 | \Phi \rangle = \mu N_\alpha \sum_{i=1}^{i=A} \left(\frac{\sigma_i(r) - \sigma_0(r)}{\sigma_0(r)}\right)^2, \tag{21}$$

where N_α is the number of alpha clusters and μ is a parameter. $\sigma_i(r)$ is the variance of the nucleon radius and $\sigma_0(r)$ is the average variance of nucleons forming an alpha particle.

Thus, this correction to the Hamiltonian describes an increase of the energy caused by different sizes of clusters and it should vanish when the wave functions for all the clusters correspond to alpha particles in their ground states. It seems that this correction could be related to the bosonic character of alpha particles. Both corrections are

Table 3: Estimated parameters of corrections (18), and (21) to the Hamiltonian (16) coming from interactions between alpha clusters.

μ	ν	d_0	κ
(MeV)	[1]	(fm)	(MeV/fm^2)
1.0	9.0	3.77	5.2

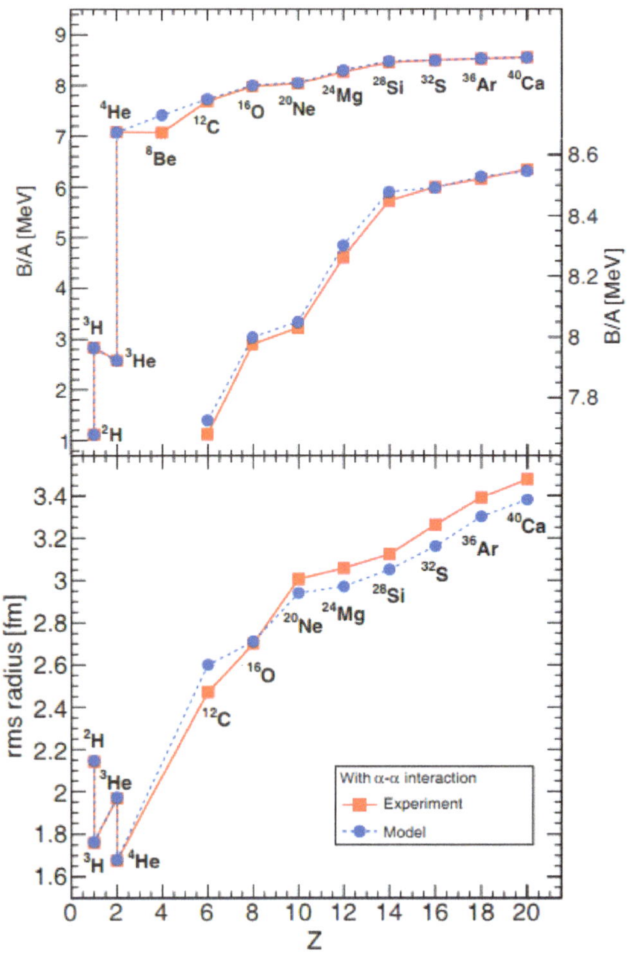

Figure. 7: Binding energies per nucleon and rms radii for the Hamiltonian with the corrections (18) and (21). The right scale of the binding energies per nucleon in the top panel refers to the magnified view for nuclei from ^{12}C to ^{40}Ca. The solid (dashed) line connects the experimental (model) points.

positive and in this way the binding energy decreases. Assuming that parameters of the main part of the Hamiltonian (14) are established by properties of light nuclei d, t, 3He and α, the idea now is to define four additional parameters in order to reproduce the binding energies and sizes of nuclei starting from 8Be up to 40Ca by imposing the structural constraints. The adopted values of these parameters are summarized in table 3. The results for the full Hamiltonian, including the corrections due to the alpha-alpha interactions are presented in fig. 7. The upper part of the figure shows the binding energies as a function of the atomic number Z. The bottom part presents the rms radii. Much better agreement between the model results and the experiment can be observed now, as compared to fig. 5.

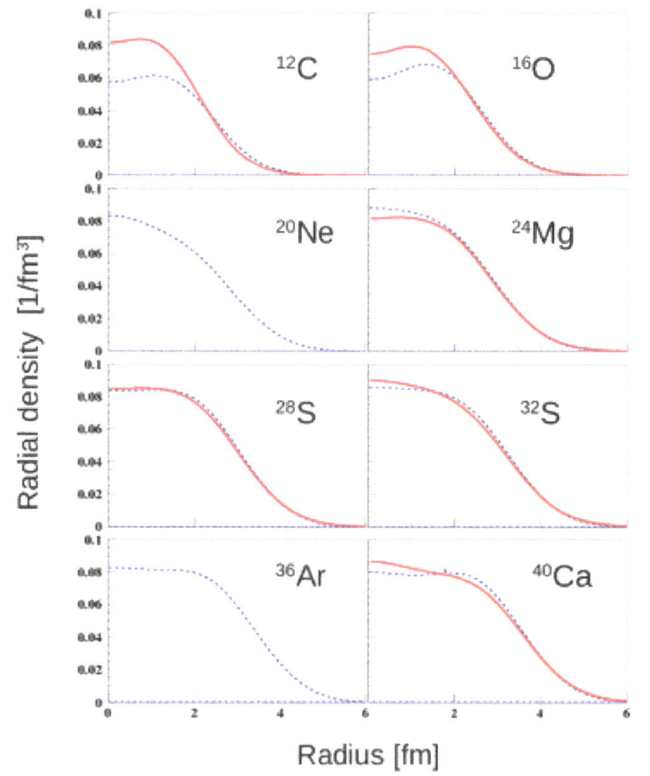

Figure. 8: Experimental (solid, red line) and model (dashed, blue line) density profiles.

Some discrepancies can be still noticed especially for ^{8}Be for which the distance between the alpha particles is pretty small and much smaller than d_0 in the harmonic potential description. Again, the properties of particles up to 4He are reproduced perfectly well, because they served as the constraints to fix the EoS parameters. Figure 8 presents a comparison of the model and experimental density profiles for the selected nuclei. The full (red) lines represent the experimental profiles and the broken (blue) lines result from the model calculations. Experimental density profiles were taken from [27]; however for ^{20}Ne and ^{36}Ar reliable experimental data were not available.

Figure 9 shows the isodensity contour plots for the model ground-state configurations of ^{8}Be, ^{12}C, ^{16}O and ^{20}Ne. Note that apart from the ^{16}O, which has a tetrahedral shape, all the other nuclei have a planar geometry (cf. fig. 4). A similar shape for ^{12}C has been obtained within the FMD [28] and AMD [29] simulations. In particular, the 20Ne case shows that the central α-like cluster is more diffused than the surface ones. More quantitatively, the standard deviations of the density distributions of surface alphas are of the order of 1.02fm, while for the core alpha they are of the order of 1.28fm.

Figure 10 presents a distribution of distances between alpha clusters for the nuclei in question. The distances between clusters from the ground-state configurations obtained with the liquid-drop–like Hamiltonian without corrections are represented by open circles. The distances obtained from the full Hamiltonian calculations are represented by closed ones. As can be seen, in the latter case the distances are grouping around 3.5fm. This can denote that these clusters behave more like rigid spheres.

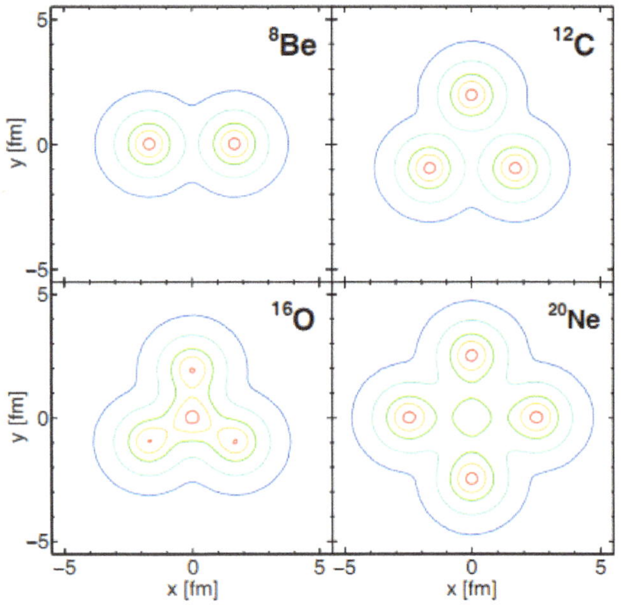

Figure. 9: Shapes of the model nuclei. The isodensity lines correspond to 1, 15, 50, 75 and 95% of the maximum density

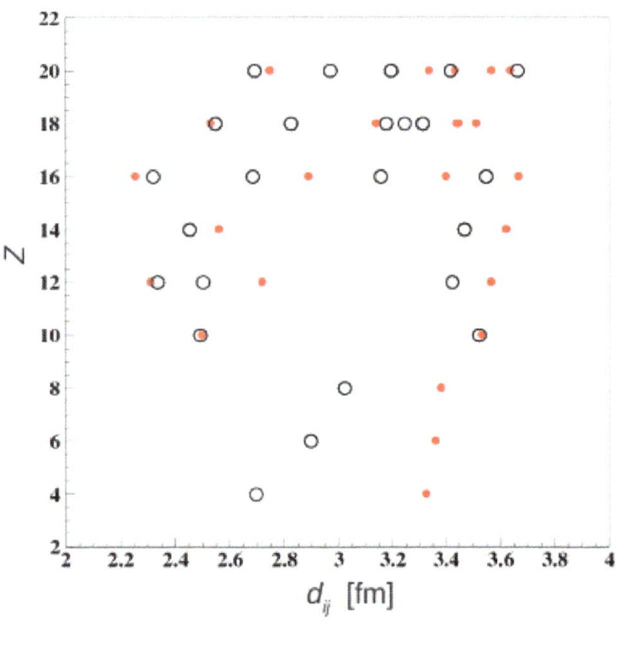

Figure. 10: The distribution of the distances between alpha clusters as a function of Z. The open (closed) circles represent the distances without (with) the correction for the alpha-alpha interaction in the Hamiltonian.

SUMMARY AND CONCLUSIONS

A simple model of nuclei formed out of alpha-like clusters in which interactions lead to strong correlations between nucleons has been presented. The model interactions have been defined using a cubic approximation of the nuclear EoS. The proposed EoS contains additional terms which take into account the spin and isospin polarizations. These additional terms cause that nucleonic clusters, with an enlarged binding energy, appear in the ground-state configurations. These clusters have full spin-isospin symmetry similarly as in [8] and in [7]. For n-alpha nuclei these clusters can be identified as alpha-like structures. When these clusters are formed in regions of smaller density, i.e. at the nuclear surface, then their sizes measured as the variance of the density distribution become comparable to the alpha-particle size. On the other hand, the formation of the clusters in the central part of nuclei leads to much bigger spreads of their density distributions and they can hardly be identified as alpha particles. It has been shown that by taking into account the liquid-drop–like form of the Hamiltonian in the minimization process it was not possible to describe the subtle changes of the binding energies of nuclei. Therefore, the liquid drop part of the Hamiltonian has been supplemented by the corrections taking into account the clustercluster interactions. The ground-state configurations obtained from the minimization of the full Hamiltonian with the imposed structural constraints were found to reproduce very well the experimental binding energies and sizes of the n-alpha nuclei and the available density profiles. In particular, the model with the cluster corrections has been able to reproduce the "fine structure" of the ground-state energy distribution (inset of fig. 7). In this respect it seems to be able to mimic the shell effects without explicit spinorbit term. The interactions have been defined using a 12-parameter EoS approximation. The thirteenth parameter has been added to control the variance of the field for finite systems. The values of these parameters and correlations among them have been constrained using the experimental binding energies and sizes of d, t, 3He and α particles. Four additional parameters have been introduced in the corrections to the liquid drop part of the Hamiltonian in order to account for the interactions between the alpha-like clusters and to substantially improve the model predictions. The differences between the calculated and measured ground-state energies for nuclei with $Z > 8$ are about an order of magnitude smaller than those obtained using the liquid drop model. Besides, the obtained parameters describe the properties of the t, ^{3}He and α very accurately (practically without error). It has also been demonstrated that correlations between nucleons in the phase space can play an important role in the description of the ground states, particularly of the ground states of nuclei with the same and even number of neutrons and protons.

REFERENCES

1. H.A. Bethe, R.F. Bacher, Rev. Mod. Phys. 8, 82 (1936).
2. L.R. Hafstad, E. Teller, Phys. Rev. 54, 681 (1938).
3. http://nz21-33.ifj.edu.pl/clusters/.
4. M. Freer, Scholarpedia 5, 9652 (2010) doi:10.4249/scholarpedia.9652.
5. W. von Oertzen, Eur. Phys. J. A 29, 133 (2006).
6. Z. Sosin, J. Kallunkathariyil, Acta Phys. Pol. B 45, 925 (2014).
7. H. Feldmeier, Nucl. Phys. A 515, 147 (1990).
8. A. Ono et al., Prog. Theor. Phys. 87, 1185 (1992).
9. Y. Kanada-En'yo et al., Prog. Theor. Phys. 2012, 01A202 (2012).
10. L. Wilets et al., Nucl. Phys. A 282, 341 (1977).
11. C. Dorso, S. Duarte, J. Randrup, Phys. Lett. B 188, 287 (1987).
12. H. Horiuchi, A. Ohnishi, T. Maruyama, in Proceedings of the 6th International Conference on Nuclear Reaction Mechanisms, Varenna, June 10–15, 1991, edited by E.

Gadioli, Ricerca Scientifica ed Educazione Permanente, Suppl. No. 84 (Univerist`a degli Studi di Milano, 1991).

13. D.H. Boal, J.N. Glosli, Phys. Rev. C 38, 1870 (1988).
14. T. Maruyama et al., Phys. Rev. C 53, 297 (1996).
15. J. Fink et al., Z. Phys. A 323, 189 (1986).
16. J. Aichelin, Phys. Rep. 202, 233 (1991).
17. P. Dirac, The Principles of Quantum Mechanics, Oxford Science Publications Series (Oxford University Press, 1981).
18. Z. Sosin, Int. J. Mod. Phys. E 19, 759 (2010).
19. Bao-An Li et al., Int. J. Mod. Phys. E 07, 147 (1998).
20. P. Danielewicz et al., Science 298, 1592 (2002).
21. J. Piekarewicz, Phys. Rev. C 69, 041301 (2004).
22. G. Col`o et al., Phys. Rev. C 70, 024307 (2004).
23. C. Sturm et al., Phys. Rev. Lett. 86, 39 (2001).
24. Ch. Hartnack et al., Phys. Rev. Lett. 96, 012302 (2006).
25. Bao-An Li, Xiao Han, Phys. Lett. B 727, 276 (2013).
26. M. Centelles et al., Phys. Rev. Lett. 102, 122502 (2009).
27. H. De Vries, C.W. De Jager, C. De Vries, At. Data Nucl. Data Tables 36, 495 (1987).
28. H. Feldmeier, J. Schnack, Prog. Part. Nucl. Phys. 39, 393 (1997).
29. Y. Kanada-En'yo, H. Horiuchi, Prog. Theor. Phys. Suppl. 142, 205 (2001).

CHAPTER 4

SCATTERING OF NUCLEONS IN THE CLASSICAL SKYRME MODEL

David Foster and Nicholas S. Manton

INTRODUCTION

The Skyrme model [1] is a nonlinear theory of pion fields that has topological soliton solutions. Encouragingly, Witten identified the Skyrme model as a low energy effective model of QCD [2] and [3]. The conserved topological charge is interpreted as the baryon number B, and the minimal energy static solutions for each integer B are called Skyrmions. They can be treated as rigid bodies, free to rotate in space and also in isospace (the three-dimensional space of the pion fields). When the rotational motion is quantised, the Skyrmions are models for nucleons and nuclei. In particular the B=1Skyrmions, quantised with spin and isospin half, model protons and neutrons.

A full quantum mechanical treatment of the interaction of two B=1 Skyrmions is difficult. Each Skyrmion has three position coordinates and three orientational coordinates, so two-Skyrmion dynamics involves twelve coordinates [4]. A truncation to ten coordinates has been useful for modelling the deuteron bound state [5], but no proper discussion of two-nucleon scattering is possible with this truncation.

An alternative is a classical approach, using the idea that a classically spinning Skyrmion is a reasonable model of a nucleon [6] and [7]. The classical angular velocity of the Skyrmion is fixed to match the quantised spin and isospin of the nucleon. Scattering of non-spinning Skyrmions was first performed using an axially symmetric ansatz [8], the first full-field simulation was reported in [9], and multi-charge Skyrmion scattering was considered in [10]. The scattering of spinning Skyrmions representing nucleons was considered by Gisiger and Paranjape [6], and they obtained analytical formulae for the scattering angle for large impact parameters. There has also been some recent work on multi-charge, purely isospinning Skyrmions [12].

A systematic, numerical investigation of Skyrmion scattering without spin, at moderate and small impact parameters [11], confirmed that matter is exchanged between the Skyrmions when the Skyrmions are close, and also showed that a substantial rotation of the Skyrmion's orientation can occur. Here, we numerically investigate the scattering of two classically spinning B=1 Skyrmions to model proton–proton, neutron–neutron and proton–neutron scattering at various impact parameters, including head-on collisions. We are particularly interested in the change in the polarisation states of the nucleons when they scatter, and also in finding evidence, using our classical approximation, for the deuteron and for the known two-nucleon resonance states.

The format of this paper is as follows. We first introduce the Skyrme model and its calibration. Then we discuss how to identify classically spinning Skyrmions as protons or neutrons. The next section presents the results of our numerical scattering of Skyrmions, and our identification of the dineutron/diproton resonances, the deuteron bound state, and the excited dibaryon state. The concluding section summarises and discusses our results.

Citation: David Foster, Nicholas S. Manton, Scattering of nucleons in the classical Skyrme model, Nuclear Physics B, Volume 899, October 2015, Pages 513-526, ISSN 0550-3213, http://dx.doi.org/10.1016/j.nuclphysb.2015.08.012.

THE SKYRME MODEL

The Skyrme model is defined by the Lagrangian density

$$\mathcal{L} = -\frac{F_\pi^2}{16}\mathrm{Tr}(R_\mu R^\mu) + \frac{1}{32e^2}\mathrm{Tr}([R_\mu, R_\nu][R^\mu, R^\nu]) + \frac{m_\pi^2 F_\pi^2}{8}\mathrm{Tr}(U - I_2), \tag{1}$$

where the Skyrme field U(t,$\boldsymbol{x}$) is an SU(2)-valued scalar and $R_\mu=\partial_\mu UU^\dagger$ is itssu(2)-valued current. F_π, e and m_π are parameters, which are fixed by comparison with experimental data. Their values will be discussed later. It is convenient for us to work in dimensionless Skyrme units. One Skyrme length unit corresponds to $\frac{2}{eF_\pi}$ in inverse MeV, and one Skyrme energy unit corresponds to $\frac{F_\pi}{4e}$ MeV. Conversion of inverse MeV to fm is as usual through ħ=197.3 MeV fm. The dimensionless pion mass in Skyrme units is

$$m = \left(\frac{2}{eF_\pi}\right) m_\pi . \tag{2}$$

In Skyrme units, the energy of a static field is

$$E = \int \left(-\frac{1}{2}\mathrm{Tr}(R_i R_i) - \frac{1}{16}\mathrm{Tr}([R_i, R_j][R_i, R_j]) + m^2\mathrm{Tr}(I_2 - U)\right) d^3x .$$

It is often convenient, especially in numerical simulations, to express U in terms of a triplet of pion fields $\boldsymbol{\pi}=(\pi_1,\pi_2,\pi_3)$ and an additional auxiliary field σ as

$$U(t, \boldsymbol{x}) = \sigma(t, \boldsymbol{x})\, I_2 + i\boldsymbol{\pi}(t, \boldsymbol{x}) \cdot \boldsymbol{\tau} , \tag{3}$$

where $\boldsymbol{\tau}=(\tau_1,\tau_2,\tau_3)$ are the three Pauli matrices, with the constraint $\sigma^2+\boldsymbol{\pi}\cdot\boldsymbol{\pi}=1$ so that U∈SU(2).

At fixed time, U(t,$\boldsymbol{x}$) is a map $U:\mathbb{R}^3\rightarrow SU(2)$, where $U\rightarrow I_2$ at spatial infinity. This boundary condition compactifies $\mathbb{R}^3\cup\{\infty\}$ to S^3. The group manifold of SU(2) is S^3, so a finite energy configuration U extends to a map $U:S^3\rightarrow S^3$, and then belongs to a class of $\pi_3(S^3)=Z$ indexed by an integer B∈Z, called the baryon number. B is also the degree of the map U which can be explicitly calculated as

$$B \equiv \int \mathcal{B}(\boldsymbol{x})\, d^3x = -\frac{1}{24\pi^2}\int \varepsilon_{ijk}\mathrm{Tr}(R_i R_j R_k)\, d^3x , \tag{4}$$

where B($\boldsymbol{x}$) is the baryon density. Skyrmions are the static field configurations of minimal energy for each value of B . In the figures we plot level-sets of baryon densityB($\boldsymbol{x}$).

It is well known that the B=1 Skyrmion can be found using the so-called Hedgehog ansatz [1]

$$U_H(\boldsymbol{x}) = \cos f(r)\, I_2 + i \sin f(r)\, \hat{\boldsymbol{x}} \cdot \boldsymbol{\tau} , \tag{5}$$

where r=|$\boldsymbol{x}$| and $\hat{\boldsymbol{x}}$ is the radial unit vector. f(r) is a real radial profile function satisfying f(0)=π,f(∞)=0. This produces a solution with rotationally symmetric energy and baryon density. Fig. 1 shows a B=1 Hedgehog Skyrmion coloured as in[7]. The centres of the white and black regions are where $\pi_1^2+\pi_2^2=0$ and $\pi_3>0,\pi_3<0$ respectively. The centres of the red, green and blue regions are where $\pi_3=0$and $\tan^{-1}\left(\frac{\pi_2}{\pi_1}\right)=0, \frac{2\pi}{3}, \frac{4\pi}{3}$ respectively. This is the colouring scheme used throughout this paper.

Figure. 1: B = 1 Hedgehog Skyrmion. (For a colour image, we refer the reader to the web version of the article.)

An important feature of the Hedgehog ansatz is that an isorotation, U($\boldsymbol{x}$)→AU($\boldsymbol{x}$)A†, which acts as a rotation among the pion fields, is equivalent to a rotation in space,U($\boldsymbol{x}$)→U(D(A)$\boldsymbol{x}$), where $D(A)_{ij}=\frac{1}{2}\mathrm{Tr}(\tau_i A\tau_j A^\dagger)$ is the standard SO(3) rotation matrix corresponding to the SU(2) matrix A. This can be understood in terms of the coloured Skyrmions, where a spatial rotation can be 'undone' by a reordering of colours, i.e. an isorotation of the pion fields. A further use of the colouring is that it shows when two Hedgehog Skyrmions are in the attractive channel [13]. This is when the colours on the nearest, facing sides of two Skyrmions match.

The B=2 Skyrmion plays an important role in the dynamics of two B=1 Skyrmions. This Skyrmion is toroidal, and is shown in Fig. 2.

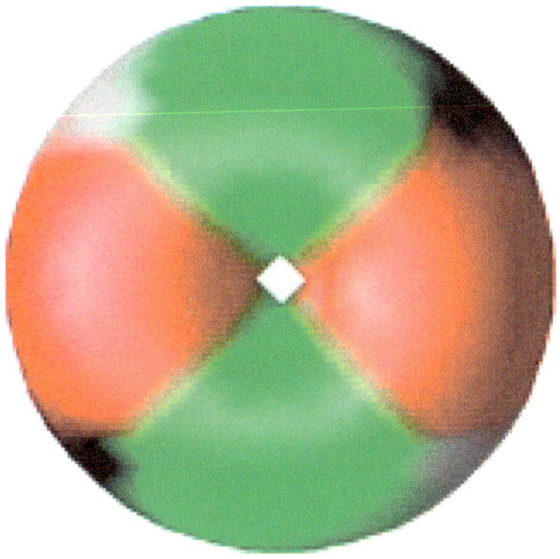

Figure. 2: B = 2 Skyrmion.

To relate the Skyrme model to nuclear physics, Skyrmions should be quantised. As the Skyrmions are energy minima they can be treated as rigid bodies rotating in space and isospace. For general Skyrmions, semi-classical quantisation is the quantisation of the time-dependent space and isospace rotations A_2 and A_1 in the rigid-body ansatz

$$U(t, x) = A_1(t)U_0(D(A_2(t))x)A_1(t)^\dagger, \quad (6)$$

where $U_0(x)$ is a static solution. $A_1(t)$ and $A_2(t)$ are SU(2) matrices, and$D(A_2(t))$ is, as before, the SO(3) matrix corresponding to $A_2(t)$. The spherical symmetry of the Hedgehog ansatz (5) implies that the rotational motion of the B=1Skyrmion only depends on the combined SU(2) matrix $A(t)=A_1(t)A_2(t)$.

Under rigid rotation the B=1 Skyrmion has body-fixed angular momentum$L_i=\lambda(-a_i+b_i)$ and body-fixed isospin angular momentum $K_i=\lambda(a_i-b_i)$, where $a_j = -i\,\mathrm{Tr}(\tau_j A_1^\dagger \dot{A}_1)$ is the angular velocity in isospace, $b_j = i\,\mathrm{Tr}(\tau_j \dot{A}_2 A_2^\dagger)$ is the angular velocity in space and λ is the moment of inertia [14]. From these one obtains the space-fixed angular momentum $J_i = -D(A_2)^T_{ij} L_j$ and the "space-fixed" isospin angular momentum $I_i=-D(A_1)_{ij}K_j$. The space-fixed angular momenta correspond to physical spin and isospin, and these can be expressed purely in terms of A(t) and its time derivative. Nucleons have both spin and isospin half. A proton is an isospin-up state, and a neutron is an isospin-down state.

Due to the spherical symmetry of the Hedgehog ansatz, $L_i + K_i = 0\ (i = 1, 2, 3)$. Something similar is true for the B=2 toroidal Skyrmion where $L_3+2K_3=0$[15]. We shall make use of these properties later. For an in-depth discussion of Skyrmion quantisation we point the reader to [14], [15], [16] and [17].

Skyrmion Calibration

For the model to be relevant to nuclear physics it requires calibration. Adkins and Nappi[16] calibrated the quantised B=1 Skyrmion against the masses of the spin $\frac{1}{2}$ nucleons, the spin $\frac{3}{2}$ delta resonances, and the pions, finding $F_\pi=108$ MeV, e=4.84 and$m_\pi=138$ MeV (so m=0.526). This calibration assumes that rigid rotation is a good approximation, but the surface fields of the delta are rotating close to the speed of light and pion radiation is very strong. Therefore, for the delta, the approximation is not reliable and this calibration is not valid for higher charge Skyrmions. Recently, Manton and Lau calibrated the Skyrme model against the states of Carbon-12 [18] and found m=0.7 to be optimal. For the B=1 Skyrmion this leads to an energy (Skyrmion mass)M=159.89, λ=55.89 and an rms matter radius $\langle r^2\rangle^{1/2}=1.006$ in Skyrme units. The calibration requires $F_\pi=117.5$ MeV and e=3.93. Planck›s constant in Skyrme units is always $\hbar=2e^2$ so for this calibration ħ=30.8. These are the values we use for our simulations and throughout this paper: m=0.7 and ħ=30.8.

The resulting nucleon mass is the static B=1 Skyrmion energy M plus the spin-energy contribution, $M + \frac{3}{8\lambda}\hbar^2$, which in physical units is

$$M_N = M\frac{F_\pi}{4e} + \frac{3}{8\lambda}e^3 F_\pi\,. \quad (7)$$

With our calibration $M_N=1243$ MeV, about 30% larger than the physical nucleon mass $M_N=939$ MeV. Also the rms matter radius is $\langle r^2\rangle^{1/2}=0.87$ fm.

The B=2 toroidal Skyrmion can also be quantised as a rigid body. The two lowest-energy states are a spin 1, isospin 0 state representing the deuteron, and a spin 0, isospin 1 state representing the low-energy resonance known as the diproton/dineutron. With our calibration, and following the quantisation procedure in [17], we find the mass for the deuteron to be 2324 MeV, which is similarly larger than the experimental value of1876 MeV. We also find the rms matter radius to be 1.08 fm, which is less than half the experimentally found rms radius of the deuteron ~2.14 fm, showing that rigid-body quantisation of the static B=2 Skyrmion gives a too tightly bound representation of the loosely bound deuteron. A better representation of the deuteron in the Skyrme model was obtained in [5]. Using a Yang–Mills instanton ansatz for the Skyrme field [19], an additional, radial degree of freedom could be included, allowing the two B=1 Skyrmions to separate. In the deuteron state, this radial degree of freedom oscillates with the zero point motion of a vibrational ground state. We will see later that modest oscillations of the radial degree of freedom are generated when spinning Skyrmions collide classically.

Although this calibration is not optimal for the proton, neutron or deuteron, we have found that the processes discussed later are not sensitive to small changes of calibration.

SPINNING SKYRMIONS AS PROTONS OR NEUTRONS

It was discovered by Adkins, Nappi and Witten [20] that the properly normalised wavefunctions for the proton and neutron in spin-up and spin-down states, along the z-axis, are

$$p^\uparrow = \frac{1}{\pi}(a_1 + ia_2), \quad p^\downarrow = -\frac{i}{\pi}(a_0 - ia_3), \quad (8)$$

$$n^\uparrow = \frac{i}{\pi}(a_0 + ia_3), \quad n^\downarrow = -\frac{1}{\pi}(a_1 - ia_2), \quad (9)$$

where $A=a_0 I_2+ia_i\tau_i$ is the SU(2) matrix that controls the B=1 Skyrmion›s orientation and $a_0^2+a_1^2+a_2^2+a_3^2=1$. It has been proposed that these wavefunctions can be used to identify protons and neutrons as classically spinning Skyrmions[6] and [7]. This is because each wavefunction has maximal magnitude on a great circle in SU(2) and motion along the great circle is equivalent to a 4π rotation of the Skyrmion in space. The phase of the wavefunction changes by 2π around the circle, so the spin is $\frac{1}{2}$.

The wavefunctions of $p^{\downarrow}$ and $n^{\uparrow}$ are maximal when the Hedgehog Skyrmion (initially in its standard orientation) is simply rotated about the z -axis, whereas the wavefunctions of$p^{\uparrow}$ and $n^{\downarrow}$ are maximal when the Skyrmion is flipped over, and then rotated about the z-axis.

A Hedgehog Skyrmion spinning with angular frequency ω about the z -axis is obtained by acting with the isorotation matrix $A(t) = \exp(\frac{i\omega t}{2}\tau_3) = \cos(\frac{\omega t}{2}) + i\sin(\frac{\omega t}{2})\tau_3$. This isospinning Skyrmion is simultaneously spinning about its white–black axis in space. It has spin and isospin projections $J_3=\lambda\omega$, $I_3=-\lambda\omega$, which are opposite. An anticlockwise spin, $\omega>0$, corresponds to an $n^{\uparrow}$ state and a clockwise spin, $\omega<0$, corresponds to a $p^{\downarrow}$ state.

The isorotation matrix $A(t)(i\tau_2)$ flips the Skyrmion by π about the y-axis, and rotates it with angular frequency ω about the z -axis. Such an isospinning Skyrmion has $J_3=-\lambda\omega$and $I_3=-\lambda\omega$, so the spin and isospin projections are now equal. An anticlockwise spin,$\omega>0$, corresponds to an $n^{\downarrow}$ state and a clockwise spin, $\omega<0$, corresponds to a $p^{\uparrow}$state.

The replacement $A(t)\rightarrow A(t)(i\tau_2)$ produces the spin flips $n^{\uparrow}\rightarrow n^{\downarrow}$ and $p^{\downarrow}\rightarrow p^{\uparrow}$because $D_{33}(A)\rightarrow -D_{33}(A)$. This leads to the identification that a Skyrmion spinning anticlockwise about its white–black axis is always a neutron and a Skyrmion spinning clockwise about the same axis is always a proton, as shown in Fig. 3. This is regardless of the orientation of the axis in space.

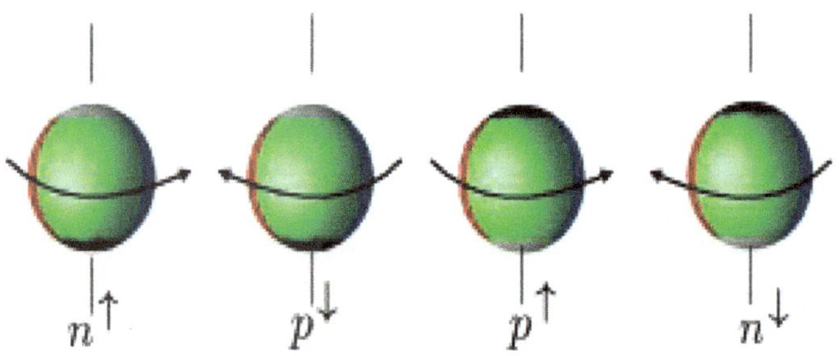

Figure. 3: Identifying Skyrmions as neutrons and protons.

To classically model a nucleon whose projected spin has magnitude $\frac{1}{2}$ we require $\lambda|\omega| = \frac{1}{2}\hbar$, so

$$|\omega| = \frac{\hbar}{2\lambda} = 0.28,$$

as $\hbar=30.8$ and $\lambda=55.89$. We use $|\omega|=0.28$ throughout this paper. With this angular velocity, the surface fields of the Skyrmion are not rotating close to the speed of light.

There are classical models of the two lowest-energy quantum states of the B=2 toroidal Skyrmion too. In the deuteron state, the torus spins around an axis orthogonal to the symmetry axis of the torus, with spin 1. The colour orientation is arbitrary and unchanging in time, as the deuteron has isospin zero. In the dineutron/diproton state the torus is coloured as in Fig. 2, and the colours circulate steadily, so that the isospin is 1. The black and white regions on the circular edge of the torus are fixed. The torus is not spatially rotating, as the dineutron/diproton has spin zero. The torus does not rotate about its symmetry axis in either state. If it did, the constraint $L_3+2K_3=0$ would imply that the state had non-zero spin and non-zero isospin.

SKYRMION SCATTERING AND ITS INTERPRETATION

Spinning Skyrmion Collisions

We are interested in classically replicating nucleon scattering, namely proton–proton, neutron–neutron and proton–neutron scattering. To do this we identify protons and neutrons as spinning Skyrmions, as discussed above. Instead of colliding non-spinning Skyrmions as in some previous work [11], we numerically collide spinning Skyrmions. We initiate a collision by forming two configurations U_1 and U_2 out of spinning and translated Hedgehog Skyrmions,

$$U_1(t,\boldsymbol{x})=U_H(D_1(t)(\boldsymbol{x}-\boldsymbol{X})), \quad (10)$$

$$U_2(t,\boldsymbol{x})=U_H(D_2(t)(\boldsymbol{x}+\boldsymbol{X})),$$

where $D_1(t)$ and $D_2(t)$ are two SO(3) time-dependent rotation matrices and $\pm\boldsymbol{X}$are the locations of the Skyrmions. We choose $\boldsymbol{X}=(b,Y,0)$, where Y is large compared with the Skyrmion radius. Then, using the product ansatz, we create an initial configuration of two Skyrmions boosted towards each other, at velocities $\pm v$, by

$$U(t,\boldsymbol{x}) = U_1(x,\gamma(y+vt),z)U_2(x,\gamma(y-vt),z), \quad (11)$$

where $\gamma = 1/\sqrt{1-v^2}$ is the usual Lorentz boost factor.

Each Skyrmion is initially moving parallel to the y-axis. b is the impact parameter chosen to give the desired orbital angular momentum $l\hbar=2Mvb$, where M is the Skyrmion mass. We define transverse polarisation to be where the Skyrmion›s white–black axis is aligned parallel to the z-axis and linear polarisation to be where the Skyrmion's white–black axis is aligned parallel to the y-axis.

We numerically evolve these initial configurations using the field equations obtained from the Skyrme Lagrangian density (1). We use a finite difference leap-frog numerical algorithm on a suitably large lattice. Leap-frog was chosen because it is a symplectic integrator.

All of the simulations are presented as videos at http://www.bristoltheory.org/people/david.foster/skyrmevideos/; we urge the reader to view these videos.

The Dineutron/Diproton

There are no physical bound states in the 2-proton or 2-neutron systems. However, a 2-proton low energy resonance is observed in proton–proton collisions and is called the diproton. It has spin $J=0$ and isospin $I=1$. The dineutron is the corresponding 2-neutron resonance. Neutron–neutron scattering is less easy to achieve than proton–proton scattering, but the dineutron resonance is observed as a decay product of the neutron-rich nucleus beryllium-16, ^{16}Be [21], where it is energetically favourable for ^{16}Be to decay to ^{14}Be by simultaneously ejecting two neutrons. The trajectories of the two detected neutrons trace back to the same point, where they form a state of lifetime $\sim 10^{-22}$ s. The ^{16}Be and ^{14}Be ground states are both $\boldsymbol{J}$=0, hence the dineutron is a $\boldsymbol{J}$=0 state. The relative energy of the two neutrons is found to be ≈0.25 MeV. So, assuming no other radiation, the neutrons separate with initial speed v=0.02.

There is no significant electrostatic repulsion between two neutrons, and since we have not included Coulomb effects in our Skyrme model, our simulations of Skyrmion scattering are likely to better model the dineutron rather than the diproton. So we collide two Skyrmions which are spinning as anti-aligned neutrons (illustrated in Fig. 3), giving a$\boldsymbol{J}$=0, $\boldsymbol{I}$=1,I_3=−1 state, with initial speeds v=0.02. The Skyrmions are either transversely polarised (parallel to the z-axis) or linearly polarised. Both types of scattering are shown in Fig. 4.

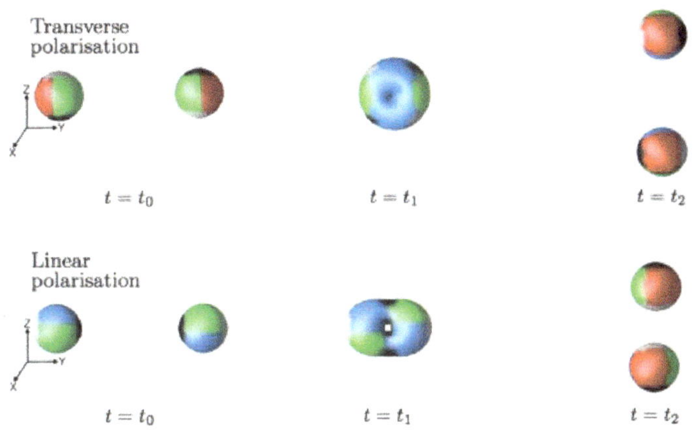

Figure. 4: Level-sets of baryon density B($\boldsymbol{x}$) for the $\boldsymbol{J}$ = 0 scattering of two neutrons. Three time-snapshots are shown here. The transverse polarisation scattering and the linear polarisation scattering videos are shown online.

It is seen that no bound state is produced, but the B=2 torus configuration is present for a short time. This torus forms in the (y,z)-plane and it does not spin, but it has isospin because the colouring continues to circulate at all times. It has the quantum numbers$\boldsymbol{J}$=0 and $\boldsymbol{I}$=1, I_3=−1 which is consistent with the short-lived dineutron state observed in the decay of ^{16}Be. For both polarisations, the black and white regions alternate around the circular edge of the torus.

The scattering shown in Fig. 4 is for a head-on collision of two neutrons. For both polarisations, the outgoing Skyrmions are moving at right angles to the incoming Skyrmions, and they are spinning as neutrons too, because of isospin conservation. Similar right angle scattering was observed for non-spinning Skyrmions in [11].

In the case of transverse polarisation the outgoing direction of motion is precisely the direction of the initial polarisation axis. There is another interesting phenomenon: the polarisation axis, the white–black axis, jumps by 90 degrees during the scattering, from being parallel to the z-axis to being parallel to the y-axis. This is possible because of the way the Skyrmions merge and break up, such that each outgoing Skyrmion is a combination of half of each incoming Skyrmion.

From our numerical simulations we can observe the sense of the polarisation, before and after the collision. For example, the neutron incoming from the negative y-direction (the left) has its polarisation in the positive z-direction, and the neutron that is outgoing in the positive z-direction has its polarisation in the negative y-direction. The result can be stated in a more invariant way, both for 2-neutron and 2-proton right angle scattering. Each particle, incoming or outgoing, has a momentum **p** and classical spin **s**. The two incoming particles and the two outgoing particles all have the same value for the vector**p**×**s**, so the scattering is accompanied by spin rotation.

In the case of linear polarisation the outgoing direction of motion can be any direction at right angles to the y-axis, and is determined by the precise initial colour orientations. The scattering shown in Fig. 4 has been selected to be from the y-direction to the z -direction. Again there is a 90 degree jump in the polarisation axis, so that after scattering the neutrons are still linearly polarised. There is however a flip in the sense of the polarisation. In our example, for both incoming neutrons the polarisation is outwards, but for the outgoing neutrons it is inwards. Helicity, **p**·**s**, is therefore the same for incoming and outgoing particles, and this result is more general, holding in the diproton case too.

Although the scattering here is classical, the results suggest that quantised 2-neutron and 2-proton scattering depends strongly on polarisations. For Skyrmions transversely polarised in the z-direction, the outgoing motion breaks the axial symmetry around the y-axis, because the outgoing particle motion is concentrated in the z-direction. Also, the outgoing polarisations are completely determined. It would be interesting to compare these classical simulations with physical proton–proton scattering in which the polarisations of both incoming particles and both outgoing particles were measured.

Deuteron Formation

The deuteron is a stable, two-baryon bound state comprised of a proton and a neutron. It has been understood as the quantised ground state of the B=2 Skyrmion [5],[15] and [17] with spin $\boldsymbol{J}$=1 and isospin $\boldsymbol{I}$=0.

Here we are interested in classically modelling deuteron formation, so we numerically collide two spinning Skyrmions, representing a proton and neutron with transverse polarisation parallel to the z-axis. The deuteron total angular momentum projected along the z -direction has three contributions, the individual spins J_p and J_n of the proton and neutron, which are $\pm\frac{1}{2}$, and the orbital angular momentum, an integer l , combining to give J_3=J_p+J_n+l[22]. Quantum models predict that the l=0 state dominates, and the deuteron spin is accounted for by the proton and neutron spins being aligned. We can replicate the deuteron state by colliding two Skyrmions head-on, where one Skyrmion is spinning as a neutron with $J_n = \frac{1}{2}$ and the other Skyrmion has the opposite orientation of the white–black axis and is spinning as a proton with $J_p = \frac{1}{2}$. The impact parameter is zero, so l=0.

The deuteron has slightly lower energy than a proton plus a neutron, so we cannot use a simple energetic argument to choose an initial collision speed. Whatever the initial speed, some energy needs to be dissipated in order to truly produce a deuteron. It is therefore a considerable surprise that our collisions do produce a configuration similar to what is expected for a deuteron in a classical Skyrmion model.

There is, however, a problem to be overcome. If we minimise the energy by choosing initial speed v=0, and the initial relative orientations arbitrary, then averaged over time the Skyrmions repel. This is because the Skyrmions are spinning, and most of the time the colours do not match, producing a repulsion. Only briefly do the colours match, and the Skyrmions attract. Similar behaviour was observed for two spinning baby Skyrmions in 2-dimensions [23]. The repulsion of spinning Skyrmions initially at rest, which is not a Coulomb effect, is shown in Fig. 5.\

Figure. 5: The repulsion of a Skyrmion proton and a Skyrmion neutron with zero initial speed. The dynamics is shown online.

Therefore, to model deuteron formation we need to collide the Skyrmions at a positive speed, and we choose v=0.2. There is now enough kinetic energy to overcome the repulsive barrier. The dynamic B=2 Skyrmion that forms is stable even with this extra energy. The collision along the y-axis of two Skyrmions spinning about the z-direction is shown in Fig. 6, from two viewpoints.

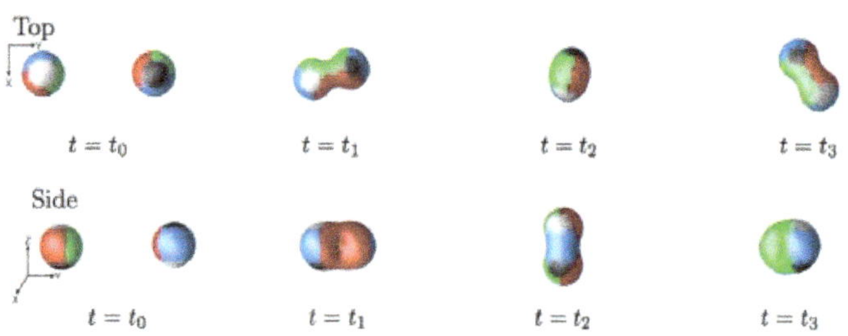

Figure. 6: Formation of a deuteron bound state in a spinning Skyrmion collision with $\boldsymbol{J} = 1$, $\boldsymbol{I} = 0$. This is shown dynamically online.

As expected [24], the B=2 torus forms in the (y,z)-plane and spins about the z -axis, which is an axis orthogonal to the symmetry axis of the torus. Therefore there is no projection of spin or isospin along the symmetry axis, and the constraint $L_3+2K_3=0$ is satisfied.

The important feature of this collision is that the B=1 Skyrmions do not scatter, but merge into a long-lived spinning B=2 Skyrmion, with superimposed oscillations. The colours are not independently spinning. When the oscillation is at its maximal amplitude, two B=1 Skyrmions can be identified. They represent two nucleons with parallel spins. This classical bound state models the physical, quantised deuteron [15], with the correct spin, and zero isospin. The oscillation is similar to that of the B=2 Skyrmion in its quantised ground state obtained using the instanton ansatz [5], but the amplitude of oscillation is somewhat greater, and the energy somewhat larger. The classical oscillation produces an oscillating matter radius, which is shown in Fig. 7 as a function of time. The time-averaged matter radius is comparable with the experimental value for the deuteron.

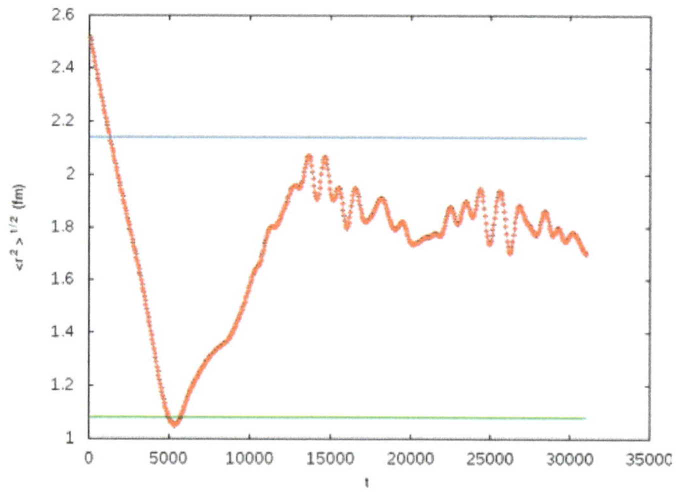

Figure. 7: Skyrmion matter radius in the classical simulation of a neutron–proton collision, as a function of time t. The upper line is the experimental deuteron matter radius and the lower line is the matter radius of the static B = 2 toroidal Skyrmion.

We have also attempted to simulate deuteron formation by colliding Skyrmions spinning as a neutron and proton, both with linear polarisation. This is shown in Fig. 8. The process does not produce a classical bound state, even though the spin and isospin are correct for a deuteron. The initial state can be understood as Skyrmions approaching in the attractive channel with two white faces being closest. The Skyrmions scatter at right angles and the initial spin is converted to orbital angular momentum, the Skyrmions separating with some non-zero impact parameter. They also spin slowly about a coloured axis as they separate – this is inferred from a scattering of Skyrmions with larger initial spin. Skyrmions spinning like this should be interpreted as a superposition of a neutron and a proton.

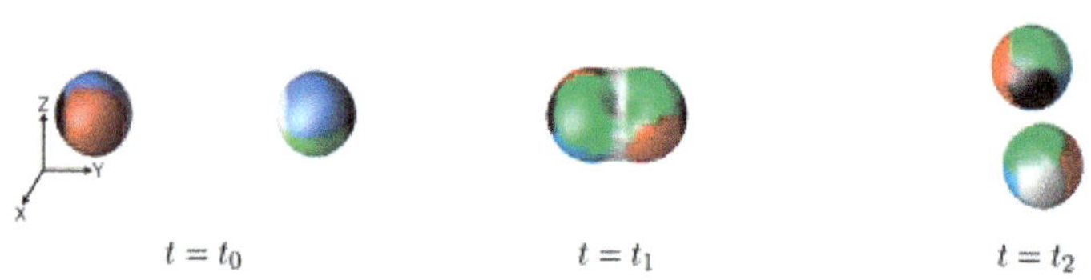

Figure. 8: Head-on collision of linearly polarised proton and neutron with $\boldsymbol{J} = 1$, $\boldsymbol{I} = 0$. Note: at $t=t_1$ the torus is spinning about its black–black axis. The dynamics is shown online.

Although the torus forms briefly in the linearly polarised case, it breaks apart. This is because Skyrmions in the attractive channel accelerate towards each other and the excess energy quickly breaks the torus up. In the transversely polarised case the excess energy is less and goes into the vibrational mode. This shows that polarisation has implications for proton–neutron fusion.

Excited Dibaryon

Recently an excited proton–neutron state has been discovered [25], called the dibaryon. It can be viewed as an excited spin 3 state of the spin 1 deuteron, with mass$M_{Dib} \approx 2380$ MeV and quantum numbers $\boldsymbol{J}=3$, $\boldsymbol{I}=0$.

To create the Skyrmion analogue requires colliding two Skyrmions each with kinetic energy $T = \frac{1}{2}(M_{\text{Dib}} - 2M_{\text{N}})$; this is achieved with initial, relativistic speeds of v=0.6. The$\boldsymbol{J}=3$, $\boldsymbol{I}=0$ state is replicated by colliding a Skyrmion spinning as a neutron and an oppositely orientated Skyrmion spinning as a proton, with the spins parallel and with orbital angular momentum l=2, such that $J_3 = \frac{1}{2} + \frac{1}{2} + 2$. The orbital angular momentum is achieved with an impact parameter b=0.32 at this speed. This impact parameter is the same order of magnitude as half the spatial width of the B=2 toroidal Skyrmion. The scattering is shown in Fig. 9, with the initial motion parallel to the y-axis, and the proton and neutron polarisations parallel to the z -axis. The dibaryon appears briefly in the form of the B=2 toroidal Skyrmion spinning with ***J***=3 about the z-axis. The spin axis is at right angles to the torus's symmetry axis, as for the deuteron.

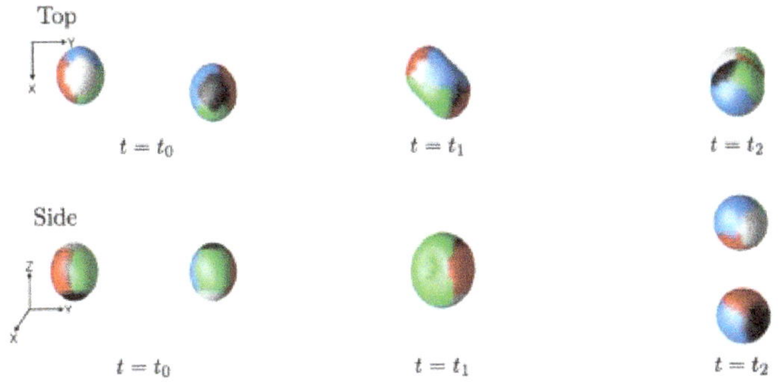

Figure. 9: Skyrmion scattering modelling proton–neutron scattering with ***J*** = 3, ***I*** = 0 at the energy of the dibaryon resonance. The orbital angular momentum is l = 2, requiring an impact parameter b = 0.32. Note: at $t=t_2$the Skyrmions are spinning about the red–cyan axis. The dynamical scattering is shown online.

There is a small amount of excess energy and as a result the B=2 Skyrmion almost breaks up into two rapidly spinning B=1 Skyrmions moving back-to-back along the z -axis. These have close to spin $\frac{3}{2}$ each, but rapidly lose energy, presumably by pion radiation, although this is hard to see in the simulation. The spinning Skyrmions stop and return to form a spinning B=2 Skyrmion again. This replicates the experimental result that two pions are emitted, and a deuteron remains.

The transient production of spin $\frac{3}{2}$ particles matches other theoretical models of the dibaryon resonance involving two deltas with spin $\frac{3}{2}$[26]. Our simulation shows that (in the proton–neutron centre-of-mass frame) the deltas are emitted transversely to the line of collision, but parallel to the proton and neutron polarisation axes. Each delta is linearly polarised, and because its spin axis is an axis perpendicular to the white–black axis, it is in a superposition of states with different charges. This superposition is not surprising, because the average charge emerging along the positive or negative z-axis is half the proton charge.

We have also attempted to produce a ***J***=3, ***I***=0 dibaryon state by colliding a proton Skyrmion and a neutron Skyrmion with the spin polarisations transverse but opposite, so that the spin angular momenta cancel. The orbital angular momentum is l=3, which requires the same initial velocities, but an impact parameter b=0.48. This scattering is shown in Fig. 10.

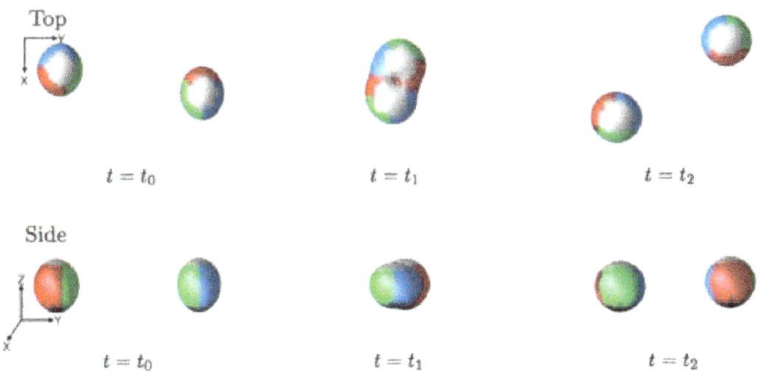

Figure. 10: Modelling dibaryon production in proton–neutron scattering, with orbital angular momentum l = 3, and impact parameter b = 0.48. This is shown dynamically online.

Here, the toroidal Skyrmion also appears but is not an excited deuteron, because it forms in the plane normal to the initial orbital angular momentum vector. In this case, one might anticipate the torus to be spinning about its symmetry axis, but it would then have non-zero isospin, which is not allowed. On closer inspection one observes that the B=2 torus is not spinning, but each B=1 Skyrmion can be interpreted as a wave propagating along the outer edge, such that the colours do not rotate. This curious dynamical configuration is not a dibaryon.

In summary, dibaryon formation can be observed in classical Skyrmion scattering, but only if the initial polarisations of the proton and neutron are transverse and parallel.

CONCLUSION

In this work we have classically scattered spinning Skyrmions to model nucleon–nucleon scattering. The initial spin angular momenta are fixed to be $\frac{1}{2}\hbar$. Interestingly, we have found short-lived states modelling the dineutron/diproton and the dibaryon. We have also found that in a collision with the quantum numbers of the deuteron, the Skyrmions do not scatter, but merge into a spinning and oscillating form of the B=2 toroidal Skyrmion. So, even without quantisation, the Skyrme model usefully captures important physical features of two-nucleon dynamics. This is gratifying but a little surprising, especially when it is recalled that our long-lived deuteron state was made from two Skyrmions colliding with a larger kinetic energy than that needed to produce the short-lived dineutron/diproton.

Our most valuable observations concern the polarisations. The classically spinning Skyrmions must be polarised along some axis, so we inevitably have complete polarisation information for the nucleons before and after scattering. Such complete information is hard to obtain experimentally, and also difficult to determine in more traditional nucleon–nucleon potential models. Classically we also fix the impact parameter. Several of our simulations lead to right angle scattering of the Skyrmions. Experimentally one cannot fix the impact parameter, but one can focus on scattering events where there is right angle scattering (in the centre-of-mass frame), and can hope that these are modelled by our collisions. For such events, our Skyrmion simulations make strong predictions for the polarisations. For example, in our simulations of a proton–neutron collision leading to the dibaryon, with total spin 3, we have seen the transient production of two linearly polarised deltas each with spin close to $\frac{3}{2}$.

We have observed that in a 2-proton or 2-neutron head-on collision, with total angular momentum zero, if the particles are transversely polarised initially then they are transversely polarised finally. If they are linearly polarised initially they are linearly polarised finally. In both cases the polarisation axes rotate through a right angle. Moreover, there is a fixed plane containing the initial and final momenta and also the polarisations. The sense of the polarisations is determined by conservation of $\mathbf{p}\times\mathbf{s}$ for each particle in the transverse polarisation case, and conservation of $\mathbf{p}\cdot\mathbf{s}$ in the linear polarisation case. It would be very interesting if something similar was observed experimentally. Such an advanced experiment would require the development of new experimental techniques, but it would help understand the structure of nuclei.

We intend to extend our simulations to scattering of polarised Skyrmions with a range of impact parameters, and from the results deduce a differential cross section. This cross section would be easier to compare with existing data. We also plan to extend our numerical simulations to model nucleon–deuteron and deuteron–deuteron collisions, possibly forming Helium-3, Tritium or Helium-4 bound states. The physical spins must be correctly modelled in the initial data, and the results will differ from those obtained with non-spinning Skyrmions.

ACKNOWLEDGEMENTS

D.F. would like to thank Prof. David Jenkins, Dr. Steffen Krusch and Dr. Mareike Haberichter for useful conversations. D.F. acknowledges the Leverhulme Trust Program Grant: Scientific Properties Of Complex Knots.

REFERENCES

1. T.H.R. Skyrme, A nonlinear field theory, Proc. R. Soc. Lond. A 260 (1961) 127.
2. E. Witten, Global aspects of current algebra, Nucl. Phys. B 223 (1983) 422.
3. E. Witten, Current algebra, baryons, and quark confinement, Nucl. Phys. B 223 (1983) 433.
4. M.F. Atiyah, N.S. Manton, Geometry and kinematics of two Skyrmions, Commun. Math. Phys. 153 (1993) 391.
5. R.A. Leese, N.S. Manton, B.J. Schroers, Attractive channel Skyrmions and the deuteron, Nucl. Phys. B 442 (1995) 228.
6. T. Gisiger, M.B. Paranjape, Recent mathematical developments in the Skyrme model, Phys. Rep. 306 (1998) 109.
7. N.S. Manton, Classical Skyrmions – static solutions and dynamics, Math. Methods Appl. Sci. 35 (2012) 1188.
8. J.J.M. Verbaarschot, T.S. Walhout, J. Wambach, H.W. Wyld, Scattering of Skyrmions in an axially symmetric system, Nucl. Phys. A 461 (1987) 603.
9. A.E. Allder, S.E. Koonin, R. Seki, H.M. Sommermann, Dynamics of Skyrmion collisions in 3+1 dimensions, Phys. Rev. Lett. 59 (1987) 2836.
10. R.A. Battye, P.M. Sutcliffe, Multi-soliton dynamics in the Skyrme model, Phys. Lett. B 391 (1997) 150.
11. D. Foster, S. Krusch, Scattering of Skyrmions, Nucl. Phys. B 897 (2015) 697–716, arXiv:1412.8719 hep-th..
12. R.A. Battye, M. Haberichter, S. Krusch, Classically isospinning Skyrmion solutions, Phys. Rev. D 90 (2014) 125035.
13. B.J. Schroers, Dynamics of moving and spinning Skyrmions, Z. Phys. C 61 (1994) 479.
14. R.A. Battye, N.S. Manton, P.M. Sutcliffe, S.W. Wood, Light nuclei of even mass number in the Skyrme model, Phys. Rev. C 80 (2009) 034323.
15. E. Braaten, L. Carson, Deuteron as a toroidal Skyrmion, Phys. Rev. D 38 (1988) 3525.
16. G.S. Adkins, C.R. Nappi, The Skyrme model with pion masses, Nucl. Phys. B 233 (1984) 109.
17. O.V. Manko, N.S. Manton, S.W. Wood, Light nuclei as quantized Skyrmions, Phys. Rev. C 76 (2007) 055203.
18. P.H.C. Lau, N.S. Manton, States of carbon-12 in the Skyrme model, Phys. Rev. Lett. 113 (2014) 232503.
19. M.F. Atiyah, N.S. Manton, Skyrmions from instantons, Phys. Lett. B 222 (1989) 438.
20. G.S. Adkins, C.R. Nappi, E. Witten, Static properties of nucleons in the Skyrme model, Nucl. Phys. B 228 (1983) 552.
21. A. Spyrou, et al., First observation of ground state dineutron decay: 16Be, Phys. Rev. Lett. 108 (2012) 102501.
22. K.S. Krane, Introductory Nuclear Physics, Wiley, New York, Chichester, 1987.
23. B.M.A.G. Piette, B.J. Schroers, W.J. Zakrzewski, Dynamics of baby Skyrmions, Nucl. Phys. B 439 (1995) 205.
24. J.L. Forest, et al., Femtometer toroidal structures in nuclei, Phys. Rev. C 54 (1996) 646.
25. WASA-at-COSY Collaboration, P. Adlarson, et al., Evidence for a new resonance from polarized neutron–proton scattering, Phys. Rev. Lett. 112 (2014) 202301.
26. A. Gal, H. Garcilazo, Three body calculation of the dibaryon candidate D03(2370), Phys. Rev. Lett. 111 (2013) 172301.

CHAPTER 5

DYNAMICAL EFFECTS OF SPIN-DEPENDENT INTERACTIONS IN LOW- AND INTERMEDIATE-ENERGY HEAVY-ION REACTIONS

Jun Xu, Bao-An Li, Wen-Qing Shen, Yin Xia

Keywords: Heavy-ion collisions, transport model, spin–orbit interaction, tensor force, polarization

INTRODUCTION

Understanding novel features of the fundamental nuclear forces and properties of strongly interacting matter under extreme conditions of density, temperature, spin, and isospin is among the main goals of nuclear physics. Heavy-ion collision (HIC) experiments play an important role in achieving these goals. Indeed, great achievements have been made using HICs at various beam energies from the sub-Colomb barrier to the highest energy available at the Large Hadron Collider. In particular, terrestrial experiments using intermediate-energy HICs have led to strong constraints on the equations of state of hadronic matter [1] and neutron-rich nucleonic matter [2, 3].

Theoretical studies have shown recently that some spin-sensitive observables of HICs can be used to explore the in-medium properties of noncentral nuclear forces. The spin-dependent nuclear interactions are important for explaining several interesting features of nuclear structure [4], such as the varying magic numbers and the shell evolution with the isospin asymmetry of finite nuclei. However, the strength, density, and isospin dependence of the nuclear spin–orbit coupling are still uncertain (see Section 2.1). Moreover, the tensor force can modify the magic number of nuclei and is an important source of the nucleon–nucleon short-range correlation. The latter is related to many interesting phenomena in nuclear physics (see Section 2.2). More studies on inmedium properties of the spin–orbit coupling and tensor force are thus very much needed. HICs provide flexible ways of adjusting the conditions of the nuclear medium and may also lead to new spin-dependent phenomena. For example, the so-called spin Hall effect [5–7], which affects the dynamics of spin-up and spin-down particles differently as a result of the spin–orbit coupling, is expected to be a general feature in any spin transport process. It thus might be interesting to test whether such a phenomenon can also occur in HICs. Considerable efforts using the time-dependent Hartree–Fock (TDHF) model, the spin- and isospindependent Boltzmann–Uehling–Uhlenbeck (SIBUU) transport model, and the quantum molecular dynamics (QMD) model have been devoted to exploring the spin dynamics in HICs (see Section 3). Indeed, some interesting phenomena were found. For example, it was found that the inclusion of the spin-dependent nuclear interaction may affect the fusion threshold, generate the spin twist during the collision process, and lead to the spin splitting of nucleon collective flows (see Section 3). Future comparisons with relevant experimental data may help extract properties of the in-medium spindependent nuclear force. Here we review briefly the main physics motivations and recent findings of studying the spin-dependent dynamics and observables in low- and intermediate-energy HICs. A major goal of this article is to stimulate more experimental work in this direction.

SPIN-RELATED NUCLEAR FORCE

Based on the one-boson-exchange picture [8], the nuclear force can be understood by exchanging mesons between nucleons. Exchanging the scalar σ meson and vector ω meson leads to, respectively, attractive and repulsive central nuclear forces as well as the spin–orbit interaction, while exchanging the π meson and ρ meson leads to, respectively, long-range and short-range nuclear tensor forces. Although in free space

Citation: Jun Xu, Bao-An Li, Wen-Qing Shen, Yin Xia, Dynamical effects of spin-dependent interactions in low- and intermediate-energy heavy-ion reactions, DOI 10.1007/s11467-015-0479-8.

the bare nuclear force is well constrained by the nucleon–nucleon scattering data, the in-medium nuclear interactions, especially the nuclear spin–orbit interaction and tensor force, are still quite uncertain. The in-medium nuclear interactions can be studied by using microscopic many-body theories or phenomenological models, such as the nonrelativistic Skyrme–Hartree–Fock (SHF) model and the relativistic mean-field (RMF) model. In the following subsections, we will discuss the effective spin-dependent nuclear force based on the energy-density functional in the phenomenological approach.

Nuclear Spin–Orbit Interaction

The nuclear spin–orbit interaction was first introduced to explain the magic numbers of nuclei [9, 10]. Nuclei with numbers of neutrons or protons equal to the magic numbers are more stable, and this reflects the special shell structure of a nucleus. Although even a simple harmonic potential leads to the shell structure of nucleon energy levels inside nuclei, the spin–orbit coupling is essential to reproduce the correct magic number. In the Skyrme interaction, the effective spin–orbit force between two nucleons at positions r1 and r2 can be expressed as [11]

$$V_{so} = \mathrm{i}W_0(\boldsymbol{\sigma}_1 + \boldsymbol{\sigma}_2) \cdot \boldsymbol{k} \times \delta(\boldsymbol{r}_1 - \boldsymbol{r}_2)\boldsymbol{k}', \tag{1}$$

where W_0 is the spin–orbit coupling constant, σ_1 and σ_2 are the Pauli matrices for the two nucleons, $k = -i(\nabla_1 - \nabla_2)/2$ is the relative momentum operator acting on the right side with ∇_1 and ∇_2 acting on the first and second nucleon, respectively, and k is its complex conjugate acting on the left. From the conventional HartreeFock method, the spin–orbit single-particle potential can be obtained based on the above effective spin–orbit force as

$$U_q^{so} = \boldsymbol{W}_q \cdot (-\mathrm{i}\nabla \times \boldsymbol{\sigma}), \tag{2}$$

where

$$\boldsymbol{W}_q = \frac{W_0}{2}(\nabla\rho + \nabla\rho_q) \tag{3}$$

is the form factor of the spin–orbit potential, with q = n or p being the isospin index and ρ being the nucleon number density. By taking the operator −i∇ as the momentum p, the right-hand side of Eq. (2) has the form of (r × p) · σ with W_q playing the role of r, and this is why it is called the spin–orbit potential. By solving the Schr¨odinger equation with the single-nucleon Hamiltonian

$$h_q = -\nabla \cdot \left(\frac{1}{2m_q^*}\nabla\right) + U_q + U_q^{so}, \tag{4}$$

With m_q^* being the effective nucleon mass and U_q being the spin-independent potential, the single-nucleon spectrum in a spherical closed-shell nucleus can be obtained. In the RMF model, one solves the Dirac equation, where the spin of the nucleon is treated explicitly with nucleon wave functions for different spin states [12]. Studies of SHF and RMF models on nuclear structure were reviewed in Ref. [13], and here we compare the effective spin–orbit potentials from both the relativistic and nonrelativistic approaches. With nonrelativistic expansion of the Dirac equation, the form factor of the nucleon effective spin–orbit potential in the RMF model can be expressed in the form [14, 15]

$$\boldsymbol{W}_{\mathrm{RMF}} = \frac{1}{(2m - C_{eff}\rho)^2}C_{eff}\nabla\rho, \tag{5}$$

where m is the nucleon mass and the coefficient C_{ef} f is related to the coupling strength and mass of the scalar σ meson and the vector ω meson, i.e.,

$$C_{eff} = \frac{g_\sigma^2}{m_\sigma^2} + \frac{g_\omega^2}{m_\omega^2}. \tag{6}$$

The form factors of the spin–orbit potential in the SHF model [Eq. (3)] and the RMF model [Eq. (5)] are different. First, the spin–orbit coupling strength is a constant in the SHF model, but the effective coupling strength depends on the density in the RMF model. Implementing an additional density-dependent effective nucleon– nucleon spin–orbit interaction with a coupling constant W1, the authors of Ref. [16] got additional contributions to the form factor as

$$\boldsymbol{W}_q^{\rho} = \frac{W_1}{2}[c\rho\nabla(\rho - \rho_q) + (2+c)(2\rho_q)^c\nabla\rho_q] + \frac{W_1}{4}c\rho^{c-1}(\rho - \rho_q)\nabla\rho, \tag{7}$$

with c mimicking the density dependence. The above form was tested in Ref. [16] in semi-infinite nuclear matter with parameters fitted to the RMF interaction. It was found that the general features of the RMF model were then reproduced with this nonrelativistic densitydependent spin–orbit interaction. Nevertheless, the density dependence of the spin–orbit coupling is still largely unknown so far, and it is related to many interesting phenomena in nuclear structure studies [17–19]. Second, the spin–orbit couplings from the SHF and RMF approaches have different isospin dependence; i.e., in the SHF approach the spin–orbit coupling is stronger for nucleons of the same isospin, whereas in the RMF approach the coupling strength is the same for neutrons and protons. This feature impacts descriptions of properties of neutron-rich nuclei, e.g., the kink in the evolution of the charge radii for lead isotopes. It was shown that the weak isospin dependence of the spin–orbit coupling in the RMF approach can better explain the kink than can the conventional SHF functional. However, if the form factor in the latter approach was modified to [20, 21]

$$\boldsymbol{W}_q = \frac{W_0}{2}(1 + \chi_w)\nabla\rho_q + \frac{W_0}{2}\nabla\rho_{q'} \quad (q \neq q'), \tag{8}$$

a similar kink can be reproduced with $\chi_w \approx 0.1$ [21], corresponding to the case with a very small Fock contribution of the spin–orbit interaction. Similar efforts were made by using a modified SHF functional to reproduce the isospin dependence of the spin–orbit field in semiinfinite nuclear matter with different neutron excesses [22] and in neutron-rich nuclei [23] from a relativistic approach. In Ref. [24], the isospin dependence of the spin– orbit coupling was compared in light drip line nuclei from the RMF theory and the nonrelativistic Skyrme model. Furthermore, it was observed that the commonly used Skyrme functional of the spin–orbit splitting overestimated the central density and the spin–orbit splitting of neutron drops [25], calling for new functionals of the spin–orbit coupling. The proton energy splitting of $h_{11/2}$ and $g_{7/2}$ outside the Z = 50 closed shell increases with neutron excess, corresponding to the decreasing strength of the nuclear spin–orbit interaction [26]. The studies so far seem to favor a weak isospin dependence of the spin– orbit coupling. However, since the isospin dependence of the spin–orbit coupling, which is important in nuclear surfaces, is often coupled with its density dependence, it is still not well settled yet.

Based on the above discussion, we proposed the following general form of the form factor of the spin–orbit coupling by taking both the density and isospin dependence into account:

$$\boldsymbol{W}_q = \frac{W_0}{2}\left(\frac{\rho}{\rho_0}\right)^{\gamma}(a\nabla\rho_q + b\nabla\rho_{q'}) \quad (q \neq q'). \tag{9}$$

The above form is artificially constructed and, for simplicity, includes only the main physics. In the above form, γ is used to mimic the density dependence of the spin– orbit coupling while fixing its strength at saturation density ρ0 to be W_0. a and b are parameters to vary the isospin dependence of the spin–orbit coupling, with a = 2 and b = 1 corresponding to the case of the standard SHF approach and a = b corresponding to the case of the RMF approach. The values of γ, a, and b are still uncertain according to the above discussion. For the strength of the spin–orbit coupling W_0, efforts have been made to extract its information from ground-state properties of various nuclei. Recent studies have shown that the spin–orbit coupling and the tensor force, which will be discussed in the next subsection, should be considered simultaneously to describe the spin–orbit splitting and single-nucleon spectra of nuclei. Based on the Skyrme functional and by taking the uncertainties of the tensor force into account, the strength of the spin–orbit coupling is ≈80–150 MeV fm^5, from fitting the properties of light to heavy nuclei [27–29]. The single-nucleon Hamlitonian of Eq. (4) is adequate to describe the ground-state properties of spherical closed-shell nuclei. For open-shell nuclei, one needs to consider an additional spin-dependent potential using the spin-current density J from Eq. (1):

$$U_q^J = -\frac{W_0}{2}\nabla\cdot(\boldsymbol{J} + \boldsymbol{J}_q). \tag{10}$$

J is actually the vector component of the spin-current density tensor $J_{\mu\nu}$ (see, e.g., Ref. [27] for a more detailed discussion). For deformed nuclei, not only the time-even potentials [Eqs. (2) and (10)] but also the time-odd potentials should be considered [30]. Starting from the effective nucleon–nucleon spin–orbit interaction [Eq. (1)] and taking Eq. (9) into consideration, one can obtain the general form of the time-even and time-odd spindependent potentials as

$$U_q^{s\text{-even}} = -\frac{W_0}{2}\left(\frac{\rho}{\rho_0}\right)^{\gamma}[\nabla\cdot(a\boldsymbol{J}_q + b\boldsymbol{J}_{q'})]$$
$$+\frac{W_0}{2}\left(\frac{\rho}{\rho_0}\right)^{\gamma}(a\nabla\rho_q + b\nabla\rho_{q'})\cdot(\boldsymbol{p}\times\boldsymbol{\sigma}), \tag{11}$$

$$U_q^{s\text{-odd}} = -\frac{W_0}{2}\left(\frac{\rho}{\rho_0}\right)^{\gamma}\boldsymbol{p}\cdot[\nabla\times(a\boldsymbol{s}_q + b\boldsymbol{s}_{q'})]$$
$$-\frac{W_0}{2}\left(\frac{\rho}{\rho_0}\right)^{\gamma}\boldsymbol{\sigma}\cdot[\nabla\times(a\boldsymbol{j}_q + b\boldsymbol{j}_{q'})] \quad (q \neq q'), \tag{12}$$

where p = −i∇ is the momentum operator, and s and j are spin density and current density, respectively. We note that the time-odd potentials play an important role in the dynamics of heavy-ion reactions, which will be discussed in Section 3.

Nuclear Tensor Force

The first piece of strong evidence for the nuclear tensor force comes from studying properties of deuterons. In the nonrelativistic approach, the nuclear tensor force between two nucleons at positions r1 and r2 is often expressed with the tensor operator written as

$$S_{12} = 3\frac{(\boldsymbol{\sigma}_1\cdot\boldsymbol{r})(\boldsymbol{\sigma}_2\cdot\boldsymbol{r})}{r^2} - (\boldsymbol{\sigma}_1\cdot\boldsymbol{\sigma}_2), \tag{12}$$

where $r = r_1 - r_2$ is the relative position vector. One thus sees that whether the tensor force is attractive or repulsive depends on the relative direction between the spin and the relative position vector; i.e., $S_{12} > 0$ for $\sigma_1(2)$ parallel to r and $S_{12} < 0$ for $\sigma_1(2)$ perpendicular to r.

The tensor term in the effective nuclear interaction was first included in the Skyrme force [31], but afterward it was neglected owing to its complex form. Recently, it has attracted renewed interest. It has been found by Otsuka et al. that the nuclear tensor force may affect the shell structure or even modify the magic number of nuclei [32– 34]. The combined effects of the spin–orbit coupling and the nuclear tensor force sometimes hamper our understanding of both of them [35, 36]. Based on the random phase approximation, the effects of the tensor force on the multipole response of magic nuclei have been studied, and a large effect on the magnetic dipole states was observed [37]. Moreover, it was proposed that the spinisospin excitation of finite nuclei may serve as a useful observable to assess the strength of the tensor force [38, 39]. The existence of the tensor force may also open a shell gap for large neutron numbers, having a consequent implication for the synthesis of neutron-rich superheavy elements [40]. Although the nuclear tensor force has no effect on the equation of state of spin-saturated nuclear matter based on the studies at the mean-field level, it affects the properties of nuclear matter from many-body calculation beyond the mean field. It was found that the repulsive central force and the tensor force are two important sources of the nucleon–nucleon short-range correlation [41]. The high-momentum tail of the nucleon distribution in nuclear matter as well as in finite nuclei was observed even at zero temperature based on these studies [42–45]. Great efforts have been made in measuring the short-range nucleon–nucleon correlation and extracting the ratio of nucleons in the high-momentum tail experimentally [46– 50] and theoretically [51, 52] (see, e.g., Ref. [53] for a review). In particular, it was found that the neutronproton correlation is much stronger than the correlation between neutron-neutron and proton-proton pairs [48– 50], and this is mainly due to the nuclear tensor force. The isospin dependence of the short-range correlation can lead to interesting consequences, such as the reduction of the kinetic contribution to the nuclear symmetry energy [54–56] compared to the free Fermi gas scenario. Because the symmetry energy at saturation density is constrained to be around 30 MeV from many analyses (see, e.g., Ref. [57]), the isospin-dependent short-range correlation effectively increases the potential contribution to the symmetry energy and thus the symmetry potential effect, which may lead to enhanced isospin effects in intermediate-energy HICs [58, 59].

SPIN IN NUCLEAR REACTIONS

The spin Hall effect was first predicted by Dyakonov and Perel in 1971 [6, 7], while the term spin Hall effect was coined by

Hirsch in 1999 [5]. Given the transport of spinup and spin-down particles with spin–orbit coupling, i.e., $U^{so} = -L \cdot \sigma$ with L being the angular momentum and σ being the particle spin, the spin-up (spin-down) particles tend to turn left (right) to couple with the angular momentum and lower the energy, leading to the splitting of final observables for different spin states, as shown in Fig. 1. In this section, we will discuss similar effects in low- and intermediate-energy HICs based on the framework of TDHF, BUU, and QMD, with more complicated forms of spin–orbit coupling from spin-dependent nuclear interactions.

TDHF Model Study

The mean-field dynamics of nucleons in the TDHF model is described by

$$i\frac{\partial}{\partial t}\phi_i = h\phi_i, \tag{14}$$

where φ_i is the wave function of the ith nucleon and the single-nucleon Hamlitonian is given by

$$h\phi_i = \frac{\delta E}{\delta \phi_i^\star}, \tag{15}$$

with E being the energy functional of the nuclear system from the Hartree-Fock calculation. The single-nucleon Hamiltonian is generally a function of nucleon number density ρ, spin density s, current density j, spin-current density J, and so on, and their definitions in terms of the nucleon wave function are

$$\rho = \sum_i \phi_i^\star \phi_i, \tag{16}$$

$$s = \sum_i \sum_{\sigma,\sigma'} \phi_i^\star \langle\sigma|\sigma|\sigma'\rangle \phi_i, \tag{17}$$

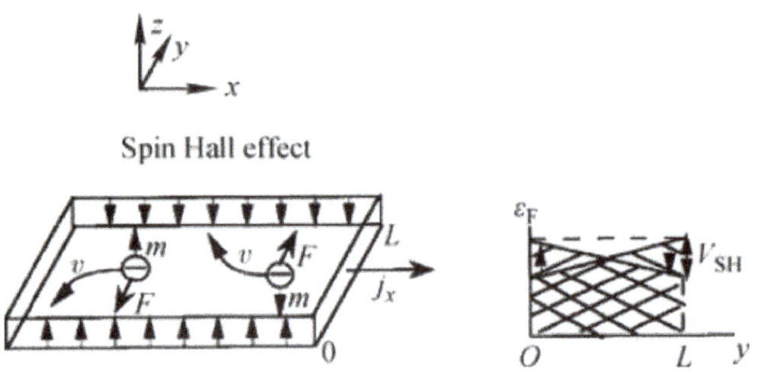

Figure. 1: Schematic of the spin Hall effect. Figure taken from Ref. [5].

$$j = \frac{1}{2i} \sum_i (\phi_i^\star \nabla \phi_i - \phi_i \nabla \phi_i^\star), \tag{19}$$

$$J = \frac{1}{2i} \sum_i \sum_{\sigma,\sigma'} (\phi_i^\star \nabla \phi_i - \phi_i \nabla \phi_i^\star) \times \langle\sigma|\sigma|\sigma'\rangle, \tag{20}$$

with <σ|σ|σ > being the Pauli matrix element. Numerically, these densities can be calculated on the coordinate space grid and Eq. (14) can be solved with a fixed time step. For more details, we refer the reader to Refs. [60, 61]. The TDHF framework works well for low-energy heavy-ion reactions and in studying resonances dynamics. In the old calculations, the single-nucleon Hamiltonian was generally calculated from the SHF model without the spin–orbit interaction and time-odd terms. The spin– orbit force was first introduced to the TDHF framework in Refs. [62–64]. It is interesting to see that the spin–orbit force enhances the dissipation in the fusion reaction and transforms the relative motions of the two nuclei into the internal excitations. The fusion threshold energy in the $O^{16} + O^{16}$ reaction is increased by about a factor of 2 [62, 63], as shown in Fig. 2 with three different parameterizations of the Skyrme force. The fusion cross section obtained from the TDHF calculation was increased after including the spin–orbit force [62].

With only a time-even contribution of the spin–orbit interaction, i.e., the spin–orbit potential [Eq. (2)] and the potential with spin-current density J [Eq. (10)], spurious spin twist can be generated in a free moving nucleus, as a result of spin–orbit coupling. Obviously, this phenomenon is not reasonable as it depends on the reference frame. By considering that all kinds of collision geometry can be realized in HICs, the time-odd terms were further introduced in the TDHF calculation in Refs. [65, 66] to satisfy the invariance under Galilei transformations. It is

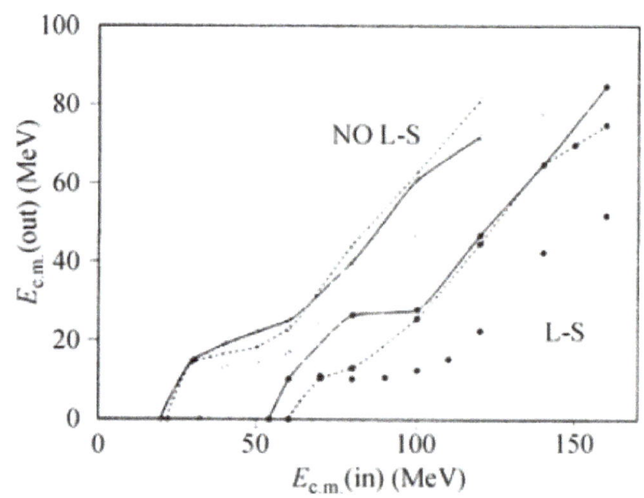

Figure. 2: Relation between the outgoing energy and incoming energy in the $O^{16} + O^{16}$ reaction from the TDHF study. The filled (open) circles are results with (without) the spin–orbit force, and the solid, dashed, and dotted curves represent results from three different Skyrme forces. Figure taken from Ref. [63].

seen from Fig. 3 that there is no such spurious spin and that the kinetic energy is a constant before 50 fm/c when the nuclei are moving freely, as a result of a suppression effect on the time-even terms from the time-odd terms. During the reaction process, the real spin twist appears owing to the overwhelming effect of the time-odd terms on the time-even terms, as shown in Fig. 4. At the end of the reaction, the energy of the outgoing nuclei is smaller with the time-odd terms, as shown in Fig. 3, indicating a stronger dissipation. Besides the spin excitation, it was found that the fusion description was further improved with the time-odd terms and the spin-current pseudotensor contribution [66]. A more detailed study on this topic done recently [67] showed that the dissipation is dominated by the time-even contribution of the spin–orbit force at lower energies but by the time-odd terms at higher energies. Besides the spin–orbit force, the additional contribution of the spin-current density J was introduced in the TDHF calculation to represent the contribution from the tensor force in Ref. [68].

It was found that the dissipation effect from the tensor force is small compared with that from the spin–orbit force [68, 69]. However, the spin

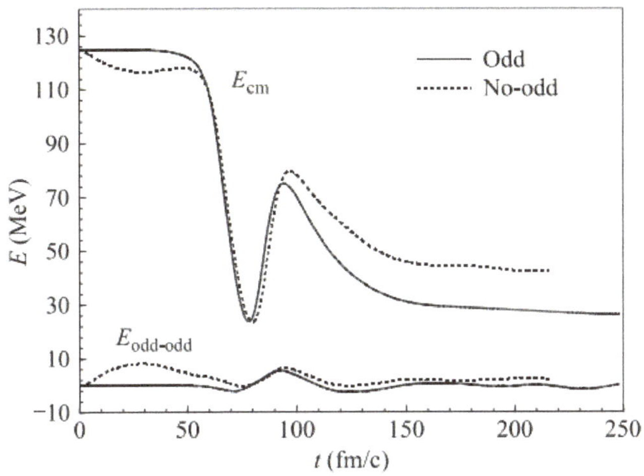

Figure. 3: Center-of-mass energy evolution in central O^{16} + O^{16} reactions with (solid lines) and without (dashed lines) time-odd contributions from the TDHF calculation. The time window of the reaction process is from about 50 to 120 fm/c. Figure taken from Ref. [65].

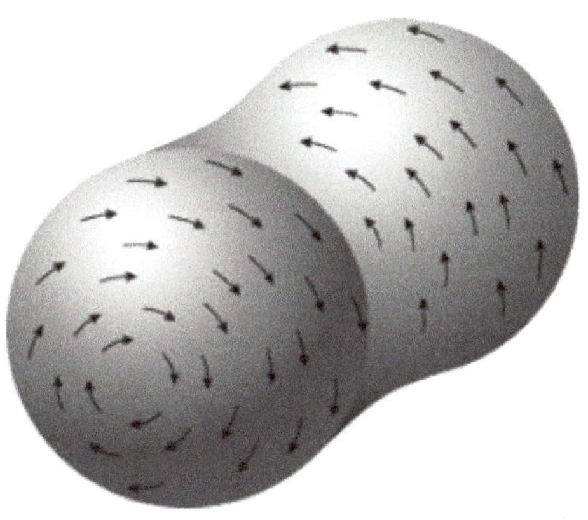

Figure. 4: Spin excitation in central O^{16} + O^{16} reactions from the TDHF study with both time-even and time-odd terms. Figure taken from Ref. [65].

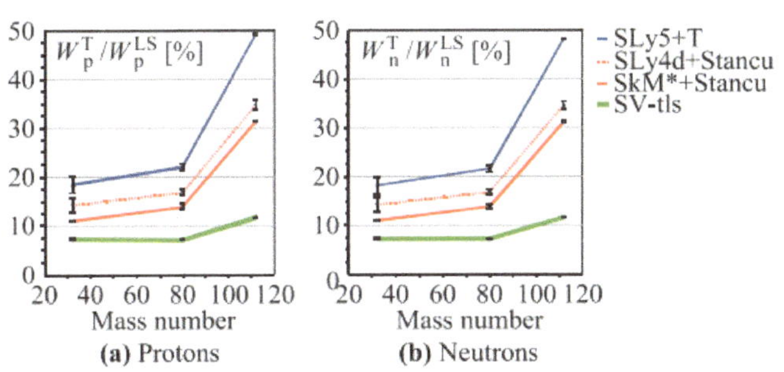

Figure. 5: Ratios of the spin mean field from the spin-current density representing the tensor force to that from the spin–orbit force for protons (a) and neutrons (b) as a function of the mass of the reaction nuclei. Different parameterizations of the Skyrme force are used in the study. Figure taken from Ref. [68].

mean field can be enhanced with the nuclear tensor force, and the enhancement becomes important with the increasing mass of the colliding nuclei, as shown in Fig. 5, depending on the parameterization of the Skyrme force. A more complete study including the full Skyrme functional as well as the tensor force in the TDHF calculation was done very recently in Ref. [70]. It was found that the Skyrme tensor force has non-negligible effects on low-energy heavy-ion dynamics and the fusion threshold energy.

BUU Model Study

The TDHF model works well in low-energy HICs, but the particle emission and nucleon–nucleon scattering are still lacking. To describe these effects in intermediate-energy HICs, BUU models and QMD models are suitable candidates. In the BUU framework, the Boltzmann equation is solved with the test particle method [71, 72]. In previous studies, an isospin-dependent BUU (IBUU) transport model has been used to describe the isospin dynamics in intermediate-energy HICs [3]. Recently, the spin degree of freedom of nucleons and the spin–orbit interaction were incorporated into the IBUU model, and the new model is dubbed as the spin- and isospin-dependent BUU (SIBUU) model [73–77]. In this section, we summarize the main results published originally in Refs. [73–77]. In the SIBUU model, each nucleon is assigned randomly a unit vector representing the expectation value of its spin. In this way, the spin projection of each nucleon in an arbitrary direction can be easily calculated. In the transport simulation, the z direction is set as the beam momentum and the x direction is for the impact parameter. Since the total angular momentum in noncentral HICs is in the y direction perpendicular to the reaction plane, i.e., the x-O-z plane, it is reasonable to study the spin polarization in the y direction. We thus determine the nucleons with spin projection on +y (−y) direction as the spin-up (spin-down) nucleons. Considering the general form of the time-even and time-odd spin-dependent potentials in Eqs. (11) and (12), one can describe the time evolutions of the coordinate, momentum, and spin degrees of freedom by

$$\frac{d\boldsymbol{r}}{dt} = \frac{\boldsymbol{p}}{m} + \frac{W_0}{2}\left(\frac{\rho}{\rho_0}\right)^{\gamma} \boldsymbol{\sigma} \times (a\nabla\rho_q + b\nabla\rho_{q'}) - \frac{W_0}{2}\left(\frac{\rho}{\rho_0}\right)^{\gamma} \nabla \times (a\boldsymbol{s}_q + b\boldsymbol{s}_{q'}), \tag{20}$$

$$\frac{d\boldsymbol{p}}{dt} = -\nabla U_q - \nabla U_q^{s-even} - \nabla U_q^{s-odd}, \tag{21}$$

$$\frac{d\boldsymbol{\sigma}}{dt} = W_0\left(\frac{\rho}{\rho_0}\right)^{\gamma} [(a\nabla\rho_q + b\nabla\rho_{q'}) \times \boldsymbol{p}] \times \boldsymbol{\sigma} - W_0\left(\frac{\rho}{\rho_0}\right)^{\gamma} [\nabla \times (a\boldsymbol{j}_q + b\boldsymbol{j}_{q'})] \times \boldsymbol{\sigma}. \tag{22}$$

One sees that the three degrees of freedom couple with each other. The number density ρ, the spin density s, the current density j, and the spin-current density J are calculated from the test particle method [71, 72, 74]. Because the mixing of the long-range Fock contribution and the spin interaction is a complex problem, the momentum dependence is not included in the spin-independent potential Uq for the moment. In addition, the spins of nucleons are randomized after nucleon–nucleon scattering, by approximately taking the spin flip effect into consideration [78, 79]. The time evolutions of the relevant density contours from the SIBUU calculation are displayed in Fig. 6. The gradient of number density, ∇ρ, and the curl of the current density, j, show the strength of the time-even and time-odd spin-dependent potential, respectively, and both of them are closely related to the evolution of the number density shown in the first row of Fig. 6. The nucleon spin tends to be parallel to p × ∇ρ from the time-even potential [Eq. (11)], whereas it tends to be parallel to ∇ × j from the time-odd potential [Eq. (12)]. The contributions from the time-even and time-odd potentials are opposite to each other. One sees that before the two nuclei touch each other there is no spin polarization as a result of the cancellation of the time-even and time-odd potentials, consistent with the findings from TDHF studies. During the collision process, the participant is polarized in the +y direction, i.e., in the direction of the total angular momentum, following the preference direction of the time-odd potential. It is seen that the direction of the spin polarization is consistent with that in Fig. 4 from the TDHF calculation with both time-even and time-odd potentials.

Transverse flow is one of the most important observ ables for extracting the equation of state of produced matter and studying the nuclear interaction in HICs [1, 72, 80]. The left panel of Fig. 7 displays the transverse flow of spin-up and spin-down nucleons as a function of reduced rapidity y_r/y_r^{beam}. We note that the target (projectile) nucleus is in the +x (−x) direction in Fig. 6, which is different from the conventional initialization, leading to the negative slope of the transverse flow. However, this does not prevent the reader from seeing the obvious splitting of transverse flow between spin-up and spin-down nucleons. With a detailed orientation analysis, one can find that again the time-odd potential dominates the effect, giving the spin-up (spin-down) nucleons an attractive (repulsive) potential. This can be understood in a naive picture in which the spin-up (spin-down) nucleons are parallel (antiparallel) to the direction of total angular momentum and thus feel an attractive (repulsive) potential. One can further define the spin up-down

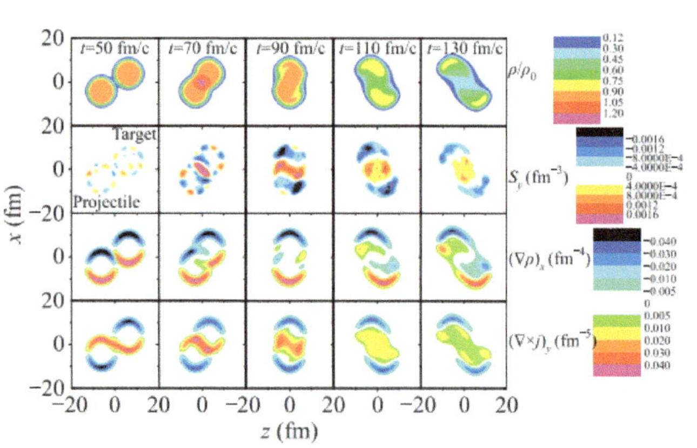

Figure. 6: Time evolution of contours of the reduced number density, ρ/ρ0, y component of the spin density, sy, x component of the number density gradient, (∇_ρ)x, and y component of the curl of the current density, $(\nabla \times_j)$y, in noncentral Au + Au collisions at a beam energy of 50 MeV. Figure taken from Ref. [73].

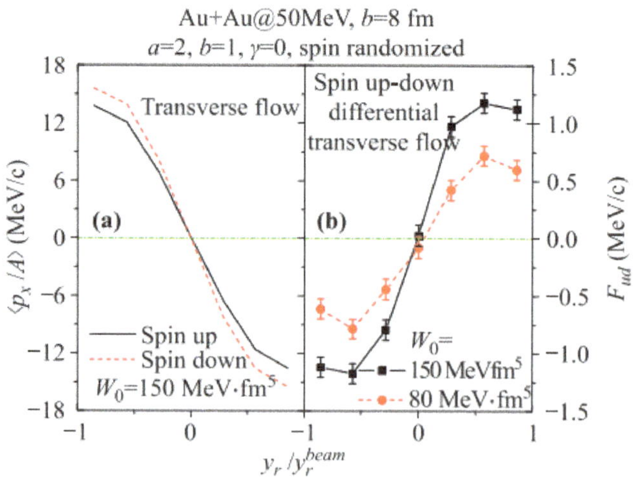

Figure. 7: Transverse flow of spin-up and spin-down nucleons (a) and the spin up-down differential transverse flow (b) with different strength of the spin–orbit coupling in noncentral Au + Au collisions at a beam energy of 50 MeV. Figure taken from Ref. [73].

differential transverse flow as follows:

$$F_{ud}(y_r) = \frac{1}{N(y_r)} \sum_{i=1}^{N(y_r)} \sigma_i (p_x)_i, \tag{23}$$

where σ_i is 1 for spin-up nucleons and −1 for spin-down nucleons, and $N(y_r)$ is the number of nucleons at rapidity y_r. The above spin up-down differential transverse flow largely cancels the effect from the spin-independent nuclear interaction while preserving the information of the spin-dependent potential. Indeed, the slope of Fud increases with increasing spin–orbit coupling constant, indicating that it is a good probe of the nuclear spindependent interaction. The spin up-down differential transverse flow was further analyzed in detail in Ref. [74]. Figure 8 displays the dependence of F_{ud} on the beam energy and the centrality. At higher beam energies, the angular momentum is larger while the nucleon–nucleon scattering is more violent, with the former enhancing the spin-dependent potential and the latter washing out part of the information of spin dynamics. The competition leads to a maximum slope of F_{ud} at a beam energy of ~100 MeV, as shown in the left panel of Fig. 8. Since the spin-dependent potential is related to the density gradient and is thus a surface effect, the slope of F_{ud} increases with the increasing value of the impact parameter, as shown in the right panel of Fig. 8.

In the neutron-rich collision system where the relevant neutron densities are larger than proton densities, the difference of the spin up-down differential transverse flow of neutrons and protons can be useful to probe the isospin dependence of the spin–orbit coupling in HICs. The analysis was performed with a stronger isospin-like coupling (a = 2, b = 1) and a stronger isospin-unlike coupling (a = 1, b = 2), and the resulting F_{ud} values were calculated at different beam energies as shown in Fig. 9. A stronger isospin-like spin–orbit coupling, which

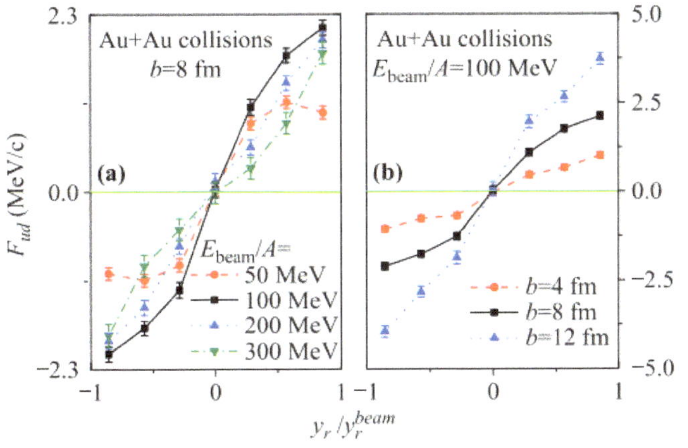

Figure. 8: Dependence of the spin up-down differential transverse flow on the beam energy (a) and the impact parameter (b) in noncentral Au + Au collisions. Figure taken from Ref. [74].

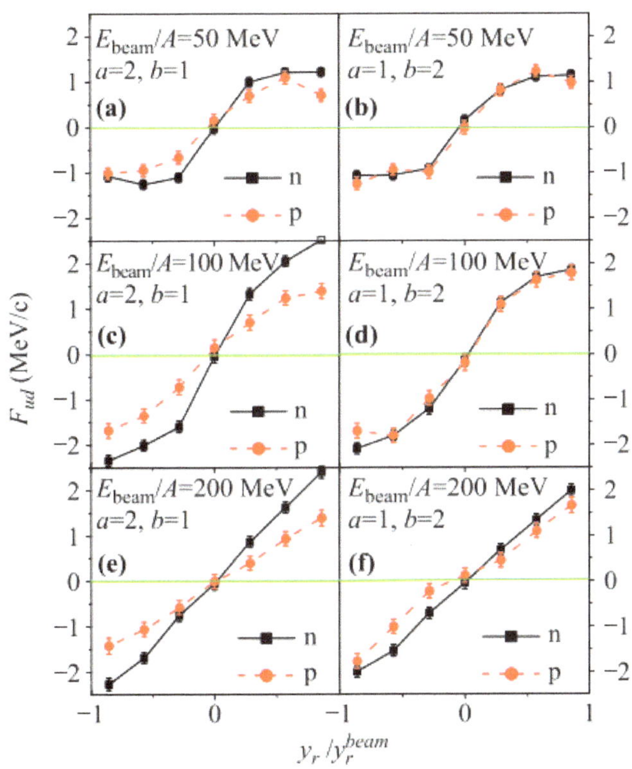

Figure. 9: Spin up-down differential transverse flow of neutrons and protons at different beam energies and with two typical isospin dependencies of the spin–orbit coupling. Figure taken from Ref. [74].

is exactly the case of the SHF interaction, leads to a larger F_{ud} for neutrons than for protons, while a stronger isospin-unlike coupling gives opposite predictions or similar F_{ud} for neutrons and protons. The effect is appreciable from beam energies of 50 to 200 MeV, with a beam energy of 100 MeV being optimal because it gives the largest magnitude of F_{ud}.

The density dependence of the spin–orbit coupling has bothered many nuclear physicists and hampered our understanding of the nuclear spin–orbit interaction in nuclear structure studies. Because HICs enable us to construct systems with designed density, isospin, and momentum current, they might be helpful in extracting useful information of the density dependence of the spin– orbit coupling. As is known, nucleons of high transverse momentum (p_T) are emitted early from the high-density phase in HICs, and the density of the high-density phase increases with increasing beam energy. This feature can be used to extract the density dependence of the spin– orbit coupling, as illustrated in Fig. 10. Without a highpT cut, the slope of F_{ud} can hardly be distinguished, as shown in the left panel of Fig. 10, because nucleon emission from the low-density phase, which is similar at different beam energies, dominates the results. With a high-p_T cut, the slope of F_{ud} is smaller at lower collision energies but larger at higher collision energies from a linearly increasing spin–orbit coupling strength, compared to the case with a constant one. In this way, the strength and the density dependence of the spin–orbit coupling can be disentangled.

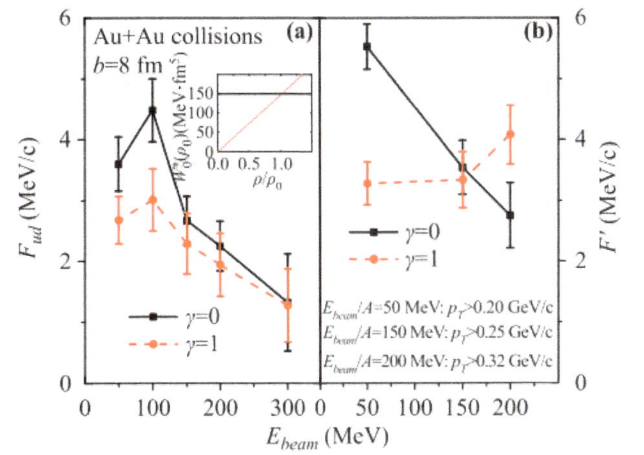

Figure. 10: Slope parameter of the spin up-down differential transverse flow F without (a) and with (b) high transverse momentum cut from different density dependencies of the spin–orbit coupling at different beam energies. Figure taken from Ref. [74].

In noncentral HICs, the azimuthal distribution of emitted nucleons can always be expressed as

$$E\frac{d^3N}{dp^3} = \frac{d^2N}{2\pi p_T dp_T dy_r}[1 + 2v_1(y_r, p_T)\cos(\phi) + 2v_2(y_r, p_T)\cos(2\phi) + \cdots] \quad (24)$$

with $\phi = \arctan(p_y/p_x)$ being the azimuthal angle, and $v_1 = \langle\cos(\phi)\rangle$ and $v_2 = \langle\cos(2\phi)\rangle$ being called the directed flow and elliptic flow, respectively. The directed flow is similar to the transverse flow, but it depends on the flow angle rather than magnitude. The elliptic flow is positive at lower energies, is negative at intermediate energies, and becomes positive again at higher energies. Positive elliptic flow means more particles move in-plane than out-of-plane as a result of hydrodynamics, whereas negative elliptic flow is a result of the squeeze-out effect on the expansion of participant matter by the spectator nucleons [1]. Despite the complicated dynamics, the elliptic flow serves as a useful probe of the properties of nuclear matter formed in HICs and the nuclear interaction. The transverse momentum dependence of v_2 of spin-up and spin-down nucleons at mid-rapidity is displayed in Fig. 11. Except for the different behaviors of v_2 at different beam energies, a larger elliptic flow of spin-up nucleons than spin-down nucleons is observed, especially at higher transverse momentum, as a result of the stronger spin–orbit coupling for energetic nucleons. At the energy range considered, a more attractive mean- field potential leads to a larger v_2 in peripheral HICs, consistent with the effect of a spin-dependent potential on the spin splitting of transverse flow discussed above. The above observables are for free nucleons. Experimentally, it is easier to detect charged particles rather than neutrons, leading to difficulties of measuring the spin splitting of transverse flows for protons and

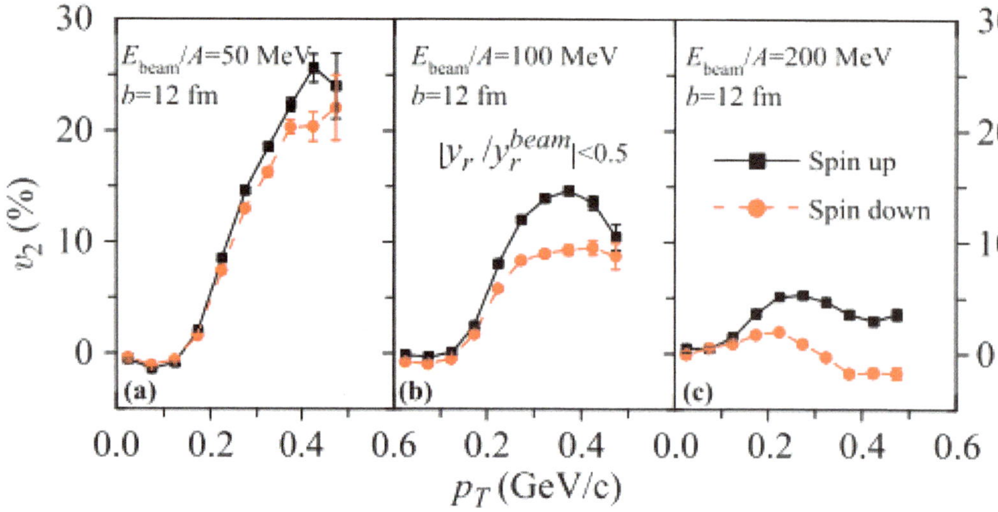

Figure. 11: Transverse momentum dependence of the elliptic flow of mid-rapidity nucleons in peripheral Au + Au collisions at different beam energies. Figure taken from Ref. [75].

neutrons and identifying the isospin dependence of the spin–orbit coupling. Of course, the spin measurement poses another challenge, as will be discussed in the next section. Once the corresponding detectors are set up, the spin splitting of observables for charged light clusters may be more easily measured. For transport models with point-like particles, the dynamical coalescence approach has been shown to be successful in studying the hadronization in relativistic HICs [81, 82] and light cluster formation in intermediate-energy HICs [83, 84]. In this approach, the probability for nucleons to form a light cluster is proportional to the nucleon Wigner function of the light cluster [83, 84], and the proportional constant is the statistical factor determined by the spinisospin degeneracy. For example, with explicit knowledge of the isospin of nucleons, the statistical factor for a neutron and a proton to form a deuteron is 3/8, whereas that for one neutron and two protons to form a ^{3}He nucleus is 1/12. Because the spin of each nucleon is also explicitly known, the dynamical coalescence can be further improved by considering the antisymmetrization of the product of spin and isospin wave function. For example, the statistical factor for a spin-up neutron and a spin-up proton to form a spin-up deuteron is 1/2, whereas for a spin-up neutron, a spin-up proton, and a spin-down proton to form a spin-up ^{3}He nucleus it is 1/2. This improvement has been applied to study spin splitting observables for deuterons, tritons, and ^{3}He nuclei [85]. It has been checked that, after spin averaging, the results reproduce those without an explicit spin treatment.

Figure 12 displays the spin splitting of the directed flows for deuterons, tritons, and ^{3}He nuclei in noncentral Au + Au collisions at a beam energy of 100 MeV. The directed flow of spin-down clusters is larger than that of spin-up ones. The spin splitting of the directed flow is largest for the deuteron owing to its large spin quantum number, i.e., S = 1. The spin splitting observables of tritons and ^{3}He nuclei might be more easily measurable for extracting the isospin dependence of the spin–orbit

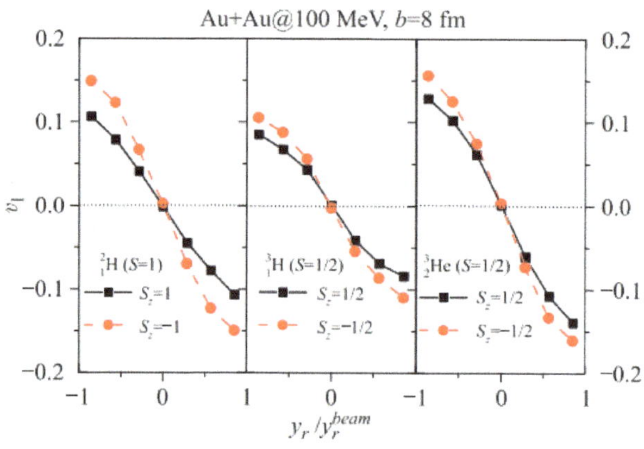

Figure. 12: Directed flow of deuterons, tritons, and ^{3}He nuclei of different spin states in noncentral Au + Au collisions at a beam energy of 100 MeV. s_z represents the spin state projecting on the y direction perpendicular to the reaction plane. Figure taken from Ref. [85].

coupling experimentally. The elliptic flow of different spin states of deuterons in noncentral Au + Au collisions at a beam energy of 100 MeV is illustrated in Fig. 13. It is seen that the elliptic flow of spin-down deuterons is more negative at mid-rapidity, but it is slightly positive at large rapidity, indicating an obvious spin splitting even with the statistical error taken into account. Again, the magnitude of v_2 as well as its spin splitting for deuterons is larger than that of free nucleons according to Ref. [75], and it might serve as a better spin-dependent observable. A further preliminary calculation with the full Skyrme functional has been done. A standard Skyrme functional with MSL0 parameterization [86] has been used in the calculation. The detailed derivation and expression of the full Skyrme functional with both time-even and time-odd terms can be found in Refs. [27, 29, 30]. The resulting spin up-down differential transverse flow is shown in the left panel of Fig. 14. One can see the similar sensitivity of F_{ud} to the spin–orbit coupling strength, although the

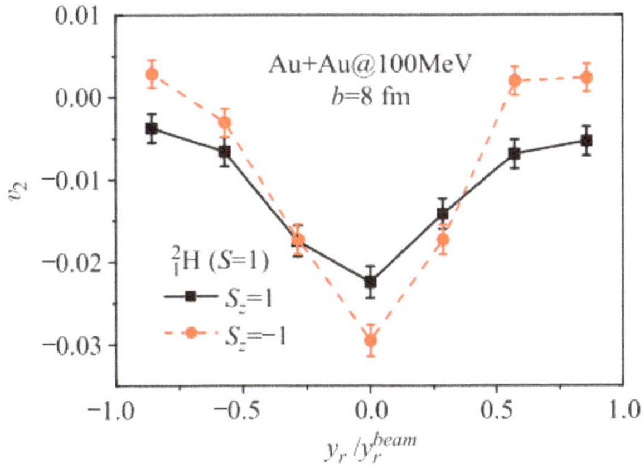

Figure. 13: Elliptic flow of spin-up and spin-down deuterons in noncentral Au + Au collisions at a beam energy of 100 MeV. s_z represents the spin state projecting on the y direction perpendicular to the reaction plane. Figure taken from Ref. [85].

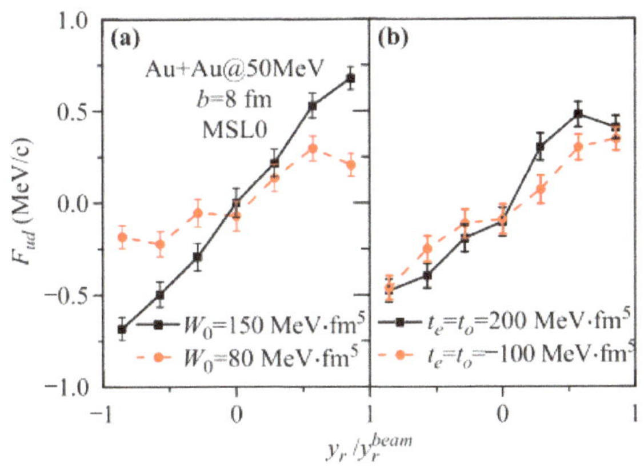

Figure. 14: Spin up-down differential transverse flow from full Skyrme calculation using the MSL0 force without (a) and with the tensor force (b). Figure taken from Ref. [85].

magnitude is a little smaller, compared to the result shown in Fig. 7 where only the spin–orbit coupling is applied. To investigate the effect of the nuclear tensor force on the spin dynamics of intermediate-energy HICs, a zero-range tensor force of the form

$$v_t(\boldsymbol{r}) = \frac{t_e}{2}\{[3(\boldsymbol{\sigma}_1\cdot\boldsymbol{k}')(\boldsymbol{\sigma}_2\cdot\boldsymbol{k}') - (\boldsymbol{\sigma}_1\cdot\boldsymbol{\sigma}_2)k'^2]\delta(\boldsymbol{r}) + \delta(\boldsymbol{r})[3(\boldsymbol{\sigma}_1\cdot\boldsymbol{k})(\boldsymbol{\sigma}_2\cdot\boldsymbol{k}) - (\boldsymbol{\sigma}_1\cdot\boldsymbol{\sigma}_2)k^2]\} + t_o[3(\boldsymbol{\sigma}_1\cdot\boldsymbol{k}')\delta(\boldsymbol{r})(\boldsymbol{\sigma}_2\cdot\boldsymbol{k}) - (\boldsymbol{\sigma}_1\cdot\boldsymbol{\sigma}_2)\boldsymbol{k}'\cdot\delta(\boldsymbol{r})\boldsymbol{k}] \quad (25)$$

is incorporated into the full Skyrme transport model calculation, where $r = r_1 - r_2$ is the relative coordinate, k and k are the relative momentum operator and its complex conjugate, respectively, and t_e and to are the triplet-even and triplet-odd strength parameters, respectively. The energy-density functional derived from the above tensor force can be found in Refs. [27, 29], where the corresponding terms (such as the spin-current density J) are non-negligible only when local spin polarization is produced. The resulting spin up-down differential transverse flow is shown in the right panel of Fig. 14. It is seen that the slope of F_{ud} is not very sensitive to the values of t_e or to unless an extremely large coupling constant is used. This feature is consistent with the TDHF study where the spin dynamics is dominated by the spin–orbit coupling. However, one would expect that, with a spinpolarized beam or target, the tensor force effect can be much enhanced.

QMD Model Study

In the QMD framework, the Wigner function of each nucleon is treated as a Gaussian wave packet in both coordinate and momentum space [87, 88], and the twonucleon interaction is related to the effective two-body interaction and the overlap of their wave functions. The equation of motion in the QMD model is given by the semiclassical canonical equation, i.e.,

$$\frac{d\boldsymbol{r}}{dt} = \nabla_p H,$$
$$\frac{d\boldsymbol{p}}{dt} = -\nabla_r H, \quad (26)$$

where r and p are, respectively, the central coordinate and momentum of the wave packet, and H is the Hamiltonian of the system including the kinetic and potential energy. In a recent study, the nuclear spin–orbit interaction was incorporated into the ultra-relativistic QMD (UrQMD) model. The potential energy contribution of the spin–orbit interaction is expressed as [89]

$$U_s = \int u_s d^3 r, \quad (27)$$

where the spin-dependent potential u_s consists of timeeven and time-odd contributions written as

$$u_s^{even} = -\frac{W_0}{2}(\rho\nabla\cdot\boldsymbol{J} + \rho_n\nabla\cdot\boldsymbol{J}_n + \rho_p\nabla\cdot\boldsymbol{J}_p), \quad (28)$$

$$u_s^{odd} = -\frac{W_0}{2}[\boldsymbol{s}\cdot(\nabla\times\boldsymbol{j}) + \boldsymbol{s}_n\cdot(\nabla\times\boldsymbol{j}_n) + \boldsymbol{s}_p\cdot(\nabla\times\boldsymbol{j}_p)], \quad (29)$$

where W0 represents the spin–orbit coupling strength, and ρ, s, j, and J are the number, spin, current, and spin-current densities, which can be calculated from the local Wigner function of the nucleon [89]. The spin dynamics was analyzed based on the above framework. Similar spin splittings of the directed flow and the elliptic flow were observed in noncentral Au + Au collisions at a beam energy of 150 MeV, as shown in Fig. 15. It was argued that the net spin-dependent

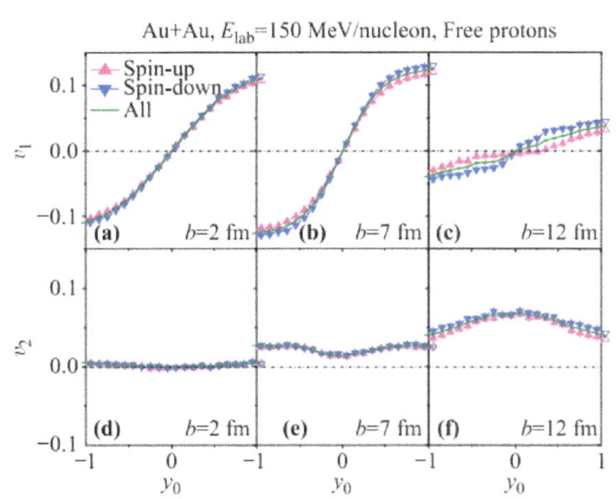

Figure. 15: The directed flow (upper panels) and elliptic flow (lower panels) for spin-up and spin-down protons in noncentral Au + Au collisions at a beam energy of 150 MeV from the QMD calculation. Figure taken from Ref. [89].

potential is attractive for spin-up protons and repulsive for spin-down protons, leading to a larger directed flow for spin-down protons than spin-up protons. The spin splitting of pT -integrated elliptic flow was found to be small and only visible in peripheral collisions, and it was found that v_2 for spin-down protons is slightly larger than that for spin-up ones. Since the conventional initial direction of the target and projectile is used as shown in Fig. 3 of Ref. [89], the spin-up (spin-down) nucleons correspond to the spin-down (spin-up) ones in the SIBUU study [73–77]. Although the spin splitting of final collective flow is a robust phenomenon in both models, further studies are needed to understand the relative sign of the splitting. Further analysis was done on the beam energy dependence of the flow splitting. By defining κ_{up} and κ_{down} as the slope parameter of the directed flow of spin-up and spin-down protons, respectively, the slope difference is shown to increase with increasing impact parameter, as shown in the upper panel of Fig. 16; this is qualitatively consistent with SIBUU studies. In noncentral Au + Au collisions, it was found that the slope difference first increases then decreases with increasing beam energy, and the maximum difference appears at a beam energy of 150 MeV, similar to the finding in the SIBUU model where

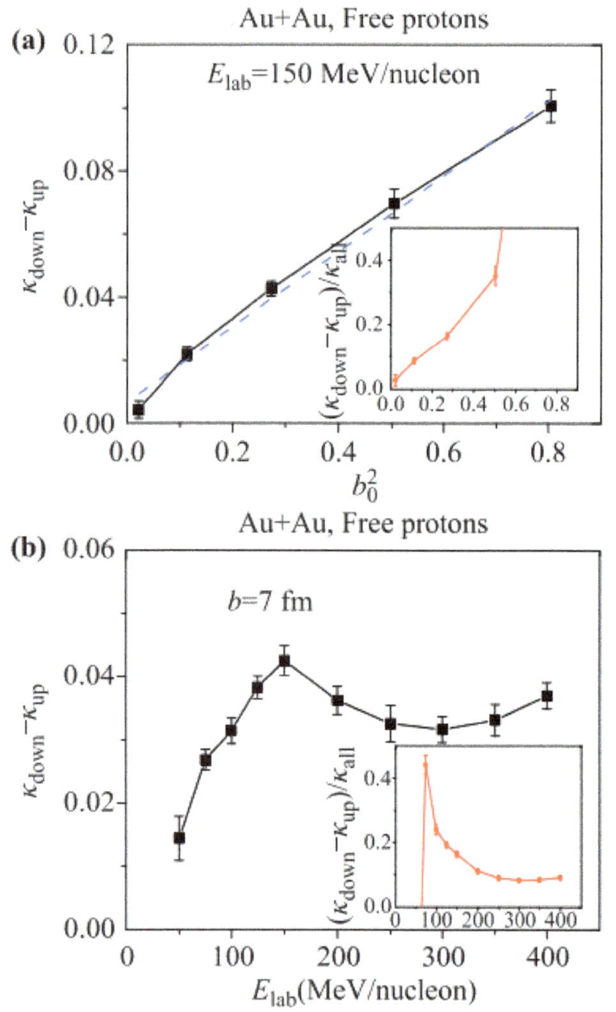

Figure. 16: Difference of the slope of directed flow between spin-up and spin-down protons as a function of impact parameter (upper panel) and beam energy (lower panel). The dashed line in the upper panel is a linear fit, while the inset in the lower panel shows the relative difference. Figure taken from Ref. [89].

the maximum slope of the spin up-down differential transverse flow appears at a beam energy of about ~100 MeV. It was further emphasized in Ref. [89] that the spinaveraged flow results do not change after including the spin–orbit interaction, as a result of the cancellation of spin-up and spin-down nucleons. In addition, the spin splitting of the flow slope caused by the spin–orbit interaction is comparable to the isospin splitting caused by the nuclear symmetry energy, especially for neutrons. These findings are all consistent with the observations in SIBUU studies [73–77].

EXPERIMENTAL STATUS

Owing to the difficulties of spin measurement in HIC experiments, the main focus in the past has been mainly on the spin-averaged observables, and information on spin dynamics has been neglected. Because of the great efforts made by experimental nuclear physicists, making measurements of the spin of free nucleons and light clusters has now become possible. Although the detailed experimental status will be presented in another review of this issue, here we briefly mention two related experiments that might be relevant in analyzing the probes discussed above. One of them is the spin-polarized beam that can be produced through pick-up or removal reactions at Rikagaku Kenkyusho (RIKEN) [90, 91], Gesellschaft f'ur Schwerionenforschung mbH (GSI) [92], the National Superconducting Cyclotron Laboratory (NSCL) [93], and the Grand Acc´el´erateur National d'Ions Lourds (GANIL) [94–96]. It is expected that the effects of spin dynamics with a spin-polarized beam will be much enhanced, providing a better system for extracting information on the spin-dependent nuclear force, especially the nuclear tensor force. For the spin-excitation state of heavy clusters, the spin polarization and alignment can be measured via the angular distribution of its γ or β decay (see, e.g., Ref. [97] for a review). Making use of the analyzing power of a nucleus might be the most promising way of identifying the spin of free nucleons or light clusters experimentally. The analyzing power indicates the left-right scattering asymmetry of an incident polarized nucleon on the target nucleus. The spin-dependent scattering is a result of the interference of the electromagnetic interaction and the hadronic force [98], and the spin flip is observed between not only charged-charged scattering but also between charged-neutral scattering. It is noteworthy that at certain energies and scattering angles the analyzing power can be as large as 100% [99]. Experimental efforts are thus encouraged by using the selected nucleus as a "detector" whose analyzing power is known a priori. In this way the spin of corresponding particles can be measured and the probes discussed in the previous sections can be analyzed.

SUMMARY

In summary, we outlined the major physics motivations for investigating in-medium spin-dependent nuclear interactions, i.e., the spin–orbit interaction and the nuclear tensor force, and summarized some recent efforts in exploring the spin dynamics in low- and intermediateenergy heavy-ion collisions. In particular, studies on the strength, the density, and the isospin dependence of the spin–orbit interaction as well as the short-range correlation induced by the tensor force are highlighted. In TDHF studies, it has been found that the spin–orbit interaction can enhance the dissipation in low-energy heavy-ion reactions and increase the fusion threshold. Incorporating both the time-even and time-odd contribution of the spin–orbit interaction can lead to nontrivial spin polarization, while the tensor force slightly enhances the spin field compared to the spin–orbit interaction. In studies using HICs at intermediate energies, both the spin- and isospin-dependent BUU models and QMD model predict the spin splitting of the nucleon collective flow, which seems to be a robust phenomenon. In the BUU model studies, efforts have been made in extracting the isospin dependence of the spin–orbit coupling and disentangling its strength and density dependence. Preliminary results on spin splitting of observables related to the light clusters and those from the full Skyrme calculation with a nuclear tensor force have also been discussed in the BUU model studies. We hope that the findings summarized in this review will soon stimulate more experimental work in this direction.

ACKNOWLEDGEMENTS

This work was supported by the Major State Basic Research Development Program (973 Program) of China under Contract Nos. 2015CB856904 and 2014CB845401, the National Natural Science Foundation of China under Grant Nos. 11320101004, 11475243, and 11421505, the "100-Talent Plan" of Shanghai Institute of Applied Physics under Grant Nos. Y290061011 and Y526011011 from the Chinese Academy of Sciences, the "Shanghai Pujiang Program" under Grant No. 13PJ1410600, the U.S. National Science Foundation under Grant No. PHY-1068022, the U.S. Department of Energy Office of Science under Award No. DE-SC0013702, and the CUSTIPEN (China–U.S. Theory Institute for Physics with Exotic Nuclei) under the U.S. Department of Energy Grant No. DE-FG02-13ER42025.

REFERENCES

1. P. Danielewicz, R. Lacey, and W. G. Lynch, Determinationof the equation of state of dense matter, Science 298(5598), 1592 (2002)
2. V. Baran, M. Colonna, V. Greco, and M. Di Toro, Reaction dynamics with exotic nuclei, Phys. Rep. 410(5–6), 335 (2005)
3. B. A. Li, L. W. Chen, and C. M. Ko, Recent progress and new challenges in isospin physics with heavy-ion reactions, Phys. Rep. 464(4–6), 113 (2008)
4. H. Liang, J. Meng, and S. G. Zhou, Hidden pseudospin and spin symmetries and their origins in atomic nuclei, Phys. Rep. 570, 1 (2015)
5. J. E. Hirsch, Spin Hall effect, Phys. Rev. Lett. 83(9), 1834 (1999)
6. M. I. Dyakonov and V. I. Perel, Possibility of orientating electron spins with current, Sov. Phys. JETP Lett. 13, 467 (1971)
7. M. I. Dyakonov and V. I. Perel, Current-induced spin orientation of electrons in semiconductors, Phys. Lett. A 35(6), 459 (1971)
8. Lecture Notes of R. Machleidt, CNS Summer School, University of Tokyo, Aug. 18–23, 2005
9. M. G. Mayer, On closed shells in nuclei, Phys. Rev. 74(3), 235 (1948) M. G. Mayer, On closed shells in nuclei (II), Phys. Rev. 75(12), 1969 (1949)
10. O. Haxel, J. H. D. Jensen, and H. E. Suess, On the "magic numbers" in nuclear structure, Phys. Rev. 75(11), 1766 (1949)
11. D. Vautherin and D. M. Brink, Hartree–Fock calculations with Skyrme's interaction (I): Spherical nuclei, Phys. Rev. C 5(3), 626 (1972)
12. P. Ring, Relativistic mean field theory in finite nuclei, Prog. Part. Nucl. Phys. 37, 193 (1996)
13. M. Bender, P. Heenen, and P.-G Reinhard, Self-consistentmean-field models for nuclear structure, Rev. Mod. Phys. 75, 121 (2003)
14. P. G. Reinhard, The relativistic mean-field description of nuclei and nuclear dynamics, Rep. Prog. Phys. 52(4), 439 (1989)
15. A. Sulaksono, T. B¨urvenich, J. A. Maruhn, P. G. Reinhard, and W. Greiner, The nonrelativistic limit of the relativistic point coupling model, Ann. Phys. 308(1), 354 (2003)
16. J. M. Pearson and M. Farine, Relativistic mean-field theory and a density-dependent spin–orbit Skyrme force, Phys.Rev. C 50(1), 185 (1994)
17. B. G. Todd-Rutel, J. Piekarewicz, and P. D. Cottle, Spinorbitsplitting in low-j neutron orbits and proton densitiesin the nuclear interior, Phys. Rev. C 69, 021301 (2004)
18. M. Grasso, L. Gaudefroy, E. Khan, T. Nikˇsi´c, J. Piekarewicz,O. Sorlin, N. V. Giai, and D. Vretenar, Nuclear "bubble" structure in Si34, Phys. Rev. C 79(3), 034318 (2009)
19. O. Sorlin and M. G. Porquet, Evolution of the N = 28 shell closure: A test bench for nuclear forces, Phys. Scr. T152, 014003 (2013)
20. P. G. Reinhard and H. Flocard, Nuclear effective forces and isotope shifts, Nucl. Phys. A 584(3), 467 (1995)
21. M. M. Sharma, G. Lalazissis, J. K¨onig, and P. Ring, Isospindependence of the spin–orbit force and effective nuclear potentials, Phys. Rev. Lett. 74(19), 3744 (1995)
22. M. Onsi, R. C. Nayak, J. M. Pearson, H. Freyer, and W. Stocker, Skyrme representation of a relativistic spin–orbit field, Phys. Rev. C 55(6), 3166 (1997)
23. J. M. Pearson, Skyrme Hartree–Fock method and the spin– orbit term of the relativistic mean field, Phys. Lett. B 513(3– 4), 319 (2001)
24. G. A. Lalazissis, D. Vretenar, W. P¨oschl, and P. Ring, Reductionof the spin–orbit potential in light drip-line nuclei,Phys. Lett. B 418(1–2), 7 (1998)
25. B. S. Pudliner, A. Smerzi, J. Carlson, V. R. Pandharipande, S. C. Pieper, and D. G. Ravenhall, Neutron drops and Skyrme energy-density functionals, Phys. Rev. Lett. 76(14), 2416 (1996)
26. J. P. Schiffer, S. J. Freeman, J. A. Caggiano, C. Deibel, A. Heinz, C.L. Jiang, R. Lewis, A. Parikh, P. D. Parker, K. E. Rehm, S. Sinha, and J. S. Thomas, Is the nuclear spin–orbit interaction changing with neutron excess? Phys. Rev. Lett. 92(16), 162501 (2004) [Erratum: Phys. Rev. Lett., 110,169901 (2013)]

27. T. Lesinski, M. Bender, K. Bennaceur, T. Duguet, and J.Meyer, Tensor part of the Skyrme energy density functional: Spherical nuclei, Phys. Rev. C 76(1), 014312 (2007)
28. M. Zalewski, J. Dobaczewski, W. Satula, and T. R. Werner, Spin–orbit and tensor mean-field effects on spin–orbit splitting including self-consistent core polarizations, Phys. Rev.C 77(2), 024316 (2008)
29. M. Bender, K. Bennaceur, T. Duguet, P. H. Heenen, T.Lesinski, and J. Meyer, Tensor part of the Skyrme energydensity functional (II): Deformation properties of magic and semi-magic nuclei, Phys. Rev. C 80(6), 064302 (2009)
30. Y. M. Engel, D. M. Brink, K. Goeke, S. J. Krieger, and D. Vautherin, Time-dependent Hartree–Fock theory with Skyrme's interaction, Nucl. Phys. A 249(2), 215 (1975)
31. T. H. R. Skyrme, The effective nuclear potential, Nucl. Phys.9(4), 615 (1958)
32. T. Otsuka, T. Suzuki, R. Fujimoto, H. Grawe, and Y. Akaishi, Evolution of nuclear shells due to the tensor force, Phys. Rev. Lett. 95(23), 232502 (2005)
33. T. Otsuka, T. Matsuo, and D. Abe, Mean field with tensorforce and shell structure of exotic nuclei, Phys. Rev. Lett.97(16), 162501 (2006)
34. T. Otsuka, T. Suzuki, M. Honma, Y. Utsuno, N. Tsunoda,K. Tsukiyama, and M. Hjorth-Jensen, Novel features of nuclear forces and shell evolution in exotic nuclei, Phys. Rev. Lett. 104(1), 012501 (2010)
35. L. Gaudefroy, O. Sorlin, D. Beaumel, Y. Blumenfeld, Z. Dombr´adi, S. Fortier, S. Franchoo, M. G´elin, J. Gibelin, S. Gr´evy, F. Hammache, F. Ibrahim, K. W. Kemper, K. L. Kratz, S. M. Lukyanov, C. Monrozeau, L. Nalpas, F. Nowacki, A. N. Ostrowski, T. Otsuka, Y.E. Penionzhkevich, J. Piekarewicz, E. C. Pollacco, P. Roussel-Chomaz, E. Rich, J. A. Scarpaci, M. G. St. Laurent, D. Sohler, M. Stanoiu, T. Suzuki, E. Tryggestad, and D. Verney, Reduction of the spin–orbit splittings at the N= 28 shell closure, Phys. Rev. Lett. 97(9), 092501 (2006)
36. G. Col`o, H. Sagawa, S. Fracasso, and P. F. Bortignon, Spin– orbit splitting and the tensor component of the Skyrme interaction, Phys. Lett. B 646(5–6), 227 (2007) [Erratum: Phys. Lett. B, 668, 457 (2008)].
37. L. G. Cao, G. Col`o, H. Sagawa, P. F. Bortignon, and L. Sciacchitano, Effects of the tensor force on the multipole response in finite nuclei, Phys. Rev. C 80(6), 064304 (2009)
38. C. L. Bai, H. Q. Zhang, H. Sagawa, X. Z. Zhang, G. Col´o, and F. R. Xu, Effect of the tensor force on the charge exchange spin-dipole excitations of Pb208, Phys. Rev. Lett. 105(7), 072501 (2010)
39. C. L. Bai, H. Q. Zhang, H. Sagawa, X. Z. Zhang, G. Col´o, and F. R. Xu, Spin-isospin excitations as quantitative constraints for the tensor force, Phys. Rev. C 83(5), 054316 (2011)
40. E. B. Suckling and P. D. Stevenson, The effect of the tensor force on the predicted stability of superheavy nuclei, Eur. Phys. Lett. 90(1), 12001 (2010)
41. H. A. Bethe, Theory of nuclear matter, Annu. Rev. Nucl. Part. Sci. 21(1), 93 (1971)
42. V. R. Pandharipande, Variational calculation of nuclear matter, Nucl. Phys. A 181(1), 33 (1972)
43. S. Fantoni and V. R. Pandharipande, Momentum distribution of nucleons in nuclear matter, Nucl. Phys. A 427(3), 473 (1984)
44. S. C. Pieper, R. B. Wiringa, and V. R. Pandharipande, Variational calculation of the ground state of O16, Phys. Rev. C 46(5), 1741 (1992)
45. C. Ciofi degli Atti and S. Simula, Realistic model of the nucleon spectral function in few- and many-nucleon systems, Phys. Rev. C 53(4), 1689 (1996)
46. A. Tang, J. W. Watson, J. Aclander, J. Alster, G. Asryan, Y. Averichev, D. Barton, V. Baturin, N. Bukhtoyarova, A. Carroll, S. Gushue, S. Heppelmann, A. Leksanov, Y. Makdisi, A. Malki, E. Minina, I. Navon, H. Nicholson, A. Ogawa, Y. Panebratsev, E. Piasetzky, A. Schetkovsky, S. Shimanskiy, and D. Zhalov, n−p short-range correlations from (p, 2p+n) measurements, Phys. Rev. Lett. 90(4), 042301 (2003)
47. K. S. Egiyan, et al. (CLAS Collaboration), Measurement of two- and three-nucleon short-range correlation probabilities in nuclei, Phys. Rev. Lett. 96(8), 082501 (2006)
48. E. Piasetzky, M. Sargsian, L. Frankfurt, M. Strikman, and J. W. Watson, Evidence for strong dominance of protonneutron correlations in nuclei, Phys. Rev. Lett. 97(16), 162504 (2006)
49. R. Subedi, R. Shneor, P. Monaghan, B. D. Anderson, K. Aniol, J. Annand, J. Arrington, H. Benaoum, F. Benmokhtar, W. Boeglin, et al., Probing cold dense nuclear matter, Science 320(5882), 1476(2008)
50. O. Hen, et al. (Jefferson Lab CLAS Collaboration), Momentum sharing in imbalanced Fermi systems, Science 346(6209), 614 (2014)
51. R. Schiavilla, R. B. Wiringa, S. C. Pieper, and J. Carlson, Tensor forces and the ground-state structure of nuclei, Phys. Rev. Lett. 98(13), 132501 (2007)
52. M. Alvioli, C. Ciofi degli Atti, and H. Morita, Protonneutron and proton-proton correlations in medium-weight nuclei and the role of the tensor force, Phys. Rev. Lett. 100(16), 162503 (2008)
53. J. Arrington, D. W. Higinbotham, G. Rosner, and M. Sargsian, Hard probes of short-range nucleon–nucleon correlations, Prog. Part. Nucl. Phys. 67(4), 898 (2012)
54. I. Vida˜na, A. Polls, and C. Providencia, Nuclear symmetry energy and the role of the tensor force, Phys. Rev. C 84, 062801(R) (2011)
55. A. Carbone, A. Polls, and A. Rios, High-momentum components in the nuclear symmetry energy, Eur. Phys. Lett. 97(2), 22001 (2012)
56. C. Xu, A. Li, and B. A. Li, Delineating effects of tensor force on the density dependence of nuclear symmetry energy, J. Phys.: Conf. Ser. 420, 012090 (2013)
57. K. Pomorski and J. Dudek, Nuclear liquid-drop model and surface-curvature effects, Phys. Rev. C 67(4), 044316 (2003)

58. O. Hen, B. A. Li, W. J. Guo, L. B. Weinstein, and E. Piasetzky, Symmetry energy of nucleonic matter with tensor correlations, Phys. Rev. C 91(2), 025803 (2015)
59. B. A. Li, W. J. Guo, and Z. Z. Shi, Effects of the kinetic symmetry energy reduced by short-range correlations in heavyion collisions at intermediate energies, Phys. Rev. C 91(4), 044601 (2015)
60. P. Hoodbhoy and J. W. Negele, Solution of Hartree-Fock equations in coordinate space for axially symmetric nuclei, Nucl. Phys. A. 288(1), 23 (1977)
61. K. T. R. Davies and S. E. Koonin, Skyrme-force timedependent Hartree–Fock calculations with axial symmetry, Phys. Rev. C 23(5), 2042 (1981) [Erratum: Phys. Rev. C, 24, 1820 (1981)]
62. A. S. Umar, M. R. Strayer, and P. G. Reinhard, Resolution of the fusion window anomaly in heavy-ion collisions, Phys. Rev. Lett. 56(26), 2793 (1986)
63. P. G. Reinhard, A. S. Umar, K. T. R. Davies, M. R. Strayer, and S. J. Lee, Dissipation and forces in time-dependent Hartree-Fock calculations, Phys. Rev. C 37(3), 1026 (1988)
64. A. S. Umar, M. R. Strayer, P. G. Reinhard, K. T. R. Davies, and S. J. Lee, Spin–orbit force in time-dependent HartreeFock calculations of heavy-ion collisions, Phys. Rev. C 40(2), 706 (1989)
65. J. A. Maruhn, P.G. Reinhard, P. D. Stevenson, and M. R. Strayer, Spin-excitation mechanisms in Skyrme-force timedependent Hartree-Fock calculations, Phys. Rev. C 74(2), 027601 (2006)
66. A. S. Umar and V. E. Oberacker, Three-dimensional unrestricted time-dependent Hartree–Fock fusion calculations using the full Skyrme interaction, Phys. Rev. C 73(5), 054607 (2006)
67. G. F. Dai, L. Guo, E. G. Zhao, and S. G. Zhou, Dissipation dynamics and spin–orbit force in time-dependent Hartree– Fock theory, Phys. Rev. C 90(4), 044609 (2014)
68. Y. Iwata and J. A. Maruhn, Enhanced spin-current tensor contribution in collision dynamics, Phys. Rev. C 84(1), 014616 (2011)
69. G. F. Dai, L. Guo, E. G. Zhao, and S. G. Zhou, Effect of tensor force on dissipation dynamics in time-dependent Hartree–Fock theory, Science China-Physics, Mechanics & Astronomy 57(9), 1618 (2014)
70. P. D. Stevenson, E. B. Suckling, S. Fracasso, M. C. Barton, and A. S. Umar, The Skyrme tensor force in heavy ion collisions, arXiv: 1507.00645 [nucl-th]
71. C. Y. Wong, Dynamics of nuclear fluid (VIII): Timedependent Hartree–Fock approximation from a classical point of view, Phys. Rev. C 25(3), 1460 (1982)
72. G. F. Bertsch and S. Das Gupta, A guide to microscopic models for intermediate energy heavy ion collisions, Phys. Rep. 160(4), 189 (1988)
73. J. Xu and B. A. Li, Probing in-medium spin–orbit interaction with intermediate-energy heavy-ion collisions, Phys. Lett. B 724(4–5), 346 (2013)
74. Y. Xia, J. Xu, B. A. Li, and W. Q. Shen, Spin–orbit coupling and the up-down differential transverse flow in intermediateenergy heavy-ion collisions, Phys. Rev. C 89(6), 064606 (2014)
75. Y. Xia, J. Xu, B. A. Li, and W. Q. Shen, The spin-splitting of collective flows in intermediate-energy heavy-ion collisions, arXiv: 1411.3057 [nucl-th]
76. J. Xu, B. A. Li, Y. Xia, and W. Q. Shen, Spin Effects in Intermediate-energy Heavy-ion Collisions, Nucl. Phys. Rev. 31, 306 (2014)
77. J. Xu, Y. Xia, B. A. Li, and W. Q. Shen, Spin-orbit coupling in intermediate-energy heavy-ion collisions, Nucl. Technol. 37, 100513 (2014) (in Chinese)
78. G. G. Ohlsen, Polarization transfer and spin correlation experiments in nuclear physics, Rep. Prog. Phys. 35(2), 717 (1972)
79. W. G. Love and M. A. Franey, Effective nucleon–nucleon interaction for scattering at intermediate energies, Phys. Rev. C 24(3), 1073 (1981); W. G. Love and M. A. Franey, Erratum: Effective nucleon–nucleon interaction for scattering at intermediate energies, Phys. Rev. C 27(1), 438 (1983)
80. P. Danielewicz and G. Odyniec, Transverse momentum analysis of collective motion in relativistic nuclear collisions, Phys. Lett. B 157(2–3), 146 (1985)
81. V. Greco, C. M. Ko, and P. L´evai, Parton coalescence and the antiproton/pion anomaly at RHIC, Phys. Rev. Lett. 90(20), 202302 (2003)
82. V. Greco, C. M. Ko, and P. L´evai, Partonic coalescence in relativistic heavy ion collisions, Phys. Rev. C 68(3), 034904 (2003)
83. L. W. Chen, C. M. Ko, and B. A. Li, Light cluster production in intermediate energy heavy-ion collisions induced by neutron-rich nuclei, Nucl. Phys. A 729(2–4), 809 (2003)
84. L. W. Chen, C. M. Ko, and B. A. Li, Light clusters production as a probe to nuclear symmetry energy, Phys. Rev. C 68(1), 017601 (2003)
85. Y. Xia, J. Xu, B. A. Li, and W. Q. Shen, in preparation.
86. L. W. Chen, C. M. Ko, B. A. Li, and J. Xu, Density slope of the nuclear symmetry energy from the neutron skin thickness of heavy nuclei, Phys. Rev. C 82(2), 024321 (2010)
87. C. Hartnack, L. Zhuxia, L. Neise, G. Peilert, A. Rosenhauer, H. Sorge, J. Aichelin, H. St¨ocker, and W. Greiner, Quantum molecular dynamics a microscopic model from UNILAC to CERN energies, Nucl. Phys. A. 495(1–2), 303 (1989)
88. J. Aichelin, "Quantum" molecular dynamics — a dynamical microscopic n-body approach to investigate fragment formation and the nuclear equation of state in heavy ion collisions, Phys. Rep. 202(5–6), 233 (1991)
89. C. C. Guo, Y. J. Wang, Q. F. Li, and F. S. Zhang, Effect of the spin–orbit interaction on flows in heavy-ion collisions at intermediate energies, Phys. Rev. C 90(3), 034606 (2014)
90. K. Asahi, M. Ishihara, N. Inabe, T. Ichihara, T. Kubo, M. Adachi, H. Takanashi, M. Kouguchi, M. Fukuda, D. Mikolas, D. J. Morrissey, D. Beaumel, T. Shimoda, H. Miyatake, and N. Takahashi, New aspect of intermediate

energy heavy ion reactions. Large spin polarization of fragments, Phys. Lett. B 251(4), 488 (1990)

91. H. Okuno, K. Asahi, H. Sato, H. Ueno, J. Kura, M. Adachi, T. Nakamura, T. Kubo, N. Inabe, A. Yoshida, T. Ichihara, Y. Kobayashi, Y. Ohkubo, M. Iwamoto, F. Ambe, T. Shimoda, H. Miyatake, N. Takahashi, J. Nakamura, D. Beaumel, D. J. Morrissey, W. D. Schmidt-Ott, and M. Ishihara, Systematic behavior of ejectile spin polarization in the projectile fragmentation reaction, Phys. Lett. B 335(1), 29 (1994)
92. W.-D. Schmidt-Ott, K. Asahi, Y. Fujita, H. Geissel, K.-D. Gross, T. Hild, H. Irnich, M. Ishihara, K. Krumbholz, V. Kunze, A. Magel, F. Meissner, K. Muto, F. Nickel, H. Okuno, M. Pf¨utzner, C. Scheidenberger, K. Suzuki, M. Weber, C. Wennemann, Spin alignment of 43Sc produced in the fragmentation of 500 MeV/u 46Ti, Z. Phys. A 350(3), 215 (1994)
93. D. E. Groh, P. F. Mantica, A. E. Stuchbery, A. Stolz, T. J. Mertzimekis, W. F. Rogers, A. D. Davies, S. N. Liddick, and B. E. Tomlin, Spin polarization of K37 produced in a singleproton pickup reaction at intermediate energies, Phys. Rev. Lett. 90(20), 202502 (2003)
94. K. Asahi, M. Ishihara, T. Ichihara, M. Fukuda, T. Kubo, Y. Gono, A. C. Mueller, R. Anne, D. Bazin, D. GuillemaudMueller, R. Bimbot, W. D. Schmidt-Ott, and J. Kasagi, Observation of spin-aligned secondary fragment beams of B14, Phys. Rev. C 43(2), 456 (1991)
95. D. Borremans, J. M. Daugas, S. Teughels, D. L. Balabanski, N. Coulier, F. de Oliveira Santos, G. Georgiev, M. Hass, M. Lewitowicz, I. Matea, Y. E. Penionzhkevich, W. D. SchmidtOtt, Y. E. Sobolev, M. Stanoiu, K. Vyvey, and G. Neyens, Spin polarization of 27Na and 31Al in intermediate energy projectile fragmentation of 36S, Phys. Rev. C 66(5), 054601 (2002)
96. K. Turz´o, P. Himpe, D. L. Balabanski, G. B´elier, D. Borremans, J. M. Daugas, G. Georgiev, F. O. Santos, S. Mallion, I. Matea, G. Neyens, Y. E. Penionzhkevich, C. Stodel, N. Vermeulen, and D. Yordanov, Spin polarization of Al34 fragments produced by nucleon pickup at intermediate energies, Phys. Rev. C 73(4), 044313 (2006)
97. Y. Ichikawa, H. Ueno, Y. Ishii, T. Furukawa, A. Yoshimi, D. Kameda, H. Watanabe, N. Aoi, K. Asahi, D. L. Balabanski, R. Chevrier, J. M. Daugas, N. Fukuda, G. Georgiev, H. Hayashi, H. Iijima, N. Inabe, T. Inoue, M. Ishihara, T. Kubo, T. Nanao, T. Ohnishi, K. Suzuki, M. Tsuchiya, H. Takeda, and M. M. Rajabali, Production of spin-controlled rare isotope beams, Nat. Phys. 8(12), 918 (2012)
98. N. H. Buttimore, E. Gotsman, and E. Leader, Spindependent phenomena induced by electromagnetic-hadronic interference at high energies, Phys. Rev. D 18(3), 694 (1978)
99. A. Zelenski, G. Atoian, A. Bogdanov, D. Raparia, M. Runtso, and E. Stephenson, Precision, absolute proton polarization measurements at 200 MeV beam energy, J. Phys.: Conf. Ser. 295, 012132 (2011)

CHAPTER 6

THE GROUND STATE OF THE POMERON AND ITS DECAYS TO LIGHT MESONS AND PHOTONS

A. A. Godizov

INTRODUCTION

In Regge phenomenology [1], the Pomeron is introduced as an even Reggeon with vacuum quantum numbers and the intercept of its Regge trajectory higher than unity. Any Reggeon is defined as an analytic continuation of some set of bound states and resonances to the region of complex values of the spin. Many meson Reggeons emerging in the Quark Model have already got their identification via association with experimentally observed hadron states [2]: f -Reggeon ($f_2(1270)$ and $f_4(2050)$ mesons), ρ-Reggeon (ρ(770) and $\rho_3(1690)$ mesons), a-Reggeon, ω-Reggeon, φ-Reggeon, π-Reggeon, etc. All meson Regge trajectories have intercepts noticeably lower than unity. As a consequence, high-energy diffractive scattering of hadrons is dominated by Pomeron interaction. In particular, the representation of elastic scattering amplitude in terms of Pomeron exchanges yields a satisfactory description of the nucleon–nucleon diffractive pattern at ultrahigh collision energies and low momentum transfers [3]. However, being one of the most important objects of strong interaction physics (the main carrier of the strong interaction in the high-energy diffraction domain), the Pomeron has not been discovered yet. The question "which of the observed $0^+(2^{++})$ states, if any, is the ground state of the Pomeron?" still waits for proper answer. From the practical standpoint, true identification of the Pomeron is possible only through estimation of the partial decay widths and branching ratios of its real states (even-spin resonances). Presumably, the Pomeron resonance decays to "hadron–antihadron" pairs are driven by the same functions as the Pomeron coupling to the corresponding hadrons in high-energy diffractive scattering. If we knew the analytic structure of these functions, we would be able to calculate the partial decay widths of the Pomeron real states. Unfortunately, at its present stage of development, QCD does not help to solve the problem. Nevertheless, some relevant information can be extracted from the available experimental data on meson–proton scattering. In this paper, we will make an attempt to identify the ground state of the Pomeron, searching for interrelation between the high-energy hadron diffraction data and the measured characteristics of observed $0^+(2^{++})$ resonances. For example, in the framework of the DL model [4], the ratio of the kaon–proton and pion–proton total cross sections is approximately equal (up to secondary Reggeon contributions) to the ratio of the Pomeron couplings to the corresponding mesons in forward scattering. At $\sqrt{s}$ = 24.1 GeV (the highest reached energy for K p scattering), the impact of secondary Reggeons on the quantity [5]

$$\frac{\sigma_{\rm tot}^{K^+p}}{\sigma_{\rm tot}^{\pi^+p}} \approx 0.844 \qquad (1)$$

is minimal. As the partial decay width of the Pomeron ground state to pair of scalar mesons is proportional to the squared absolute value of the corresponding meson–Pomeron coupling, so, using (1), we obtain a qualitative estimation

$$\frac{\Gamma_{{\rm P}\to K^+K^-}}{\Gamma_{{\rm P}\to\pi^+\pi^-}} \sim 0.844^2 \approx 0.71. \qquad (2)$$

This value is consistent with the $f_2(1950)$ resonance branching ratio

$$\frac{\Gamma_{f_2(1950)\to K^+K^-}}{\Gamma_{f_2(1950)\to\pi^+\pi^-}} = 0.565^{+0.19}_{-0.27} \qquad (3)$$

Citation: A. A. Godizov, The ground state of the Pomeron and its decays to light mesons and photons, DOI 10.1140/epjc/s10052-016-4229-z.

extracted from the Belle Collaboration data on the quantities $\Gamma(\gamma\gamma)\Gamma(\pi^0\pi^0)/\Gamma_{tot}$ [6] and $\Gamma(\gamma\gamma)\Gamma(K^+K^-)/\Gamma_{tot}$ [7].

On the one hand, such a proportion between the partial decay widths could be considered characteristic for any SU(3)f singlet. On the other hand, in the t–J plane, $f_2(1950)$ resonance is located very close to the DL Pomeron Regge trajectory $\alpha_P^{DL}(t) = 1.08 + 0.25t$ [8,9], widely used in phenomenology. Therefore, it may be considered as a promising candidate for the Pomeron ground state. However, $0^+(2^{++})$ hadrons different from $f_2(1950)$ should not immediately be excluded from the consideration. Of course, the observed smallness of the branching ratio $\frac{\Gamma_{f_2(1270)\to K\bar{K}}}{\Gamma_{f_2(1270)\to\pi\pi}}$ and $\frac{\Gamma_{f_2'(1525)\to\pi\pi}}{\Gamma_{f_2'(1525)\to K\bar{K}}}$ [2] allows one to consider resonances $f_2(1270)$ and f_2'(1525) as mesons of certain quark–antiquark content and to relate them to secondary Reggeons, but the current interpretation of other well-established $0^+(2^{++})$ resonances (namely, $f_2(2010)$, $f_2(2300)$, and $f_2(2340)$ [2]) is not so clear. To distinguish reliably the Pomeron spin-2 state among various vacuum resonances of the same spin, we need more information on the Pomeron dynamics in both the spacelike and the domains. First, let us estimate the Pomeron coupling to light mesons in the diffractive scattering regime.

THE SPACELIKE POMERON COUPLING TO LIGHT MESONS

Elastic Scattering of Scalar Mesons on Protons

In the leading approximation, the angular distributions in π^+ p and K^+ p elastic scattering can be calculated in the same way as for pp scattering (see [3] and references therein):

$$\frac{d\sigma}{dt} = \frac{|T(s,t)|^2}{16\pi s^2},$$
$$T(s,t) = 4\pi s \int_0^\infty db^2\, J_0(b\sqrt{-t})\, \frac{e^{2i\delta(s,b)} - 1}{2i},$$
$$\delta(s,b) = \frac{1}{16\pi s}\int_0^\infty d(-t)\, J_0(b\sqrt{-t})\, \delta_P(s,t)$$
$$\approx \frac{1}{16\pi s}\int_0^\infty d(-t)\, J_0(b\sqrt{-t})\, g_{hhP}(t)\, g_{ppP}(t)$$
$$\times\left(i + \tan\frac{\pi(\alpha_P(t) - 1)}{2}\right)\pi\alpha_P'(t)\left(\frac{s}{2s_0}\right)^{\alpha_P(t)}, \quad (4)$$

where s and t are the Mandelstam variables, b is the impact parameter, $s_0 = 1\ \text{GeV}^2$, $\alpha_P(t)$ is the Regge trajectory of the Pomeron, $g_{pp}P(t)$ is the Pomeron coupling to the proton, and $g_{hh}P(t)$ is the Pomeron coupling to the incoming meson. At $t < 0$, the Pomeron Regge trajectory and the Pomeron coupling to the proton are approximated by the same test functions as in nucleon–nucleon scattering [3],

$$\alpha_P(t) = 1 + \frac{\alpha_P(0) - 1}{1 - \frac{t}{\tau_a}}, \quad g_{ppP}(t) = \frac{g_{ppP}(0)}{(1 - a_g t)^2}, \quad (5)$$

Table 1: The parameter values for (5) fitted to the nucleon–nucleon elastic scattering data

Parameter	Value
$\alpha_P(0) - 1$	0.109 ± 0.017
τ_a	$(0.535 \pm 0.057)\ \text{GeV}^2$
$g_{ppP}(0)$	(13.8 ± 2.3) GeV
a_g	$(0.23 \pm 0.07)\ \text{GeV}^{-2}$

where the free parameters take on the values presented in Table 1. Such a choice of parametrization for $\alpha_P(t)$ is in part specified by the QCD-motivated asymptotic behavior of the Pomeron Regge trajectory [10],

$$\lim_{t\to-\infty} \alpha_P(t) = 1, \quad (6)$$

and by the conditions

$$\frac{d^n\alpha(t)}{dt^n} > 0 \quad (n = 1, 2\ldots;\ t < 0), \quad (7)$$

which originate from the dispersion relations for Regge trajectories [1] and are expected to be valid for any Reggeon.1 Fixing

$$g_{\pi\pi P}(t) = g_{\pi\pi P}(0) = 8.0\ \text{GeV} \quad (8)$$

and

$$g_{KKP}(t) = g_{KKP}(0) = 7.1\ \text{GeV}, \quad (9)$$

we come to the description presented in Fig. 1 (the solid lines). The secondary Reggeon interaction influence on the π^+ p and K^+ p total cross sections can be roughly taken into account via straight addition of the corresponding DL model terms [4] (see the dashed lines in Fig. 1):

$$\Delta\sigma_{tot}^{\pi^+p} = 27.56\left(\frac{s}{s_0}\right)^{-0.4525} \text{mb},$$
$$\Delta\sigma_{tot}^{K^+p} = 8.15\left(\frac{s}{s_0}\right)^{-0.4525} \text{mb}. \quad (10)$$

The deviation of the model differential cross section curves from the experimental data cannot be removed so easily, since it requires introduction of extra unknown functions of t. The main problem is that, in view of the rather high slopes of the secondary Regge trajectories in the resonance region,

$$\alpha_R'(t) \sim 1\ \text{GeV}^{-2} \quad (t > 0), \quad (11)$$

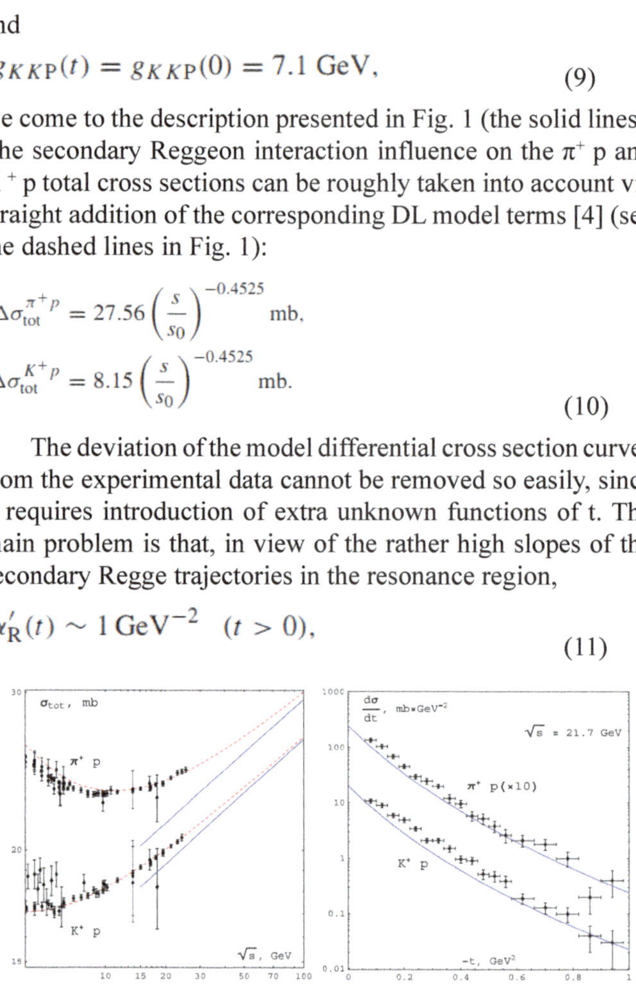

Figure. 1: The model description of the π^+ p and K^+ p scattering observables [5,11]. The solid lines correspond to the Pomeron exchange eikonal approximation (4), (5) of the amplitude. The dashed lines are obtained by adding the secondary Regge pole terms (10) of the DL model for total cross sections [4]

their intercept values

$$\alpha_R(0) \sim 0.5, \tag{12}$$

their asymptotic behavior in the deeply Euclidean domain [12,13],

$$\lim_{t \to -\infty} \alpha_R(t) = 0, \tag{13}$$

and conditions (7), they are expected to be essentially nonlinear in the diffractive scattering range. Hence, t-behavior of the corresponding Regge residues turns out strongly related to the nontrivial t-behavior of the factors $\alpha'_R(t)$ which originate from the Reggeon resonance propagators. As a consequence, reliable determination of the secondary Reggeon exchange contributions into the eikonal (Born amplitude) at nonzero t is possible only via combined fitting of all the unknown functions to the overall set of available data on nucleon–nucleon and meson–proton scattering, including charge exchange reactions [14–16]. Such a statistically satisfactory description is not achieved yet. It still remains one of the most crucial challenges in Regge phenomenology.

Nonetheless, the relative divergence between the presented model curves and the data is limited. The underestimation of $|T(s,t)|$ at $\sqrt{s} = 21.7$ GeV does not exceed 16 and 25 % for the $\pi^+ p$ and $K^+ p$ scattering, respectively. This fact might be interpreted as a hint to rather weak tdependence of the Pomeron coupling to light scalar mesons in the diffractive scattering regime. Unfortunately, the contamination by secondary Reggeon exchanges (though subdominant) does not allow one to establish this t-behavior explicitly in the framework of approximation (4). To elucidate on this matter, we need to analyze available data on the high-energy scattering of vector mesons on protons.

Exclusive Photoproduction of Vector Mesons on Protons

The vector meson photoproduction data [17–24] are available in the kinematic range wherein the influence of secondary Reggeon exchanges is negligible. According to the hypothesis of s-channel helicity conservation (SCHC), con- firmed experimentally [25–28], the cross sections of reactions $\gamma + p \to V + p$ are dominated by the non-flip helicity amplitudes. In the leading approximation, these amplitudes can be represented as [29,30]

$$T_{\gamma p \to Vp}(W^2, t) \approx \sum_{V'} C_T^{V'}(0)\, T_{V'p \to Vp}(W^2, t), \tag{14}$$

where W is the collision energy and $C_T^V(0)$ are the vector dominance model (VDM) coefficients:

$$C_T^V(Q^2) = \sqrt{\frac{3\Gamma_{V \to e^+e^-}}{\alpha_e M_V}}\frac{M_V^2}{M_V^2 + Q^2} \tag{15}$$

(M_V is the vector meson mass and $Q^2 \ll M_V^2$ is the photon virtuality). It is well known that at small values of t the Pomeron diagonal coupling to hadrons is much stronger than the off-diagonal one. Otherwise, say, the cross section of the low-mass diffractive excitation of proton in high-energy pp collisions would be comparable to the elastic cross section. Thus, in the range W > 30 GeV,

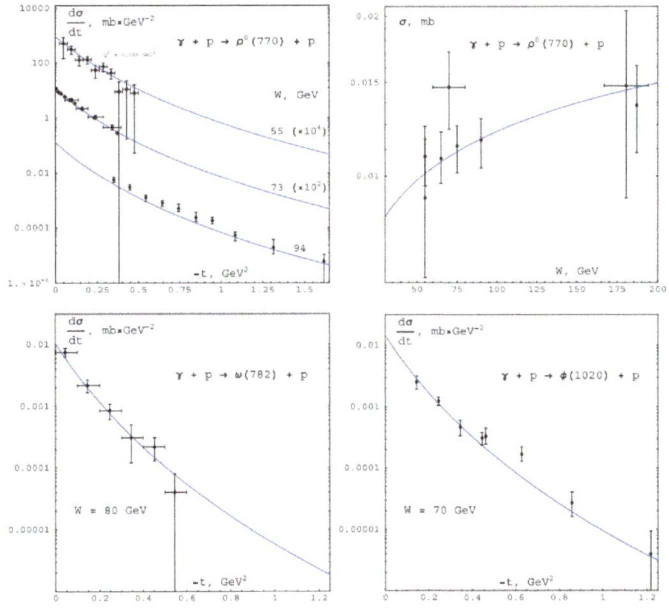

Figure. 2: The model description of the vector meson photoproduction observables [17–24]

$$T_{\gamma p \to Vp}(W^2, t) \approx C_T^V(0)\, T_{Vp \to Vp}(W^2, t), \tag{16}$$

Where $T_{Vp \to Vp}(W^2, t)$ is calculated with the help of (4). Fixing

$$g_{\omega\omega P}(t) = g_{\rho\rho P}(t) = g_{\rho\rho P}(0) = 7.07\ \text{GeV},$$
$$g_{\phi\phi P}(t) = g_{\phi\phi P}(0) = 6.7\ \text{GeV}, \tag{17}$$

we come to the description presented in Fig. 2 and Table 2. The predicted exponential slope of dσ/dt for $\gamma + p \to \rho^0(770) + p$ at $W = 73$ GeV and $0.073\ \text{GeV}^2 < -t < 0.4\ \text{GeV}^2$, $B^{\text{model}} \approx 9.1\ \text{GeV}^{-2}$ is compatible with the measured value, $B^{\text{exp}} = (9.8 \pm 1.36)\ \text{GeV}^{-2}$ [20].

The deviation of the $\rho^0(770)$ and $\phi(1020)$ photoproduction curves from the data at $-t > 0.4\ \text{GeV}^2$ could be explained by the 15 % normalization uncertainty [22] not included in the error bars. If we multiply the $\rho^0(770)$ photoproduction data set [22] by factor 0.85 and exclude two outlying points, at $t = -0.019\ \text{GeV}^2$ and $t = -0.371\ \text{GeV}^2$, from data set [21], the description quality becomes much better: $\chi^2/N_{DoF} \approx 1.17$ (37 points, one free parameter). Fitting the t-dependent test expressions of the $\rho^0(770)$- Pomeron coupling,

Table 2: The quality of description of the experimental angular distributions in the vector meson exclusive photoproduction on protons a $g_{\omega\omega P}(t) = g_{\rho\rho P}(t) = g_{\rho\rho P}(0) = 7.07$ GeV and $g_{\phi\phi P}(t) = g_{\phi\phi P}(0) = 6.7$ GeV

Data set	W, GeV	Number of points	$\chi^2_{[i]}$
[19] (ρ)	55	10	6.5
[20] (ρ)	73	4	7.9
[21] (ρ)	73	9	36.4
[22] (ρ)	94	10	19.0
[23] (ω)	80	6	4.0
[24] (ϕ)	70	4	1.8
[22] (ϕ)	94	4	8.6

$$g_{\rho\rho P}(t) = \frac{g_{\rho\rho P}(0)}{1 - bt} \tag{18}$$

or

$$g_{\rho\rho P}(t) = g_{\rho\rho P}(0)\, e^{bt}, \tag{19}$$

we reach the minimum of χ^2 at $g_{\rho\rho P}(0) = 7.08$ GeV and $b = 0.020$ GeV^{-2} for both parametrizations. However, the description quality becomes even worse, since the slight decrease of χ^2 is overcompensated by the decrease of NDoF due to introduction of extra parameter. Such a result can be interpreted as a manifestation of extremely weak t-dependence of $g_{\rho\rho P}(t)$ and $g_{\omega\omega P}(t)$ in the diffraction domain.

The obtained values of $g_{\rho\rho P}(0) = g_{\omega\omega P}(0)$ and $g_{\varphi\varphi P}(0)$ allow one to estimate the Pomeron coupling to real photons.

High-Energy γ p Scattering

Owing to the optical theorem: $W^2\sigma_{tot}^{\gamma p}(W) = \mathrm{Im}\, T_{\gamma p\to\gamma p}(W^2, 0)$. In the range $W > 30$ GeV, the diagonal fluctuations of photons to mesons $\rho^0(770)$, $\omega(782)$, and $\phi(1020)$ provide just about 80 % of $\sigma_{tot}^{\gamma p}$ (see the dashed line in Fig. 3). Other contributions can be roughly taken into account via introduction of the pole-like term A(W) $= g_A(0)\, g_{\rho\rho P}(0)\, \pi\alpha'_P(0) \left(\frac{W^2}{2s_0}\right)^{\alpha_P(0)}$, where $g_A(0) = 0.007$ GeV, so that the full amplitude is approximated by (the solid line in Fig. 3)

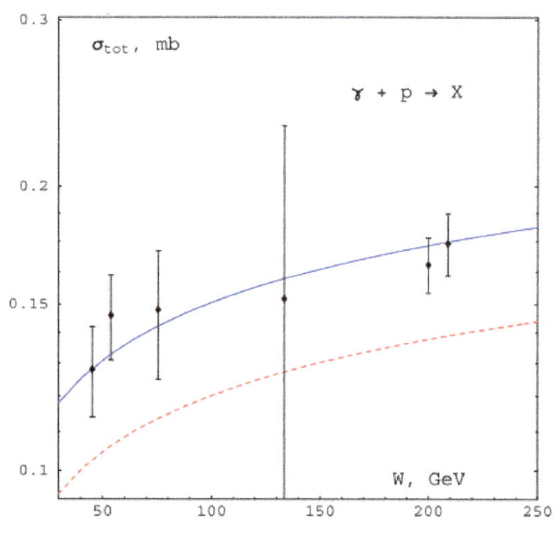

Figure. 3: The total cross sections of γ p scattering [5]. The dashed line is the contribution of the fluctuations $\gamma \to \rho^0(770), \omega(782), \phi(1020) \to \gamma$

$$\mathrm{Im}\, T_{\gamma p\to\gamma p}(W^2, 0) \approx \sum_{V=\rho,\omega,\phi} [C_T^V(0)]^2\, \mathrm{Im}\, T_{Vp\to Vp}(W^2, 0) + A(W). \tag{20}$$

The total contribution of absorptive corrections to the imaginary part of the ρ^0 p scattering forward amplitude is negative. In the interval 40 GeV < W < 210 GeV, it does not exceed 16 % of the corresponding Born amplitude. Hence, we do not expect that the underestimation of the Pomeron coupling to photon which emerges under fitting gA(0)to the data is higher than 20 % of this parameter, since the Pomeron interaction of the heavier vector mesons is not expected to be stronger than of the $\rho^0(770)$ meson. Thus, we obtain the following estimation of $g_{\gamma\gamma P}(0)$:

$$g_{\gamma\gamma P}(0) = \sum_{V=\rho,\omega,\phi} [C_T^V(0)]^2\, g_{VVP}(0) + \kappa\, g_A(0) \quad (1 < \kappa < 1.2). \tag{21}$$

THE POMERON DECAYS TO LIGHT MESONS AND PHOTONS

Reggeon coupling to light mesons in the diffraction and resonance domains is driven by the same structure functions. Although these formulas are initially derived for spinless particles, they can be adapted to vector mesons and photons. Let us formulate the comparability assumption, which underlies the further discussion:

The Pomeron coupling to light mesons in the resonance region is comparable or even approximately equal to its coupling to the corresponding particles in the diffractive scattering regime. In particular $|g_{hhP}(2;\, M_P^2, m_h^2, m_h^2)| \sim |g_{hhP}(0)| \equiv g_{hhP}(\alpha_P(0);\, 0, m_h^2, m_h^2)$, where M_P is the Pomeron ground state mass.

Though this assumption is highly nontrivial (due to the high sensitivity of the resonance partial decay width to the value of its coupling to the decay products), it does not seem to be exceptional. For example, as applied to ρ-mesons [2], Eq. (A.4) yields

$$\frac{|g_{\pi\pi\rho}(3;\, M^2_{\rho 3(1690)}, m_\pi^2, m_\pi^2)|}{|g_{\pi\pi\rho}(1;\, M^2_{\rho(770)}, m_\pi^2, m_\pi^2)|} = 1.1 \pm 0.05. \tag{22}$$

From the physical standpoint, such a weak t-dependence just implies that the corresponding Reggeon perceives pions (and, possibly, some other mesons) as pointlike particles. Regarding the Pomeron spin-2 resonance decays to pairs of light vector mesons or photons, the comparability assumption supplemented by the above-mentioned SCHC hypothesis (which is valid in the diffraction region) implies the dominance of the only orbital mode: L = 0. Other orbital modes are related to helicity amplitudes conjugated to the spin-flip vertices in Pomeron exchanges. Hence, they are expected to be dynamically suppressed on the same reasons as the spin-flip amplitudes in the high-energy diffractive scattering. Consequently, in the case of Pomeron interaction, Eq. (A.4) is applicable, as well, to the decays to pairs of vector mesons or photons (of course, the spin-2 resonance decays to pairs of scalar mesons take place in the L = 2 orbital mode only). Let us try to describe the f2(1950) resonance observables with the help of the above-stated assumption. For example, the constant related to the Pomeron ground state decay to two real photons of different polarization can be estimated as

$$|g_{\gamma\gamma P}(2;\, M_P^2, 0, 0)| = g_{\gamma\gamma P}(0) = (39.7 \pm 0.7)\,\text{MeV}. \tag{23}$$

Inserting $M_P = M_{f2(1950)} = (1.944 \pm 0.012)$ GeV [2] into (A.4), we come to the following result:

$$\Gamma_{f_2(1950)\to\gamma\gamma} = (960 \pm 50)\,\text{eV}. \tag{24}$$

In addition, with the help of estimations (8) and (9), we predict

$$\Gamma_{f_2(1950)\to\pi^0\pi^0} = (37 \pm 1)\,\text{MeV},$$
$$\Gamma_{f_2(1950)\to K^+K^-} = (29 \pm 1)\,\text{MeV}, \tag{25}$$

and, then, in view of $\Gamma_{f_2(1950)\to X} = (472 \pm 18)$ MeV [2], obtain

$$\frac{\Gamma_{f_2(1950)\to\gamma\gamma}\Gamma_{f_2(1950)\to\pi^0\pi^0}}{\Gamma_{f_2(1950)\to X}} = (75 \pm 5)\ \text{eV},$$
$$\frac{\Gamma_{f_2(1950)\to\gamma\gamma}\Gamma_{f_2(1950)\to K^+K^-}}{\Gamma_{f_2(1950)\to X}} = (59 \pm 4)\ \text{eV}, \tag{26}$$

while the corresponding experimental values are [6,7]

$$\frac{\Gamma_{f_2(1950)\to\gamma\gamma}\Gamma_{f_2(1950)\to\pi^0\pi^0}}{\Gamma_{f_2(1950)\to X}} = 54^{+23}_{-14}\ \text{eV},$$
$$\frac{\Gamma_{f_2(1950)\to\gamma\gamma}\Gamma_{f_2(1950)\to K^+K^-}}{\Gamma_{f_2(1950)\to X}} = (61 \pm 2 \pm 13)\ \text{eV}. \tag{27}$$

The same procedure applied to $f_2(2300)$ resonance ($M_{f_2(2300)} = (2.297 \pm 0.028)$ GeV, $\Gamma_{f_2(2300)\to X} = (149 \pm 40)$ MeV [2]) yields the estimation

$$\frac{\Gamma_{f_2(2300)\to\gamma\gamma}\Gamma_{f_2(2300)\to K\bar{K}}}{\Gamma_{f_2(2300)\to X}} = (1300 \pm 370)\ \text{eV}, \tag{28}$$

which catastrophically diverges from the Belle data [31]:

$$\frac{\Gamma_{f_2(2300)\to\gamma\gamma}\Gamma_{f_2(2300)\to K\bar{K}}}{\Gamma_{f_2(2300)\to X}} = 3.2^{+0.5+1.3}_{-0.4-2.2}\ \text{eV}. \tag{29}$$

This inconsistency allows one to exclude $f_2(2300)$ resonance from the candidates for the Pomeron ground state. Unfortunately, analogous data for $f_2(2010)$ and $f_2(2340)$ resonances [2] are currently unavailable.

DISCUSSION

Thus, direct propagation of the Pomeron couplings $g_{K\,KP}$ and $g_{V\,VP}$ from the diffraction to resonance region yields a correct estimation of $\Gamma_{f_2(1950)\to\gamma\gamma}\Gamma_{f_2(1950)\to K^+K^-}/\Gamma_{f_2(1950)\to X}$. The model and experimental estimations of $\Gamma_{f_2(1950)\to\gamma\gamma}\Gamma_{f_2(1950)\to\pi^0\pi^0}/\Gamma_{f_2(1950)\to X}$ overlap as well, but the significant systematic uncertainties [6] not included into the presented value (27) make this result not so impressive. A quite independent argument in favor of our reasoning is the $f_2(1950)$ resonance location in the t–J plane (see Fig. 4), which allows a smooth linkage between the linear Chew–Frautschi plot and the nonlinear Regge trajectory of the Pomeron (5) extracted from the nucleon–nucleon scattering data. This coincidence encourages to consider more seriously the possibility of the $f_2(1950)$ resonance correspondence to the Pomeron ground state, since other observed $0^+(2^{++})$ resonances do not seem to be so good candidates. Of course, at present, the discussed coincidences should be qualified just as some noteworthy indications. For a firm conclusion, we need new and, if possible, more accurate data on both the $f_2(1950)$ resonance observables and the πp and K p scattering at higher energies. Another prediction which should be checked is the opposite polarization of the photons in $f_2(1950) \to \gamma\,\gamma$ decays which follows from the dominance of the L = 0 orbital mode predicted above.

In fact, the possible physical consequences of the presumed comparability are much deeper than just a practical feasibility of the Pomeron identification. In the Euclidean domain, the behavior like $\{\alpha_P(t) > 1, \lim_{t\to-\infty}\alpha_P(t) = 1\}$ is characteristic for glueball Reggeons (for details, see[10,32–37]). If the assumption is correct, then the Pomeron is expected to keep its glueball nature in the resonance region as well. In other words, if the predictions, regarding the $K\bar{K}$, $\pi\pi$, and $\gamma\gamma$ decays, are confirmed, we would have enough grounds to consider $f_2(1950)$ resonance as not only the Pomeron ground state, but the lightest tensor glueball.

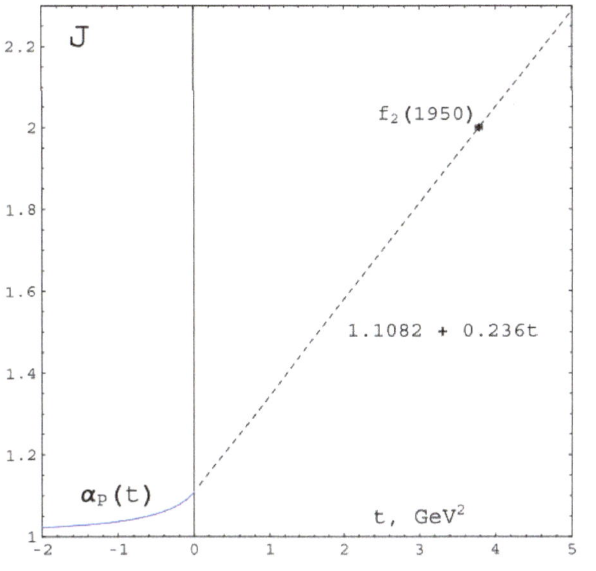

Figure. 4: Expected behavior of the Chew–Frautschi plot for the Pomeron. The solid line is the Pomeron Regge trajectory (5) fitted to the nucleon– nucleon scattering data. The dashed line corresponds to $\alpha_P^{lin}(t) = 1.1082 + 0.236\,t$

The idea that $f_2(1950)$ resonance as a glueball is, certainly, not new. Although its mass is significantly lower than the predictions by lattice QCD [38,39], it is discussed as a glueball candidate from time to time, because of the high value of its decay width [40], and some features of the highenergy p+ p → p+ $f_2(1950)$+ p differential cross sections [41,42].

ACKNOWLEDGMENTS

The author is indebted to his colleagues from the IHEP Division of Theoretical Physics for helpful discussions.

REFERENCES

1. P.D.B. Collins, An Introduction to Regge Theory and High Energy Physics (Cambridge University Press, Cambridge, 1977)
2. Particle Data Group, http://pdg.lbl.gov/2015/tables/rpp2015-sum-mesons.pdf
3. A.A. Godizov, Eur. Phys. J. C 75, 224 (2015)
4. A. Donnachie, P.V. Landshoff, Phys. Lett. B 296, 227 (1992)
5. Particle Data Group, http://pdg.lbl.gov/2015/hadronic-xsections/ hadron.html
6. The Belle Collaboration (S. Uehara et al.), Phys. Rev. D 79, 052009 (2009)
7. The Belle Collaboration (K. Abe et al.), Eur. Phys. J. C 32, 323 (2004)

8. A. Donnachie, P.V. Landshoff, Nucl. Phys. B 267, 690 (1986)
9. S. Donnachie, G. Dosch, P. Landshoff, O. Nachtmann, Pomeron Physics and QCD (Cambridge University Press, Cambridge, 2002)
10. P.D.B. Collins, P.J. Kearney, Z. Phys. C 22, 277 (1984)
11. NA22 Collaboration (M. Adamus et al.), Phys. Lett. B 186, 223 (1987)
12. J. Kwiecinski, Phys. Rev. D 26, 3293 (1982)
13. R. Kirschner, Z. Phys. C 67, 459 (1995)
14. A.V. Barnes et al., Phys. Rev. Lett. 37, 76 (1976)
15. O.I. Dahl et al., Phys. Rev. Lett. 37, 80 (1976)
16. V.D. Apokin et al., Nucl. Phys. B 255, 253 (1985)
17. ZEUS Collaboration (M. Derrick et al.), Z. Phys. C 63, 391 (1994)
18. ZEUS Collaboration (M. Derrick et al.), Z. Phys. C 69, 39 (1995)
19. H1 Collaboration (S. Aid et al.), Nucl. Phys. B 463, 3 (1996)
20. ZEUS Collaboration (M. Derrick et al.), Z. Phys. C 73, 253 (1997)
21. ZEUS Collaboration (J. Breitweg et al.), Eur. Phys. J. C 2, 247 (1998)
22. ZEUS Collaboration (J. Breitweg et al.), Eur. Phys. J. C 14, 213 (2000)
23. ZEUS Collaboration (M. Derrick et al.), Z. Phys. C 73, 73 (1996)
24. ZEUS Collaboration (M. Derrick et al.), Phys. Lett. B 377, 259 (1996)
25. ZEUS Collaboration (S. Chekanov et al.), Nucl. Phys. B 695, 3 (2004)
26. ZEUS Collaboration (S. Chekanov et al.), Nucl. Phys. B 718, 3 (2005)
27. ZEUS Collaboration (S. Chekanov et al.), PMC Phys. A 1, 6 (2007)
28. H1 Collaboration (F.D. Aaron et al.), JHEP 1005 (2010) 032
29. J.J. Sakurai, Currents and Mesons (The University of Chicago Press, Chicago, 1969)
30. V.A. Petrov, Mod. Phys. Lett. A 30, 1550164 (2015)
31. The Belle Collaboration (S. Uehara et al.), Prog. Theor. Exp. Phys. 2013 123C01
32. F.E. Low, Phys. Rev. D 12, 163 (1975)
33. S. Nussinov, Phys. Rev. D 14, 246 (1976)
34. D. Heckathorn, Phys. Rev. D 18, 1286 (1978)
35. H. Cheng, T.T. Wu, Phys. Rev. Lett. 22, 666 (1969)
36. R. Kirschner, L.N. Lipatov, Z. Phys. C 45, 477 (1990)
37. A.A. Godizov, V.A. Petrov, JHEP 0707, 083 (2007)
38. Y. Chen et al., Phys. Rev. D 73, 014516 (2006)
39. E. Gregory et al., JHEP 1210, 170 (2012)
40. F. Brünner, D. Parganlija, A. Rebhan, Phys. Rev. D 91, 106002 (2015)
41. WA102 Collaboration (D. Barberis et al.), Phys. Lett. B 474, 423 (2000)
42. A. Kirk, Phys. Lett. B 489, 29 (2000)

CHAPTER 7 TWO-PARTICLE RAPIDITY CORRELATIONS BETWEEN RELATIVISTIC PARTICLES IN CENTRAL COLLISIONS OF 197AU NUCLEI IN EMULSION AT 11.6 A GEV/C

U. U. Abdurakhmanov, K. G. Gulamov, V. Sh. Navotny

INTRODUCTION

Interest to the study of relativistic nucleus-nucleus collisions is caused by many reasons, such as the possibility to study the nuclear fragmentation processes or the processes of propagation of newly born particles ("in statu nascendi") through the nuclear medium. Particular attention is paid to issues related to the collective properties of the particles produced. Since the energy density can reach high values, the question arises on the possibility of an ensemble of produced particles to manifest some collective behaviour (the quark-gluon plasma is the primary goal here), especially in light of the fact that in such collisions the bounce-off effects are clearly observed [1]. It can be assumed that the collective properties of an ensemble of produced particles may also demonstrate themselves in two-particle and multiparticle correlations. Recent publications [2–4] suggest that rapidity correlations may be considered as a useful tool for distinguishing between different mechanisms of multiparticle production in relativistic heavy ion collisions. This paper presents experimental data on two-particle correlations among pseudorapidities of relativistic particles produced in central interactions of ^{197}Au nuclei with emulsion nuclei at 11.6 AGeV/c.

EXPERIMENTAL DATA

The experimental data are accumulated in the framework of E-863 experiment of the EMU-01 collaboration [5,6]. Emulsions were irradiated by 197Au nuclei at 11.6 AGeV/c on the AGS accelerator of the Brookhaven National Laboratory (Upton, United States). A beam of gold ions with density 5×10^3 nuclei/cm^2 with an admixture of foreign ions of no more than 2 % was used for these purposes.

For the analysis the events of inelastic incoherent interactions of gold nuclei in emulsion were collected, the events of electromagnetic nature were excluded from consideration. In accordance with emulsion technique, the secondary charged particles in events were divided into different groups: black or b-particles, mainly consisting of protons from the target nuclei with momenta $p \leq 0.2$ GeV/c and also heavier nuclear fragments; gray or g-particles corresponding to protons with the momenta $0.2 \leq p \leq 1$ GeV/c; they mainly consist of protons – fragments of the target nuclei, contribution of slow pions does not exceed several percent. Black and gray particles may be combined into a group of strongly ionizing h-particles with a multiplicity $n_h = n_b + n_g$;

shower ors-particles – singly charged particles with speed $\beta \geq 0.7$, mainly consisting of produced particles and the singly charged fragments of the projectile nucleus; projectile fragments – fast particles with charge $Z \geq 2$ and ionization $I/I_0 \geq 4$, not changing at long distances from the point of interaction in emulsion, where I0 is the ionization of a singly charged relativistic particle. For all the above types of particles their multiplicity and the emission angles were determined. For the analysis of angular distributions of s-particles we use pseudorapidity:

$$\eta = -\log\tan\frac{\theta}{2} \qquad (1)$$

Citation: U. U. Abdurakhmanov, K. G. Gulamov, V. Sh. Navotny, Two-particle rapidity correlations between relativistic particles in central collisions of 197Au nuclei in emulsion at 11.6 A GeV/c ,DOI 10.1140/epjc/s10052-016-4174-x.

Table 1: The average multiplicities of s-particles for events with $n_h \geq 8$ and different n_α

n_α	0–2	3–4	≥5
$\langle n_s \rangle$	147.92 ± 12.76	127.94 ± 7.88	113.86 ± 3.94

where θ is the emission angle of the s-particle. For pions pseudorapidity is related with true rapidity by a simple equation:

$$\sinh \eta = \frac{m_T}{p_T} \sinh y, \quad \text{where} \quad m_T^2 = m^2 + p_T^2. \tag{2}$$

Statistics of the experiment consists of 1057 incoherent inelastic interactions of gold nuclei in emulsion. More information about the experiment is given in [7]. As a measure of the centrality of interactions with respect to the target nucleus we use conventional in emulsion technique criterion $n_h \geq 8$, which means that we effectively select collisions of gold nuclei with heavy emulsion nuclei. The number of such events is equal to 450 from a total number of 1057 inelastic interactions collected As a measure of centrality of collisions with respect to the projectile, we use the number n_α of relativistic doubly charged particles - projectile fragments, consisting mostly of α-particles. All central events considered by us were divided into three groups according to the number of doubly charged particles in them. Table 1 presents the average multiplicity of s-particles in these groups. One can see that the average multiplicity of shower particles in these groups increases with the decreasing number of doubly charged particles from a projectile nucleus, indicating that the number of doubly charged particles can indeed be considered as a measure of the centrality of interactions considered. From this point of view events with $n_h \geq 8$ and $n_\alpha = 0 - 2$ belong to the most central collisions of 197Au nuclei with heavy emulsion nuclei considered by us.

CORRELATION FUNCTIONS AND MODELS

For study of pseudorapidity correlations between produced particles we use the technique of inclusive correlation functions. Two-particle correlation function is defined as

$$C_2(\eta_1, \eta_2) = \rho_2(\eta_1, \eta_2) - \rho_1(\eta_1)\rho_1(\eta_2), \tag{3}$$

where $\rho_2(\eta_1, \eta_2)$, $\rho_1(\eta)$ are the one- and two-particle pseudorapidity distributions, respectively

$$\rho_1(\eta_1) = \frac{1}{\sigma}\frac{d\sigma}{d\eta}, \quad \rho_2(\eta_1, \eta_2) = \frac{1}{\sigma}\frac{d^2\sigma}{d\eta_1 d\eta_2}. \tag{4}$$

In our analysis we use the normalized two-particle correlation function

$$R_2(\eta_1, \eta_2) = \frac{\rho_2(\eta_1, \eta_2)}{\rho_1(\eta_1)\rho_1(\eta_2)} - 1, \tag{5}$$

which is less sensitive to the details of one-particle inclusive distributions. It is well-known (see, e.g. [8–10] and references therein) that the magnitude of correlation functions depends not only on the strength of true dynamic correlations between particles in an event, but also on more general factors, such as the shape of multiplicity distribution of produced particles and/or the dependence of inclusive distributions on the multiplicity, i.e. factors related with inhomogeneous nature of ensembles of inclusive events. For example, it follows from normalization of inclusive distributions that

$$\iint C_2(\eta_1, \eta_2) d\eta_1 d\eta_2 = \langle n(n-1) \rangle - \langle n \rangle^2. \tag{6}$$

For nucleus-nucleus interactions the inhomogeneous nature of ensembles of events is obvious even for the final states with the same n_s because nuclear interactions with different geometry and the therefore different number of intranuclear nucleon-nucleon collisions may lead to the same multiplicity n_s of produced particles. Therefore, when analysing experimental data on correlation functions, it is important to evaluate contributions of both trivial and truly dynamic effects to their magnitude and shape, which is impossible to do in a model independent way.

The experimental data of the present paper were compared with results of Monte-Carlo simulations in the framework of two models. In the framework of a phenomenological model of independent emission (IEM) of s-particles [10–12] we assume that: (i) multiplicity (n_s) distributions of simulated events in each one of subensembles (with different n_α, for example) reproduce the experimental distributions in the real group (subensemble) of events; (ii) one-particle pseudorapidity distributions of s-particles in each one of simulated subensembles of events (within, for instance, the fixed range of n_s and n_α) reproduce the experimental "semiinclusive" distributions for the same n_s and n_α; (iii) emission angles of s-particles in each one of simulated events are statistically independent.

Thus, events simulated in the framework of IEM reproduce experimental multiplicity and angular distributions of s-particles for each one of considered group (subensemble) of events but the emission angles of these particles are statistically independent. The model by definition does not introduce any dynamic correlations among s-particles. It can be used to evaluate the contribution of effects related to multiplicity and angular distributions to the correlation functions. Of course, the model does not take into account the energy-momentum conservation which influence on correlation functions was found [10] to be negligibly small for high-multiplicity interactions. In practice we have simulated artificial events in which emission angles of s-particles were generated on the basis of the experimental angular distributions for the given multiplicities n_α (and n_s). Then events with different n_s were mixed up with statistical weights corresponding to the experimental multiplicity distribution of s-particles. Altogether we have simulated more than 10^5 events so that the statistical errors are much less than the experimental ones.

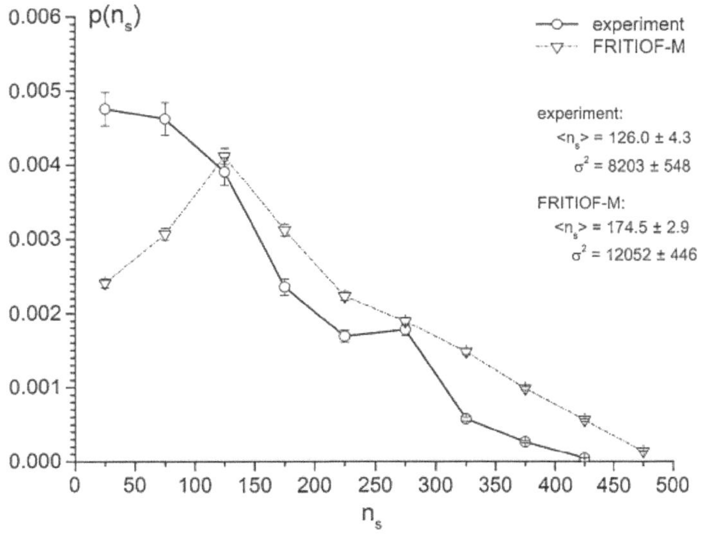

Figure. 1: Multiplicity distribution for central events

The experimental data were also compared with results of Monte-Carlo calculations performed in the framework of a modified FRITIOF-M model [13]. In this model the first stage of collisions is simulated following the phenomenological approach used in the well-known original FRITIOF [14,15], which integrates phenomenological string model for nucleon-nucleon collisions with the Glauber theory for nucleus-nucleus interactions. The next stage of an interaction related with disintegration of residual nuclei is considered in this approach following the statistical nuclear multifragmentation model [16]. No cascade processes are allowed in the model. Like original FRITIOF, the FRITIOF-M takes into account production and decay of intermediate resonances mostly in the central region of nucleus-nucleus collisions. These resonances can be a source of short-range rapidity correlations between s-particles of the final state. As regards residual nuclei, they undergo statistically isotropic decays in their own rest frames, but the model does not introduce any additional dynamic correlations. More than 5000 events generated in the framework of this model for interactions of gold nuclei in emulsion at 11.6 A GeV/c were processed following the same rules and criteria as the experimental ones. It may be noticed that the statistical errors for model calculations are considerably less than for the experimental data.

EXPERIMENTAL RESULTS

In Fig. 1 we show multiplicity distribution of s-particles in central ($n_h \geq 8$) collisions of gold nuclei with heavy emulsion nuclei at 11.6 A GeV/c in comparison with corresponding results for the FRITIOF-M model. We see that the experimental n_s-distribution is very broad and it cannot be reproduced by the FRITIOF-M model [13]. In fact, the FRITIOFM represents by itself an attempt to combine the original FRITIOF, which describes satisfactorily the data on produced particles in hadron-hadron and hadron-nucleus interactions, with multifragmentation model, which describes a number of characteristics of nuclear fragmentation processes. One can conclude that this attempt is not very successful, at least for interactions considered.

In Fig. 2 we show the values of the correlation function $R_2(\eta_1, \eta_2)$ at $\eta_1 = \eta_2$, reflecting the magnitude of shortrange correlations between s-particles, for central collisions of 197Au nuclei with heavy emulsion nuclei with different numbers of doubly charged projectile fragments in the final state. We see, first of all, that $R_2(\eta_1, \eta_2)$ at $\eta_1 = \eta_2$ shows almost no dependence on the pseudorapidities of s-particles for all three groups of events and the value of R_2 is almost the same in fragmentation and central regions of heavy ions collisions. For the groups of more central interactions with $n_\alpha = 0 - 2$ and $n_\alpha = 3 - 4$ the experimental values of $R_2(\eta_1, \eta_2)$ are approximately equal to zero, which corresponds to the absence of any significant short-range correlations between produced particles in the experiment. For more peripheral with respect to the projectile collisions with $n\alpha \geq 5$ the value of $R_2(\eta_1, \eta_2)$ is less than zero, but also does not depend on η.

The general behaviour of $R_2(\eta_1, \eta_2)$ in heavy-ions collisions considered is quite different from that in high-energy hadron-nucleon and hadron-nucleus collisions. For example, in high-energy hadron-nucleon interactions correlation functions demonstrate clearly the short-range order, they have a bump in the central region and tend to become zero only in the fragmentation regions [9,17]. Of course, it is necessary to note that at energies considered by us the central region cannot be separated from fragmentation regions of colliding nuclei. Probably this is one of the reasons why the correlation function $R_2(\eta_1, \eta_2)$ demonstrates flat η-dependencies.

Comparison of calculations performed in the framework of FRITIOF-M model with the experimental data shows that the agreement is observed for events with $n_\alpha \geq 5$, while for central events with $n_\alpha = 0 - 2$ and $n_\alpha = 3 - 4$ the model underestimates the value of R_2 at $\eta_1 = \eta_2$.

Model of independent emission of s-particles well describes the experimental data for all three groups of events. In other words, dynamic correlations between produced particles do not reveal themselves in experimental data on correlation functions, and deviations of the value of R_2 at $\eta_1 = \eta_2$ from zero can be attributed to the influence of the multiplicity and angular distributions of produced particles taken into account in the IEM.

The experimental values of the correlation function $R_2(\eta_1, \eta_2)$ for fixed $\eta_1 = 2.0$ are shown in Fig. 3 in com-

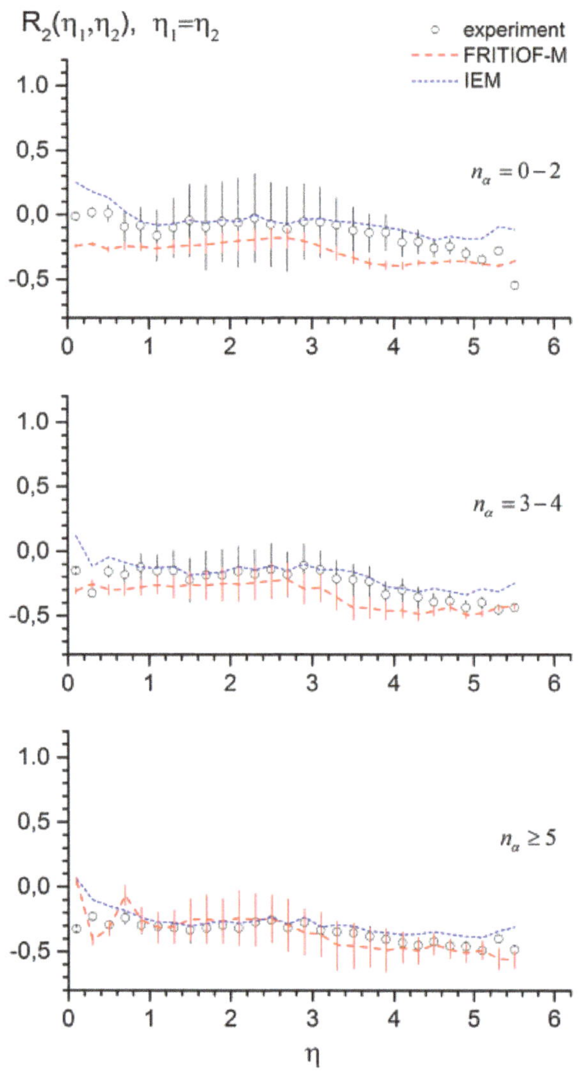

Figure. 2: $R_2(\eta_1, \eta_2)$ for three groups of central events at $\eta_1 = \eta_2$

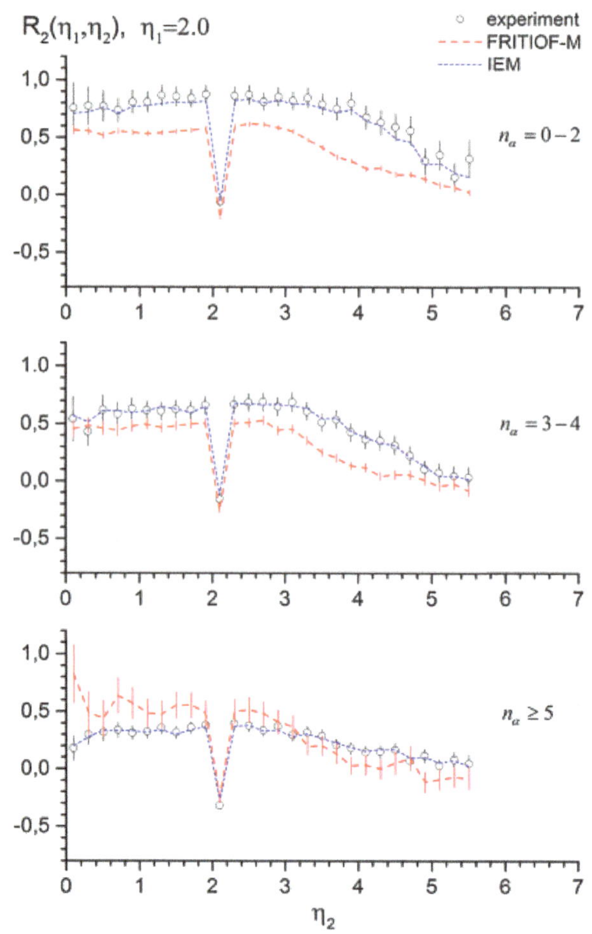

Figure. 3: $R_2(\eta_1, \eta_2)$ for three groups of central events at $\eta_1 = 2.0$

parison with results of calculation in the framework of the FRITIOF-M and IEM. Here the magnitude of $R_2(\eta_1, \eta_2)$ can be considered as a measure of long-range correlations between s-particles for all values of pseudorapidities, except for values of $\eta_1 = \eta_2 = 2.0$. We see that for all of the considered groups of events the experimental values of $R_2(\eta_1, \eta_2)$ demonstrate the presence of significant correlations of the long-range nature. Of course, for $\eta_1 = \eta_2 = 2.0$ the value of $R_2(\eta_1, \eta_2)$ becomes close to zero or even negative in agreement with what we have seen before for short-range correlations in Fig. 2. In other words, in this experiment, two-particle correlations between pseudorapidities of produced particles are of the long-range nature, whereas the short-range correlations are suppressed considerably.

The values of $R_2(\eta_1, \eta_2)$ for the FRITIOF-M model also show the existence of long-range correlations, although the calculated values are below of the experimental ones for the groups of events with $n_\alpha = 0 - 2$ and $n_\alpha = 3 - 4$, while for events with $n_\alpha \geq 5$ they generally describe the data.

Independent emission model well describes the experimental data for all three groups of events. This implies that the long-range correlations observed are not of the dynamic nature but linked to the fluctuations of s-particles' multiplicities and angular distributions.

CONCLUSIONS

Analysis of the two-particle pseudorapidity correlations for relativistic particles shows that short-range correlations are strongly suppressed in central collisions of ^{197}Au nuclei with heavy emulsion nuclei in comparison with high-energy hadron-nucleon interactions. In fact the experimental data do not reveal any noticeable short-range correlations between pseudorapidities of s-particles. At the same time the data on the normalized correlation functions demonstrate the presence of considerable long-range correlations. The enhanced contribution of long-range correlations between pseudorapidities of particles produced in hadron-nucleus interactions in comparison with hadron-nucleon collisions observed earlier [11] was linked to an additional source of multiplicity fluctuations associated with multiple collisions of a projectile inside the target nucleus. The same reasoning can be applied to the case of relativistic heavy-ions interactions.

The comparison of experimental data with calculations based on the independent emission model shows that the source of these correlations is not of a dynamic origin, and most probably is related with fluctuations in the number of intranuclear nucleon-nucleon collisions which influences the multiplicity of produced particles. This conclusion is confirmed by the fact that the long-range correlations also appear in the FRITIOF-M model. Taking into account the nuclear geometry FRITIOF-M introduces heterogeneity of events which is in turn reflected in multiplicity and pseudorapidity distributions of final state particles. Numerical deviations of the FRITIOF-M model from the experiment may be the result of the fact that the model does not well describe multiplicity distributions of produced particles for the central collisions of gold nuclei with heavy emulsion nuclei. We are thankful to all members of the EMU-01 collaboration with whom the experimental data were collected and analyzed.

REFERENCES

1. M.I. Adamovich, M.M. Aggarwal et al., Bounce—off in 197Auinduced collisions with Ag(Br) nuclei at 11.6 A GeV/c. Eur. Phys. J. A 2, 61–67 (1998)
2. Monnai, B. Schenke, Pseudorapidity correlations in heavy ion collisions from viscous fluid dynamics. Phys. Lett. B 752, 317–321 (2016)
3. P. Bozek, W. Broniowski, A. Olszewski, Two-particle correlations in pseudorapidity in a hydrodynamic model. Phys. Rev. C 92, 054913 (2015)
4. V.V. Vechernin, Forwardbackward correlations between multiplicities in windows separated in azimuth and rapidity. Nucl. Phys. A 939, 21–45 (2015)
5. E. Stenlund et al., Quark matter '93. Proceedings, 10th International Conference on Ultrarelativistic Nucleus-Nucleus Collisions, Borlaenge, Sweden, June 20–24, 1993. Nucl.Phys. A 566 (1994)
6. M.I. Adamovich, M.M. Aggarval et al., Complex analysis of gold interactions with photoemulsion nuclei at 10.7 GeV/nucleon within the framework of cascade and FRITIOF models. Z. Phys. A 358, 337–351 (1997)
7. A.M.Mashkov, Interaction of gold ions in nuclear emulsion at the primary energy 10.7 GeV/nucleon. Thesis, submitted for the ful-fillment of the requirements for the degree of candidate of physicaland mathematical sciences (Tashkent, 1999)
8. R. Slansky, Correlations, short-range order and clustering. Nucl. Phys. B 73, 477–496 (1974)
9. K. Eggert, H. Frenzel, W. Thome et al., Angular correlationsbetween the charged particles produced in pp collisions at ISRenergies. Nucl. Phys. B 86, 201–215 (1975)
10. S.A. Azimov, K.G. Gulamov, V.Sh. Navotny, V.I. Petrov, L.P. Chernova,U.M.Chernov, Multiparticle correlations in multiple particle production in the range 20–400 GeV. in Multiparticle production inhadron-nucleon and hadron-nucleus interactions (Fan publishers, Tashkent, 1985) pp. 124–167
11. K.G. Gulamov, S.A. Azimov, A.I. Bondarenko, G.M. Chernov et al., Two particle rapidity correlations among produced particles in proton-nucleus interactions at high energies. Z. Phys. A 280,107–109 (1977)
12. S.A. Azimov, G.M. Chernov, K.G. Gulamov, VSh Navotny, V.I.Petrov, Inclusive two- and three particle rapidity correlations in high-energy hadron-nucleus collisions. Nucl. Phys. B 178, 457– 476 (1981)
13. V.V. Uzhinskii, Modified code FRITIOF. User's Guide (JiNR, E2- 96-192, Dubna, 1996)
14. B. Andersson, G. Gustafson, B. Nilsson-Almqvist et al., A modelfor low-pT hadronic reactions with generalizations to hadronnucleus and nucleus-nucleus collisions. Nucl. Phys. B 281, 289– 309 (1987)
15. B. Nilsson-Almqvist, E. Stenlund, Interactions between hadronsand nuclei: the Lund Monte Carlo—FRITIOF version 1.6-. Comput. Phys. Commun. 43, 387–397 (1987)
16. J.P. Bondorf, A.S. Botvina, et al., Statistical multifragmentation of nuclei, Phys. Rep. 257, 133–221 (1995)
17. W. Ko, Rapidity correlations between negative pions produced in the K + p interactions. Phys. Rev. Lett. 28, 935–939 (1972). [Err: Phys. Rev. Lett. 28, 1010 (1972)]

CHAPTER 8

EFFECTIVE TRANSVERSE RADIUS OF NUCLEON IN HIGH-ENERGY ELASTIC DIFFRACTIVE SCATTERING

A. A. Godizov

INTRODUCTION

Elastic diffractive scattering of hadrons is one of the most interesting and important areas of high-energy hadron physics: in pp collisions the fraction of elastic diffraction events in the total number of events is very high (from more than 15 % at the ISR to about 25 % at the LHC). However, the main problem related to this sector of strong interaction physics is the fact that the characteristic distances for diffractive interaction of hadrons are of order 1 fm, so that perturbative QCD is inapplicable. Hence, one has to search for some approaches not related to pQCD directly, to provide an at least qualitative description of the corresponding high-energy observables. One of the most natural theoretical frameworks which helps to deal with the nonperturbative sector of hadron physics is Regge theory, wherein the interaction of hadrons is described in terms of exchanges by reggeons (off-mass-shell and off-spin-shell composite particles). In Ref. [1] a simple Regge-eikonal model for high-energy elastic diffraction of nucleons was examined. In this model, the standard eikonal representation of the nonflip scattering amplitude [2],

$$T_{el}(s,t) = 4\pi s \int_0^\infty db^2 J_0(b\sqrt{-t}) \frac{e^{2i\delta(s,b)} - 1}{2i},$$
$$\delta(s,b) = \frac{1}{16\pi s}\int_0^\infty d(-t) J_0(b\sqrt{-t})\delta(s,t) \quad (1)$$

(here s and t are the Mandelstam variables and b is the impact parameter), is exploited together with the single-reggeonexchange approximation to the eikonal (Born amplitude) in the kinematic range $s \gg \{m_p^2, |t|\}$:

$$\delta(s,t) = \delta_P(s,t) = \left(i + \mathrm{tg}\frac{\pi(\alpha_P(t)-1)}{2}\right)\Gamma_P{}^2(t)\left(\frac{s}{s_0}\right)^{\alpha_P(t)}, \quad (2)$$

where $s_0 = 1$ GeV^2 and $\alpha_P(t)$ is the Regge trajectory of a pomeron (a C-even reggeon, which absolutely dominates over other reggeons at the SPS, Tevatron, and LHC energies). In [1] the unknown functions $\alpha_P(t)$ and $\Gamma_P(t)$ are considered independent and treated with the help of some simple test parametrizations which are fitted to the data. Formally, the used parametrizations have provided a satisfactory description of the available experimental data at $\sqrt{s} > 500$ GeV and $-t < 2$ GeV^2. However, a question emerges about possible correlation between the behavior of $\alpha_P(t)$ and $\Gamma_P(t)$ at low t, since both functions have a strong impact on the t-behavior of the scattering amplitude. Such a correlation exists and can be taken into account explicitly. As a consequence, it becomes possible to extract from the data valuable information concerning the effective transverse radius of nucleon in the nonperturbative regime of high-energy diffractive scattering.

STRUCTURE OF THE POMERON REGGE RESIDUE

First of all, we should note that in the framework of the Regge-eikonal approach the pomeron exchange contribution into the eikonal of nucleon–nucleon elastic scattering appears as

$$\delta_P(s,t) = \left(i + \mathrm{tg}\frac{\pi(\alpha_P(t)-1)}{2}\right) g_P^2(t)\, \pi\alpha'_P(t)\left(\frac{s}{2s_0}\right)^{\alpha_P(t)}. \quad (3)$$

Where $\alpha'_P(t)$ originates from the pomeron propagator and, in general, the factor $2^{-\alpha_P(t)}\pi\alpha'_P(t)$ is not related to the

Citation: A. A. Godizov, Effective transverse radius of nucleon in high-energy elastic diffractive scattering, DOI 10.1140/epjc/s10052-015-3450-5.

pomeron–nucleon coupling. In the literature this factor is usually included into the corresponding Regge residue [2]: $g_P^2(t)2^{-\alpha_P(t)}\pi\alpha'_P(t) \equiv \Gamma_P{}^2(t)$. The real part of the eikonal has simple poles at those values of t where the pomeron Regge trajectory takes on even integers: $\alpha_P(M_j^2 - iM_j\Gamma_j) = j = 2n$ (here j, M_j, and Γ_j j are the glueball spin, mass, and decay width). The existence of such poles provides the corresponding glueball resonance contributions into the t-channel amplitude.

If to consider $g_p(t)$ as a nontrivial unknown function of t, then expressions (2) and (3) are equivalent from the standpoint of description of data. However, the replacement $g_P(t) \to \Gamma_P(t)$ degrades the physical transparency of the model, since the shortened form (2) ignores the evident correlation between the behavior of the pomeron Regge residue and the pomeron Regge trajectory. Moreover, if the t-dependence of $g_p(t)$ is weak at low t, then usage of (2) instead of (3) leads to the loss of physical information.

Treating the high-energy elastic scattering of nucleons in the same way as the lepton-proton elastic scattering, we define the quantity $g_p(0)$ as the effective "pomeron charge" of nucleon, while the ratio $g_p(t)/g_p(0)$ should be considered as the nucleon "pomeron form factor". By analogy with the extraction of the proton charge radius from the proton charge form factor, one can extract the effective transverse "pomeron radius" of the nucleon from $g_p(t)$. Hence, a possible weak t-dependence of this quantity at low t could be interpreted as the effective transverse (quasi-)pointlikeness of the nucleon in the high-energy diffractive scattering regime.

Let us recall that although QCD itself does not predict the behavior of $\alpha_p(t)$ in the diffraction domain ($0 < -t < 2$ GeV^2), this function is expected to satisfy the following conditions [2,3]:

$$\frac{d^n\alpha_P}{dt^n} > 0 \quad (n = 1, 2, \ldots;\ t < 0), \qquad \lim_{t\to-\infty} \alpha_P(t) = 1. \tag{4}$$

The first condition originates from the dispersion relations for Regge trajectories (if not more than one subtraction is needed), and the second one follows from the natural presumption that at high values of the transferred momentum the exchange by a pomeron turns, due to asymptotic freedom, into the exchange by two noninteracting gluons, which can be considered in the same way as the exchange by two photons. At high energies such Born amplitudes behave as ~ s^1 [4].

Thus, if the effective transverse radius of nucleon is small, then, in view of restrictions (4), the t-dependence of eikonal (3) is determined mainly by the Herglotz function $\alpha_p(t)$. Consequently, the quality of the description of the differential cross section $\frac{d\sigma}{dt} = \frac{|T(s,t)|^2}{16\pi s^2}$ becomes extremely sensitive to the quantitative behavior of $\alpha_p(t)$ at low negative t. Such a sensitivity implies that the procedure of fitting $\alpha_p(t)$ and $g_p(0)$ to experimental angular distributions in a wide enough kinematic range could be considered as an implicit extraction of these quantities from the data.

FITTING TO THE EXPERIMENTAL DATA

Let us reconsider the model from [1], having singled out the factor $2^{-\alpha_P(t)}\pi\alpha'_P(t)$ in the Regge residue, as in (3):

$$\alpha_P(t) = 1 + \frac{\alpha_P(0) - 1}{1 - \frac{t}{\tau_a}}, \quad \Gamma_P(t) \to g_P(t) = \frac{g_P(0)}{(1 - a_g t)^2}. \tag{5}$$

The results of fitting $\alpha_p(t)$ and $g_p(t)$ to the experimental differential cross sections at $\sqrt{s} > 500$ GeV and 0.005 GeV^2 $< -t <$ 2 GeV^2 [5–14] are presented in Tables 1, 2, and 3 and Fig. 1. The deviation of the model predictions from the pp elastic scattering data in the dip region at √s = 62.5 GeV [15] can be explained by the noticeable contribution of secondary reggeons into the real part of the eikonal. A detailed discussion of this matter can be found in [1].

The D0 data [16] were not included into the fitting procedure, since they have a normalization uncertainty about 14.4 %. If to multiply them by a factor 0.92, the quality of the description becomes much better (see Table 4). The same can be said regarding the unrenormalized CDF data at √s = 1800 GeV [17], which are inconsistent with the E-710 data.

Table 1: The parameter values obtained via fitting to the high-energy differential cross-section data

Parameter	Value
$\alpha_P(0) - 1$	0.109 ± 0.017
τ_a	(0.535 ± 0.057) GeV^2
$g_P(0)$	(13.8 ± 2.3) GeV
a_g	(0.23 ± 0.07) GeV^{-2}

Table 2: The quality of description of the data [5–14] on the angular distributions of nucleon–nucleon scattering

$\sqrt{s}$ (GeV)	Number of points	χ^2
546 ($\bar{p}p$; UA1, UA4, CDF)	231	253
630 ($\bar{p}p$; UA4)	17	11
1800 ($\bar{p}p$; E710)	51	16
7000 (pp; TOTEM, ATLAS)	201	188
Total	500	468

Table 3: Predictions for the pp total and elastic cross sections and the forward logarithmic slope of the corresponding differential crosssections

$\sqrt{s}$ (GeV)	σ_{tot} (mb)	σ_{el} (mb)	B (GeV^{-2})
62.5	43.0 ± 4.4	7.4 ± 1.2	14.5 ± 0.8
200	53.3 ± 3.8	10.2 ± 1.2	16.0 ± 0.8
546	63.8 ± 3.3	13.2 ± 1.0	17.3 ± 0.9
1800	78.5 ± 3.4	17.8 ± 1.0	19.1 ± 1.0
7000	98.4 ± 5.4	24.5 ± 1.8	21.4 ± 1.1
8000	100.5 ± 5.7	25.3 ± 1.9	21.6 ± 1.1
13000	108.6 ± 6.9	28.2 ± 2.5	22.5 ± 1.2
14000	109.9 ± 7.1	28.6 ± 2.5	22.6 ± 1.2
32000	124.9 ± 9.7	34.1 ± 3.7	24.2 ± 1.4
100000	148.0 ± 14.1	42.8 ± 5.6	26.6 ± 1.7

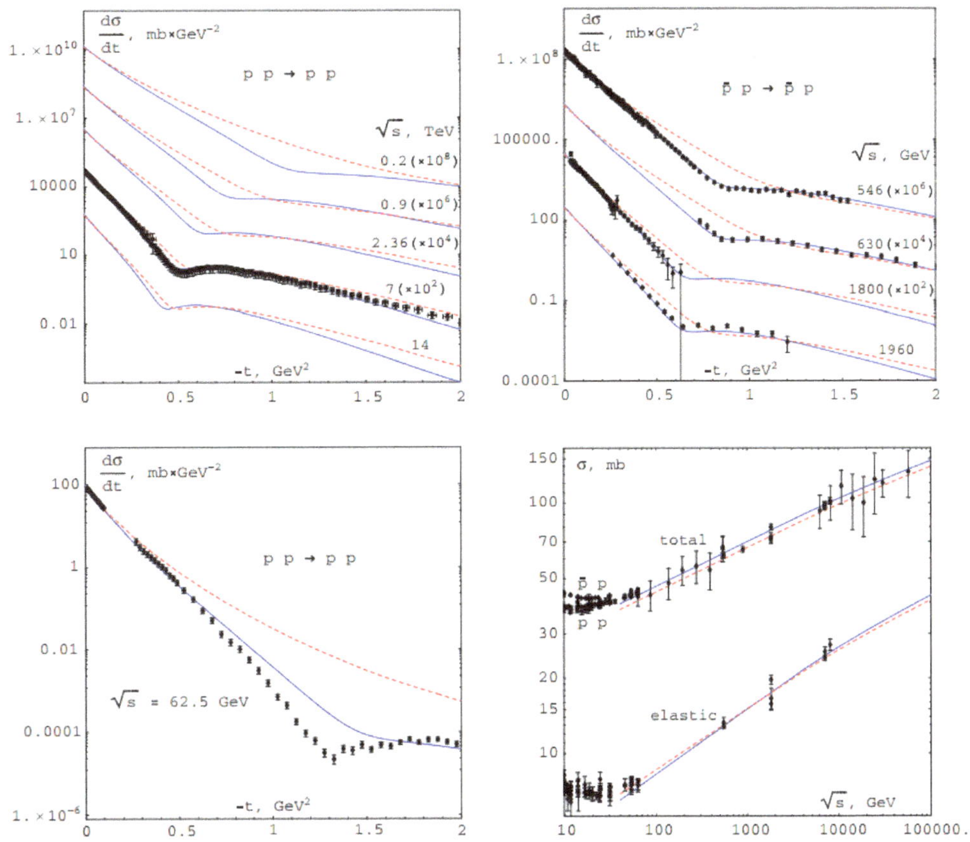

Figure. 1: The high-energy evolution of the nucleon–nucleon elastic scattering observables. The dashed lines correspond to the value $a_g = 0$ instead of $a_g = 0.23$ GeV^{-2}

JUSTIFICATION FOR USAGE OF THE PARAMETRIZATION FOR $A_P(T)$

Before drawing any physical conclusions from the fitting results, we should discuss our choice of parametrization for $\alpha_P(t)$. First, the essential nonlinearity of $\alpha_P(t)$ from (5), which, in fact, is not related to this concrete expression but follows from the restrictions (4), seems to be in contradiction with the observed approximate linearity of the Chew– Frautschi plots for secondary reggeons. However, we would like to point out that such a linear behavior of secondaries takes place at t > 0 and, in principle, is not guaranteed for t < 0. Moreover, it was determined in the framework of the BFKL approach [18,19] that secondary Regge trajectories behave as $(t) = \left(\frac{8\alpha_s(\sqrt{-t})}{3\pi}\right)^{1/2} + O(\alpha_s^{5/6}(\sqrt{-t}))$ at high negative t, where $\alpha_s(\mu)$ is the QCD running coupling. Hence, in view of the rather high slopes of the corresponding Chew–Frautschi plots in the resonance region, we have a very simple alternative: either secondary Regge trajectories are essentially nonlinear in the diffraction domain, or they are not monotonic functions in the interval $-\infty < t < 0$. A detailed discussion of this matter can be found in [20]. Second, we should ascertain that the obtained smallness of a_g (see Table 1) is not related directly to the chosen specific parametrization for $\alpha_P(t)$, since this is crucial for the main conclusion of the paper. If we consider a simple generalization of (5),

Table 4: The quality of description of the data [16,17] not included into the fitting procedure

$\sqrt{s}$ (GeV)	Number of points	χ^2
1960 ($\bar{p}p$; D0)	17	55
1960 ($\bar{p}p$; D0, multiplied by 0.92)	17	29
1800 ($\bar{p}p$; CDF)	26	178
1800 ($\bar{p}p$; CDF, multiplied by 0.88)	26	45

$$\alpha_P^{(k)}(t) = 1 + \frac{\alpha_P^{(k)}(0) - 1}{\left(1 - \frac{t}{\tau_a^{(k)}}\right)^k}, \quad g_P^{(k)}(t) = \frac{g_P^{(k)}(0)}{(1 - a_g^{(k)} t)^2}, \tag{6}$$

where k takes on integer and half-integer values [the parametrization (5) corresponds to k = 1], then it is possible to provide a satisfactory description of the data in a few cases of this series (see Table 5). The description for k = 1/2 and k ≥ 3 is unsatisfactory. For any k > 1, we find that $a_g^{(k)} < a_g^{(1)}$ Thus, the conclusion we draw below is not related to the specific form (5) of $\alpha_P(t)$ only. At the very end of this section we would like to note that although in this paper we restricted ourselves to the simplest test parametrizations, the quantities $\alpha_P(t)$ and gP(t) should, in general, be treated as unknown functions in the framework of the considered model. Namely, any expression for $\alpha_P(t)$ could be used which is analytic at t < 0 and satisfies the conditions (4). Certainly, it should provide a satisfactory description of the available data as well.

DISCUSSION

Now let us analyze the outcomes produced. The description of the nucleon–nucleon diffractive pattern in the considered kinematic range is satisfactory. The replacement of the fitted value of a_g by $a_g = 0$, which implies the neglect of the nucleon shape, disfigures the differential cross sections (see the dashed lines in Fig. 1). However, this distortion decreases with energy and does not become catastrophic already at the LHC energies. It is a consequence of the fact that, although we did not fix the nucleon transverse radius, it has turned out rather small. Indeed, the form factor $g_P(t)/g_P(0) = (1 - a_g t)^{-2}$ corresponds to a certain effective transverse distribution in the impact parameter representation: $f(b) = (4\pi a_g^3)^{-1} b K_1(b/a_g)$, where $K_1(x)$ is the modified Bessel function. The effective transverse radius of the nucleon obtained via an average over this distribution, $\sqrt{\langle b^2 \rangle} \sim 0.2 \div 0.3$ fm, is noticeably smaller than the effective transverse size of the diffractive interaction region (in the considered interval of the collision energy $\sqrt{2B} > 1$ fm, B is the forward logarithmic slope of dσ/dt, which increases with energy). Such a difference could be interpreted as if the pomeron were coupled to a very small zone inside nucleon. Hence, we arrive at the main conclusion:

Table 5: The fitting results for different variants of parametrization (6)

Parameter	$k = 3/2$	$k = 2$	$k = 5/2$
$\alpha_P^{(k)}(0) - 1$	0.108 ± 0.017	0.108 ± 0.017	0.108 ± 0.017
$\tau_a^{(k)}$	(0.708 ± 0.079) GeV^2	(0.874 ± 0.100) GeV^2	(1.040 ± 0.125) GeV^2
$g_P^{(k)}(0)$	(13.1 ± 2.2) GeV	(12.6 ± 2.0) GeV	(12.3 ± 2.0) GeV
$a_g^{(k)}$	(0.167 ± 0.075) GeV^{-2}	(0.120 ± 0.077) GeV^{-2}	(0.085 ± 0.080) GeV^{-2}
χ^2/DoF	0.915	0.93	0.97

• The quantitative evolution of the nucleon–nucleon elastic diffractive scattering observables at ultrahigh energies is determined mainly by the behavior of the pomeron Regge trajectory $\alpha_P(t)$ and the value of the effective "pomeron charge" $g_P(0)$ of nucleon and rather weakly depends on the nucleon shape.

The examined single-reggeon-exchange eikonal approximation is expected to be valid at least in the interval 0.2 TeV $\leq \sqrt{s} \leq$ 14 TeV. Therefore, the forthcoming TOTEM measurements (as well as desirable analogous measurements at the RHIC) could decrease the uncertainties of the model degrees of freedom and, thus, improve the phenomenological estimations as for the nucleon effective transverse radius, and therefore for the pomeron Regge trajectory and the pomeron charge of nucleon.

ACKNOWLEDGMENTS

The author thanks V. A. Petrov and R. A. Ryutin for discussion.

REFERENCES

1. A.A. Godizov, Phys. Lett. B 735, 57 (2014)
2. P.D.B. Collins, An Introduction to Regge Theory and High Energy Physics (Cambridge University Press, Cambridge, 1977)
3. P.D.B. Collins, P.J. Kearney, Z. Phys. C 22, 277 (1984)
4. H. Cheng, T.T. Wu, Phys. Rev. Lett. 22, 666 (1969)
5. J.R. Cudell, A. Lengyel, E. Martynov. http://www.theo.phys.ulg. ac.be/~cudell/data. Accessed 19 June 2012
6. UA1 Collaboration, G. Arnison et al., Phys. Lett. B 128, 336 (1983)
7. UA4 Collaboration, M. Bozzo et al., Phys. Lett. B 147, 385 (1984)
8. UA4 Collaboration, M. Bozzo et al., Phys. Lett. B 155, 197 (1985)
9. UA4 Collaboration, D. Bernard et al., Phys. Lett. B 198, 583 (1987)
10. UA4 Collaboration, D. Bernard et al., Phys. Lett. B 171, 142 (1986)
11. E-710 Collaboration, N. Amos et al., Phys. Lett. B 247, 127 (1990)
12. The TOTEM Collaboration, Europhys. Lett. 101, 21002 (2013)
13. ATLAS Collaboration, Nucl. Phys. B 889, 486 (2014)
14. Particle Data Group. http://pdg.lbl.gov/2014/hadronic-xsections/ hadron.html. Accessed 22 Mar 2015
15. U. Amaldi, K.R. Schubert, Nucl. Phys. B 166, 301 (1980)
16. D0 Collaboration, Phys. Rev. D 86, 012009 (2012)
17. CDF Collaboration, F. Abe et al., Phys. Rev. D 50, 5518 (1994)
18. J. Kwiecinski, Phys. Rev. D 26, 3293 (1982)
19. R. Kirschner, Z. Phys. C 67, 459 (1995)
20. A.A. Godizov, V.A. Petrov, JHEP 0707, 083 (2007)
21. L. Van Hove, Phys. Lett. B 24, 183 (1967)
22. F. Carlson, Dissertation, Uppsala, Sweden, 1914
23. G.N. Watson, Proc. R. Soc. 95, 83 (1918)
24. A. Zommerfeld, Partial Differential Equations in Physics (Academic Press, New York, 1949)

CHAPTER 9

PROBING SUPERFAST QUARKS IN NUCLEI THROUGH DIJET PRODUCTION AT THE LHC

Adam J. Freese, Misak M. Sargsian, Mark I. Strikman

INTRODUCTION

The dynamics of quantum chromodynamics (QCD) in the nuclear medium is one of the most interesting areas of modern nuclear physics. Many aspects of it are currently being investigated, including the formation of collective quark-gluon states such as quark-gluon plasma, the shadowing of the small x parton densities in nuclei, the hadronization of quarks and gluons within the nucleus, and the medium modification of partonic distribution functions in nuclei. Another interesting aspect of nuclear QCD is the possibility for quarks to carry a light cone momentum fraction higher than that of a free nucleon at rest. Deep inelastic scattering (DIS) from such a parton will correspond to Bjorken $x > 1$ (where x is normalized to run from 0 to the nuclear mass number A), and we will hereafter refer to quarks with such a light cone momentum fraction as superfast quarks.

Due to the short-range nature of strong interactions, detecting a superfast quark in a nucleus requires probing the nucleus at extremely short distance scales. The characteristic space-time distances in nuclei become shorter with an increase in Q^2 for fixed x owing to a property of QCD evolution, namely that a probed parton at high Q^2 will have come from a parent quark with a higher light cone momentum fraction. Thus, the theoretical expectation is that superfast quarks at large Q^2 will allow one to probe unprecedented small space-time distances in nuclei, on the order of 1/xmN [1,2].

Our current understanding of the dynamics of nuclei at short distances is extremely limited. Due to the short range nature of the strong interaction, one expects that they will be dominated by multi-nucleon short range correlations (SRCs) [3], which may include non-nucleonic degrees of freedom (such as and N N^* components), followed by the transition from baryonic to quark-gluon degrees of freedom.

There has been considerable progress recently made in studies of two-nucleon SRCs in inclusive and semi-inclusive nuclear processes, which have been dominated by quasielastic (QE) scattering of a high-energy probe (either an electron or a proton) off of a nucleon in the SRC [4–16]. These studies unambiguously established the existence of two-nucleon SRCs and measured the probability of a nucleon existing in one for the given nuclei. Due to the quasi-elastic nature of the scattering processes, however, it is much more difficult to reach kinematics dominated by three-nucleon SRCs [7]. Moreover, probing the transition to quark and gluon degrees of freedom has so far remained elusive. The possibility of probing superfast quarks in hard nuclear processes provides a new venue in studies of SRCs. From the theoretical point of view, as was mentioned above, the QCD evolution of quark distributions gives an additional handle on the ability to probe exceedingly small space-time distances through an increase in Q^2 (at the scale of Ioffe time $1/m_N\, x_B$). From the experimental point of view, deep inelastic processes do not restrict the phase space of the final states (as is the case for quasielastic processes), and hence one can use closure to express the cross section through the parton distribution functions of the nucleus. Measurement of nuclear parton distributions at $x \geq 1$ allows one to explore different aspects of nuclear short-range phenomena.

One of the conventional methods for probing superfast quarks in nuclei is deep inelastic scattering (DIS) from nuclei at Bjorken $x > 1$. A number of attempts have been made over the years. The attempts to measure quarks at $x \gtrsim 1$ were undertaken at CERN using a muon beam [17] and at FNAL

Citation: Adam J. Freese, Misak M. Sargsian, Mark I. Strikman, Probing superfast quarks in nuclei through dijet production at the LHC, doi:10.1140/epjc/s10052-015-3755-4.

using a neutrino beam [18], with mutually contradictory results. Measurements using electron beams were taken for $x \sim 1$ at SLAC [19] with $Q^2 \leq 10\,\text{GeV}^2$ and at Jefferson Lab for $x > 1$ and $Q^2 = 7\,\text{GeV}^2$ 2 [20]. However, this x range gets a significant contribution from higher-twist quasieleastic scattering up to fairly large $Q^2(\sim 15\,\text{GeV}^2)$. and at Jefferson Lab for For example, the nucleus/deuteron cross section ratio is reduced for x = 1 and $Q^2 = 10\,\text{GeV}^2$ by a factor of 2–2.5due to QE contributions [4]. The only way to avoid this is to probe larger values of $Q^2 \geq 30\,\text{GeV}^2$, for which the quasieleastic contribution will be a small correction [2]. Such experiments are currently included in the physics program for the 12 GeV upgrade of Jefferson Lab [21], and the first experimental data will be available within the next few years [22]. In this work, we propose a new approach for probing superfast quarks by considering dijet production in protonnucleus collisions at LHC kinematics. This approach is based on the possibility of relating the light cone momentum fractions of the initial partons to the measured kinematics of dijets; by selecting transverse jet momenta and pseudo-rapidities, one can isolate scattering off the superfast quarks within the nucleus. We develop a theoretical framework for calculating this reaction, which requires addressing several issues, such as modeling the high-momentum (short range) properties of the nuclear wave function, and calculating the medium modification of parton distributions within the nucleus and evolving this modification to the large Q^2 values relevant to the LHC. Within the framework we develop, we calculate the absolute cross section for the reaction and study its sensitivity to two- and three-nucleon SRCs in nuclei. The article is organized as follows: In Sect. 2 we review the formalism of the dijet production reaction, including its kine matics and the cross section formula. The cross section for dijet production depends on the nuclear parton distribution functions (PDFs), which are discussed in Sect. 3. Section 3 also discusses short range correlations and medium modifications, and how they factor into and affect the nuclear PDFs. This section contains a description of the structure of the elusive three-nucleon SRCs, and provides a good fit of EMC effect data using the color screening model of medium modifications. Section 4 discusses the hard subprocesses that contribute to dijet production and justifies the use of the leading order of QCD in calculating them. In Sect. 5, we present numerical estimates for the cross section, and find that at a characteristic p A luminosity reached at the LHC, rates are high enough to measure $x \geq 1$. Additionally, we find high rates for kinematics corresponding to the EMC effect, which at LHC energies allows the EMC effect to be studied with negligible higher-twist contributions. In this section we also discuss potential issues related to the accuracy of jet reconstruction. Conclusions and outlook are given in Sect. 6.

BASIC FORMALISM

The reaction we consider is the production of a dijet from a proton-nucleus collision,

$$p + A \rightarrow \text{dijet} + X, \tag{1}$$

where the kinematics of the two outgoing jets are fully determined. The reaction is treated at the leading order (LO) in perturbative QCD (pQCD), meaning that the jets are produced by a two-body parton-parton scattering process, in which one parton originates in the proton and the other in the nucleus. In our derivation, we consider the nuclear parton to have originated within a bound nucleon. This scenario is presented in Fig. 1, where we also define the kinematic variables of the reaction (1).

Jet Kinematics

We consider a reference frame where the incoming proton moves in the +z direction and the heavy nucleus, with charge number Z and mass number A, moves in the −z direction. The four-momenta are described using light cone coordinates, namely

$$p^\mu \equiv (p^+, p^-, \mathbf{p}_T), \tag{2}$$

where $p^\pm = E \pm p_z$ and pT is the two-component transverse momentum. Using the on-mass shell condition and the fact

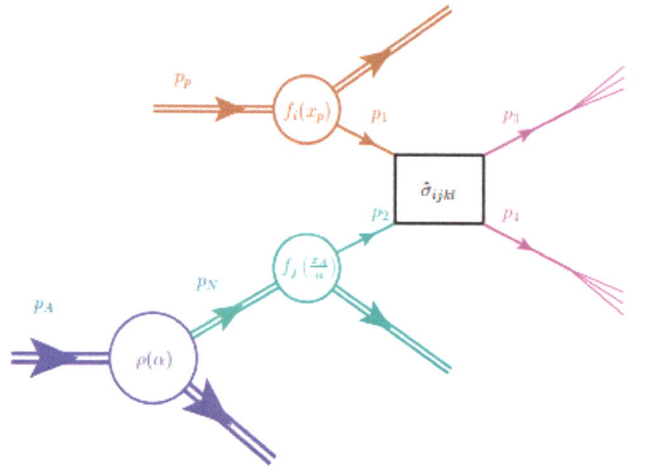

Figure. 1: Diagram of reaction

that the energies of the proton and nucleus greatly exceed their masses, in the collider reference frame one has

$$p_p^\mu = \left(p_p^+, \frac{m_p^2}{p_p^+}, \mathbf{0}_T\right) = (2E_0, 0, \mathbf{0}_T) = \left(\sqrt{\frac{As_{NN}^{\text{avg.}}}{Z}}, 0, \mathbf{0}_T\right) \tag{3}$$

$$p_A^\mu = \left(\frac{M_A^2}{p_A^-}, p_A^-, \mathbf{0}_T\right) = (0, 2ZE_0, \mathbf{0}_T) = \left(0, \sqrt{AZs_{NN}^{\text{avg.}}}, \mathbf{0}_T\right), \tag{4}$$

where E_0 is the beam energy per proton in the same reference frame, and $s_{NN}^{\text{avg.}} = 4\frac{Z}{A}E_0^2$ is the square of the average center-of-mass energy per nucleon. As an example, a leadproton collision with a beam energy of 4 TeV per proton would have an average center-of-mass energy per nucleon of $\sqrt{s_{NN}^{\text{avg.}}} \approx 5.02$ TeV. However, since the motion of the bound nucleon inside the nucleus will in general be variable, the actual center-of-mass energy per nucleon $\sqrt{s_{NN}}$ is not a fixed parameter, though it will be equal to $\sqrt{s_{NN}^{\text{avg.}}}$. N N in the limit where the nucleons do not interact and all move forward with equal momenta. At leading order, the collision results in an interaction between two partons, one each from the proton and the nucleus. Their respective four-momenta are labeled p_1 and p_2. We use a collinear approximation, in which the initial partons are treated as having zero transverse momentum. Moreover, they are treated as massless and on-shell, so

$$p_1 = (p_1^+, 0; \mathbf{0})$$
$$p_2 = (0, p_2^-; \mathbf{0}).$$

The light cone momentum fractions are defined for each parton as

$$x_p = \frac{p_1^+}{p_p^+} = \sqrt{\frac{Z}{A}} \frac{p_1^+}{\sqrt{s_{NN}^{\text{avg.}}}} \quad (5)$$

$$x_A = A\frac{p_2^-}{p_A^-} = \sqrt{\frac{A}{Z}} \frac{p_2^-}{\sqrt{s_{NN}^{\text{avg.}}}}. \quad (6)$$

Note that x_A is scaled by a factor of A. The rationale behind this is that the parton from the nucleus is found within one of its nucleons, and in a limiting case where the nucleons are all non-interacting and carry equal momenta, their light cone momentum is equal to $p_{N,\text{avg}}^- = \frac{p_A^-}{A}$, meaning $x_A \leq 1$ in this case. In reality, however, the bound nucleons do interact and it is possible that $p_2^- > p_{N,\text{avg}}^-$. In this situation, the nuclear parton originates from a nucleon which has a larger-than-average momentum $(p_N^- > p_{N,\text{avg}}^-)$. This indicates that finding an exceedingly large $x_A > 1$ will identify a bound nucleon with momentum significantly larger than average.

The parton momentum fractions cannot be directly measured. However, they can be related to the kinematic parameters of the jets. The jets from the proton and the nucleus are respectively ascribed four-momenta p_3 and p_4. At leading order, they come from the fragmentation of two partons (with the same momenta), which are treated as massless and on-shell. From energy-momentum conservation, it follows that

$$p_1 + p_2 = p_3 + p_4, \quad (7)$$

and due to the assumed collinear approximation, $p_{3T} = -p_{4T} \equiv p_T$. Using this relation, and neglecting the masses of the produced jets, we obtain:

$$p_3^+ p_3^- = p_4^+ p_4^- = p_T^2. \quad (8)$$

To proceed, we define the rapidity η as

$$\eta = \frac{1}{2}\log\left(\frac{p^+}{p^-}\right), \quad (9)$$

and use this, with the massless limit (in which $p^+p^- = p_T^2$) to obtain

$$p^\pm = p_T e^{\pm\eta}. \quad (10)$$

Applied to the jets of reaction (1), this results in $p_3^\pm = p_T e^{\pm\eta_3}$ and $p_4^\pm = p_T e^{\pm\eta_4}$. Using these relations, and energy momentum conservation, viz. Eq. (7), the light cone

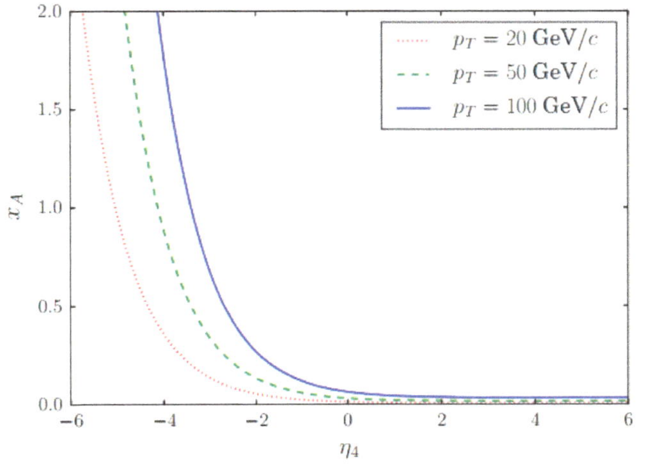

Figure. 2: Dependence of x_A on rapidity η_4 of parton from nucleus, for several transverse jet momenta. $\eta_3 = 0$ and $\sqrt{s_{NN}^{\text{avg.}}} = 5.02$ TeV in this plot

momenta of the initial partons can be expressed through jet kinematics in the following form:

$$p_1^+ = p_3^+ + p_4^+ = p_T\left(e^{\eta_3} + e^{\eta_4}\right)$$
$$p_2^- = p_3^- + p_4^- = p_T\left(e^{-\eta_3} + e^{-\eta_4}\right).$$

These relations can be used to express the momentum fractions x_p and x_A in terms of jet observables, namely

$$x_p = \sqrt{\frac{Z}{A}} \frac{p_T}{\sqrt{s_{NN}^{\text{avg.}}}}\left(e^{\eta_3} + e^{\eta_4}\right) \quad (11)$$

$$x_A = \sqrt{\frac{A}{Z}} \frac{p_T}{\sqrt{s_{NN}^{\text{avg.}}}}\left(e^{-\eta_3} + e^{-\eta_4}\right). \quad (12)$$

The main question that concerns us is whether partons with $x_A > 1$ can be seen in proton-nucleus collisions at the LHC. Equation (12) suggests that we should look for three conditions: large p_T, small (or somewhat negative) η_3, and small (or highly negative) η_4. Negative η_3 will require that the parton originating from the proton reverse its direction before fragmenting into a jet, which is highly unlikely owing to the large momentum transfer necessary to effect this. Thus, for a given p_T, the most plausible scenario is to keep η_3 small (~ 0), and to look for jets with high rapidity in the nucleus beam direction. Such a situation is presented in Fig. 2, where the dependence of x_A on η_4 is given at $\eta_3 = 0$ for different values of p_T. More practically, as will be discussed in Sect. 5,one can integrate over an η_3 range centered around $\eta_3 = 0$ in order to increase the cross section. In addition to the momentum fractions, we introduce a characteristic "hardness" scale Q^2, defined through the invariant momentum transfer:

$$Q^2 \equiv -t = -(p_1 - p_3)^2 = 2(p_1 \cdot p_3) = p_1^+ p_3^-$$
$$= p_T^2\left(1 + e^{-y_3+y_4}\right) \approx p_T^2.$$

This scale is used as both a renormalization and factorization scale.

Dijet Cross Section

The factorization formula for the hadronic cross section can be presented in terms of the partonic cross section as follows:

$$\sigma_{pA}=\sum_{ij}\int_0^1 \mathrm{d}x_p\int_0^A \mathrm{d}x_A f_{i/p}(x_p,Q^2)f_{j/A}(x_A,Q^2)\sigma_{\text{hard}},$$

where $f_{i/p}(x_p,Q^2)$ and $f_{j/A}(x_A,Q^2)$ are parton distribution functions (PDFs) for the proton and nucleus, respectively. This is similar in form to the standard factorization formula for pp collisions [23,24], and in fact reduces to it in the free nucleon limit. Since, at leading order, the momenta of the outgoing partons coincide with the jet momenta, we can integrate out the transverse momentum of one of the jets and express differentials through the rapidities of the outgoing jets. For this purpose, we present the differential form of Eq. (13) as

$$\begin{aligned}\mathrm{d}\sigma_{pA}=&\sum_{ijkl} f_{i/p}(x_p,Q^2)f_{j/A}(x_A,Q^2)\frac{1}{4(p_1\cdot p_2)}\\ &\times\frac{\overline{|\mathcal{M}_{ij\to kl}|^2}}{1+\delta_{kl}}(2\pi)^4 2\delta^{(1)}(p_1^+-p_3^+-p_4^+)\\ &\times\delta^{(1)}(p_2^- - p_3^- - p_4^-)\delta^{(2)}(\mathbf{p}_{3T}+\mathbf{p}_{4T})\\ &\times\frac{\mathrm{d}^3\mathbf{p}_3}{2E_3(2\pi)^3}\frac{\mathrm{d}^3\mathbf{p}_4}{2E_4(2\pi)^3}\mathrm{d}x_p\mathrm{d}x_A,\end{aligned} \tag{14}$$

where we have expressed four-momentum conservation through the light cone momenta. Here, the indices i, j, k, and l indicate parton types, and $M_{ij\to kl}$ is the invariant Feynman amplitude for the hard partonic scattering process.

Using the definitions of the light cone momenta fractions in Eqs. (5, 6), we can express the delta functions for $p^{\pm}$ conservation as:

$$\begin{aligned}2\delta^{(1)}(p_1^+-p_3^+-p_4^+)\delta^{(1)}(p_2^--p_3^--p_4^-)=&\frac{2A}{p_p^+p_A^-}\\ \times\delta^{(1)}\left(x_p-\frac{p_3^++p_4^+}{p_p^+}\right)\delta^{(1)}&\left(x_A-A\frac{p_3^-+p_4^-}{p_A^-}\right).\end{aligned} \tag{15}$$

Inserting this relation into Eq. (14), one can integrate out p_{4T}, x_p and x_A, resulting in

$$\begin{aligned}\mathrm{d}\sigma_{pA}=&\sum_{ijkl}\frac{A}{16\pi}\frac{1}{p_p^+p_A^-}\frac{1}{2(p_1\cdot p_2)}f_{i/p}(x_p,Q^2)\\ &\times f_{j/A}(x_A,Q^2)\frac{\overline{|\mathcal{M}_{ij\to kl}|^2}}{1+\delta_{kl}}\frac{\mathrm{d}p_{3z}}{E_3}\frac{\mathrm{d}p_{4z}}{E_4}\mathrm{d}p_T^2.\end{aligned}$$

The p_z elements can be rewritten as rapidities, since $d_\eta = \frac{dp_z}{E}$. In addition, as $2(p_1\cdot p_2)=\frac{x_p x_A}{A}p_p^+p_A^-$, we have

$$\begin{aligned}\mathrm{d}\sigma_{pA}=&\sum_{ijkl}\frac{1}{16\pi}\left(\frac{A}{p_p^+p_A^-}\right)^2\frac{f_{i/p}(x_p,Q^2)}{x_p}\frac{f_{j/A}(x_A,Q^2)}{x_A}\\ &\times\frac{\overline{|\mathcal{M}_{ij\to kl}|^2}}{1+\delta_{kl}}\mathrm{d}\eta_3\mathrm{d}\eta_4\mathrm{d}p_T^2.\end{aligned}$$

Since the square of the average center-of-mass energy per nucleon is

$$s_{NN}^{\text{avg.}}=\frac{p_p^+p_A^-}{A}, \tag{16}$$

the differential cross section can be written in the standard form:

$$\frac{\mathrm{d}^3\sigma}{\mathrm{d}\eta_3\mathrm{d}\eta_4\mathrm{d}p_T^2}=\sum_{ijkl}\frac{1}{16\pi(s_{NN}^{\text{avg.}})^2}\frac{f_{i/p}(x_p,Q^2)}{x_p}\frac{f_{j/A}(x_A,Q^2)}{x_A}\times\frac{\overline{|\mathcal{M}_{ij\to kl}|^2}}{1+\delta_{kl}}. \tag{17}$$

NUCLEAR PDFS

The main theoretical issue to be worked out is the nuclear PDFs $f_{j/A}(x_A,Q^2)$. Currently, there is a lack of strong experimental constraints on nuclear PDFs. Phenomenological parametrizations exist for several nuclei, and are based on experimental measurements of the nuclear structure function $F_2^{(A)}(x,Q^2)$) in deep inelastic scattering (DIS) experiments in a restricted range of x and Q^2. (cf. e.g. Refs. [25–30].) Most treatments of the nuclear PDF parametrize the ratio between the nucleus and the nucleon, which only makes sense for treating x_A up to 1. Reference [26] instead relates the nuclear and nucleonic PDFs though a Mellin convolution, and parametrizes the function that is convoluted with the nucleonic PDF. This approach has the possibility of accounting for $x_A > 1$, but owing to a lack of experimental constraint, Ref. [26] did not treat this region. In order to describe $x_A > 1$ in the absence of robust empirical parametrizations of the nuclear PDFs, we must account for them in terms of the nucleonic PDF theoretically. To this end, we must look at the theoretical relationship between the nuclear and nucleonic parton distributions.

Treating the nucleus strictly as a collection of unmodi- fied nucleons with a light cone distribution normalized by the baryon sum rule leads to a contradiction with experimental data—in particular for the ratio of nuclear and deuteron DIS cross sections. (See Sect. 3.2 for details.) However, the introduction of a priori unknown bound nucleon PDFs $f_{j/N}^{(b)}$ allows for the nuclear PDF to be expressed as a convolution between these PDFs and the nuclear light cone fraction distribution [1,31,32]. In the Bjorken limit, where the photon virtuality Q^2 and energy q0 both go to infinity at a fixed $x_A=\frac{AQ^2}{2M_A q_0}$, the convolution formula takes the following form:

$$f_{i/A}(x_A, Q^2) = \sum_N \int_x^A \frac{d\alpha}{\alpha} \int d^2\mathbf{p}_T f_{N/A}(\alpha, \mathbf{p}_T) \times f_{i/N}^{(b)}\left(\frac{x_A}{\alpha}, \alpha, \mathbf{p}_T, Q^2\right), \quad (18)$$

which can be derived as an impulse approximation. Here, $f_{N/A}(\alpha, p_T)$ is the light cone fraction distribution of a nucleon N in the nucleus A, described in Sect. 3.1; α is a scaled light cone momentum fraction given in Eq. (20) of the same section, and p_T is the transverse momentum of the nucleon. The (b) in the superscript of $f_{i/N}^{(b)}\left(\frac{x_A}{\alpha}, \alpha, \mathbf{p}_T, Q^2\right)$ signifies that this is an bound nucleonic PDF, which differs from the free nucleon PDF owing to modifications from the nuclear medium. In particular, it is a function of α and p_T in addition to $x_N = \frac{x_A}{\alpha}$ and the factorization scale Q^2 because of medium modifications. Equation (18) indicates that to construct nuclear PDFs, one needs to address two theoretical issues: the nuclear light cone fraction distribution $f_{N/A}(\alpha, p_T)$, and possible medium modification effects on the bound nucleon PDFs.

Light Cone Distribution and SRCS

We formally define the light cone fraction distribution in terms of the nuclear wave function as:

$$f_{N/A}(\alpha, \mathbf{p}_T) = \int \prod_{j=1}^{A} \frac{d\alpha_j d^2\mathbf{p}_{j,T}}{\alpha_j} \psi^\dagger(\alpha_1, \mathbf{p}_{T1}, \ldots, \alpha_A, \mathbf{p}_{TA}) \times \psi(\alpha_1, \mathbf{p}_{T1}, \ldots, \alpha_A, \mathbf{p}_{TA}) \times \delta^{(1)}\left(A - \sum_{i=1}^{A}\alpha_i\right)\delta^{(2)}\left(\sum_{i=1}^{A}\mathbf{p}_{iT}\right) \times \left\{\sum_{i=1}^{A}\delta^{(1)}(\alpha - \alpha_i)\delta^{(2)}(\mathbf{p}_T - \mathbf{p}_{iT})\right\} \quad (19)$$

where the (scaled) light cone momentum fraction α given by

$$\alpha = A\frac{p_N^+}{p_A^+}, \quad (20)$$

and p_T is the transverse momentum of the nucleon. The light cone fraction distribution is normalized to obey two sum rules, namely, the baryon number and momentum sum rules:

$$\int_0^A d\alpha \int d^2\mathbf{p}_T f_{N/A}(\alpha, \mathbf{p}_T) = 1 \quad (21)$$

$$\int_0^A d\alpha \int d^2\mathbf{p}_T \alpha f_{N/A}(\alpha, \mathbf{p}_T) = 1. \quad (22)$$

The light cone distribution thus defined can be related to the nuclear light cone density matrix $\rho_{N/A}(\alpha, \mathbf{p}_T)$ (cf. e.g. [1,3, 33]) in the following way:

$$f_{N/A}(\alpha, \mathbf{p}_T) = \frac{1}{\alpha}\rho_{N/A}(\alpha, \mathbf{p}_T). \quad (23)$$

We employ the light cone fraction distribution $f_{N/A}(\alpha, p_T)$, however, because it represents a direct analogy to PDFs, although for the distribution of nucleons in nuclei rather than of partons in hadrons. For our calculations, we expand the light cone fraction distribution as a sum of contributions from the nuclear mean field and multi-nucleon short range correlations (SRCs). The decomposition takes the form:

$$f_{N/A}(\alpha, \mathbf{p}_T) = f_{N/A}^{(\mathrm{MF})}(\alpha, \mathbf{p}_T) + \sum_{j=2}^{A} f_{N/A}^{(j)}(\alpha, \mathbf{p}_T), \quad (24)$$

where $f_{N/A}^{(\mathrm{MF})}(\alpha, \mathbf{p}_T)$ is the mean field part, and $f_{N/A}^{(j)}(\alpha, \mathbf{p}_T)$ is the distribution of j-nucleon SRCs.

Mean Field Distribution

The mean field part of the light cone fraction distribution describes how the nucleons in the nucleus would be distributed if they were only acted upon the mean field generated by the (A − 1) other nucleons. The mean field distribution can be related to the wave function of the nucleus, which is calculated in the non-relativistic limit since the relevant momenta are smaller than the typical Fermi momentum of heavy nuclei. The simplest mean field model considers the heavy nucleus as a degenerate Fermi gas, and the typical Fermi momentum is around 250 MeV/c [34]. More sophisticated momentum distributions (calculated based on, for example, Hartree-Fock approximations) still fall off sharply above the Fermi momentum, so the leading order relativistic corrections are at most on the order of the mean field momentum distribution strength above k_F, which $\Psi_{\mathrm{MF}}^{(N)}(p)$ (which carries an index of (N) for isospin since the wave function will in general be different for protons and neutrons) is likewise normalized to unity, so we equate

$$\left|\Psi_{\mathrm{MF}}^{(N)}(p)\right|^2 d^3\mathbf{p} = f_{N/A}^{(\mathrm{MF})}(\alpha, \mathbf{p}_T)d\alpha d^2\mathbf{p}_T.$$

In the non-relativistic limit, we can write, in the nuclear center of mass frame,

$$\alpha = A\frac{E + p_z}{m_A} \approx A\frac{m_N + p_z}{m_A}$$

$$p_z \approx \frac{m_A}{A}\alpha - m_N$$

$$dp_z \approx \frac{m_A}{A}d\alpha.$$

Consequently, we can identify

$$f_{N/A}^{(\mathrm{MF})}(\alpha, \mathbf{p}_T) = \frac{m_A}{A}\left|\Psi_{\mathrm{MF}}^{(N)}(p)\right|^2. \quad (25)$$

One modification must be made when accounting for SRCs. It is not the mean field part of the light cone distribution, but the light cone distribution taken as a whole that is normalized to satisfy the sum rules of Eqs. (21, 22). Therefore, $f_{N/A}^{(\mathrm{MF})}(\alpha, \mathbf{p}_T)$ must be scaled down by a factor of $a_1^{(N)}(A)$, which is defined as

$$a_1^{(N)}(A) = 1 - \sum_{j=2}^{A} \int d\alpha d^2\mathbf{p}_T f_{N/A}^{(j)}(\alpha, \mathbf{p}_T), \tag{26}$$

i.e. it subtracts off the probability that the nucleon is in a short range correlation. For instance, if a nucleon is in a short range correlation 25 % of the time, then $a_1^{(N)}(A)$ will be 0.75. In general, however, $a_1^{(N)}$ will be different for protons and neutrons. For numerical estimates, we will use the momentum distributions calculated in Ref. [35] for the mean field.

Two-Nucleon Correlations

Around 25 % of the time, a nucleon in a heavy nucleus is in a two-nucleon short range correlation [4]. A short range correlation occurs when two nucleons are separated by a distance on the order of 1 fm and consequently have a relative momentum larger than k_F . The analysis of recent experiments indicates [12] that in the momentum range from k_F to ~ 650 MeV/c, the nucleus is well-described by two-nucleon SRCs. Due to the short distance between the nucleons, the dynamics of the nucleons are primarily influenced by their mutual interaction rather than by the mean field. The most immediate consequences of two-nucleon SRCs are a large tail in the momentum distribution above the Fermi momentum, and large excitation energies of the residual system produced by the removal of the fast nucleons, which are not reproduced by mean field models [14].

Within the last decade, there has been considerable experimental and theoretical effort put into studying twonucleon SRCs and their properties. There are a fair number of triple-coincidence experiments demonstrating their existence [10,11,16] and demonstrating that two-nucleon SRCs primarily form between a proton and a neutron [9–11]. The fact that most two-nucleon SRCs are predominantly pn pairs has an important implication for momentum distributions (and thus for the 2N light cone fraction distribution): for neutron rich nuclei, a given proton is more likely to be in a short range correlation than a given neutron [36–39]. This means that $f_{p/A}^{(2)}(\alpha, \mathbf{p}_T) > f_{n/A}^{(2)}(\alpha, \mathbf{p}_T)$, and consequently (for the sum rules Eqs. (21, 22) to be satisfied), $a_1^{(p)} < a_1^{(n)}$. Another important aspect of two-nucleon SRCs is the universal high-momentum tail that they introduce to the momentum distribution of nuclei [1,37,40,41]. Most significantly, because of the dominance of spin-one, isosinglet SRCs, the high-momentum tail behaves to a good approximation like a scaled version of the high-momentum tail in the deuteron momentum distribution. Based on the arguments discussed above, a model of the two-nucleon SRC contribution to the nuclear light cone fraction distribution should incorporate two main properties: first, the universality of the form of the high momentum tail and its proportionality to the deuteron momentum distribution, and second, different high-momentum sharing between protons and neutrons in asymmetric nuclei. To account for universality, we use the light cone approximation of Refs. [1,3,33], which uses the requirement of rotational invariance to relate the light cone deuteron wave function to the non-relativistic wave function by using the internal p_n light cone momentum

$$k = \sqrt{\frac{m^2(\alpha-1)^2 + p_T^2}{\alpha(2-\alpha)}}, \tag{27}$$

which is obtained by the definitions $k_T = p_T$ and

$$\alpha = 1 + \frac{k_z}{\sqrt{m^2 + k^2}}, \tag{28}$$

and which also uses high momentum scaling to relate $f_{N/d}(\alpha, \mathbf{p}_T)$ to $f_{N/A}^{(2)}(\alpha, \mathbf{p}_T)$. To account for isopsin-singlet dominance, we use the model of Ref. [38,42], in which the high-momentum part of the nuclear momentum distribution is inversely proportional to the relative fraction of protons or neutrons. The model additionally includes a scaling factor $a_2(A)$, which is extracted from SRC studies in inclusive eA processes [4,7,8].

Accounting for these two effects leads to the two-nucleon SRC light cone fraction distribution:

$$f_{N/A}^{(2)}(\alpha, \mathbf{p}_T) = \frac{a_2(A)}{2\chi_N} \frac{\overline{|\psi_d(k)|^2}}{\alpha(2-\alpha)} \Theta(k - k_F), \tag{29}$$

where the extra factor of $\alpha(2-\alpha)$ is from the phase space of the struck and spectator nucleons. The details of the derivation of Eq. (29) Here, $\psi_d(k)$ is the relativistic, light-cone deuteron wave function, related to the non-relativistic wave function by

$$|\psi_d(k)|^2 = \sqrt{m^2 + k^2}\, |\psi_{NR}(k)|^2, \tag{30}$$

and χ_N is the relative abundance of nucleons of type N in the nucleus, i.e. $\chi_p = \frac{Z}{A}$ and $\chi_n = \frac{A-Z}{A}$. The factor of χ_N is present in Eq. (29) results in the equality

$$\sum_p f_{p/A}^{(2)}(\alpha, \mathbf{p}_T) = \sum_n f_{n/A}^{(2)}(\alpha, \mathbf{p}_T).$$

This relation is confirmed by quantum Monte-Carlo calculations for light nuclei up to A = 12, and for medium to heavy nuclei it was confirmed using a correlated basis calculation of the nuclear momentum distributions in a non-relativistic approach [43]. Such a relation is also in agreement with the recent analysis of momentum sharing in heavy nuclei [39].

In the present model we neglect by center of mass motion of the SRC.[2] As a result, the two-nucleon SRC contribution to the nuclear light cone distribution sets in as soon as $k > k_F$, which is signified by the step function $\Theta(k - k_F)$ in Eq. (29).

The remaining factor in Eq. (29), $a_2(A)$, is a scaling factor that describes how large the high-momentum tail of the nucleus is relative to the high-momentum tail of the deuteron. It is determined experimentally by examining ratios of quasielasitc inclusive cross sections of A(e, e X) and ^{2}H(e, e X) reactions:

$$a_2(A) = \frac{2\sigma_{eA}(x, Q^2)}{A\sigma_{ed}(x, Q^2)} \quad \text{at} \quad x \gtrsim 1.4 \quad \text{and} \quad Q^2 \gtrsim 1.5\,\text{GeV}^2,$$

where a roughly flat plateau in the ratio of the quasi-elastic cross sections is observed [4,7,8,13,15]. The values of $a_2(A)$ for various nuclei are well-constrained by experiment [4,8, 13] and it is equal to roughly 5.6 for medium nuclei such as iron-56. In this work, we will also use $a_2(^{208}Pb) = 5.6$, as no experimental data for this quantity exist for lead.

For numerical estimates, we use the Paris potential for parameterizing the non-relativistic deuteron wave function [44] in Eq. (30).

Three-Nucleon Correlations

When the light cone momentum fraction $\alpha > 2$ (and probably already for $\alpha \sim 1.6$ or 1.7), one expects the nuclear light cone distribution to be dominated by three-nucleon short range correlations. We consider a three-nucleon SRC to occur when the center of mass momentum of the 3N system is small, and when the struck nucleon, as well as the other nucleons in the SRC, have large ($> k_F$) internal momentum. Such a configuration can naturally generate a light cone momentum fraction exceeding 2.

There are two distinct mechanisms that could generate three-nucleon correlations. The first is a sequence of shortrange two-nucleon interactions, and the second an irreducible three-nucleon interaction. The second contributes a much larger removal energy part to the nuclear spectral function than the first mechanism [45]. Since the light cone momentum distribution is an integral over removal energies, one expects it to be dominated by the first mechanism, which occurs at lower values of removal energies than the irreducible 3N interaction. Therefore, neglecting the irreducible 3N interactions, we develop a model of three-nucleon SRCs where the struck nucleon obtains its momentum from a sequence of two-nucleon short range interactions. This model is based on the collinear approximation of Ref. [3], where the three nucleons in the SRC are moving collinearly prior to the interactions that generate the SRC. In the present work, we develop the model further using the recently observed dominance of p_n SRCs. This allows us to express the three-nucleon SRC as occurring through a sequence of p_n interactions, and thus the expressing 3N SRC part of the nuclear light cone fraction distribution as a convolution of two 2N SRC light cone momentum distributions.

$$f^{(3)}_{N/A}(\alpha,\mathbf{p}_T)=\{a_2(A)\}^2\frac{1}{\alpha}\int\frac{d\alpha_3 d^2\mathbf{p}_{3T}}{\alpha_3(3-\alpha-\alpha_3)}\left\{\frac{3-\alpha_3}{2(2-\alpha_3)}\right\}^2$$

$$\times\,\overline{|\psi_d(k_{12})|^2}\,\Theta(k_{12}-k_F)\,\overline{|\psi_d(k_{23})|^2}\,\Theta(k_{23}-k_F). \qquad (31)$$

Here, k_{12} and k_{13} are relative light cone momenta between pairs of nucleons in the three-nucleon SRC. Their functional forms are given by Eqs. (B_{34}, B_{26}). α_3 and p_{3T} are, respectively, the light cone momentum fraction and the transverse momentum of one of the spectator nucleons in the threenucleon SRC. Since the three-nucleon SRC is generated through a sequence of two short-range p_n interactions, it can only occur if the conditions under which the individual p_n interactions would occur are satisfied. For this reason, the threshold for three-nucleon SRCs to appear is that the relative light cone momenta of the pairs should each satisfy the threshold condition for which short range two-nucleon interactions occur, namely they should both be above the Fermi momentum k_F. It is worth noting that k_{12} and k_{23} are both functions of the light cone momentum fraction α3 of a spectator nucleon, which is integrated over. Because of this, the discontinuity of the $\Theta(\,)$ function is smeared out and $f^{(3)}_{N/A}(\alpha,\mathbf{p}_T)$) is itself a smooth function. The factor $\{a_2(A)\}^2$ appearing in Eq. (31) is a consequence of the fact that three-nucleon SRCs arise from a sequence of short-range p_n interactions. In such a scenario, for heavy nuclei, the probability of having a three-nucleon correlation should be proportional to the square of the probability of having a two-nucleon correlation Equation (31) may also be written in terms of a scaling factor $a_3(A)$, similar to $a_2(A)$ the scaling factor for twonucleon SRCs. If it is assumed that three-nucleon SRCs are universal in their behavior, the parameter $a_3(A)$ would be possible to extract in a similar way to $a_2(A)$ in the case of 2N SRCs. In particular, as is the case for $a_2(A)$, it should be possible to extract $a_3(A)$ through the ratio of quasielastic cross sections for eA and e^3He processes at high $x > 2$ and Q^2, viz.

$$a_3(A)=\frac{3\sigma_{eA}(x,Q^2)}{A\sigma_{e^3\mathrm{He}}(x,Q^2)}\quad : \quad x\gtrsim 2.4, \qquad (32)$$

provided that a 3N SRC plateau is actually observed in this ratio. In principle, the observation of $a_3(A)$ may provide an experimental test of the three-nucleon SRC model presented in this work. By definition, $a_3(^3He) = 1$, and for Eq. (31)

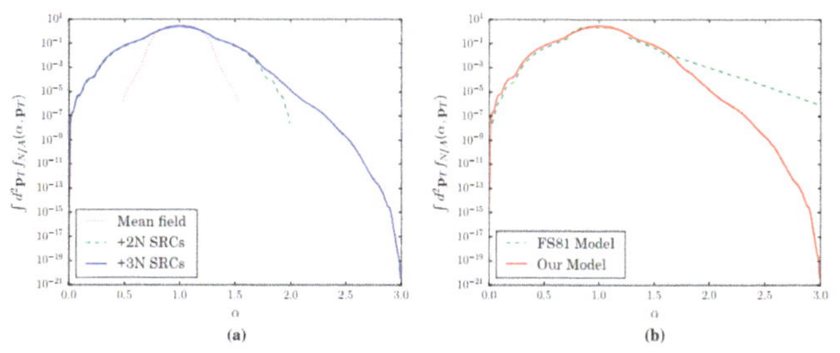

Figure. 3: The α dependence of the nuclear light cone distribution for ^{208}Pb. a Compares the mean field by itself to the light cone distribution with 2N and 3N SRCs. b Compares our model to the model in Chapter 5 of Ref. [3]

and Eq. (32) to be consistent would require that a3(A) ∝ $\left(\frac{a_2(A)}{a_2(^3\mathrm{He})}\right)^2$. Consequently, we have

$$f^{(3)}_{N/A}(\alpha,\mathbf{p}_T)=\left\{a_2(^3\mathrm{He})\right\}^2\frac{a_3(A)}{\alpha}\int\frac{d\alpha_3 d^2\mathbf{p}_{3T}}{\alpha_3(3-\alpha-\alpha_3)}$$

$$\times\left\{\frac{3-\alpha_3}{2(2-\alpha_3)}\right\}^2\overline{|\psi_d(k_{12})|^2}$$

$$\times\,\Theta(k_{12}-k_F)\overline{|\psi_d(k_{23})|^2}\,\Theta(k_{23}-k_F). \qquad (33)$$

The experimental status of a three-nucleon scaling plateau is, however, ambiguous. While observation of an $x > 2$ plateau was reported in Ref. [8] and in the analysis of SLAC data [1], a later experiment at similar kinematics and better resolution did not observe the plateau [13]. The discrepancy may be due to a resolution issue, as argued in Ref. [46]. On the theoretical side, one still needs to investigate whether the kinematics covered by these experiments correspond to high enough nucleon momentum that universality should be expected to set in. In lieu of experimental data for $a_3(A)$, we calculate $f^{(3)}_{N/A}(\alpha,\mathbf{p}_T)$ in this work using Eq. (31), which requires knowing only the experimentally known quantity $a_2(A)$. In Fig. 3, we present a calculation of the nuclear light cone distribution for ^{208}Pb. In Fig. 3a, we compare the light cone distribution using Eq. (24), considering only the mean field, and two-nucleon and three-nucleon correlations in addition. As our calculations show, 3N SRCs dominate starting only at $\alpha \gtrsim 1.7$.

In Fig. 3b, we have compared our model for the nuclear light cone density to a model in Ref. [3]—in particular, in its Eq. (5.11)—which we call the FS81 model. In this model, the relative weight of $j \geq 3$ SRCs was fixed based on a fit of $p + A \rightarrow p + X$ data with 400 GeV protons [47]. Since this model contains $j \geq$ 4-nucleon SRCs, it is expected to exceed our model, as can be seen in Fig. 3b.

Medium Modifications

Our next goal is to calculate PDFs for a bound nucleon, $f_{i/N}^{(b)}\left(\frac{x}{\alpha}, \alpha, \mathbf{p}_T, Q^2\right)$, which enter into Eq. (18). In doing so we should take into account that the nuclear medium is strongly interacting, and each nucleon has a high probability (20–30 %) of being in a short-range correlation where the nucleons themselves overlap. As a result, it should be expected that the parton distributions for nucleons immersed in the nuclear medium should be modified in some way. Deviations of the nuclear PDFs from expectations based on describing the nucleus as a system of unmodified nucleons was first observed by the European Muon Collaboration [48] in measurements of the ratio of cross sections for deeply inelastic muon scattering for nuclear and deuteron targets, i.e. in

$$\mathcal{R}_{\mathrm{EMC}}(x, Q^2) = \frac{2}{A}\frac{\sigma_{eA}}{\sigma_{ed}} \approx \frac{2}{A}\frac{F_2^{(A)}(x, Q^2)}{F_2^{(d)}(x, Q^2)}. \quad (34)$$

The ratio was measured in DIS kinematics which allowed the cross sections to be related to the structure function $F_2(x, Q^2)$. It was originally expected that, except for smearing from Fermi motion at high x, the structure functions should just be the structure functions for free nucleons, and thus the ratio should be 1. However, there is a dip below 1 in the range $0.3 < x < 0.7$ that cannot be reproduced by nucleonic motion alone [49,50]. This phenomenon is commonly referred to as the EMC effect.

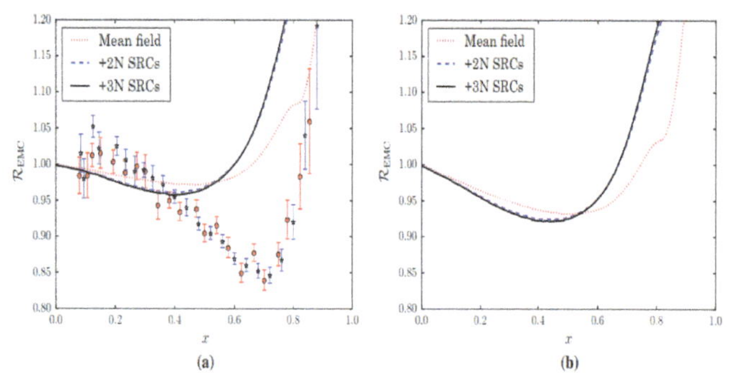

Figure. 4: EMC ratios at $Q^2 = 10$ GeV^2, without medium modifications. Data are from SLAC experiments: Stars (blue) are from [58]; circles (red) are from [59]. The nucleon structure functions are computed with the Bodek-Ritchie parametrization [60–62]. a EMC ratio for ^{56}Fe. b EMC ratio for ^{208}Pb

This dip in the ratio REMC is ubiquitous throughout nuclei, and roughly saturates for large A. Additionally, it is proportional to the local density of the nucleus considered [51], suggesting that it is an effect of the nuclear medium. Since the structure function $F_2(x, Q^2)$is directly related to the parton distribution functions, through

$$F_2(x, Q^2) = \sum_i e_i^2 x f_i(x, Q^2), \quad (35)$$

the EMC effect is commonly interpreted to be due to a modifi- cation of the PDF of a bound nucleon in the nuclear medium. Even though there is a consensus about this, the existing theoretical models based on modification of the bound nucleon PDFs differ due to different assumptions made about the nature of nuclear medium effects (for reviews of the EMC effect see [52,53]). It was argued in Refs. [54–56] that the EMC effect for the bound nucleon should be proportional to the first approximation of its kinetic energy, or more precisely to the off-shellness of the bound nucleon. This indicates that more modification should occur in the high momentum part of the nuclear wave function. Since our goal is to study the effects of SRCs on the reaction (1), and since SRCs dominate the high momentum part of the nuclear wave function, it will be necessary to account for the medium modifications that produce the EMC effect when studying this reaction. In particular, the medium modi- fication effects will be prominent in the nuclear parton distribution $f_{i/A}(x, Q^2)$ present in Eq. (17). Numerical estimates of the EMC effect will be based on the color screening model [1,54], which satisfactorily describes the phenomenology of the EMC effect observed in inclusive DIS reactions. (See also Ref. [57].)

Before proceeding with a specific medium modification model, we present estimates of the ratio given in Eq. (34) in the absence of medium modifications in order to assess the extent of the modifications that must occur. Here, $F_2^{(A)}(x, Q^2)$ is calculated using

$$F_2^{(A)}(x, Q^2) = \sum_N \int_x^A d\alpha \int d^2\mathbf{p}_T\, f_{N/A}(\alpha, \mathbf{p}_T) \times F_2^{(N,b)}\left(\frac{x}{\alpha}, \alpha, \mathbf{p}_T, Q^2\right), \quad (36)$$

which follows from Eq. (35) and the PDF convolution formula Eq. (18), and a phenomenological parametrization for free nucleons is used to estimate $F_2^{(N,b)}\left(\frac{x}{\alpha}, \alpha, \mathbf{p}_T, Q^2\right)$.[4] It should be noted that, in the absence of medium modifications, the nucleonic structure function $F_2^{(N,\mathrm{free})}\left(\frac{x}{\alpha}, Q^2\right)$ is not a function of α (except through $x_N = \frac{x}{\alpha}$) or $\mathbf{p}_T$. The nuclear light cone distribution $f_{N/A}(\alpha, \mathbf{p}_T)$ in Eq. (36) is calculated using the formalism of Sect. 3.1. As Fig. 4 shows, there is a significant discrepancy between the calculation and the data in the SRC-dominated region $x > 0.6$, indicating that nucleons in SRCs are more strongly modified by the nuclear medium than nucleons in the mean field. This is consistent with our expectation that modification should increase with nucleon virtuality, since SRCs occur at significantly larger momenta.

Correctly Defining X

Up until now, data from DIS experiments have been plotted against the kinematic Bjorken scaling variable $x_N = \frac{Q^2}{2m_p\nu}$,

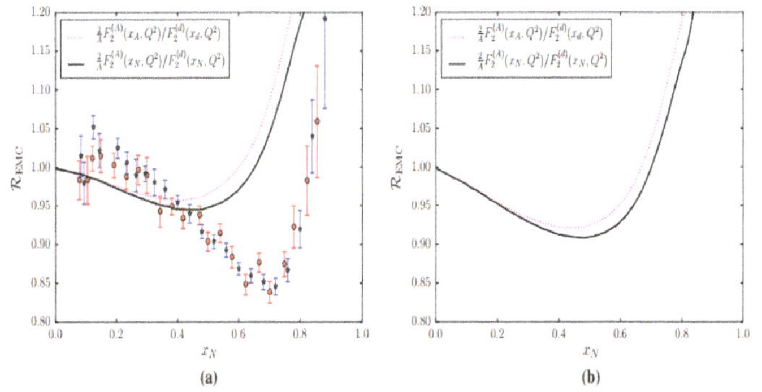

Figure. 5: EMC ratios at Q^2 = 10 GeV^2, without medium modifications. 2N and 3N correlations are included in theory curves. The dotted (magenta) lines compare structure functions as functions of the dynamical QCD scaling variables x_A and x_d, while solid (black) lines compare structure functions as functions of the kinematical variable x_N. Data and nucleonic structure function parametrizations are as in Fig. 4. a EMC ratio for ^{56}Fe. b EMC ratio for ^{208}Pb

where m_p is the mass of the proton. However, the Bjorken scaling variable that enters into the dynamics of QCD, including the convolution formulas of Eqs. (18, 36), is instead $x_A = \frac{AQ^2}{2M_A\nu}$. Thus, it is as arguments of xA and xd rather than of xN that the ratio of Eq. (34) should be presented in order to see dynamical QCD effects. By contrast, presenting data as a function of x_N instead of $x_{A/d}$ artificially shifts the arguments of the nuclear and deuteron structure functions in Eq. (34) by different amounts, since $x_A = \frac{Am_p}{M_A}x_N$ and $x_d = \frac{2m_p}{M_d}x_N$. Since the deuteron is a loosely bound system with little binding energy, the arguments for the nuclear and deuteron structure functions are shifted by substantially different amounts when the nucleus has a high binding energy. It has been observed [30,63] that presenting structure function ratios as an argument of x_N rather than x_A and x_d introduces an artificial dip in the ratio for $x_N \gtrsim 0.5$. In Fig. 5, we present theoretical calculations for REMC(x_A, x_d) (as a function of the dynamical QCD scaling variables $x_{A/d}$) and REMC(x_N) (as a function of the kinematical Bjorken scaling variable x_N) for both ^{56}Fe and ^{208}Pb. In these plots, the use of x_N as a variable produces an artificial dip, which partially explains the EMC effect up to $x_N \sim 0.5$. However, there continues to be a discrepancy between theory and data, which requires medium modification to explain.

Color Screening Model of the EMC Effect

Any nucleon can be expected to spend some of its time in a point-like configuration (PLC), in which its constituent quarks are compressed into a volume that is much smaller than the average volume of the nucleon. A PLC is largely invisible to the color force, in analogy to neutrally charged atoms in a gas whose van der Waals forces become weaker if the atoms are compressed. Due to the color neutrality of the nucleon, any color interaction between nucleons owes to higher moments (dipole, quadrupole, etc.), which decrease with distance between the color-charged constituents. Moreover, it can be shown by the renormalizability of QCD that meson exchange between nucleons, one of which is in a PLC, is suppressed [54]. Since nucleons in an average-sized configuration (ASC) and a PLC will interact differently, the probability that the nucleon can be found in either configuration should be modi- fied by the immersion of a nucleon in the nuclear medium. In particular, PLCs are expected to be suppressed compared to ASCs since the bound nucleon will assume a configuration that maximizes the binding energy and brings the nucleus to a lower-energy ground state. The change in probability can be estimated using nonrelativistic perturbation theory, as has been done in Refs. [1, 54]. Since the PLC components in the nucleon PDFs are expected to dominate in the x ≥ 0.6 region, it was found that in this region the bound nucleon PDFs should be modified by a factor $\delta_A(k^2, x_N)$, which depends on the nucleon momentum (or virtuality) as

$$\delta_A(k^2, x_N \gtrsim 0.6) = \frac{1}{(1+z)^2} \tag{37}$$

$$z = \frac{\frac{k^2}{m_p} + 2\epsilon_A}{\Delta E_A}, \tag{38}$$

Where ϵ_A is the nuclear biding energy and ΔE_A characterizes the nucleon excitation energy in the medium. In analogy with

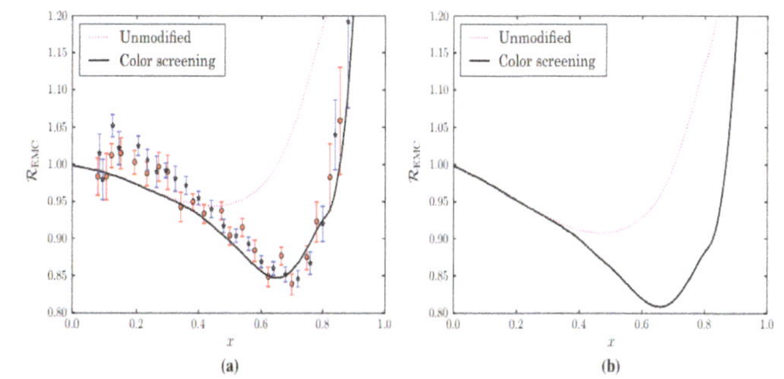

Figure. 6: EMC ratios with and without the color screening model of medium modifications. 2N and 3N correlations are included. Q^2 = 10 GeV^2, and data and nucleonic structure function parametrizations are as in Fig. 4. a EMC ratio for ^{56}Fe. b EMC ratio for ^{208}Pb

electric charge screening, this is called the color screening model of the EMC effect. We shall use the color screening model as an example for accounting for medium modifications in calculation of the dijet cross section. For this, one should first fix the parameter ΔE_A as well as extrapolate the suppression factor δ to the whole range of x. To do so we implement the color screening model in calculating the nuclear structure function $F_2^{(A)}(x, Q^2)$ and compare it with the available lepton DIS data. Such an implementation is done by substituting

$$F_2^{(N,b)}\left(\frac{x}{\alpha}, \alpha, \mathbf{p}_T, Q^2\right) = F_2^{(N)}\left(\frac{x}{\alpha}, Q^2\right)\delta\left(k^2(\alpha, \mathbf{p}_T), \frac{x}{\alpha}\right) \tag{39}$$

in the convolution formula of Eq. (36) and using the phenomenological parameterization mentioned above [60–62] for the free nucleon structure function $F_2^{(N)}(x, Q^2)$. It is worth emphasizing that this phenomenological parametrization also contains contributions from higher-twist effects, so Eq. (39) implies that higher-twist effects are modified the same way as partonic distributions. Justification for such an approximation follows from the fact that no strong Q^2 dependence has been observed for the experimentally measured range $2\,\mathrm{GeV}^2 < Q^2 < 200\,\mathrm{GeV}^2$ though errors for large Q^2 are fairly large [52,64].

In comparing the color screening predictions of Eq. (36) with the data, we assumed that no medium modification occurs at $x_N \lesssim 0.45$, and the region between $0.45 < x_N < 0.6$ is interpolated linearly[5]:

$$\delta_A(k^2, 0.45 < x_N < 0.6)$$
$$= 1 + \frac{x_N - 0.45}{0.15}\left\{\delta_A(k^2, x_N \gtrsim 0.6) - 1\right\}. \tag{40}$$

Since the nucleon excitation energy ΔE_A is a dynamical parameter depending on the spin-isopsin composition of interacting nucleons, we fitted it separately for the mean field and SRC domains of nucleon momenta. For a nucleon in the mean field of a heavy nucleus, we expect the excitation energy ΔE_A to be in the range 300–500 MeV, namely between the excitation energies of a Δ and an N^* resonance. The best fit to data appears to be $\Delta E_A \approx 500$ MeV. For the deuteron, as well as for a nucleon in a 2N SRC, we expect the lowest excited state to be a $\Delta\Delta$ configuration, giving $\Delta E_d \approx 600$ MeV. We assumed the same $\Delta E = 600$ MeV

excitation energy for a nucleon in a 3N SRC, since such a correlation in our approach is generated through a sequence of 2N correlations. We take ϵ_A to be the binding energy per nucleon of the nucleus under consideration when in the mean field and neglect it for a nucleon in an SRC, since it is expected to be negligible compared to k^2/m in an SRC.

The formula for k in terms of α and p_T also depends on whether the nucleon is in the mean field or a short range correlation. The non-relativistic approximation is adequate for describing a nucleon in the mean field, and for two-nucleon SRCs Eq. (27) is valid. For three-nucleon SRCs,

$$k^2 = 2\frac{(\alpha - 1)^2 m^2 + \mathbf{p}_T^2}{\alpha(3 - \alpha)} \tag{41}$$

is a reasonable approximation if the relative momentum between the spectator nucleons is small. As can be seen in Fig. 6, using the parameters described above results in reasonably good description of the EMC

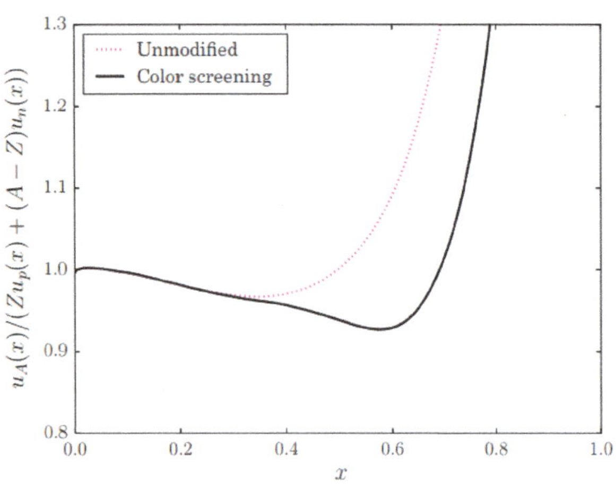

Figure. 7: Isospin-weighted ratio of the u quark distribution for ^{208}Pb and the free nucleon, both with and without the color screening model. 2N and 3N correlations are included in the ^{208}Pb distribution. Curves are at $Q^2 = 10$ GeV^2. PDF parametrization is CT10 [65]

effect. After fixing the δ factor of Eq. (38) and its xdependence, the color screening model is now applied to quark PDFs rather than structure functions in the following form:

$$f_{i/A}(x, Q^2) = \sum_{j=1}^{3}\sum_{N}\int_x^A \frac{d\alpha}{\alpha} d^2\mathbf{p}_T\, f_{N/A}^{(j)}(\alpha, \mathbf{p}_T)$$
$$\times f_{i/N}\left(\frac{x}{\alpha}, Q^2\right)\delta_A^{(j)}\left(k^2(\alpha, \mathbf{p}_T), \frac{x}{\alpha}\right). \tag{42}$$

The sum over j here indicates the sum over the mean field (j = 1) and j-nucleon SRC (j = 2, 3) parts of the light cone fraction distribution. $\delta_A^{(j)}\left(k^2(\alpha, \mathbf{p}_T), \frac{x}{\alpha}\right)$ has an index j here to indicate that the color screening model affects the mean field and SRCs differently.

To estimate the medium modification for gluon distributions we notice that the medium modifications occur predominantly at large x, where valence quarks dominate. Gluons in this region would originate as radiation from quarks, and would therefore inherit medium modifications from the quarks. For this reason, we apply Eq. (42) to gluons as well as quarks and anti-quarks.[6] The result of applying the color screening model to the nuclear PDF is demonstrated in Fig. 7.

Evolution of Medium Modifications

The EMC effect has so far been observed in a restricted range of Q^2, around $Q^2 \sim 2$–200 GeV^2, with the greatest modi- fication at $x \gtrsim 0.6$. The experimental errors, especially at large Q^2, do not allow the identification of any appreciable Q^2 dependence of the EMC ratio. However, from the point of view of QCD, one should expect at least logarithmic Q^2 dependence, owing to the evolution of high-x, lowQ^2 partons into lower x with an increase of Q^2. This will be the case with evolution to the high $Q^2 \sim 10^4$ GeV^2 relevant to the LHC. In our estimates of medium modifi- cation in dijet production, we predict how the EMC effect will manifest at $Q^2 = 10^4$ GeV^2 using DGLAP evolution [66–68].

It is necessary when considering medium modifications to use DGLAP evolution to evolve the nuclear PDFs obtained at low Q^2 to high Q^2, rather than to apply the mediummodified convolution formula Eq. (42) to a high-Q^2 nucleonic PDF. The factors $\delta_A^{(j)}(k^2, x_N)$ in the color screening model, for instance, contain x dependence in the form of a linear interpolation, preventing this term from being factored out of a DGLAP integral. The x dependence in this factor is particular to the model's purpose of explaining DIS cross section ratios at low Q^2, so cannot a priori be expected to have the same x dependence at large Q^2. Eq. (42) cannot be directly applied at high Q^2 without evolving $\delta_A^{(j)}(k^2, x_N)$, which has unknown Q^2 dependence. However, by applying DGLAP evolution to the medium-modified nuclear PDF as a whole, which is obtained from Eq. (42) at low Q^2, the correct Q^2 dependence of medium modification can be obtained.

The standard DGLAP formula is modified to account for the possibility of x > 1 inherent in dealing with nuclei; in particular, we have

$$\frac{\partial f_{i/A}(x, Q^2)}{\partial \log(Q^2)} = \frac{\alpha_s(Q^2)}{2\pi} \sum_{ij} \int_x^A \frac{dy}{y} P_{ij}\left(\frac{x}{y}\right) f_{j/A}(y, Q^2). \tag{43}$$

Here, P_{ij} are the Alterelli-Parisi splitting functions[68]. In our calculation, we solved Eq. (43) numerically within the leading order approximation, using the nuclear PDFs obtained in the previous sections as input. (Next-to-leading order corrections to the evolution of medium modifications are expected to be negligible at the high x being considered here.) The results of this calculation are shown in Fig. 8, which considers isospin-weighted ratios of the nuclear u quark distribution to the free nucleonic u distribution, since the effects of evolution are most easily seen in ratios. Figure 8a in particular compares the effects of evolution on nuclear PDFs with and without medium modification accounted for. Figure 8b shows how the modified PDFs, as functions of Q^2, evolve at several fixed values of x. At x ~ 0.5, where the PDF ratio mostly deviates from 1 due to medium modifications, the ratio is roughly constant as Q2 increases, suggesting little evolution of the medium modification itself. On the other hand, the PDF ratio at x ~ 0.7 is more strongly Q^2-dependent, showing that at high Q^2, the enhancement of the nuclear PDF relative to the nucleonic PDF due to Fermi motion and SRCs becomes more important at lower x— a fact that can also be seen in Fig. 8a, where the steep increase in the ratio occurs at a lower x when Q^2 is increased.

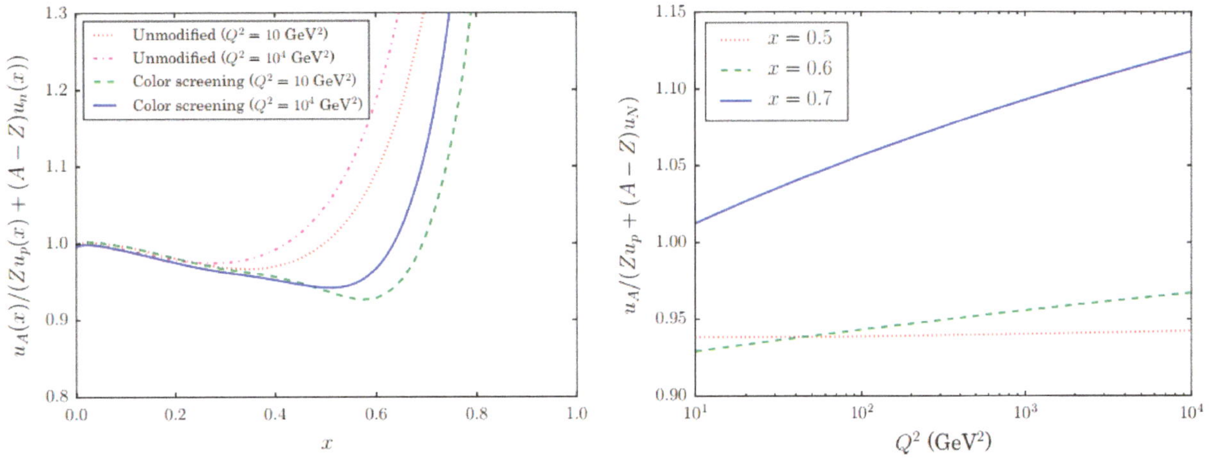

(a) **(b)**

Figure. 8: Isospin-weighted ratio of u quark distribution for ^{208}Pb to that of the free nucleon. 2N and 3N correlations are included in the ^{208}Pb distribution. a As a function of x. Unevolved lines are at Q^2 = 10 GeV^2, and evolved at Q^2 = 10,000 GeV^2. b As a function of Q^2, with the EMC effect accounted for by the color screening model. The Q^2 = 10 GeV^2 PDFs are computed using the CT10 parametrization [65], as well as the convolution formula Eq. (18)

The high-x enhancement of nuclear PDFs moves to the left in Fig. 8a because, as a parton evolves, it radiates other partons which carry off part of its forward light cone momentum, causing its light cone momentum fraction x to decrease as Q^2 increases. Accordingly, a quark with a somewhat moderate value of x which is probed at the high Q^2 characteristic of the LHC could have originated within a "primordial," low-Q^2 quark with a higher momentum fraction x. The extent to which lower-x, higher-Q^2 partons can be considered as originating within higher-x, lower-Q^2 partons is quantified by the QCD evolution trajectory, which is detailed in Ref. [69].

The QCD evolution trajectory starts at a point (x_0, Q_0^2) in x-Q^2 parameter space, and describes the evolution of a particular parton flavor i. Another point (x, Q^2) lies on the evolution trajectory if half of the PDF $f_{i/A}(x, Q^2)$ originated from evolution of the portion of the $Q^2 = Q_0^2$ PDF at $x \geq x_0$, i.e. if, given $\tilde{f}_{i/A}(x, Q^2; x_0, Q_0^2)$ defined by

$$\frac{\partial \tilde{f}_{i/A}(x, Q^2; x_0, Q_0^2)}{\partial \log(Q^2)} = \frac{\alpha_s(Q^2)}{2\pi} \sum_{ij} \int_{x_0}^A \frac{dy}{y} P_{ij}\left(\frac{x}{y}\right) \times \tilde{f}_{j/A}(y, Q^2; x_0, Q_0^2) \tag{44}$$

$$\tilde{f}_{i/A}(x, Q_0^2; x_0, Q_0^2) = f_{i/A}(x, Q_0^2)\Theta(x - x_0), \tag{45}$$

then the x for a given Q^2 on the trajectory is given by the requirement

$$\tilde{f}_{i/A}(x, Q^2; x_0, Q_0^2) = \frac{1}{2} f(x, Q^2). \tag{46}$$

Note that Eq. (45) defines an initial condition, which ensures that when its argument x < x0, the function $\tilde{f}_{i/A}(x, Q^2; x_0, Q_0^2)$ comes from the evolution of partons with x ≥ x0, and thus that Eq. (46) has the required interpretation.

With these considerations in mind, several QCD evolution trajectories for the nuclear u quark distribution in ^{208}Pb are presented in Fig. 9. Consistent with the findings of Ref. [69], the trajectories do not move very far to the left for large x0, so there is little hope of seeing primordially superfast (x > 1) quarks at moderate x values; even the trajectory which starts at $(x_0 = 1, Q_0^2 = 10\text{ GeV}^2)$ fails to reach $x = 0.95$ by the time $Q^2 = 40000\text{ GeV}^2$, corresponding to a p_T = 200 GeV/c jet.

HARD SUBPROCESSES

We have developed a model of nuclear PDFs at x > 0.1 that accounts for both short range correlations and effects from medium modification. In addition to a model for PDFs, however, we also need to calculate the squares of the invariant matrix elements $\overline{|\mathcal{M}_{ij}|^2}$ for the hard partonic subprocesses that contribute to Eq. (17). We consider two partons, labeled i and j, to exist in the initial state of the reaction—one from

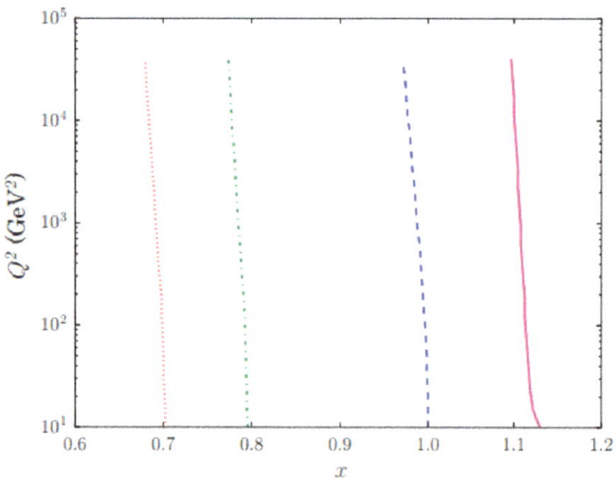

Figure. 9: Evolution trajectories for ^{208}Pb, from $Q^2 = 10$ GeV^2 to $Q^2 = 40{,}000$ GeV^2. 2N and 3N SRCs as well as color screening are considered. The $Q^2 = 10$ GeV^2 PDFs are computed using the CT10 parametrization [65], and Eq. (18)

Table 1: Squared matrix element for hard partonic processes at leading order, summed over final spins and averaged over initial spins, and divided by $g_s^4 \cdot j \neq k$ in subprocesses where both indices are present. The table is from Ref. [70].

Subprocess	$\frac{\overline{\lvert\mathcal{M}\rvert^2}}{g_s^4}$
$q_j + q_k \to q_j + q_k$	$\frac{4}{9}\frac{s^2+u^2}{t^2}$
$q_j + q_j \to q_j + q_j$	$\frac{4}{9}\left(\frac{s^2+u^2}{t^2} + \frac{s^2+t^2}{u^2}\right) - \frac{8}{27}\frac{s^2}{ut}$
$q_j + \bar{q}_j \to q_k + \bar{q}_k$	$\frac{4}{9}\frac{t^2+u^2}{s^2}$
$q_j + \bar{q}_j \to q_j + \bar{q}_j$	$\frac{4}{9}\left(\frac{s^2+u^2}{t^2} + \frac{t^2+u^2}{s^2}\right) - \frac{8}{27}\frac{u^2}{st}$
$q_j + \bar{q}_j \to g + g$	$\frac{32}{27}\frac{u^2+t^2}{ut} - \frac{8}{3}\frac{u^2+t^2}{s^2}$
$g + g \to q_j + \bar{q}_j$	$\frac{1}{6}\frac{u^2+t^2}{ut} - \frac{3}{8}\frac{u^2+t^2}{s^2}$
$q_j + g \to q_j + g$	$-\frac{4}{9}\frac{u^2+s^2}{us} + \frac{8}{3}\frac{u^2+s^2}{t^2}$
$g + g \to g + g$	$\frac{9}{2}\left(3 - \frac{ut}{s^2} - \frac{us}{t^2} - \frac{st}{u^2}\right)$

each of the proton and the heavy nucleus participating in the reaction (1). At leading order, a dijet is associated with two partons in the final state, which we label k and l. Each parton can in principle be a quark, an anti-quark, or a gluon, although what partonic subprocesses are possible is limited by quark flavor conservation. A list of the possible two-parton to twoparton subprocesses and their squared matrix elements is presented in Table 1 (cf. [70]). At next-to-leading order (NLO), a dijet can be associated either with two or with three partons in the final state, provided two of the three partons have small enough differences in both their rapidities and their azimuthal angles, so that $\Delta\phi^2 + \Delta\eta^2 < R^2$, where R is the cone radius used for defining jets. Additionally, three-parton dijets are needed to cancel the infrared divergences in the NLO contributions to the two-parton to two-parton matrix elements [71]. Finally, identifying two partons as a single jet in the NLO contribution alters the simple kinematical formula Eq. (6) for determining x_A through dijet kinematics.

These facts indicate that we should verify the adequacy of the leading order approximation. To make this verification, we perform a leading order calculation of the reaction (1) for the case of A = 1, i.e. for dijet production from protonproton collisions, for which a fair amount of experimental data in the kinematical regime of interest exist. We compare the calculation within the leading order to experimental data from Ref. [72]. The data, however, are given in terms of a two-fold differential cross section $d^2\sigma/dm_{JJ}d|y^*|$, where $m_{JJ}^2 = (p_3 + p_4)^2$ and y^* * is the rapidity of an individual jet in the dijet center of mass frame. The relationship of this two-fold differential cross section to the three-fold differential cross section of Eq. (17), and a calculation of the leading-order calculation with the data is given in Fig. 10. As the comparison shows, we have a reasonably good agreement with the data that validates the use of leading order for calculating the hard subprocesses that contribute to reaction (1).

NUMERICAL ESTIMATES

In the previous sections, we have developed all the components necessary to calculate the three-fold differential cross section of Eq. (17) for the dijet production reaction (1). It is our intention to use them to determine the sensitivity of the dijet production reaction (1) to the presence of short range correlations in the nucleus, as well as to medium modifi- cation effects. Additionally, we will look at nuclear partons with $x_A > 1$ in particular. Since x_A can be related to measurable jet kinematics via Eq. (6), it is necessary to elaborate on the kinematics we will consider. Once the kinematics considered are clarified, we will present numerical estimates of the three-fold differential cross section of Eq. (17) at these kinematics, and additionally we will estimate the partially and fully integrated cross sections to find out if events corresponding to $x_A > 1$ are frequent enough to be measurable at the LHC.

Kinematics Considered

Since one of our goals is to probe $x_A > 1$, we consider kinematics that maximize x_A. 7 According to Eq. (6), three variables are related to xA. The jet rapidities η_3 and η_4 should be as small or highly negative as possible in order to maximize x_A. Additionally, the transverse jet momentum p_T should be as large as possible. On the other hand, non-central jets are more difficult to detect and identify, and the jet cross section for large p_T is known to drop rapidly. It is necessary to find a balance between kinematics that will increase x_A and those that will produce an appreciable yield.

Since central jets are easier to detect, and since we are interested in superfast partons in the nucleus, we consider dijets with one jet produced in the central rapidity region where LHC detectors have the best resolution and the second jet moving in the direction of the nucleus beam. In particular, the rapidity ranges considered are $-2.5 < \eta_3 < 2.5$ and $3 < -\eta_4 < 5$. We also consider a p_T range from 40 to 200 GeV/c.

Three-Fold Differential Cross Section

Equation (17) can be used to obtain numerical estimates of the three-fold differential cross section $d^3\sigma/d\eta_3 d\eta_4 dp_T^2$ for the dijet production reaction (1). Numerical estimates are presented in Fig. 11, where estimates with and without short range correlations are presented for comparison, and likewise estimates with and without medium modification are presented. As can be seen in Fig. 11a, the presence of SRCs increases the cross section for superfast $(x_A > 1)$ quarks considerably. Figure 11b demonstrates the relative importance of three-nucleon SRCs in particular at large x_A. On the other hand, medium modification decreases the cross section, especially at large x where highly-modified SRCs dominate the cross section, as can be seen in Fig. 11c, d.

In addition to comparisons of the effects of SRCs and medium modifications within the model for the nuclear light cone density presented in this work, we compare in Fig. 12 the prediction for the three-fold differential cross section using our model to the one made using the FS81 model [3]. The FS81 model uses exponential decay to model the large α part of the nuclear light cone density, and likewise the nuclear PDF calculated using this model exhibits exponential decay for large x_A. It is for this reason the curve for the FS81 model appears as a straight line in the log plot of Fig. 12. Qualitatively, the model of this work can be distinguished experimentally from the FS81 model by determining whether the three-fold differential cross section of Eq. (17) falls exponentially or faster than a simple exponential.

Partially Integrated Cross Section

Next we study the sensitivity to SRCs and medium modifi- cation effects of the one-fold differential cross section $\frac{d\sigma}{dp_T}$, which is obtained by integrating Eq. (17) over the pseudorapidities of the jets. The main goal in doing so is to maximize the absolute cross section while retaining highest possible sensitivity to short-range nuclear phenomena, namely the x > 1 region. For this we choose the ranges of rapidity integrations in order to maximally preserve the sensitivity from the x > 1 domain. The ranges of η3 and η4 that are considered have been discussed above, namely the jet originating from the proton should be in the central region $-2.5 < \eta_3 < 2.5$ and the jet originating from the nucleus should be forward in the nucleus beam region, i.e. $-5 < \eta_4 < -3$. Thus, the one-fold differential cross section is:

$$\frac{d\sigma}{dp_T} = \int_{-2.5}^{2.5} d\eta_3 \int_{-5}^{-3} d\eta_4 \frac{2p_T d^3\sigma}{d\eta_3 d\eta_4 dp_T^2}. \tag{47}$$

In order to verify the sensitivity of Eq. (47) to the kinematic domain of superfast quarks (x > 1), we compare it with the one-fold differential cross section over the same range of η_3 and η_4, but with x > 1 events selected for only, viz.

$$\frac{d\sigma(x_A > 1)}{dp_T} = \int_{-2.5}^{2.5} d\eta_3 \int_{-5}^{-3} d\eta_4 \frac{2p_T d^3\sigma}{d\eta_3 d\eta_4 dp_T^2}\Theta(x_A - 1).$$
48)

By comparing the differential cross sections of Eqs. (47) and (48), we can determine the range of p_T in which the results are close, and accordingly identify the p_T region that is sensitive to short-range nuclear phenomena. In Fig. 13we present estimates of the differential cross sections of Eqs. (47, 48). Figure 13a compares the differential cross section, in the absence of medium modifications, without and with short-range correlations accounted for. Most dijets with transverse momentum up to about 100 GeV/c can be attributed to nucleons in the mean field. By contrast, dijets with high $p_T \gtrsim 150$ GeV/c can be attributed predominantly

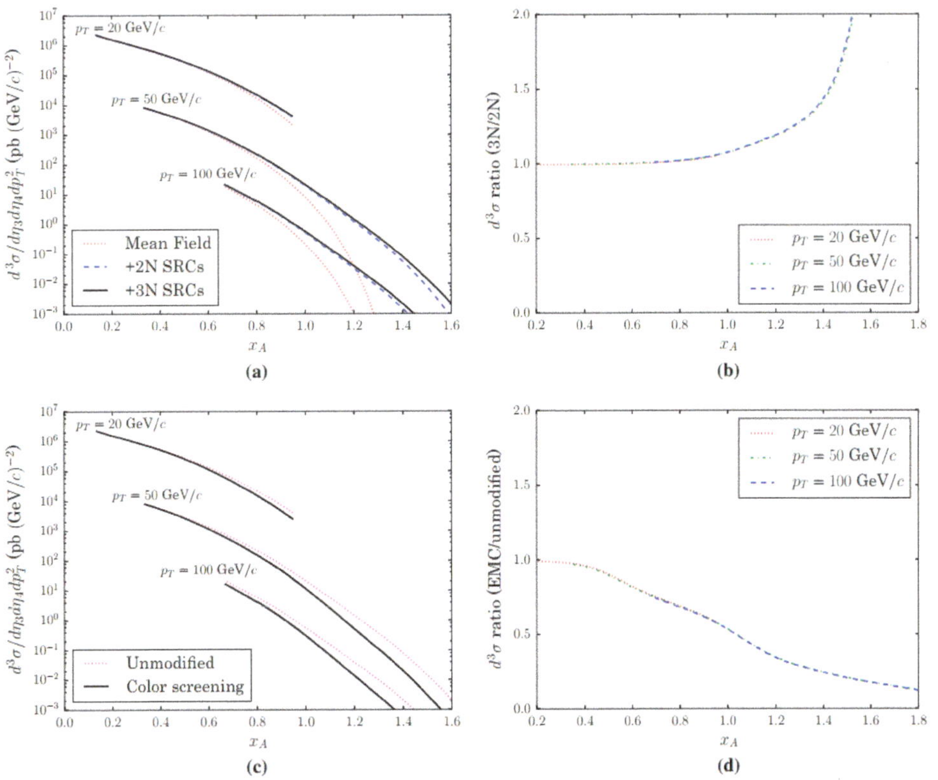

Figure. 11: Three-fold differential cross section for p+^{208}Pb→ dijet+ X according to Eq. (17). a and c are absolute cross sections. b and d are ratios. b is the ratio of the three-fold differential cross section when 2N and 3N SRCs are both considered to when only 2N SRCs are considered, b is the ratio of the three-fold differential cross section with medium modifications to without. η4 is constrained to the range (−5, −3), and η_3 = 0. Proton PDF used is CT10 [65]. a No medium modifications. b No medium modifications. c All SRCs contribute. d All SRCs contribute

to short range correlations. In addition, selecting only dijets with kinematics corresponding to an initial state nuclear parton with $x_A > 1$ produces a differential cross section that is at least an order of magnitude larger when SRCs are considered than for the mean field by itself. This confirms that $x_A > 1$ events can be attributed predominantly to the presence of SRCs in the nucleus, and also indicates that finding a large yield for $x_A > 1$ events can demonstrate the presence of SRCs.

On the other hand, since SRCs are highly modified in the nuclear medium, $x_A > 1$ events in particular are suppressed by the EMC effect, as can be seen in Fig. 13b. This means the value of the one-fold differential cross section for $x_A > 1$ dijet events experiences two effects that compete with oneanother: it is increased by the presence of SRCs, but it is at the same time suppressed by medium modifications, which most strongly affect SRCs. In addition to one-fold differential cross sections, we also present estimates for the total integrated cross section in order to show that $x_A > 1$ events should produce a measurable yield. Since jet identification at low pT is difficult, we consider the total cross section integrated from a minimum p_T value of 50 GeV/c. In Table 2 we present numerical estimates for the total cross section for different bins of x as indicated in the table. In addition to comparing the integrated cross section with and without medium modifications, we have compared the model of this work to the FS81 model [1] (with medium modifications accounted for by the color screening model). The FS81 model predicts a larger cross section than our model does at $x_A > 1$, owing to the different form of 3N correletions, as well as the implicit inclusion of 4, 5, etc. nucleon correlations in their light cone density.

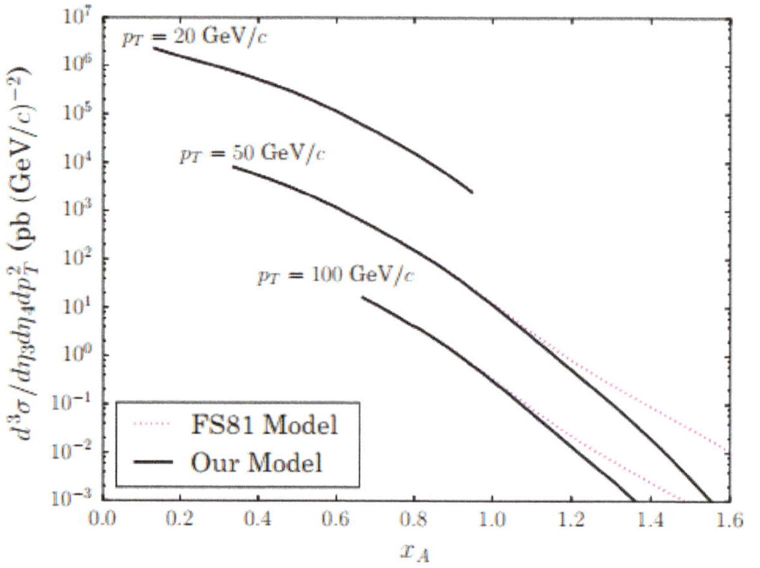

Figure. 12: The color screening model accounts for medium modifications for both models. Calculation within our model considers 2N and 3N SRCs to contribute

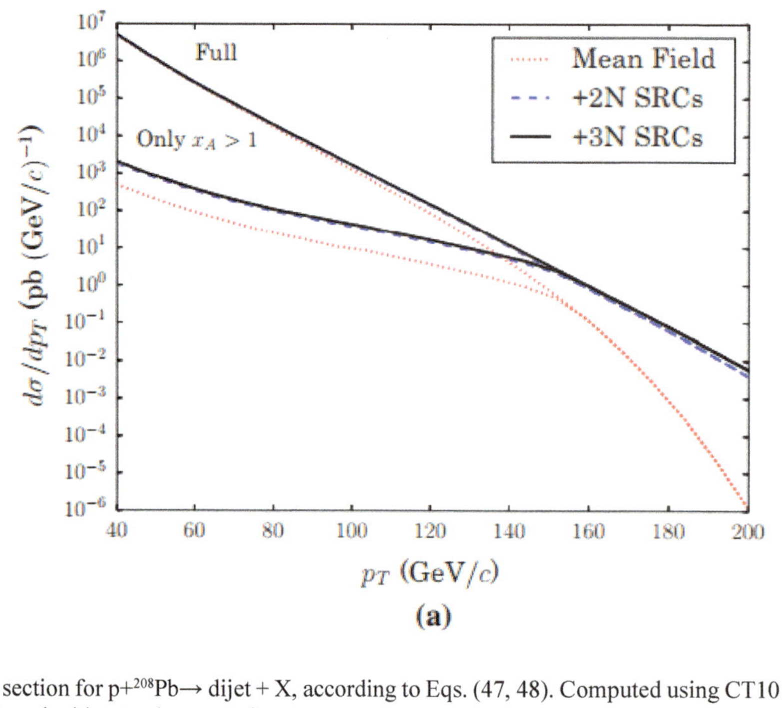

Figure. 13: Differential cross section for p+^{208}Pb→ dijet + X, according to Eqs. (47, 48). Computed using CT10 for proton PDF [65]. a Without medium modifications. b With and without color screening

Table 2: Estimates of integrated cross sections, for different bins of x_A. Calculations include two- and three-nucleon SRCs

	Unmodified	Modified (no SRCs)	Modified (SRCs)	FS81 (modified)
All x_A	7.8 μb	6.8 μb	6.9 μb	6.9 μb
$0.6 < x_A < 0.7$	0.94 μb	0.73 μb	0.72 μb	0.76 μb
$0.7 < x_A < 0.8$	0.38 μb	0.25 μb	0.27 μb	0.28 μb
$0.8 < x_A < 0.9$	0.17 μb	0.07 μb	0.08 μb	0.09 μb
$0.9 < x_A < 1$	36 nb	14 nb	21 nb	23 nb
$1 < x_A$	12 nb	2.3 nb	5.7 nb	7.3 nb

As can be seen in Table 2, medium modifications suppress the integrated cross section, especially at high x_A. This is consistent with the expectation that medium modifications affect SRCs especially strongly. However, despite the suppression by medium modification, the cross section remains large enough to be observable. In particular, with an integrated luminosity of 35.5 nb−1 (achieved in the 2013 p A run of the LHC), the expected yield for $x_A > 1$ events (with suppression from the color screening model considered) is about 200 events. Additionally, the expected yield for $x_A > 0.6$—the classical kinematical region of the EMC effect—is about 39,000 events, suggesting dijet production from p A collisions could be used to study the EMC effect at high Q^2 where higher-twist effects are negligible, but where existing experimental data have large error bars.

To use the leading order kinematic relations Eqs. (11, 12) requires unambiguous identification of two jets with equal pT , and neglects effects of jet fragmentation. There could be a number of issues related to jet production and reconstruction which may spoil these relations. Experimental measurement of the cross section of jet (or dijet) production involves using a particular algorithm for extracting information about single (di-) jet production kinematics. Currently, the anti-kt jet clustering algorithm [73] is widely used. For jets with pT ≥ 150 GeV/c, which we consider here, a cone with R = 0.2-0.4 is usually used. Numerical studies for the p + A → jet + X process with the jet produced in the proton fragmentation region, performed in [74] using anti-kt clustering algorithm and PYTHIA, find that in the kinematics studied by ATLAS [75], the value of $y = (E_{\text{jet}} + p_{z,\text{jet}})/2E_p$ determined using anti-kt clustering algorithm with R = 0.4 is very close to the x p of the colliding proton for $x_p \geq 0.2$ up to $x_p \sim 0.7$ (higher x_p were not studied in Ref. [74] due to the lack of statistics). This allows good tracking of x for the large x parton.

In principle, fluctuations in the energy deposited in the calorimeter could have affect such a determination for large x where distribution drops very steeply, leading to a large value of $\frac{\partial \ln F_2^{(A)}(x,Q^2)}{\partial x}$. However, it is easy to check that for $F_2^{(A)}(x,Q^2) \propto \exp(-bx)$ with $b \sim 8$, variation with x for x > 0.6 is weaker in the nucleus case. Moreover, detection of the central jet we envision should help to suppress the effects of large fluctuations. The observation of the central jets would additionally help to suppress contribution of multiparton interactions (MPI), in which production of two jets with e.g. $x_A \sim 0.5$ could mimic production of one $x_A \sim 1$ jet.

Another potential problem is the subtraction of the contributions of the underlying events for $x_A > 0.5$. On one hand, the underlying activity is higher in this case than in pp collisions. On the other hand, for $\eta \sim -4$ it is significantly smaller than for the central rapidities. The anti-k_t algorithm is well suited for dealing with such problems.

Obviously further simulations are necessary, which would be based on more detailed experimental information about the underlying event fluctuations in the backward kinematics than are available now. Simulations should also account for energy and angular resolutions and the acceptances of the detectors, as well as treating NLO effects like p_T inbalance between the two jets. Such a simulation would require a joint effort of experimentalists and theorists, and is beyond the scope of this paper.

CONCLUSIONS

In this work, we have shown that superfast partons inside a heavy nucleus with x > 1 are experimentally accessible at the LHC, and can be probed in dijet production reactions from proton-nucleus collisions. The value of x can be related to jet kinematics in the LHC rest frame in a straightforward way, so it is possible to select events that correspond to x > 1. Partons with x > 1 can predominantly be attributed to the presence of short range correlations between nucleons within the nucleus. Expected medium modifications suppress the cross section appreciably, but the total x > 1 cross section remains on the order of a few nanobarns, making measurement feasible. Additionally, it is possible to to perform novel studies of the EMC effect in the x < 1 region. The observability of superfast partons in the nucleus owes itself to the presence of short range correlations. This fact tells us that it is inadequate to treat the nucleus as a collection of free nucleons, moving forward with equal light cone momenta. The internal structure of the nucleus, and the relative momenta of its nucleons, are necessary to account for. We observe that even in the one-fold differential cross section for inclusive dijet production, looking at large transverse jet momenta (pT ≥ 150 GeV/c) allows one to select events originating from short range correlations in nuclei. Overall, our conclusion is that even at such high energies as those at the LHC, short range nuclear structure can be explored, and it will dominate the rate of the events with large transverse momenta and negative pseudo-rapidities in nuclear fragmentation region. We expect the formalism to have further applications beyond dijet production in proton-nucleus collisions. In particular, short range nuclear structure and medium modification can be checked in Drell-Yan processes both in protonnucleus and nucleus-nucleus collisions. Additionally, it will prove necessary to account for multi-nucleon short range correlations to correctly predict the high-x cross section in inelastic nuclear DIS experiments at high energy. The Electron Ion Collider [76] in particular will be able to carry out such experiments, where the formalism developed in this work will prove necessary. In particular, the nuclear PDF at high Q^2 will need to be obtained through QCD evolution, since the phenomenological models that exist to account for nuclear modifications are constructed using lower-Q^2 experimental data.

ACKNOWLEDGMENTS

We are thankful to Alberto Accardi, Brian Cole, Wim Cosyn, Leonid Frankfurt, Shunzo Kumano, and Gerald Miller for many useful discussions. Our special thanks are to Matteo Cacciari, Dennis Perepelitsa, Gavin Salam for discussion of the problems of extraction of nuclear quark distributions at large x_A from p A jet production data. This work is supported by US Department of Energy grants under contracts DE-FG02-01ER-41172 and DE-FG02-93ER40771.

REFERENCES

1. L.L. Frankfurt, M.I. Strikman, Phys. Rep. 160, 235 (1988)
2. M. Sargsian, J. Arrington, W. Bertozzi, W. Boeglin, C. Carlson et al., J. Phys. G29, R1 (2003).
3. L.L. Frankfurt, M.I. Strikman, Phys. Rep. 76, 215 (1981)
4. L. Frankfurt, M. Strikman, D. Day, M. Sargsian, Phys. Rev. C 48, 2451 (1993)
5. J.L. Aclander, J. Alster, D. Barton, G. Bunce, A. Carroll et al., Phys. Lett. B 453, 211 (1999)
6. Tang, J. Watson, J.L. Aclander, J. Alster, G. Asryan et al., Phys. Rev. Lett. 90, 042301 (2003).
7. K. Egiyan, et al. (CLAS Collaboration), Phys. Rev. C 68, 014313 (2003).
8. K. Egiyan, et al. (CLAS Collaboration), Phys. Rev. Lett. 96, 082501 (2006). arXiv:nucl-ex/0508026
9. E. Piasetzky, M. Sargsian, L. Frankfurt, M. Strikman, J. Watson, Phys. Rev. Lett. 97, 162504 (2006).
10. R. Shneor et al. (Jefferson Lab Hall A Collaboration), Phys. Rev. Lett. 99, 072501 (2007).
11. R. Subedi, R. Shneor, P. Monaghan, B. Anderson, K. Aniol et al., Science 320, 1476 (2008).arXiv:0908.1514
12. L. Frankfurt, M. Sargsian, M. Strikman, Int. J. Mod. Phys. A 23, 2991 (2008).arXiv:0806.4412
13. N. Fomin, J. Arrington, R. Asaturyan, F. Benmokhtar, W. Boeglin et al., Phys. Rev. Lett.108, 092502 (2012). arXiv:1107.3583
14. J. Arrington, D. Higinbotham, G. Rosner, M. Sargsian, Prog. Part. Nucl. Phys. 67, 898 (2012). arXiv:1104.1196
15. J. Arrington, A. Daniel, D. Day, N. Fomin, D. Gaskell et al., Phys. Rev. C 86, 065204 (2012). arXiv:1206.6343
16. Korover, N. Muangma, O. Hen, R. Shneor, V. Sulkosky et al., Phys. Rev. Lett. 113, 022501 (2014). arXiv:1401.6138
17. Benvenuti et al. BCDMS Collaboration, Z. Phys. C 63, 29 (1994)
18. M. Vakili et al. (CCFR Collaboration), Phys. Rev. D 61, 052003 (2000). arXiv:hep-ex/9905052 [hep-ex]
19. P.E. Bosted, A. Lung, L. Andivahis, L. Stuart, J. Alster et al., Phys. Rev. C 46, 2505 (1992)
20. N. Fomin, J. Arrington, D. Day, D. Gaskell, A. Daniel et al., Phys. Rev. Lett. 105, 212502 (2010). arXiv:1008.2713
21. J. Dudek, R. Ent, R. Essig, K.S. Kumar, C. Meyer et al., Eur. Phys. J. A 48, 187 (2012)
22. J. Arrington et al., Jefferson lab experiment 12-06-105: inclusive scattering from nuclei at x1x1 in the quasielastic and deeply inelastic regimes.http://www.jlab.org/exp_prog/proposals/06/PR12-06-105.pdf
23. J. Collins, Foundations of perturbative QCD (2011)
24. R.K. Ellis, W.J. Stirling, B.R. Webber, *QCD and collider physics*, vol. 8 (1996) pp. 1–435
25. K. Eskola, V. Kolhinen, C. Salgado, Eur. Phys. J. C 9, 61 (1999). arXiv:hep-ph/9807297[hep-ph]
26. D. de Florian, R. Sassot, Phys. Rev. D 69, 074028 (2004). arXiv:hep-ph/0311227 [hep-ph]
27. M. Hirai, S. Kumano, T.-H. Nagai, Phys. Rev. C 76, 065207 (2007). arXiv:0709.3038[hep-ph]
28. K. Eskola, H. Paukkunen, C. Salgado, JHEP 0904, 065 (2009). arXiv:0902.4154 [hep-ph]
29. D. de Florian, R. Sassot, P. Zurita, M. Stratmann, Phys. Rev. D85, 074028 (2012).arXiv:1112.6324 [hep-ph]
30. H. Honkanen, M. Strikman, V. Guzey, (2013). arXiv:1310.5879 [hep-ph]
31. M. Arneodo, Phys. Rep. 240, 301 (1994)
32. D.F. Geesaman, K. Saito, A.W. Thomas, Ann. Rev. Nucl. Part. Sci. 45, 337 (1995)
33. L. Frankfurt, M. Strikman, Nucl. Phys. B 148, 107 (1979)
34. E. Moniz, I. Sick, R. Whitney, J. Ficenec, R.D. Kephart et al., Phys. Rev. Lett. 26, 445 (1971)
35. M.V. Zverev, E.E. Saperstein, Yad. Fiz. 43, 304 (1986)
36. M. McGauley, M.M. Sargsian, (2011). arXiv:1102.3973
37. H. Feldmeier, W. Horiuchi, T. Neff, Y. Suzuki, Phys. Rev. C84, 054003 (2011).arXiv:1107.4956
38. M.M. Sargsian, Phys. Rev. C 89, 034305 (2014). arXiv:1210.3280
39. O. Hen, M. Sargsian, L. Weinstein, E. Piasetzky, H. Hakobyan et al., Science 346, 614 (2014). arXiv:1412.0138
40. M. Alvioli, C. Ciofi Degli Atti, L.P. Kaptari, C.B. Mezzetti, H. Morita, Int. J. Mod. Phys. E22, 1330021 (2013). arXiv:1306.6235
41. R. Wiringa, R. Schiavilla, S.C. Pieper, J. Carlson, (2013). arXiv:1309.3794
42. M.M. Sargsian, J. Phys. Conf. Ser. 496, 012007 (2014). arXiv:1312.2263
43. M. Vanhalst, W. Cosyn, J. Ryckebusch (2014). arXiv:1405.3814
44. M. Lacombe, B. Loiseau, J. Richard, R. Vinh,Mau, J. Cote et al., Phys. Rev. C 21, 861 (1980)
45. M. Sargsian, T. Abrahamyan, M. Strikman, L. Frankfurt, Phys. Rev. C 71, 044615 (2005).arXiv:nucl-th/0501018
46. D.W. Higinbotham, O. Hen, (2014). arXiv:1409.3069

47. Y. Bayukov, V. Efremenko, S. Frankel, W. Frati, M. Gazzaly et al., Phys. Rev. C 20, 764 (1979)
48. J. Aubert et al. European Muon Collaboration, Phys. Lett. B 123, 275 (1983)
49. G.A. Miller, J.R. Smith, Phys. Rev. C 65, 015211 (2002). arXiv:nucl-th/0107026
50. J.R. Smith, G.A. Miller, Phys. Rev. C 65, 055206 (2002). arXiv:nucl-th/0202016
51. J. Seely, A. Daniel, D. Gaskell, J. Arrington, N. Fomin et al., Phys. Rev. Lett. 103, 202301 (2009). arXiv:0904.4448
52. P. Norton, Rept. Prog. Phys. 66, 1253 (2003)
53. K. Rith, (2014). arXiv:1402.5000 [hep-ex]
54. L. Frankfurt, M. Strikman, Nucl. Phys. B 250, 143 (1985)
55. S.A. Kulagin, G. Piller, W. Weise, Phys. Rev. C 50, 1154 (1994). arXiv:nucl-th/9402015
56. W. Melnitchouk, M. Sargsian, M. Strikman, Z. Phys. A 359, 99 (1997). arXiv:nucl-th/9609048
57. M. Frank, B. Jennings, G. Miller, Phys. Rev. C 54, 920 (1996). arXiv:nucl-th/9509030
58. J. Gomez, R. Arnold, P.E. Bosted, C. Chang, A. Katramatou et al., Phys. Rev. D49, 4348 (1994)ADS
59. R. Arnold, P.E. Bosted, C. Chang, J. Gomez, A. Katramatou et al., Phys. Rev. Lett. 52, 727 (1984)
60. Bodek, M. Breidenbach, D. Dubin, J. Elias, J.I. Friedman et al., Phys. Rev. D20, 1471 (1979)ADS
61. Bodek, J. Ritchie, Phys. Rev. D 23, 1070 (1981)
62. Bodek, J. Ritchie, Phys. Rev. D 24, 1400 (1981)
63. L. Frankfurt, M. Strikman, Int. J. Mod. Phys. E 21, 1230002 (2012). arXiv:1203.5278[hep-ph]
64. J. Arrington, R. Ent, C. Keppel, J. Mammei, I. Niculescu, Phys. Rev. C73, 035205 (2006).arXiv:nucl-ex/0307012
65. J. Gao, M. Guzzi, J. Huston, H.-L. Lai, Z. Li et al., Phys. Rev. D89, 033009 (2014).arXiv:1302.6246 [hep-ph]
66. Y.L. Dokshitzer, Sov. Phys. JETP 46, 641 (1977)ADS
67. V. Gribov, L. Lipatov, Sov. J. Nucl. Phys. 15, 438 (1972)
68. G. Altarelli, G. Parisi, Nucl. Phys. B 126, 298 (1977)
69. L. Frankfurt, V. Guzey, M. Strikman, Phys. Rep. 512, 255 (2012). arXiv:1106.2091 [hep-ph]
70. B.L. Combridge, J. Kripfganz, J. Ranft, Phys. Lett. B 70, 234 (1977)
71. R.K. Ellis, J. Sexton, Nucl. Phys. B269, 445 (1986)
72. G. Aad et al. (ATLAS Collaboration), Phys. Rev. D 86, 014022 (2012). arXiv:1112.6297[hep-ex]
73. M. Cacciari, G.P. Salam, G. Soyez, JHEP 0804, 063 (2008). arXiv:0802.1189 [hep-ph]
74. D. Perepelitsa, private communication
75. G. Aad et al. (ATLAS), (2014). arXiv:1412.4092 [hep-ex]
76. Accardi, J. Albacete, M. Anselmino, N. Armesto, E. Aschenauer et al. (2012).arXiv:1212.1701
77. S.J. Brodsky, H.-C. Pauli, S.S. Pinsky, Phys. Rep. 301, 299 (1998). arXiv:hep-ph/9705477 [hep-ph]
78. Ciofi degli Atti, S. Simula, Phys.Rev. C 53, 1689 (1996). arXiv:nucl-th/9507024
79. O. Artilles, A. Freese, M. Sargsian, in progress.

CHAPTER 10

THE ROLE OF THE NEXT-TO-LEADING ORDER TRIANGLE-SHAPED DIAGRAM IN TWO-BODY HADRONIC DECAYS

J. Schneitzer, T. Wolkanowski, F. Giacosa

INTRODUCTION

The study of decays is an important subject of atomic, nuclear and particle physics [1]. Some subatomic particles possess a lifetime which is so short that they can be seen only through their decay products, and hence one usually calls them resonances. This is indeed the case for the recently discovered Higgs particle, see e.g. Refs. [2] and [3] and references therein. In the realm of the strong interactions also many hadrons were discovered via their decay processes [4]; in addition to that, decays turn out to be crucially important for the understanding of their quantum numbers and inner structure.

The main problem concerning the fundamental theory of quarks and gluons (Quantum Chromodynamics or QCD) is the fact that this theory is non-perturbative in the low-energy regime. Hence one relies on other approaches, as for instance effective models based on symmetries [5] and [6], where the physical degrees of freedom are not quarks and gluons, but composite particles, namely hadrons. Decays of hadrons have often been evaluated within such models in the lowest order approximation – in other words at tree-level, see e.g. Refs. [7], [8], [9] and [10] and references therein. In particular, in the recent work of Ref. [10] decays of various mesons up to 1.5 GeV were computed in a chirally and dilatation invariant framework and were found to be in agreement with the corresponding experimental values as provided by the PDG [11].

A two-body tree-level decay is the easiest nontrivial process in quantum field theory. It is depicted in Fig. 1a: The unstable bosonic particle S decays into two identical particles, denoted as ϕ. The decay amplitude is simply a constant in the case of scalar particles and non-derivative interactions. When derivatives and/or particles with nonzero total angular momentum J are considered, a dependence of the momenta appears in the tree-level amplitude(s).

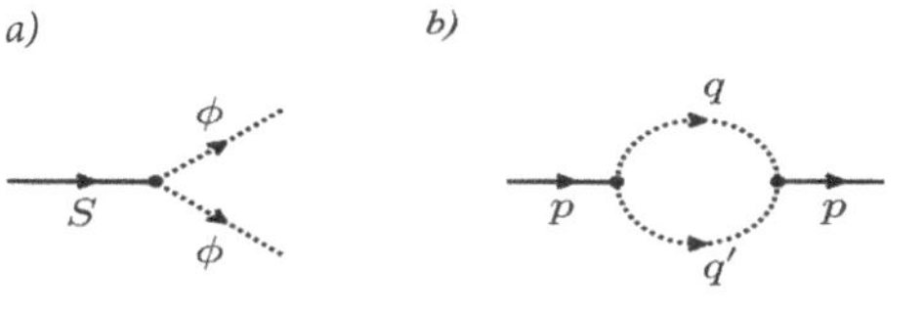

Figure. 1: (a) Decay process $S \to \phi\phi$ at leading order (tree-level). (b) Self-energy at leading order.

The next step in the context of effective models has been the study of (hadronic) loops, see for instance Refs. [12], [13], [14], [15], [16], [17], [18], [19], [20], [21] and [22] and references therein. The leading contribution to the self-energy is shown by the diagram inFig. 1b. Both the mass and the width of the decaying particle are influenced by the quantum fluctuations due to the coupling to hadronic intermediate states. The optical theorem assures that the imaginary part of the one-loop diagram from Fig. 1b coincides with the tree-level decay formula. The unstable particle is described by a spectral function (i.e., an energy distribution), which is given by the imaginary part of the one-loop resummed propagator. Alternatively, the properties of the unstable particle can also be described as a complex pole in the appropriate unphysical Riemann sheet, a procedure first proposed by Peierls [23] a long time ago. The quantum theoretical treatment of unstable particles became an object of much interest, see e.g. Refs. [24], [25],[26] and [27]. The general outcome of such studies is –

Citation: J. Schneitzer, T. Wolkanowski, F. Giacosa, The role of the next-to-leading order triangle-shaped diagram in two-body hadronic decays, Nuclear Physics B, Volume 888, November 2014, Pages 287-299, ISSN 0550-3213, http://dx.doi.org/10.1016/j.nuclphysb.2014.10.003.

disregarding problems like mixing – that when the particle is narrow-shaped, quantum fluctuations have a small influence on its properties (i.e., mass and width) but are non-negligible for broad resonances. It turns out to be the ratio 'width over mass' that is decisive here: as long as this number is smaller than ~0.3 the loop contributions have a small impact [17].

There is, however, another open issue: what is the role of the next-to-leading order (NLO) diagram for hadronic decays? We depict this kind of triangle-shaped diagram in Fig. 2; it is proportional to the third power of the coupling constant. In the context of hadronic decays in effective field theories/models it is usually not taken into account. Nevertheless, one should stress that the coupling constant in hadronic models is in general not a small number, thus there is a priori no guarantee that the NLO diagram is smaller than the tree-level one.

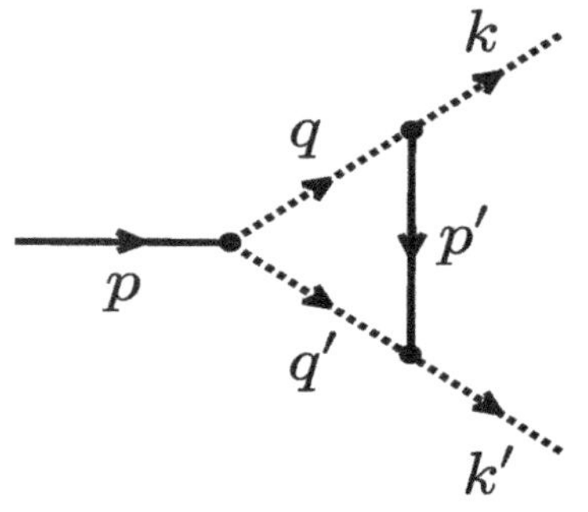

Figure. 2: Triangle-shaped NLO diagram for a two-body decay.

The aim of this work is to close this gap. To this end, we evaluate the role of the triangle diagram from Fig. 2 in the case of a simple scalar field theory without derivative interactions. We do this in plain perturbation theory (i.e., without resummation), meaning that the virtual S-particle exchanged in Fig. 2 is described by its free propagator. After discussing the analytic properties of the triangle diagram, we adopt our results to some decays of well-known scalar resonances: $f_0(500)$, $f_0(980)$, $f_0(1370)$ and$f_0(1500)$[11]. All these resonances decay predominately in two pions and are therefore a good test for our purpose. For completeness, we also look at the pion–pion and kaon–kaon decay channels of $f_0(1710)$ as well as the kaon–kaon decay of the scalar–isovector state $a_0(1450)$ [11]. Yet, our investigation is quite general and applies also to decays involving derivatives and particles with spin.

The main result of our work is that the triangle contribution is indeed negligible and one can consequently justify a posteriori all previous studies in which those types of contributions (and, in turn, higher-order contributions as well) were not taken into account. Since in the field of hadron physics there are usually other (and even larger) sources of uncertainties due to various (and sometimes subtle) approximations and simplifications, the restriction to the leading order tree-level diagram from Fig. 1a and to the (resummed) one-loop quantum corrections from Fig. 1b are reasonable and usually sufficient.

The paper is organized as follows: we present the model and some analytic aspects of the triangle diagram in Section 2, while the numerical results are shown in Section 3. The last Section 4 contains the conclusions.

THE MODEL AND THE TRIANGLE DIAGRAM

We introduce a model with the scalar fields S (with mass m_S) and ϕ (with mass m_ϕ) described by the Lagrangian

$$\mathcal{L} = \frac{1}{2}(\partial_\mu S)^2 - \frac{m_S^2}{2}S^2 + \frac{1}{2}(\partial_\mu \phi)^2 - \frac{m_\phi^2}{2}\phi^2 + gS\phi^2. \tag{1}$$

The interaction term induces the decay process $S\to\phi\phi$. The parameter g is the coupling constant (with dimension of energy). For previous studies and details of the model see Refs. [16], [17], [18], [19], [28] and [29]. The decay width in perturbation theory can be expressed as

$$\Gamma_{S\to\phi\phi} = \frac{1}{2}\frac{\sqrt{\frac{m_S^2}{4}-m_\phi^2}}{8\pi m_S^2}|-i\mathcal{M}|^2 , \tag{2}$$

where the decay amplitude $-i\mathcal{M}$ in perturbation theory is written as a sum

$$-i\mathcal{M}=-i\mathcal{M}_1-i\mathcal{M}_3+\ldots. \tag{3}$$

The term $-i\mathcal{M}_{2n+1}$ represents the contribution of order g^{2n+1} to the amplitude. It is quite remarkable that an exact solution to this problem has not yet been found (there are, however, quantum mechanical models for which this is possible, see Refs. [1],[18], [30], [31], [32], [33] and [34]).

(1) Leading order . The leading order with n=0 is given by the tree-level amplitude ofFig. 1a, for which one obtains

$$-i\mathcal{M}_1=2ig. \tag{4}$$

Thus, the tree-level decay width simply reads

$$\Gamma^{\text{tree}}_{S\to\phi\phi} = \frac{1}{2}\frac{\sqrt{\frac{m_S^2}{4}-m_\phi^2}}{8\pi m_S^2}(2g)^2. \tag{5}$$

(2) The one-loop diagram. The loop diagram shown in Fig. 1b actually does not enter directly into the expression of Eq. (2), because the latter is valid in plain perturbation theory without resummation. The role of the one-loop resummation has been widely studied and, although not directly relevant for our calculation, we recall the main features in view of its general importance for the problem that we are studying and for future works[12], [13], [14], [15], [16], [17], [18], [19] and [29]. By denoting $\Pi(p^2)$ as the loop contribution, the propagator of the field S changes upon resummation to

$$\Delta_S(p^2) = \frac{i}{p^2-m_S^2+i\epsilon} \quad\rightarrow\quad G_S(p^2) = \frac{i}{p^2-m_S^2-\Pi(p^2)+i\epsilon}. \tag{6}$$

The self-energy $\Pi(p^2)$ is linked to the tree-level decay width via the optical theorem:

$$-\operatorname{Im}\Pi(p^2) = \sqrt{p^2}\,\Gamma^{\text{tree}}_{S\to\phi\phi}(m_S^2\to p^2) = \sqrt{p^2}\frac{g^2\sqrt{\frac{p^2}{4}-m_\phi^2}}{4\pi p^2}. \tag{7}$$

The properties of the unstable particle S (i.e., its mass and decay width) are often identified with the complex pole of the full propagator $G_S(p^2)$ in the second Riemann sheet, $p^2=(m_S^{pole}-i\Gamma^{pole}_{S\to\phi\phi}/2)^2$. The (shifted) mass m_S^{pole} is given by the real part of the pole and the modified decay width $\Gamma^{pole}_{S\to\phi\phi}$ by the negative imaginary part multiplied by two. For a small coupling constant g the quantities $\Gamma^{pole}_{S\to\phi\phi}$ and $\Gamma^{tree}_{S\to\phi\phi}$ are close to each other, but then deviate when increasing g [29].

At this point it should be stressed that hadronic theories do not need to undergo a renormalization process as for theories of elementary particles, since they are only valid in a limited energy regime. (For instance, if we restrict our attention to mesons made ofu-, d-, and s -quarks, the associated 'cutoff' has a value of about ~1.5 GeV.) For this reason, a finite energy cutoff is used when evaluating the loop diagram from Fig. 1b. Different forms for the consequently needed cutoff function can be applied, but the general outcome shows just a soft dependence on the precise choice as long as convergence is guaranteed [17].

In an ideal scattering experiment of the type $\phi\phi\to\phi\phi$, the unstable state S manifests itself as an enhanced peak for a center of mass energy close to m_S. More precisely, the spectral function $d_S(E)=-\frac{2E}{\pi}\,\mathrm{Im}\,G_S(E^2)$ plays an important role: as was argued long time ago by Matthews and Salam [20], it can be interpreted as a 'mass distribution' of the unstable particle S , which can be well described by a Breit–Wigner function for narrow resonances. Namely, even if there are low-energy threshold(s) and high-energy distortions, as long as the ratio $\Gamma^{tree}_{S\to\phi\phi}/m_S$ is a small number, the role of hadronic loop contributions is small [17] and [18]. However, when this ratio becomes large one observes big deviations from a typical Breit–Wigner peak and in some cases a very peculiar phenomenon takes place, often called pole-doubling. This means that new poles can emerge in the unphysical Riemann sheet(s) [12], [14] and [15]. Although there is only one unstable state in the Lagrangian, quantum fluctuations might be able to generate two (or more) resonance poles. Such a mechanism of dynamical generation is present in the ongoing debate among hadron physicists on where some of the (known) resonances in the hadron spectrum arise from.[1] A particular interesting field of study is that of charmed mesons, see Refs. [13] and [36].

(3) The triangle diagram . We now turn our attention to the main subject of this work: the NLO (non-resummed) perturbation theory. To this end, we evaluate the triangle diagram corresponding to the amplitude $-iM_3$ as depicted in Fig. 2. Its analytic expression takes the form

$$-i\mathcal{M}_3(p\to kk')=8ig^3\int\frac{d^4q}{(2\pi)^4}\frac{1}{q^2-m_\phi^2+i\epsilon}\,\frac{1}{(q-p)^2-m_\phi^2+i\epsilon}\,\frac{1}{(q-k)^2-m_S^2+i\epsilon}. \tag{8}$$

Note, the propagator of S is taken as the free propagator $\Delta_S(p^2)$. The factor of 8 arises due to identical particles in the final state. Solving the integration over q^0 by the residue theorem and after introducing spherical coordinates the amplitude can be re-expressed as

$$\mathcal{M}_3(p\to kk')=\frac{2g^3}{\pi^2}\int_0^\infty du\,u^2\int_{-1}^{1}d\chi\,\{P_{12}(u,\chi)+P_3(u,\chi)\}, \tag{9}$$

where

$$P_{12}(u,\chi)=\frac{1}{8\sqrt{u^2+m_\phi^2(m_\phi^2+w^2(1-\chi^2))}}\times\frac{u^2-2w^2+uw\chi}{(u-p_1-i\epsilon)(u-p_2+i\epsilon)(u-p_3(\chi)-i\epsilon\xi)(u-p_4(\chi)+i\epsilon\xi)}, \tag{10}$$

$$\xi=\mathrm{sgn}(-\Delta_2-2uw\chi), \tag{11}$$

$$P_3(u,\chi)=\frac{1}{8u\sqrt{u^2+m_S^2(w^2(\chi^2-1)-m^2)(u-p_5(u,\chi)-i\epsilon\xi')}}, \tag{12}$$

$$\xi'=\mathrm{sgn}\left(\frac{m_S^2}{u}-2w\chi\right), \tag{13}$$

$$\Delta_1=m_S^2-m_\phi^2, \tag{14}$$

$$\Delta_2=2m_\phi^2-m_S^2. \tag{15}$$

Here we introduced $|\mathbf{q}|=u$, $|\mathbf{k}|=w$ and $\chi=\cos\theta$, where θ is the angle between **q**and **k**. We applied the shift $\mathbf{q}\to\mathbf{q}-\mathbf{k}$ for P_3 before the transformation to spherical coordinates. The integrand in Eq. (9) has five poles:

$$p_1=w, \tag{16}$$

$$p_2=-w, \tag{17}$$

$$p_3(\chi) = \frac{-w\chi\Delta_2 - \sqrt{u^2\chi^2\Delta_2^2 - 4w^2\Delta_1(-w^2 - m_\phi^2 + w^2\chi^2)}}{2(-w^2 - m_\phi^2 + w^2\chi^2)}$$
$$= \frac{-2w\chi\Delta_2 - wm_S\sqrt{4w^2 + m_\phi^2(3+\chi^2)}}{2(-w^2 - m_\phi^2 + w^2\chi^2)}, \tag{18}$$

$$p_4(\chi) = \frac{-w\chi\Delta_2 + \sqrt{w^2\chi^2\Delta_2^2 - 4w^2\Delta_1(-w^2 - m_\phi^2 + w^2\chi^2)}}{2(-w^2 - m_\phi^2 + w^2\chi^2)}$$
$$= \frac{-2w\chi\Delta_2 + wm_S\sqrt{4w^2 + m_\phi^2(3+\chi^2)}}{2(-w^2 - m_\phi^2 + w^2\chi^2)}, \tag{19}$$

$$p_5(\chi) = \frac{m_S^2 w\chi}{w^2(\chi^2 - 1) - m^2}. \tag{20}$$

One finds that only p_1, p_3 and p_5 are positive and located on the path of integration (where, for the latter, this in fact depends on the value of χ), and thus contribute to the imaginary part of the integral in Eq. (9). The contribution from p_1 is easy to calculate analytically, whereas the one from p_3 yields a rather complicated result and is therefore computed numerically. Notice that just the term P_{12} contains those two singularities and that the contribution of p_5 vanishes.

Before moving to our final results, two comments are in order:

- The NLO diagram in Fig. 2 is convergent and well-defined also for an infinite cutoff. However, as we explained above, a finite value of the cutoff naturally comes into a hadronic theory because of the non-elementary nature of the fields [17], [18] and [37]and the finite size of the corresponding particles [14] and [15], respectively. One should in principle evaluate this diagram by including such a cutoff even if convergence is ensured in the limit of $\Lambda\to\infty$. It turns out that for what concerns the triangle diagram the influence of the cutoff parameter is small – a value of about ~1 GeV or taking the infinity limit generates only small changes.
- Triangle-shaped diagrams were indeed studied in hadron physics, but in a rather different framework. For instance, one has studied the processes $\pi^0\to\gamma\gamma$ and$f_0\to\gamma\gamma$[38], where f_0 represents a generic scalar state (see Section 3). These decays occur via a triangle-loop of quarks and represent the leading order contributions (there is no tree-level diagram for those processes). For a case in which mesonic loops contribute to $\gamma\gamma$ emission see Ref. [39]. The mentioned investigations have similar technical aspects to our present interest, yet they could not give an answer to the question of the role of the triangle diagram as a next-to-leading order contribution.

RESULTS

General Case

We first present numerical results of our calculations without referring to any particular mesonic state. To this end, we fix the energy units in the following way:

$$[g]=[m_\varphi]=[M]=[\Gamma]=[\Lambda]=1m_S, \tag{21}$$

i.e., all dimensionful quantities are expressed in terms of the mass of the unstable state S.

In the upper panel of Fig. 3 the ratio $|M_3|/|M_1|$ of the decay amplitudes is shown for different masses of the particle ϕ in dependence of the coupling constant g . In this way the role of the triangle diagram is visualized. As expected, the larger the coupling, the larger $|M_3|/|M_1|$ and for equal g the ratio is larger for smaller masses mφ=m We denote g=g_* as the value of the coupling for which the amplitudes are equal,$|M_3|=|M_1|$, implying that then the triangle diagram is exactly as large as the tree-level one. The value g_* represents an upper limit for the validity of a tree-level calculation in particular and for a perturbative expansion in general (see Table 1 for a list of numerical values).

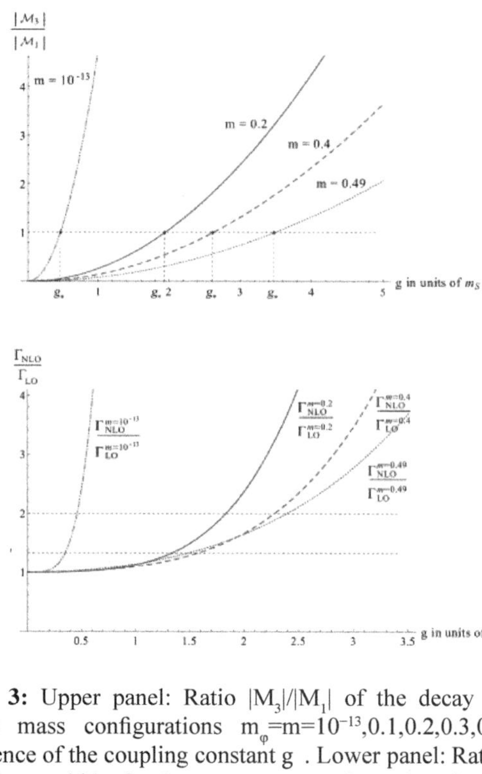

Figure. 3: Upper panel: Ratio $|M_3|/|M_1|$ of the decay amplitudes for the mass configurations m_φ=m=10^{-13},0.1,0.2,0.3,0.4,0.49 in dependence of the coupling constant g . Lower panel: Ratio Γ_{NLO}/Γ_{LO} of the decay widths for the same mass configurations in dependence of the coupling constant g.

Table 1: Specific coupling constants for different mass configurations: The case $|M_3|=|M_1|$ is obtained for $g=g_*$, the case $g=g'$ for $\Gamma_{NLO}/\Gamma_{LO}=1.33$ and the case $g=g''$ for $\Gamma_{NLO}/\Gamma_{LO}=2$. All quantities are in units of m_S.

m_φ	g_*	g'	g''
10^{-13}	0.4581	0.3459	0.4571
0.1	1.6407	1.1766	1.5894
0.2	1.9365	1.3327	1.8316
0.3	2.2155	1.4279	2.0164
0.4	2.6122	1.5672	2.2767
0.49	3.4742	1.4796	2.3908

In Fig. 3, lower panel, we also show the ratio Γ_{NLO}/Γ_{LO} for different values of m_φ as function of g , where the lowest order is the tree-level decay width $\Gamma_{LO}=\Gamma^{tree}_{S\to\phi\phi}$, meaning $-iM=-iM_1=2ig$ in Eq. (2), and the next-to-leading order is Γ_{NLO}, meaning $-iM=-iM_1-iM_3$. We also define the two specific couplings g' and g'', where g' corresponds to the ratio $\Gamma_{NLO}/\Gamma_{LO}=1.33$ (the decay width at NLO is 33% larger than the tree-level width, i.e., this marks a 'soft' limit for the validity of the tree-level calculation) and g'' corresponds to $\Gamma_{NLO}/\Gamma_{LO}=2$ (a 'hard' limit). The values of g' and g'' are reported in Table 1.

The following comments are in order:

(i) `Although the limit $m_\varphi\to 0$ is subtle because it contains infrared divergences, even for the very small value $m_\varphi=m=10^{-13}$ the NLO becomes dominant only for $g\gtrsim 0.45$. This case is however unrealistic for hadronic physics in which $m_S\sim 1$ GeV and $m_\varphi \geq m_\pi$.

(ii) For the ratio $m_\varphi/m_S \gtrsim 0.1$, which is usually fulfilled for decays of hadrons, one has $g_*\sim g''\gtrsim 1.6$ and $g'\gtrsim 1.2$, see Table 1. These values of coupling constants correspond to very large decay widths. In physical cases the value of g is usually safely smaller, showing that the NLO is subdominant, see next Subsection for explicit examples.

(iii) The fact that $g_*\sim g''$ is possible because the NLO amplitude, $-iM_3$, has a dominant imaginary part. As a consequence

$$\Gamma_{NLO}=\frac{1}{2}\frac{\sqrt{\frac{m_S^2}{4}-m_\phi^2}}{8\pi m_S^2}|-i\mathcal{M}_1-i\mathcal{M}_3|^2\simeq\frac{1}{2}\frac{\sqrt{\frac{m_S^2}{4}-m_\phi^2}}{8\pi m_S^2}(|-i\mathcal{M}_1|^2+|-i\mathcal{M}_3|^2)$$
$$=\Gamma_{LO}+\frac{1}{2}\frac{\sqrt{\frac{m_S^2}{4}-m_\phi^2}}{8\pi m_S^2}|-i\mathcal{M}_3|^2, \qquad (22)$$

which shows that Γ_{NLO} is always larger than Γ_{LO} (interference effects that involve only the real part of $-iM_3$ are small) and that, when the NLO equals the LO, the width doubles.

(iv) Although in our study one has a decay into particles with equal masses, nothing substantial would change for the decay into two particles with different masses. The numerical evaluation would be more involved because other poles could contribute.

Specific Examples

We now turn to the concrete examples of the well-known scalar resonances $f_0(500)$,$f_0(980)$, $f_0(1370)$ and $f_0(1500)$ and calculate their decays into pions [11] (for a discussion of the internal structure of these states in terms of quarks and gluons see also e.g. Refs. [7], [8], [9], [10], [12], [17], [22], [40] and [41]). We present the results in Fig. 4, plotting the decay widths as function of g . The experimental value of the width (which is $\Gamma_{f0\to\pi\pi}/3$ since only the $\pi^0\pi^0$-channel is considered) is marked on the y -axis and the error bars are indicated by a gray band. In the case of the two resonances below 1 GeVthe data was taken from the dispersive analysis of Ref. [21], while for the other two resonances the values from the PDG [11] were used. Notice that the PDG value(350±150) MeV refers to the full decay width of $f_0(1370)$. We are thus making the simplifying assumption that the $\pi\pi$-decay mode is dominant (see for instance the recent study in Ref. [42] and references therein). This choice represents an upper limit: for smaller $\pi\pi$ branching ratios, the effect of the triangle diagram will be smaller.

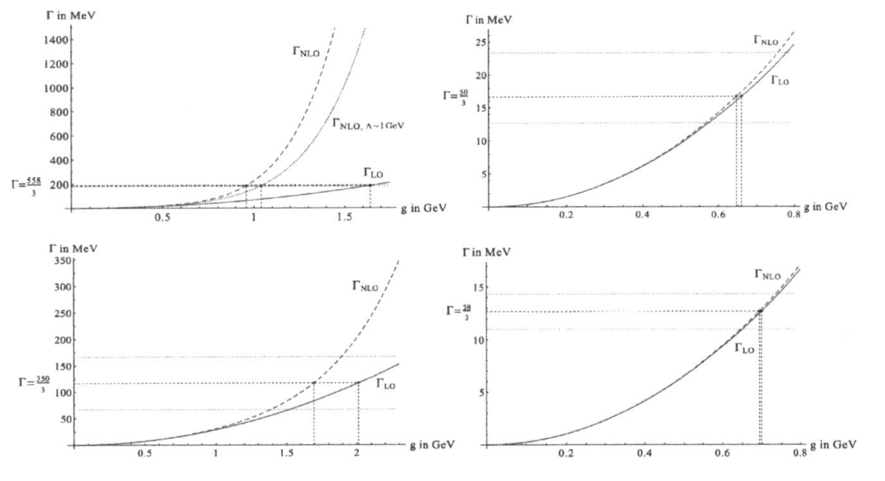

Figure. 4: Decay widths of $f_0(500)$ (upper left), $f_0(980)$ (upper right), $f_0(1370)$ (lower left) and $f_0(1500)$(lower right). The 'known' value (for the two resonances below 1 GeV taken from Ref. [21], for the other two from the PDG [11]) of the width is marked on the y -axis and error bars are indicated by a gray band. Note that here $\Gamma=\Gamma_{f0\to\pi^0\pi^0}=\Gamma_{f0\to\pi\pi}/3$. The vertical lines correspond to the determination of the coupling constant at LO and NLO, respectively.

The calculation for the resonance $f_0(500)$ was done twice: once with a cutoff $\Lambda \sim \frac{10^5}{2}$ GeV (i.e., the practical limit $\Lambda \to \infty$) and once with a physical cutoff of $\Lambda \sim 1$ GeV in order to demonstrate how the outcome is influenced by a cutoff value which is typical in hadron physics. As can be seen in Fig. 4, the NLO correction to the decay width is only important for the resonance $f_0(500)$, for which (i) the mass of the decay products is comparable to $m_{f0(500)}$ and (ii) $\Gamma^{\text{tree}}_{f_0(500)} \sim m_{f_0(500)}$: in such a case $\Gamma_{NLO} \sim 2\Gamma_{LO}$, thus the value of the coupling g corresponds roughly to g″. However, it should be stressed that the NLO process does not include all the other $\pi\pi$ -scattering contributions. In the low-energy regime in which $f_0(500)$ appears, those contributions interfere and, in virtue of chiral symmetry (see e.g. Ref. [21]), the full NLO result is expected to be smaller than what our simple model suggests.

For $f_0(980)$, $f_0(1370)$ and $f_0(1500)$ the NLO differs in a small amount with respect to the full result. (Since $f_0(1370)$ is quite broad, the NLO correction gives a small but non-negligible contribution to the width. This is so because the ratio $\Gamma^{\text{tree}}_{f_0(1370)}/m_{f_0(1370)}$ is already in the range of ~0.1. However, the large error bars do not allow to distinguish the NL and the NLO decay widths.) We also observe that a finite cutoff parameter $\Lambda \sim 1$ GeV would affect the decays of $f_0(980)$, $f_0(1370)$ and $f_0(1500)$ only marginally, since the outcome curve would lie somewhere in between Γ_{LO} and Γ_{NLO}.

Furthermore, we show in Fig. 5 analogous plots for $f_0(1710)$ going into pions and kaons, and for $a_0(1450)$ going into kaons. The experimental value of the width (which is $\Gamma_{f0\to\pi\pi}/3$ and $\Gamma_{f0/a0\to KK}/2$, respectively) is marked on the y-axis and the error bars are again indicated by a gray band. Here, data was taken from the PDG only [11]. By looking at the plots it becomes pretty clear that the NLO correction has only a small influence. Notice that there is a subtle difference in the calculation of the kaon–kaon decay due to the fact that the kaons are distinguishable particles: the factor of 2 in the LO expression of the amplitude in Eq. (4), the factor of 8 in the NLO expression in Eq. (8), as well as the symmetry factor 1/2 in Eq. (5) are replaced by the unity.

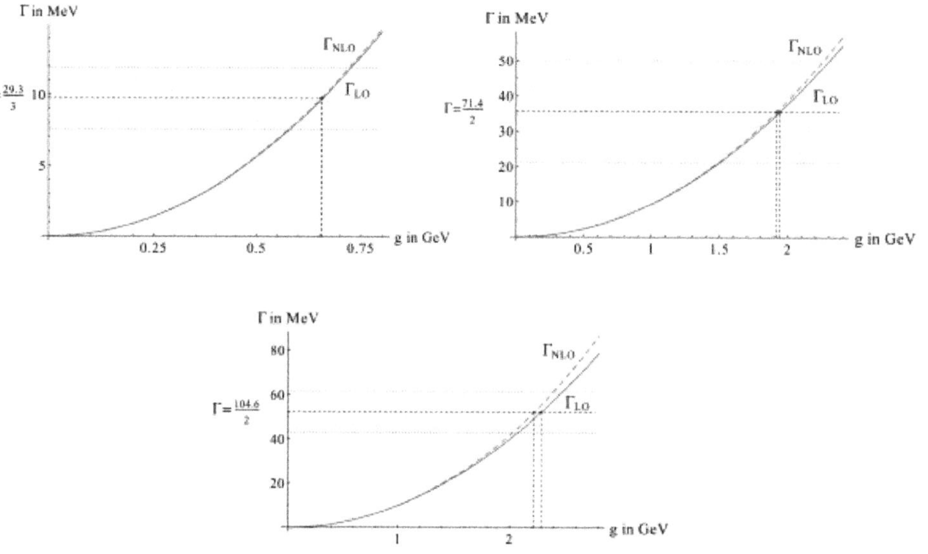

Figure. 5: Decay widths of $f_0(1710)$ going into pions (upper left) and into kaons (upper right), and $a_0(1450)$ into kaons (lower middle). The 'known' value (taken from the PDG [11]) of the width is marked on the y -axis and error bars are indicated by a gray band. Note that here $\Gamma=\Gamma_{f0\to\pi^0\pi^0}=\Gamma_{f0\to\pi\pi}/3$ and $\Gamma = \Gamma_{f_0/a_0\to K^+K^-} = \Gamma_{f_0/a_0\to K^0\bar{K}^0} = \Gamma_{f_0/a_0\to KK}/2$. The vertical lines correspond to the determination of the coupling constant at LO and NLO, respectively.

As a next step, we check if the coupling constants obtained in Fig. 4 and Fig. 5 are compatible with the ones chiral approaches deliver. In Ref. [10] the resonancesf$_0(1370)$, $f_0(1500)$ and $a_0(1450)$ were studied. The various terms in the amplitude for a given channel (i.e., terms with and without derivatives) can be summarized in a unique effective coupling constant for that channel. One gets ~1.7 GeV for $f_0(1370)$into pions, ~0.7 GeV for $f_0(1500)$ into pions, and 2.28 GeV for $a_0(1450)$ into kaons. Also, the values of the effective couplings of $f_0(1710)$ were determined in Ref.[42] to be 0.64 GeV for pions and 1.88 GeV for kaons.

All those values are well compatible with the values shown in Fig. 4 and Fig. 5. Similar comments can be made concerning the resonances below 1 GeV by regarding at the works done in Refs. [17] and [19]. These considerations confirm also that the ranges of couplings studied in Fig. 3 were realistic.

CONCLUSIONS

In this work we investigated the importance of the (usually neglected) triangle-shaped NLO contribution to two-body hadronic decays of the form $S\to\phi\phi$, see Fig. 1b. To this end, the NLO diagram was calculated analytically and numerically in the framework of a quantum field theory involving scalar fields without derivatives (avoiding unnecessary complications due to spin and/or derivatives).

We studied different cases where the mass m_ϕ of the decay products varied from nearly zero up to almost $m_S/2$, see Fig. 3. The (finite) contribution from the triangle diagram to the decay width turned out to be negligible if the mass of the decaying particle is sufficiently large, a condition which is usually met in hadron physics. We then moved to physical decays into pions of the well-known scalar resonances $f_0(500)$, $f_0(980)$,$f_0(1370)$, $f_0(1500)$ and $f_0(1710)$, as well as to the kaonic channels of $f_0(1710)$and of $a_0(1450)$. The outcome was, too, very clear, see Fig. 4 and Fig. 5: the NLO correction gave only a relevant contribution for $f_0(500)$, obviously because $\Gamma^{\text{tree}}_{f_0(500)} \sim m_{f_0(500)}$. The resonance $f_0(500)$ is however an 'extreme example': in all other cases the correction is much smaller.

Within the perturbative framework we have studied in this work, the unstable state Sexchanged in the triangle diagram (Fig. 2) has been considered as stable (i.e., the free propagator was used). Indeed, when considering the fact that S has a finite width, the contribution from the triangle diagram would be even smaller. However, the correct way of going beyond the present study is the following: Besides the one-loop diagram in Fig. 1b, one should perform the resummation of the self-energy of the unstable state S also by incorporating the NLO correction depicted in Fig. 6. Such a study is certainly nontrivial because the full propagator of S enters here – one is left with a typical problem of the Bethe–Salpeter type, see e.g. Ref. [43] and references therein. Quite interestingly, the results of our work show that the modifications coming from such a computation are likely to be in most cases negligible. Thus, the tree-level results or at most a description using the (resummed) one-loop propagator of an unstable state, give(s) a good description of unstable hadronic states.

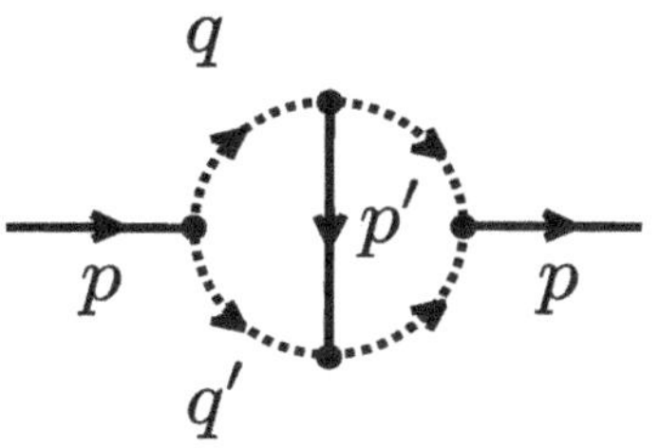

Figure. 6: NLO correction to the self-energy of the state S.

Other future studies are possible by considering different forms of the Lagrangian, including derivative interactions, particles with higher spin, fermionic fields (i.e., baryons), three-body decays and unstable states which decay in more than one channel.

ACKNOWLEDGEMENTS

The authors thank G. Pagliara, D.H. Rischke and J. Wambach for useful discussions. T.W. acknowledges financial support from HGS-HIRe, F&E GSI/GU and HIC for FAIR Frankfurt.

REFERENCES

1. L. Fonda, G.C. Ghirardi, A. Rimini, Rep. Prog. Phys. 41 (1978) 587–631.
2. A. Djouadi, Phys. Rep. 457 (2008) 1–216, arXiv:/0503172.
3. J. Ellis, T. You, J. High Energy Phys. 06 (2013) 103–120, arXiv:1303.3879
4. C. Amsler, N.A. Törnqvist, Phys. Rep. 389 (2004) 61–117; E. Klempt, A. Zaitsev, Phys. Rep. 454 (2007) 1–202, arXiv:0708.4016
5. S. Gasiorowicz, D.A. Geffen, Rev. Mod. Phys. 41 (1969) 531–573.
6. U.G. Meissner, Phys. Rep. 161 (1988) 213–361.
7. P. Ko, S. Rudaz, Phys. Rev. D 50 (1995) 6877–6894; M. Urban, M. Buballa, J. Wambach, Nucl. Phys. A 697 (2002) 338–371, arXiv:/0102260.
8. R.L. Jaffe, Phys. Rev. D 15 (1977) 267–280; R.L. Jaffe, Phys. Rev. D 15 (1977) 281–289; L. Maiani, F. Piccinini, A.D. Polosa, V. Riquer, Phys. Rev. Lett. 93 (2004) 212002, arXiv:/0407017; F. Giacosa, Phys. Rev. D 74 (2006) 014028, arXiv:/0605191; G. 't Hooft, G. Isidori, L. Maiani, A.D. Polosa, V. Riquer, Phys. Lett. B 662 (2008) 424–430, arXiv:0801.2288
9. A.H. Fariborz, R. Jora, J. Schechter, Phys. Rev. D 72 (2005) 034001, arXiv:/0506170; A.H. Fariborz, Int. J. Mod. Phys. A 19 (2004) 2095, arXiv:/0302133; M. Napsuciale, S. Rodriguez, Phys. Rev. D 70 (2004) 094043, arXiv:/0407037; F. Giacosa, Phys. Rev. D 75 (2007) 054007, arXiv:/0611388.
10. S. Gallas, F. Giacosa, D.H. Rischke, Phys. Rev. D 82 (2010) 014004, arXiv:0907.5084
11. D. Parganlija, F. Giacosa, D.H. Rischke, Phys. Rev. D 82 (2010) 054024, arXiv:1003.4934
12. S. Janowski, D. Parganlija, F. Giacosa, D.H. Rischke, Phys. Rev. D 84 (2011) 054007, arXiv:1103.3238
13. D. Parganlija, P. Kovacs, G. Wolf, F. Giacosa, D.H. Rischke, Phys. Rev. D 87 (2013) 014011, arXiv:1208.0585
14. J. Beringer, et al., Particle Data Group, Phys. Rev. D 86 (2012) 010001.
15. E. van Beveren, T.A. Rijken, K. Metzger, C. Dullemond, G. Rupp, J.E. Ribeiro, Z. Phys. C 30 (1986) 615–620, arXiv:0710.4067
16. G. Rupp, S. Coito, E. van Beveren, Acta Phys. Pol. B, Proc. Suppl. 5 (2012) 1007–1014, arXiv:1209.1475
17. N.A. Törnqvist, M. Roos, Phys. Rev. Lett. 76 (1996) 1575–1578, arXiv:/9511210v1.
18. M. Boglione, M.R. Pennington, Phys. Rev. D 65 (2002) 114010, arXiv:/0203149.
19. N.N. Achasov, A.V. Kiselev, Phys. Rev. D 70 (2004) 111901, arXiv:/0405128.
20. F. Giacosa, G. Pagliara, Phys. Rev. C 76 (2007) 065204, arXiv:0707.3594
21. F. Giacosa, Found. Phys. 42 (2012) 1262–1299, arXiv:1110.5923
22. nucl-th
23. F. Giacosa, T. Wolkanowski, Mod. Phys. Lett. A 27 (2012) 1250229, arXiv:1209.2332
24. T. Wolkanowski, F. Giacosa, Acta Phys. Pol. B, Proc. Suppl. 6 (2013) 893–898, arXiv:1305.2603
25. P.T. Matthews, A. Salam, Phys. Rev. 112 (1958) 283–287; P.T. Matthews, A. Salam, Phys. Rev. 115 (1959) 1079–1084.
26. R. García-Martín, R. Kaminski, ′ J.R. Peláez, J. Ruiz de Elvira, Phys. Rev. Lett. 107 (2011) 072001, arXiv:1107.1635
27. I. Caprini, G. Colangelo, H. Leutwyler, Phys. Rev. Lett. 96 (2006) 132001, arXiv:/0512364; F.J. Yndurain, R. García-Martín, J.R. Peláez, Phys. Rev. D 76 (2007)

074034, arXiv:/0701025; H. Leutwyler, AIP Conf. Proc. 1030 (2008) 46, arXiv:0804.3182

28. R. Kaminski, ´ R. García-Martín, P. Grynkiewicz, J.R. Peláez, Nucl. Phys. B, Proc. Suppl. 186 (2009) 318–321, arXiv:0811.4510
29. E. van Beveren, D.V. Bugg, F. Kleefeld, G. Rupp, Phys. Lett. B 641 (2006) 265–271, arXiv:/0606022.
30. R.E. Peierls, Proceedings of the Glasgow Conference on Nuclear and Meson Physics, Pergamon Press, London, 1954.
31. G. Höhler, Z. Phys. 152 (1958) 546–565.
32. M. Lévy, Nuovo Cimento 13 (1959) 115–143.
33. S. Aramaki, T. Osawa, Prog. Theor. Phys. 29 (1963) 451–460.
34. P.V. Landshoff, Nuovo Cimento 28 (1962) 123–131.
35. M.J.G. Veltman, Physica 29 (1963) 186–207.
36. T. Wolkanowski, Resonances and poles in the second Riemann sheet, Master thesis, Frankfurt am Main, 2013, arXiv:1303.4657
37. T.D. Lee, Phys. Rev. 95 (1954) 1329–1334; C.B. Chiu, E.C.G. Sudarshan, G. Bhamathi, Phys. Rev. D 46 (1992) 3508–3528.
38. A.G. Kofman, G. Kurizki, B. Sherman, J. Mod. Opt. 41 (1994) 353–384.
39. P. Facchi, H. Nakazato, S. Pascazio, Phys. Rev. Lett. 86 (2001) 2699–2703, arXiv:quant-ph/0006094; P. Facchi, S. Pascazio, Phys. Lett. A 241 (1998) 139–144, arXiv:quant-ph/9905017.
40. F. Giacosa, Phys. Rev. A 88 (2013) 052131, arXiv:1305.4467 quant-ph
41. K. Koshino, A. Shimizu, Phys. Rep. 412 (2005) 191–275, arXiv:quant-ph/0411145.
42. F. Giacosa, Phys. Rev. D 80 (2009) 074028, arXiv:0903.4481
43. N. Brambilla, S. Eidelman, B.K. Heltsley, R. Vogt, G.T. Bodwin, E. Eichten, A.D. Frawley, A.B. Meyer, et al., Eur. Phys. J. C 71 (2011) 1534, arXiv:1010.5827
44. C. Amsler, F.E. Close, Phys. Rev. D 53 (1996) 295–311, arXiv:/9507326.
45. A. Faessler, T. Gutsche, M.A. Ivanov, V.E. Lyubovitskij, P. Wang, Phys. Rev. D 68 (2003) 014011, arXiv:/0304031; M.K. Volkov, A.E. Radzhabov, V.L. Yudichev, Phys. At. Nucl. 66 (2003) 2143–2148, Yad. Fiz. 66 (2003) 2193–2198, arXiv:/0210306; F. Giacosa, T. Gutsche, V.E. Lyubovitskij, Phys. Rev. D 77 (2008) 034007, arXiv:0710.3403
46. M.D. Scadron, G. Rupp, R. Delbourgo, Fortschr. Phys. 61 (2013) 994–1027, arXiv:1309.5041
47. F. Giacosa, T. Gutsche, A. Faessler, Phys. Rev. C 71 (2005) 025202, arXiv:/0408085.
48. C. Hanhart, Y.S. Kalashnikova, A.E. Kudryavtsev, A.V. Nefediev, Phys. Rev. D 75 (2007) 074015, arXiv:/0701214; T. Branz, T. Gutsche, V.E. Lyubovitskij, Eur. Phys. J. A 37 (2007) 303–317, arXiv:0712.0354
49. D. Lohse, J.W. Durso, K. Holinde, J. Speth, Nucl. Phys. A 516 (1990) 513–548; J. Weinstein, N. Isgur, Phys. Rev. D 41 (1990) 2236–2257; O. Krehl, J. Speth, Nucl. Phys. A 623 (1997) 162–166; M.B. Voloshin, L.B. Okun, JETP Lett. 23 (1976) 333–336, Pis'ma Zh. Eksp. Teor. Fiz. 23 (1976) 369–372; K. Maltman, N. Isgur, Phys. Rev. Lett. 50 (1983) 1827–1830; V. Baru, J. Haidenbauer, C. Hanhart, Yu. Kalashnikova, A.E. Kudryavtsev, Phys. Lett. B 586 (2004) 53–61, arXiv:/0308129; T. Branz, T. Gutsche, V.E. Lyubovitskij, Eur. Phys. J. A 37 (2008) 303–317, arXiv:0712.0354
50. S. Krewald, R.H. Lemmer, F.P. Sassen, Phys. Rev. D 69 (2004) 016003, arXiv:/0307288; J.A. Oller, E. Oset, J.R. Peláez, Phys. Rev. D 59 (1999) 074001, arXiv:/9804209; J.A. Oller, E. Oset, J.R. Peláez, Phys. Rev. D 60 (1999) 099906 (Erratum); J.A. Oller, E. Oset, J.R. Peláez, Phys. Rev. D 75 (2007) 099903 (Erratum); T. Branz, T. Gutsche, V.E. Lyubovitskij, Phys. Rev. D 78 (2008) 114004, arXiv:0808.0705
51. D. Black, A.H. Fariborz, J. Schechter, Phys. Rev. D 61 (2000) 074001, arXiv:/9907516; A.H. Fariborz, R. Jora, J. Schechter, Phys. Rev. D 77 (2008) 034006, arXiv:0707.0843
52. A.H. Fariborz, R. Jora, J. Schechter, Phys. Rev. D 79 (2009) 074014, arXiv:0902.2825
53. A.H. Fariborz, R. Jora, J. Schechter, M.N. Shahid, Phys. Rev. D 84 (2011) 113004, arXiv:1106.4538
54. T.K. Mukherjee, M. Huang, Q.-S. Yan, Phys. Rev. D 86 (2012) 114022, arXiv:1203.5717
55. S. Janowski, F. Giacosa, D.H. Rischke, arXiv:1103.3238
56. R. Alkofer, L. von Smekal, Phys. Rep. 353 (2001) 281–465, arXiv:/0007355.

CHAPTER 11

ESTIMATING FISSION-BARRIER HEIGHT BY THE SPHERICAL-BASIS METHOD

Hiroyuki Koura

INTRODUCTION

In the region of heavy and superheavy nuclei, various decay modes coexist, such as α-decay, β-decay, and spontaneous fission. Spontaneous fission is a dynamical and drastic decay process in which a nucleus splits into two (sometimes three or more) nuclei. This mechanism is part of the nuclear many-body problem and is rather complicated. Nuclear fission is, of course, important due to its providing a large amount of atomic energy. In addition, nuclear fission is the main reason for the instability of heavy nuclei due to the Coulomb repulsive force, more than α-decay. Understanding the fission mechanism therefore leads to an understanding of the limit on the existence of nuclei with a high-Z (proton) number. Bohr and Wheeler [1] qualitatively described the mechanism of nuclear fission by using the liquid-drop model. Later, Strutinsky developed a method of calculating shell-correction energies by using deformed single-particle levels [2,3]. The method of using deformed single-particle potentials to obtain deformed shell energies [4,5] is often referred to as the Nilsson–Strutinsky method. We note that nuclear energy can thus be regarded as having two components: macroscopic liquiddrop energy and microscopic shell energy. The idea of calculating the fission barrier from the deformed single-particle configuration and the liquid-drop energy was elaborately extended to the macroscopic–microscopic calculation by Möller and his collaborators. They presented several versions of their model, e.g., the finite-range droplet model (FRDM) [6] and the finite-range liquid-drop model (FRLDM) [7].

In contrast, other calculations for the fission barrier have been carried out based on a microscopic approach using Skyrme-type effective interactions [8,9]. We will not describe this approach here in detail; however, we point out for discussion in this paper that it is based on a consideration of a deformed single-particle configuration with a coordinate constraint in order to obtain fission barriers, in which a nucleus is well deformed. In the above-mentioned calculations, a nuclear shape (or a constraint) is assumed and set in advance, and then a search is made for the eigenstate for its shape. In this paper, we propose a novel method of calculating a fission barrier that uses an approach that is radically different from using deformed single-particle states. We establish "spherical" single-particle states of the difference number of nucleons for one nucleus, and a deformed nuclear state is regarded as a representation of a configuration mixing of each of the spherical single-particle states. If the prepared spherical states yield a complete set in the quantum system, this assumption has validity in quantum mechanics. Several years ago, we developed a method of calculating ground-state nuclear masses for the entire region of a nuclear chart; these are the KUTY and KTUY mass formulae [10], which we also refer to as the spherical-basis method. In contrast to other methods, we do not use deformed single-particle states; instead, we express the deformed nuclear states as configuration mixtures of various spherical single-particle states (some corrections are introduced). We successfully described the ground-state shell energies of deformed nuclei; the calculated ground-state masses accurately reproduced those that were experimentally measured, and the derivatives of the masses as neutron and proton separation energies were better than those calculated by other models available at that time. The deformation parameters were also obtained, and they were comparable to experimentally measured nuclear deformations [11]. In this pa3per, we apply the spherical-basis method for calculating the heights of fission barriers. Compared with a ground-state nuclear-mass calculation, a fission-barrier calculation requires larger deformations. Using this method, we give the fission-barrier landscape in the heavier nuclear-mass chart, and discuss the origin of the appearance of nuclei with significantly higher

Citation: Hiroyuki Koura, Estimating fission-barrier height by the spherical-basis method, doi: 10.1093/ptep/ptu148.

and lower fission barriers than those of surrounding nuclei. Furthermore, we use the fission-barrier heights to estimate a border indicating the existence of nuclei. The mass asymmetry and large deformation are essentially important factors in discussing the fission phenomenon. In order to focus on their qualitative properties, however, nuclear shapes are limited to being axially and reflectionally symmetric, and only small deformations are considered in this paper. A discussion on the fission-fragment mass distribution cannot be given in this work due to this lack of asymmetric shape. This extension to asymmetric deformations will be left as a task for future work. The rest of this paper is organized as follows. In Sect. 2, we give a short explanation of the method. The results and discussion are presented in Sect. 3, and our conclusions are presented in Sect. 4.

METHOD

The spherical-basis method [10] is based on the assumption that a deformed state of a nucleus can be regarded as a superposition of "spherical states" that correspond to concentric spheres with different radii. Under this assumption, the nuclear shell energy with deformations can be expressed as weighted sums of spherical shell energies, and the weights are obtained from geometric shapes. The spherical shell energy of a nucleus for a given proton number, Z, and for a neutron number, N, are obtained by totaling the single-particle levels, subtracting the averaged bulk energy, and by making adjustment for pairing effects, such as the Bardeen–Cooper–Schrieffer (BCS) configuration, and the highly excited state components. The spherical shell energies of a nucleus with Z and N are defined for neutrons and protons, and are denoted as $E^{\rm sph}_{\rm nsh}(Z, N)$ and $E^{\rm sph}_{\rm psh}(Z, N)$. This summation and extraction is analogous to the Strutinsky method [2,3], except for the treatment of the deformation and the deduction of the bulk energy. We are able to compute the spherical shell energies of any nuclei by using the global spherical single-particle potential [14]. Suppose a nucleus is deformed into, e.g., a rugby-ball shape. In this case, the nucleus has both minimum and maximum radii from the center to the surface. Each of these radii can be assigned to a spherical nucleus with the same radius. Under this consideration, we assume that a deformed nucleus can be decomposed into a part of a spherical nuclei with various radii, and the shell energy can be expressed as a part of the spherical shell energy with a certain weight. We describe the relation between a radius and the number of protons Z or neutrons N in the spherical nucleus. We adopt $r(Z') = (Z'/Z)^{1/3}R_0$ or $r(N') = (N'/N)^{1/3}R_0$ with $R_0 = 1.2A^{1/3}$ m as the relation. The shell energy of a deformed nucleus with proton number Z and neutron number N is obtained as a weighted sum of the spherical shell energies and the macroscopic liquid-drop deformation energy as

$$E_{\rm sh}(Z, N, {\rm def.}) = \sum_{Z'} W_{\rm p}(Z'; Z, N)E^{\rm sph}_{\rm psh}(Z', N'') + \sum_{N'} W_{\rm n}(N'; Z, N)E^{\rm sph}_{\rm nsh}(Z'', N')$$
$$+ \Delta E_{\rm surf}(Z, N, {\rm def.}) - \Delta E_{\rm Coul}(Z, N, {\rm def.}) - \Delta E_{\rm pro}(Z, N, {\rm def.}). \quad (1)$$

Here, N'' and Z'' are the integers closest to $Z' \cdot N/Z$ and $N' \cdot Z/N$, respectively. This means that a charge equilibrium in a nucleus is assumed in this calculation. The meaning of the spherical proton and neutron shell energies, Esph $E^{\rm sph}_{i\rm sh}$ $(i = {\rm p, n})$, was explained earlier, and W_i $(i = {\rm p, n})$ are their mixing weights. The weight W_i is expressed by using the solid angle $\Omega(r)$ for a given distance r from the center as

$$W_{\rm p} = -\frac{1}{4\pi}\frac{d\Omega(r(Z'))}{dZ'},$$
$$W_{\rm n} = -\frac{1}{4\pi}\frac{d\Omega(r(N'))}{dN'}. \quad (2)$$

The weight depends on nuclear shapes through the solid angle.

The quantities $\Delta E_{\rm surf}$ and $\Delta E_{\rm Coul}$ in Eq. are the differences due to deformation of the macroscopic surface and the Coulomb energies. Other corrections are explained in Ref. [10]. The ground-state shell energy is obtained by minimizing the sum in Eq. (1) with respect to the deformation, denoted by "def.", and the results are shown in Refs. [10]. In this paper, we focus on larger deformations in which there are various saddle points, the highest of which can be regarded as a fission saddle, or fission barrier. In our previous work on calculating ground-state masses, it was necessary to determine spherical shell energies for nuclei with proton numbers of up to 270 and neutron numbers of up to 400, which correspond to the higher excited states caused by deformations in the model space; these are shown in Fig. 2 of Ref. [10]. This is sufficient if we only consider ground-state masses for nuclei with proton numbers ranging up to 130 and neutron numbers up to 200. However, in the case of fission barriers, we found by performing the preliminary calculations that many more spherical shell energies are necessary [12,13]. Thus, we extended these calculations to include spherical shell energies for proton numbers of up to 670 and neutron numbers of up to 998. The procedure for calculating the spherical shell energies is the same as that in Ref. [10]. We limited the deformations of nuclei to only those with axial and reflectional symmetry, parametrized by α_2, α_4, and α_6, as in the ground-state mass calculations [10]. The radius is then expressed as

$$R(\theta) = \frac{R_0}{\lambda}[1 + \alpha_2 P_2(\cos\theta) + \alpha_4 P_4(\cos\theta) + \alpha_6 P_6(\cos\theta)], \quad (3)$$

where λ is introduced to guarantee volume conservation and is defined as

$$\lambda = \left[1 + \frac{3}{4\pi}\left(\frac{4\pi}{5}\alpha_2{}^2 + \frac{4\pi}{9}\alpha_4{}^2 + \frac{4\pi}{13}\alpha_6{}^2\right)\right]^{1/3}. \quad (4)$$

In order to analyze just the qualitative properties, we expand the liquid-drop parts $\Delta_{\rm Esurf}$ and $\Delta_{\rm ECoul}$ in terms of the deformation parameters ($\alpha_{2,4,6}$, including the cross terms) and use the terms up to the fourth order. This reduces computational time by omitting the four-dimensional numerical integrals of the Coulomb energy. If we compare this fourth-order approximation with the full calculations for a few selected nuclei, we see that it is accurate to within a deformation of $\alpha_2 \leq$ 0.4–0.5. We thus focus primarily on $\alpha_2 < 0.4$.

We would like to emphasize that these parameter values are not changed or optimized for the fission-barrier height: they are the ones that best reproduced the experimental ground-state masses, as presented in the initial paper [10]. Finding a saddle point in multiparameter space is essentially difficult; see, e.g., Ref. [7]. Thus, we simply performed a 2D saddle-point search; i.e., we first calculated the shell energy with various values of α_2 and α4, with $\alpha_6 = 0$, then we found the minimum with respect to α_6. Next, we searched for a saddle point by looking for a location at which there was both positive and negative curvature in each set of orthogonal coordinates. After the search, we inspected selected cases to ascertain the validity of the search method, and we found no unreasonable results. We used a mesh size of 0.01 for α_2 and α4, and 0.001 for α_6. For the ground-state mass, a more accurate search for the minimum was carried out.

RESULTS AND DISCUSSION

Comparison of Experimental and Calculated Fission-Barrier Heights

Figure 1 shows a comparison between the experimental fission-barrier heights [15] and the heights calculated using models for radium and actinide isotopes (Z = 88–96). In the case of the experimental barrier heights, both the inner and outer barriers are indicated if they are experimentally identified. The experimental barrier heights [15] lie in the range of 4–9 MeV. Our estimated barrier heights generally matched the higher barrier from among the inner and outer experimental barriers, like Np, Pu, Am, and Cm. In the case of the Th, Pa, and U regions, however, the heights were overestimated by ~1 MeV. From the experimental data in Ref. [15], the shapes of the outer barrier for known isotopes from Th to Cm are known as mass asymmetry. In this paper, we only perform calculations for symmetric shapes, and thus we cannot make a comparison among barriers in the Th, Pa, and U regions because the shapes of these barriers appear to be asymmetric. We note that the outer fission barrier would be lower if we included an additional degree of freedom for asymmetry. If we adopt the higher barrier from among the inner and outer barriers as an experimental barrier for comparison with the calculated ones, the RMS deviation from the experimental barrier for the 52 adopted nuclei is 0.71 MeV, and the average difference from the experimental value is +0.37 MeV. For the extended Thomas–Fermi plus Strutinsky integral (ETFSI) [9], which is based on the Strutinsky approach with a two-body interaction as the Skyrme-type effective force, we adopted a β_2, β_4, and β_6 parametrization for the nuclear shape, which is essentially the same parametrization used in this study. The ETFSI fission barriers well reproduce the experimental trends, including those in

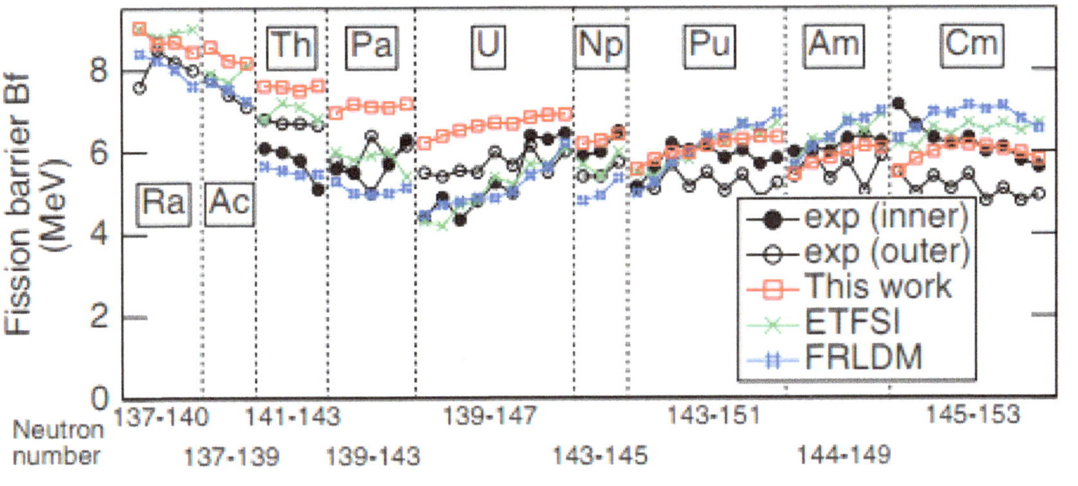

Figure. 1: Comparison between calculated and experimental fission-barrier heights. Circle: experiment [15]; open circle represents outer barrier, filled circle represents inner barrier. Square: this study. Cross: ETFSI [9]. Hash mark: FRLDM [7]. In the theoretical values, only the highest fission barriers are plotted because only the highest barriers appeared in the original papers [7,9].

the Th, Pa, and U regions. The RMS deviation for the same nuclei is 0.61 MeV, and the average difference from the experimental value is +0.05 MeV. We also plotted the FRLDM [7], which is the latest version of the macroscopic–microscopic approach and which uses 5D parametrization that includes a parameter for asymmetry. The RMS deviation is 0.77 MeV, and the average differences is −0.19 MeV

Landscape of Fission-Barrier Height

Figure 2 shows the fission-barrier heights for symmetric shapes in the heavy and superheavy nuclearmass region ranging from 76 to 148 of Z, and from 120 to 270 of N. Known nuclei are indicated by small black squares. In the known nuclear region, there is a kind of "hill" around Z ≈ 100 and N ≈ 150. The fission barrier in this region is about 6 MeV and it is higher than those of neighboring nuclei. This region includes actinide nuclei with measured fission-barrier heights as shown in Fig. 1. Another "basin" region is also found around Z ≈ 110 and N ≈ 168. The heights of these nuclei are generally lower than those of neighboring nuclei, and are in the range of 1–3 MeV. These nuclei are expected to be fissioning nuclei due to their lower fission barriers. Actually, spontaneously fissioning nuclei [16,17] were found experimentally, shown as open squares in Fig. 2, which, as expected, are in the same region.

Similar results are found for the ETFSI method, as shown in Fig. 3. The regions of higher ("hill") and lower ("basin") barriers are also seen, although the location of the "basin" is shifted to the neutron-richer side and is away from the experimentally fissioning region. Note that the fission barriers of the ETFSI method are calculated from deformed single-particle levels.

In order to clarify the origin of the significantly higher and lower barriers than those of surrounding nuclei in our method, as shown in Fig. 2, we show two cases: ^{252}Fm (Z = 100, N

= 152) as a case of a "hill" region, and ^{278}Ds (Z = 110, N = 168) as a case of a "basin" region. Figure 4 (upper) shows the potential energy surface of ^{252}Fm. The deformation parameters at the saddle point are α_2 = 0.27, α_4 = 0.03, and α_6 = 0.01. Generally, the neutron and proton numbers of a well deformed nucleus are far from the number of a closed-shell number, or a magic number. In the spherical-basis method, if a nucleus is deformed, the weighted sum of the spherical shell energies decreases, while the sum of the macroscopic deformed liquid-drop energy increases. An explanation of the lowering of the energy due to the weighted sum is given as follows. Figure 5 shows the mixing weight in this deformation and corresponding spherical shell energies in the case of a saddle-point shape. In such a well deformed shape, minimum and maximum radii exist, and the corresponding numbers of

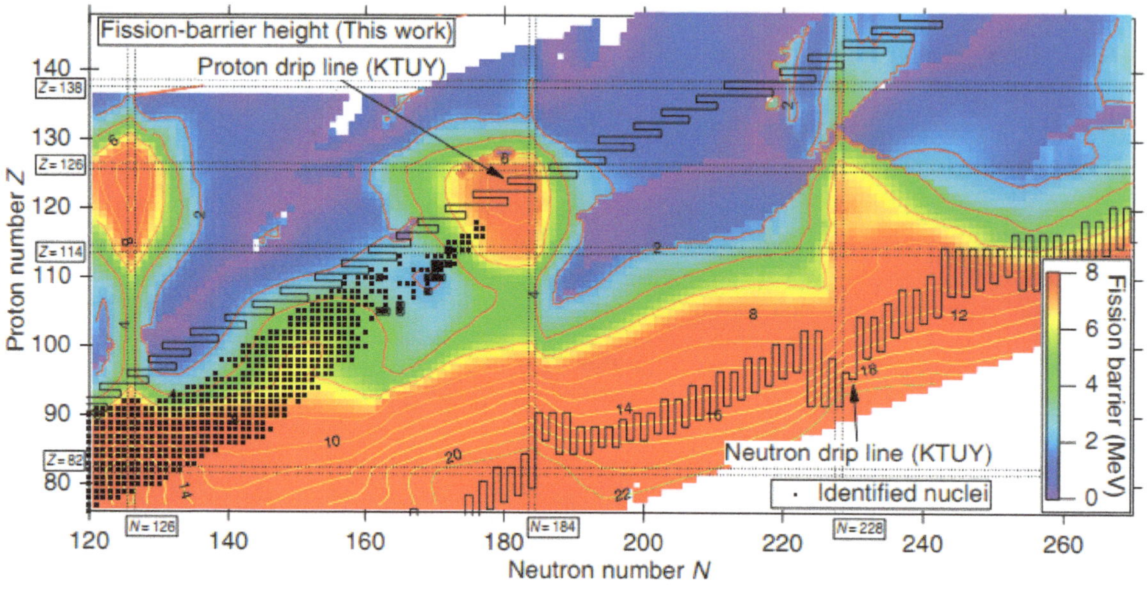

Figure. 2: Fission-barrier height in the heavy and superheavy mass region. Proton and neutron drip lines from the KTUY mass formula are shown as solid lines. Known nuclides [16,17] are also shown, as small black squares. Nuclei for which fission events were measured as occurring at a rate of 10% or more are shown as open squares (only for nuclei with neutron numbers N > 161). These are in the region Z ≈ 110 and N ≈ 168, as shown in the figure.

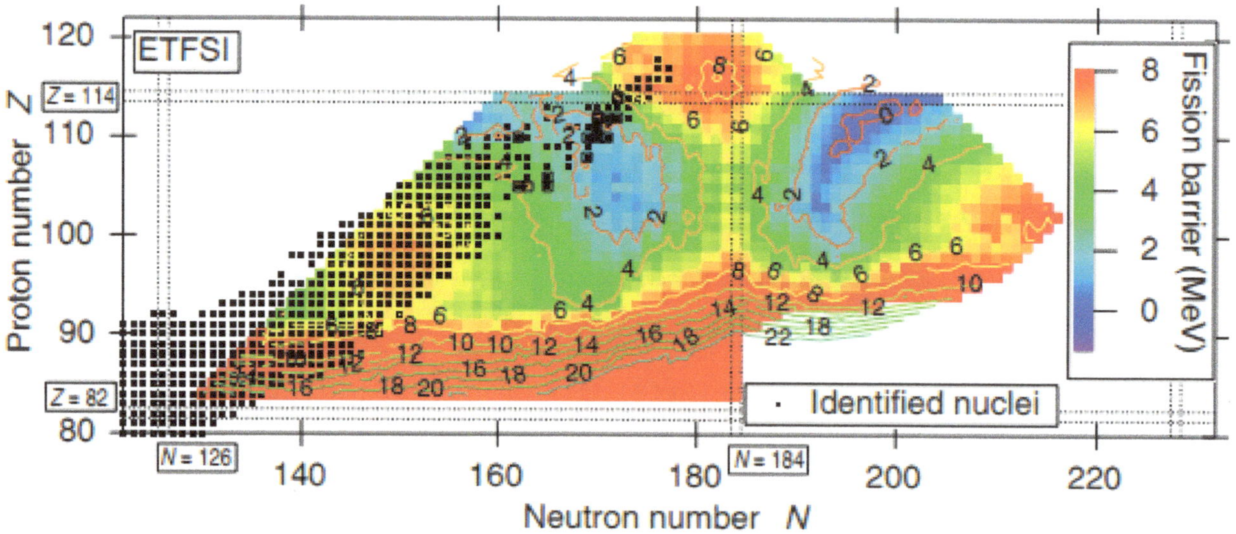

Figure. 3: Fission-barrier heights estimated by ETFSI [9].

neutrons and protons are also defined. In the case of neutrons, 97 is the corresponding number of neutrons to the minimum radius, and 232 is the corresponding number to the maximum radius. The summation is conducted through this range. In the case of protons, 64 and 152 are the smallest and largest numbers of protons. In Fig. 5, the end of the left side is the minimum number, and the end of the right side is the maximum number. Of these four numbers of nucleons, only 232 for neutrons is the closest number to the closed shell, 228. The weighted sum of spherical shell energies lowers the total energy, and the appearance of the saddle in this case comes from the 228 neutron shell closure. The other three sides of the weight (fewer neutrons and both more and fewer protons) do not reach strong shell closures, and thus do not lower the saddle point. The energy is therefore not lowered by much: ~6 MeV.

Figure 4 (lower panel) shows the potential energy surface of ^{278}Ds. The deformation parameters are $\alpha_2 = 0.18$, $\alpha_4 = -0.01$, and $\alpha_6 = 0.007$. The mixing weight and spherical shell energies of 268Ds at the saddle point are shown in Fig. 6. In this case, the smaller side of the neutron weight reaches the N = 126 neutron shell closure, and also the smaller side of the proton weight reaches the Z = 82

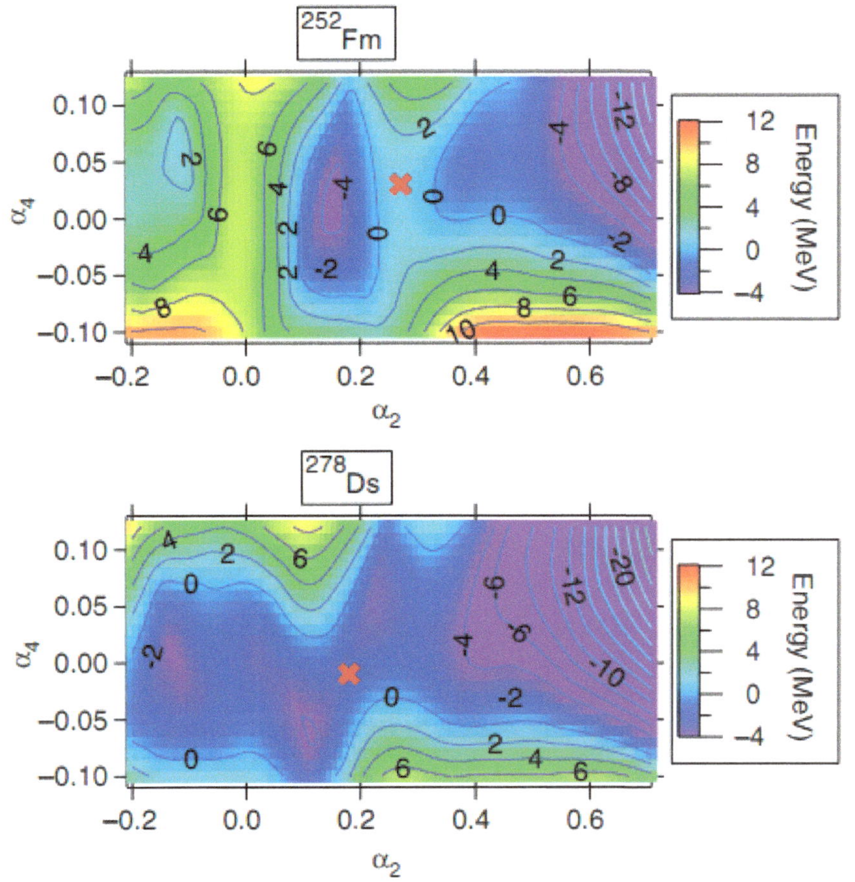

Figure. 4: Potential energy surface. Minimization by α_6 has been carried out. The highest saddle point is indicated by a cross. Upper: ^{252}Fm. Lower: ^{278}Ds.

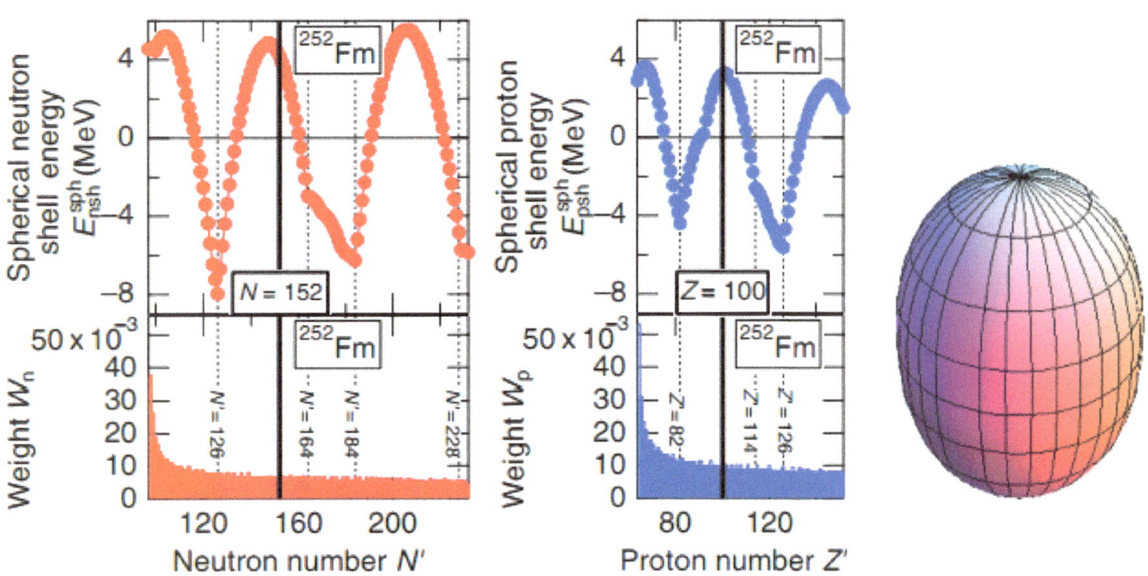

Figure. 5: (Left) Mixing weights (lower) and corresponding spherical shell energies (upper) at the saddle point for ^{252}Fm. Each quantity appears in Eq. (1). (Right) Shape at the saddle point for ^{252}Fm

proton shell closure. These two regions of weights produce the minimum of the saddle, and furthermore the weights in the smaller particle region (or inner nuclear region) have large values, as shown in the figures. In other words, this configuration can be regarded as a component of the core of ^{208}Pb (Z = 82 and N = 126) plus (collective) valence nucleons. Thus, this configuration has a saddle point with much lower values: 1–3 MeV in this mass region.

Limit of Existence of Nuclei from Systematics of Fission-Barrier Heights

Again going back to Fig. 2, we can see some isolated "islands" in the neutron-deficient heavy and superheavy mass region. The island in the lighter-mass region is found along N = 126 with Z ≈ 114

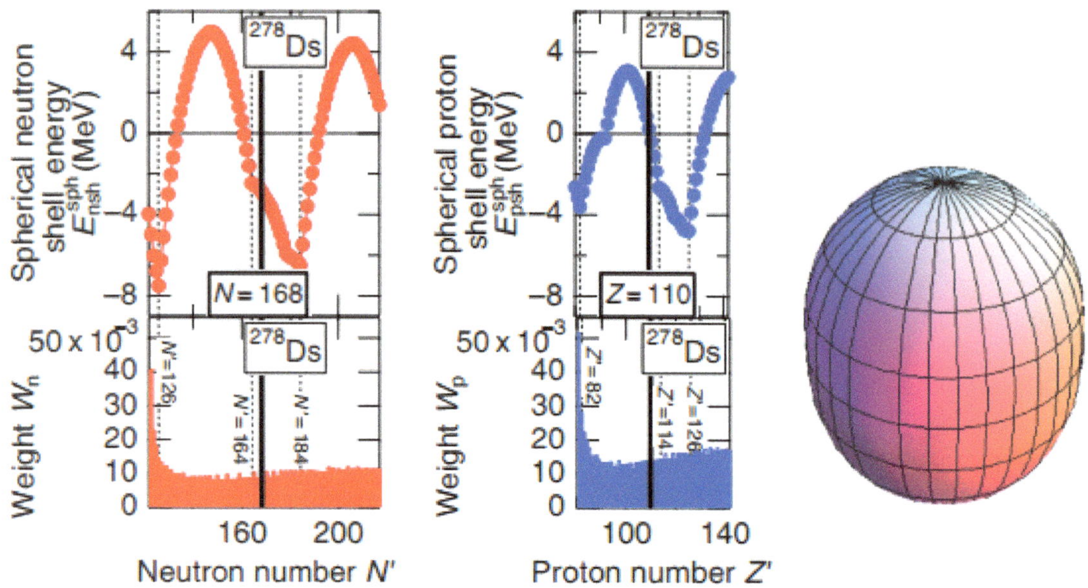

Figure. 6: (Left) Mixing weights and corresponding spherical shell energies at the saddle point for ^{278}Ds. (Right) Shape at the saddle point for ^{278}Ds. See explanation in Fig. 5.

and 126. This island is caused as a result of the strong shell closure of N = 126 and Z = 114 and 126, and the barrier heights of these nuclei are more than 8 MeV. However, the existence of nuclei in this region is unexpected because this region is far outside the proton drip line. Because of the existence of this island, the fission barriers of the neutron-deficient nuclei near N ~ 126 are enhanced, compared to the macroscopic estimation, and some nuclei such as U and Pu isotopes would be expected to have longer half-lives, and have more heavy-ion synthesis cross sections due to an enhancement of the evaporation-residue cross section. The island along N = 184 with Z between approximately 114 and 126 is expected to be an island of stability in the superheavy mass region. The estimated barrier height is more than 8 MeV. The adopted nuclear potential in this work results in doubly magic spherical single-particle levels for 298Fl (Z = 114) and ^{310}[126] [18]. The proton shell gaps of 114 for ^{298}Fl and ^{126} for ^{310}[126] are comparable (actually the gap of 126 is slightly larger than that of 114), and this property results in the creation of an island.

In the much heavier region, a peninsula appeared along N = 228 with Z between 114 and 126. This enhanced region is due to the strong shell gap of N = 228, and the reason the well defined island disappeared is that the nucleus ^{354}[126] is not a doubly magic nucleus in this nuclear potential (see Ref. [18]). Nevertheless, the barrier height of this nucleus is obtained as 6.59 MeV, comparable with those of the known actinides, as shown in Fig. 1, and α2 at the saddle point is 0.12. Furthermore, this nucleus is estimated to be β-stable [18]. Therefore, we can expect that this nucleus has a comparatively long half-life.

Borders of the Inner and Outer Fission Barriers

Figure 7 shows the deformation parameter α_2 at the saddle point. In the known actinide region (90 < Z < 103 and 120 < N < 160), there is a distinct border of $\alpha_2 \approx 0.4$. As we move from the border in the direction of higher Z, the saddle-point α_2 deformation is approximately 0.2–0.3, and the highest saddle point is the inner one if the outer barrier exists. As we move toward lower Z, the saddle-point α_2 deformation is more than 0.4, and the highest saddle point is the outer one if the inner barrier exists. If we compare these to the experimentally determined locations of such transitions, as shown in Fig. 1, we see that the calculated location of the transition is shifted to a region in which Z includes one or two more protons. Our calculations reproduced the qualitative trend of the

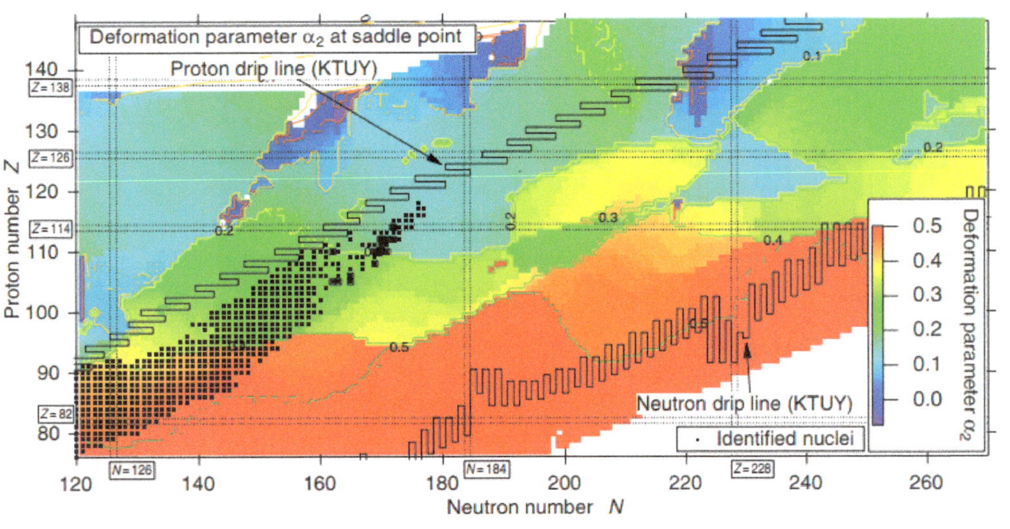

Figure. 7: α_2 deformation parameter at the fission saddle point. See explanation in Fig. 2.

experimentally determined locations of the dominant outer and inner barriers. For a more quantitative analysis, however, it will be necessary to extend our method to include larger deformations and asymmetric shapes.

CONCLUSION

We have presented a novel method for estimating the heights of fission barriers. We used the spherical-basis method instead of the Nilsson–Strutinsky framework to explain the existence of basins and hills in the superheavy mass region. In this study, however, we limited the nuclear shape to small deformations so that we could focus on the reasons for the shape of the landscape. An important area of our future work will be to apply this method to asymmetric shapes as, thus far, we have only considered shapes with axial and reflectional symmetry, but in general, especially with larger deformations, nuclear shapes are asymmetric. Recently, we have begun some calculations of potential energy surfaces for shapes with axial symmetry but without reflectional symmetry, for some selected nuclei in the same framework [19]. We will next extend our method to asymmetric shapes.

ACKNOWLEDGEMENTS

We wish to thank Prof. M. Yamada and Prof. S. Chiba for their valuable discussions. This work was partly supported by Grant-in-Aid for Young Scientists (B) 19740151, provided by the Ministry of Education, Culture, Sports, Science and Technology (MEXT) of JAPAN. The numerical calculations were mainly carried out using the PRIMERGY BX900 computer at the Center for Computational Science and E-Systems, Japan Atomic Energy Agency.

REFERENCES

1. N. Bohr and J. A. Wheeler, Phys. Rev., 56, 426 (1939).
2. V. M. Strutinsky, Nucl. Phys. A, 95, 420 (1967).
3. V. M. Strutinsky, Nucl. Phys. A, 122, 1 (1968).
4. S. G. Nilsson, Dan. Mat. Fys. Medd., 29, 16 (1955).
5. S. G. Nilsson et al., Nucl. Phys. A, 131, 1 (1969).
6. P. Möller, J. R. Nix, W. D. Myers, and W. J. Swiatecki, At. Data Nucl. Data Tables, 59, 185 (1995).
7. P. Möller, A. J. Sierk, T. Ichikawa, A. Iwamoto, R. Bengtsson, H. Uhrenholt, and S. Åberg, Phys. Rev. C, 79, 64304 (2009).
8. A. Mamdouh, J. M. Pearson, M. Rayet, and F. Tondeur, Nucl. Phys. A, 644, 389 (1999).
9. A. Mamdouh, J. M. Pearson, M. Rayet, and F. Tondeur, Nucl. Phys. A, 679, 337 (2001).
10. H. Koura, M. Uno, M. Uno, and M. Yamada, Nucl. Phys. A, 674, 47 (2000).
11. S. Raman, C. W. Nestor Jr, and P. Tikkanen, At. Data Nucl. Data Tables, 78, 1 (2001).
12. H. Koura, AIP Conf. Proc., 561, 388 (2001).
13. H. Koura, J. Nucl. Radiochem. Sci., 3, 201 (2002).
14. H. Koura and M. Yamada, Nucl. Phys. A, 671, 96 (2000).
15. T. Belgya et al., Handbook for Calculations of Nuclear Reaction Data (IAEA, Vienna, 2006), RIPL-2, IAEA = TECDOC-1506.
16. T. Tachibana, H. Koura, and J. Katakura, Chart of the Nuclides 2010 (Japanese Nuclear Data Committee and Nuclear Data Center) (Japan Atomic Energy Agency, Tokai, Ibaraki, 2010).
17. Yu. Ts. Oganessian et al., Phys. Rev. Lett., 109, 162501 (2012).
18. H. Koura and S. Chiba, J. Phys. Soc. Jpn., 82, 14201 (2013).
19. H. Koura, Meeting Abstract of the Phys. Soc. Jpn. 68, 51 (2013)

CHAPTER 12 ELASTIC SCATTERING, FUSION, AND BREAKUP OF LIGHT EXOTIC NUCLEI

J.J. Kolata, V. Guimar˜aes, and E.F. Aguilera

INTRODUCTION

This article is a review of progress made over the past decade in reactions induced by light ($A < 20$) radioactive beams at energies near to and below the Coulomb barrier ($E/A \leq$ 10MeV). Studies of elastic scattering, breakup, transfer, and fusion reactions on targets with $A > 4$ that were published between 2006 and 2015 are reported. Experimental work is emphasized, together with a selection of some of the many relevant theoretical papers on these subjects. Early publications within this time frame have already been reviewed by Keeley et al. [1,2]. In addition, a review of reactions with weakly bound nuclei by Canto et al. [3], which discusses reactions induced by both stable and radioactive projectiles and includes extensive theoretical analyses of the data, has recently appeared. All three of these publications are valuable resources for assessing the progress and future directions in this area in light of the fact that more exotic beams at higher intensity are becoming available.

REACTIONS WITH ^{6}HE

Reactions induced by ^{6}He have continued to be a fruitful area of research in the last decade, both because of the neutron-halo nature of this nuclide and the relative ease of producing intense beams of it at energies near to and below the Coulomb barrier. Over 150 articles reporting on fusion, elastic scattering, and breakup reactions, approximately 1/3 of them experimental, have appeared during this period. The experimental papers, and some of the more important theoretical work, are reviewed below.

Citation: J. J. Kolata, V. Guimarães, and E. F. Aguilera, Elastic scattering, fusion, and breakup of light exotic nuclei, doi:10.1140/epja/i2016-16123-1.

Elastic scattering of ^{6}He

Elastic scattering of ^{6}He and other light radioactive nuclei on various targets was last reviewed by Keeley et al. in 2009 [1]. Work that appeared after this publication is emphasized below.

^{6}He + ^{6,7}Li

All available elastic-scattering data on these systems are reviewed in ref. [1].

^{6}He + ^{9}Be

Elastic scattering of ^{6}He on ^{9}Be has been measured at an incident energy of 16.8MeV by Majer et al. [4], as part of a project to study ^{7}He produced via the ^{9}Be(^{6}He, ^{8}Be)^{7}He reaction. The measured angular distribution, shown in fig. 1, is compared with data for ^{6}Li + ^{9}Be scattering at 16.6MeV [5]. Optical-model (OM) and continuumdiscretized coupled-channels (CDCC) calculations are also shown. The optical-model calculations were performed by varying the real and imaginary well depths of the ^{6}Li+^{9}Be potential of Cook and Kemper [6]. They begin to differ from the data at angles $\theta_{c.m.} > 100°$ for ^{6}Li+^{9}Be and 70° for ^{6}He + ^{9}Be. A CDCC calculation coupling to the ^{6}He continuum above the $\alpha + 2n$ breakup threshold provides superior agreement with the data, illustrating the effect of the projectile breakup channel on the elastic scattering. A total reaction cross section of 1436mb was deduced. In addition, quasifree scattering of ^{6}He on an α-particle cluster in ^{9}Be was studied, together with two-neutron stripping to states in 11Be.

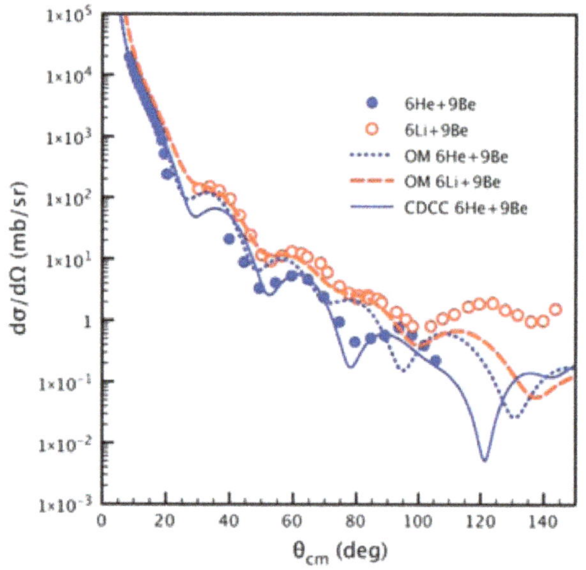

Figure 1: Elastic scattering of ^{6}He on ^{9}Be at 16.8 MeV. Figure reprinted from ref. [4] with kind permission of The European Physical Journal (EPJ).

Additional data for elastic scattering of ^{6}He on ^{9}Be were obtained at E_{lab} = 16.2 and 21.3MeV by Pires et al. [7]. The effect of target and projectile excitation was studied via coupled-channels calculations including coupling to the first three members of the ground-state rotational band in ^{9}Be and the 2^+ state in ^{6}He. The rotational couplings were found to have only a minor effect, while the coupling to the 2^+ resonance improved the agreement with experiment. Three-body and four-body CDCC calculations were also performed, and it was found that a dominant role is played by nuclear couplings to the 2^+ resonance. Surprisingly, the trivial local potentials extracted from these calculations displayed the long-range absorption effects found for heavier targets, even though the interaction is mainly nuclear rather than Coulomb in this system. An additional analysis of this same data set was presented in ref. [8]. Average total reaction cross sections of 1525(104)mb and 1527(106)mb at 16.2MeV and 21.3MeV, respectively, were deduced. This implies an enhancement of $\approx$ 25% in comparison with stable systems.

^{6}He + ^{12}C and ^{6}He + ^{27}Al

All available elastic-scattering data on these systems are reviewed in ref. [1].

^{6}He + ^{51}V

Angular distributions for ^{6}He quasielastic scattering on ^{51}V at laboratory energies of 15.4 and 23.0MeV are reported in ref. [9]. The data at 23.0MeV were compared with ^{4}He scattering data at 23.2MeV taken at the same time. Optical-model calculations showed that ^{6}He is far more absorbed than ^{4}He. The ^{6}He + ^{51}V total reaction cross sections were found to be 1836mb and 2219mb at 15.4 and 23.0MeV, respectively.

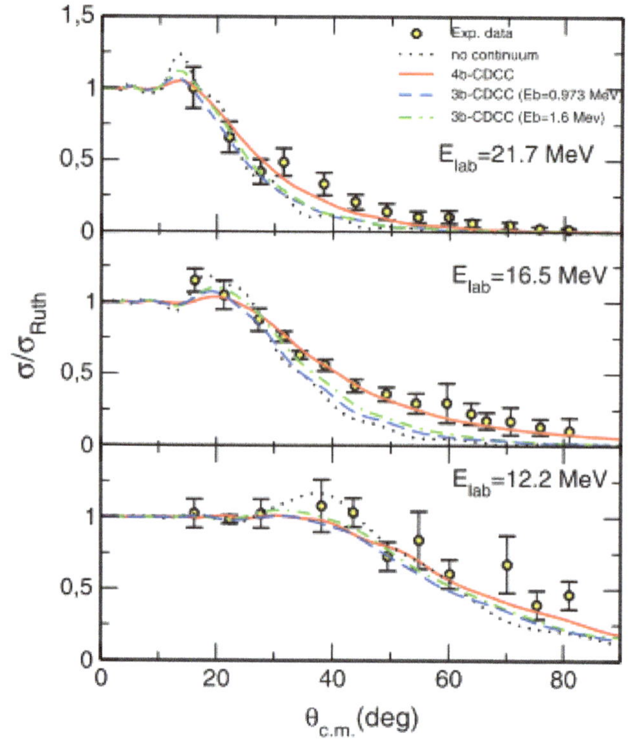

Figure 2: Elastic scattering of ^{6}He on ^{58}Ni from ref. [10].

^{6}He + ^{58}Ni

Elastic-scattering angular distributions for ^{6}He on ^{58}Ni at E_{lab} = 12.2, 16.5, and 21.7MeV were presented in ref. [10]. The data were analyzed via three-body and four-body CDCC calculations. The results are shown in fig. 2. It can be seen that the effect of coupling to the continuum is to remove cross section from the region of the Fresnel peak. Both the three-body and four-body CDCC calculations captured this effect. However, the three-body calculations failed to reproduce the experimental data in detail while the four-body calculations provided an excellent fit at all energies. These results were found to be quite stable with respect to changes in the optical-model potentials used in the calculations. The trivially-equivalent local potentials derived from the four-body calculation displayed a longrange imaginary part, indicating absorption from the combined effect of the nuclear and Coulomb potentials. The real part of the potential is repulsive in the surface region, which acts to reduce the fusion cross section. Finally, the total reaction cross sections were found to be 1185, 1833, and 2185mb at E_{lab} = 12.2, 16.5, and 21.7MeV, respectively. These values correspond to an $\approx$ 50% enhancement over total reaction cross sections for ^{6,7}Li or ^{9}Be on ^{58}Ni.

^{6}He + ^{64}Zn and ^{6}He + ^{65}Cu

All available elastic-scattering data on these systems are reviewed in ref. [1].

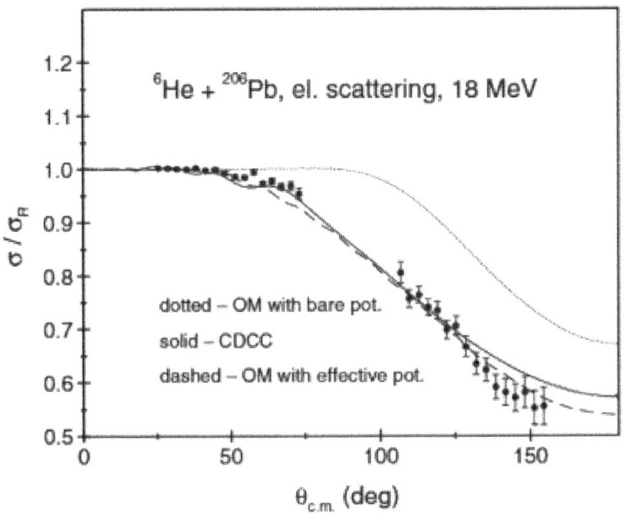

Figure 3: Elastic scattering of ^{6}He on ^{206}Pb. Figure reprinted with permission from ref. [13]. c 2013, The American Physical Society.

^{6}He + ^{120}Sn

The elastic scattering of ^{6}He on ^{120}Sn has been studied at E_{lab} = 17.40, 18.05, 19.80, and 20.5MeV [11]. The angular distributions were analyzed using the optical model, and also a four-body CDCC calculation. The latter followed the methodology laid out in ref. [12] and included coupling to the breakup channel. The total reaction cross sections extracted from these analyses were compared with those from elastic scattering of stable projectiles, both stronglyand weakly-bound, on several targets. It was observed that the reduced total reaction cross sections for 6He were greater than those of weakly-bound stable nuclei, which in turn were greater than those for strongly-bound projectiles. A preliminary discussion of these data was given in ref. [9].

^{6}He + ^{197}Au

All available elastic-scattering data on this system at energies such that the angular distributions are not pure Rutherford are reviewed in ref. [1].

^{6}He + ^{206}Pb

An elastic-scattering angular distribution for this system at E_{lab} = 18MeV has been reported in ref. [13]. The data were well-reproduced by a three-body CDCC calculation [14] that included only the effect of the breakup of 6He (fig. 3). This calculation very slightly over-predicted the differential cross section at c.m. angles ≥ 125◦, while a more sophisticated four-body CDCC calculation [15] tends to under-predict the data in the same angular region. The corresponding dynamic polarization potential is similar to that derived for ^{6}He+^{208}Pb, with a long-range, attractive real part as well as long-range absorption.

^{6}He + ^{208}Pb

Elastic-scattering data on this system are reviewed in ref. [1], including all information available up to 2009. Since then, two papers of interest have appeared. The first (ref. [16]) presents an analysis of the 18MeV data from ref. [17] in terms of the dipole polarizability of ^{6}He, which was found to be much greater than those of ^{2}H, ^{3}He, or ^{7}Li. For example, the ^{6}He dipole polarizability was found to be 1.2fm^{3}, the highest value yet observed, compared with only 0.70fm^{3} for ^{2}H. The second paper (ref. [18]) was a measurement of ^{6}He + ^{208}Pb elastic scattering at 22MeV. This system had previously been studied [17] but over a more limited angular range. Agreement between the two data sets was observed. The new data were analyzed in the context of a three-body CDCC calculation, which reproduced them fairly well.

^{6}He + ^{209}Bi

This system, which is the most-extensively studied of all those presented here, is reviewed in ref. [1]. No new information has become available since the publication of that article.

Breakup of ^{6}He

The first indication of the anomalous breakup cross section of ^{6}He at energies near the Coulomb barrier was published in 2000 [19]. In this paper, an astonishingly strong α-particle group from ^{6}He+^{209}Bi, possibly resulting from breakup of ^{6}He, was noted. However, the energy spectrum of the α-particles was consistent with that expected from Q-window calculations for 2n-transfer to ^{211}Bi. Subsequently, a series of α-n coincidence experiments [20–22] established that 2n-transfer was indeed the most important process, and that 1n-transfer and direct breakup were less important, each having only about 1/2 the cross section for 2n-transfer. In this section, further work in the last decade on ^{6}He breakup near the Coulomb barrier is reviewed. Additional data on a number of systems can be found in ref. [1].

α-particle production in ^{6}He reactions near the Coulomb barrier

Alpha-particles from the breakup of ^{6}He on a ^{120}Sn target were observed by deFaria *et al.* [23]. The α-particle energies and angular distributions led to conclusions about the reaction mechanism that agreed with those discussed in ref. [19]. A preliminary report of this work appeared in ref. [9].

Escrig *et al.* [24] observed α-particle production in the ^{6}He + ^{208}Pb reaction at energies near the barrier and developed a method to calculate 2n-transfer to continuum states in ^{210}Pb. Their analysis indicated that the α-particles emitted at backward angles in this reaction are mainly due to two-neutron transfer to weakly-bound states of the final nucleus. More recently, Fern´andez-Garc´ıa *et al.* [25] and Acosta *et al.* [18] have used this method to analyze data for the same system at a number of nearby energies and reached the identical conclusion. In all three experiments, however, the α-particle yield was somewhat under predicted.

In addition to elastic scattering, α-particle production in the ^{6}He+^{206}Pb reaction at 18 MeV was also reported in

ref. [13]. The observed yield was analyzed in the context of the DWBA. For 1n-transfer, (d,p) spectroscopic factors for transitions to known low-lying states were used. The emission angle of the resulting α-particles was assumed to be identical with that of the outgoing ^{5}He. Since the 2n-transfer Q-window is at high excitation energy, a fictitious 6S_0 dineutron state was placed at an excitation energy of 14 MeV in ^{208}Pb. Both calculated transfer processes peak at angles ≥ 100 and their sum slightly over predicts the observed cross section at backward angles. The 1n- and 2n-transfer cross sections were calculated to be of similar magnitude, though it was emphasized that these calculations are very much model dependent. The effect of transfer on the elastic channel was investigated in a Coupled Reaction Channel (CRC) calculation using the neutron-transfer parameters from the DWBA. Introduction of the transfer channels significantly reduced the predicted elastic scattering at backward angles. However, a reasonably good fit to the experimental data could be achieved by replacing the ^{6}He + ^{206}Pb optical-model potential by one having a short-range imaginary potential in order to avoid double-counting in the absorptive potential.

Effect of breakup on elastic scattering and fusion near the Coulomb barrier

As mentioned in sect. 2.1.5 above, the trivially-equivalent local potential (TELP) for ^{6}He+^{58}Ni scattering at an energy of E_{lab} = 16.5MeV has been evaluated by Morcelle et al. [10] using both 3-body and 4-body CDCC. The results, shown in fig. 4, demonstrate the effect of breakup coupling on the scattering potential. In particular, the main features are the generation of long-range absorption (due to both Coulomb and nuclear effects) and an additional repulsion in the real part of the potential in the surface region (R ≈ 6–7fm) which serves to enhance the Coulomb barrier. Similar features have been observed in lighter systems where the Coulomb potential is less dominant [7], and seem to be a general characteristic of polarization potentials derived from CDCC calculations for neutron-halo projectiles.

Sargsyan et al. [26] analyzed the effect of breakup on the fusion cross section of a number of systems, including ^{6}He incident on targets of ^{64}Zn, ^{197}Au, and ^{209}Bi. They used the quantum diffusion approach (see for example ref. [27]) in an attempt to gain an understanding of breakup systematics on fusion suppression near the barrier. Neutron transfer channels and deformation effects were taken into account within this approach. Calculated capture cross sections were compared with experimental

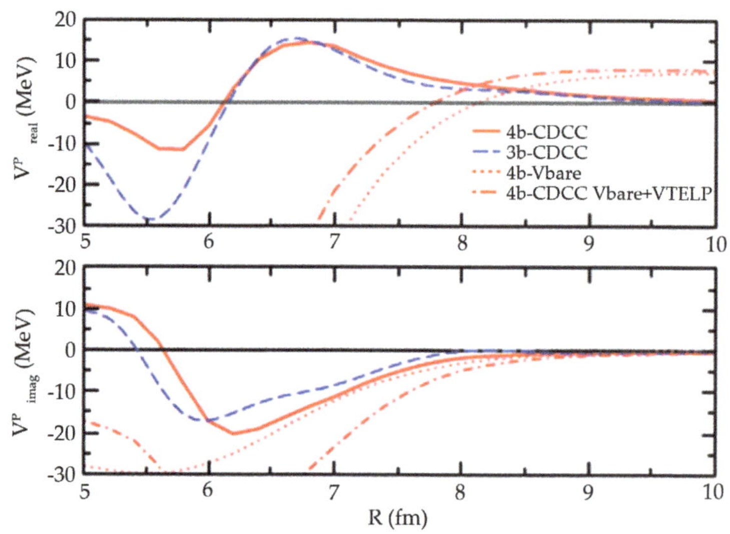

Figure 4: Trivially-equivalent local potential for ^{6}He on ^{58}Ni from ref. [10].

fusion data. Unfortunately, no systematic trends were observed. The computed fusion excitation functions reproduced the energy dependence for ^{6}He fusion, but it was necessary to normalize the calculated cross sections by factors of 0.4, 0.8, and 0.68 for ^{64}Zn, ^{197}Au, and ^{209}Bi, respectively. A similar conclusion was reached by Gomes et al. [28] in a search for systematics in the fusion of 9Be with a number of targets, using a very different method.

Kucuk [29] has analyzed the elastic scattering of ^{6}He on targets of ^{12}C, ^{27}Al, ^{58}Ni, ^{64}Zn, ^{65}Cu, ^{197}Au, ^{208}Pb, and 209Bi at energies near the Coulomb barrier, as well as on ^{12}C at 38 and 42MeV/A. A double-folding-model approach based on a No Core Shell-Model (NCSM) wave function for ^{6}He, together with a phenomenological imaginary potential, was used. All data were taken from the literature. The predictions for all targets were quite good. In addition, the ^{6}He + ^{208}Pb data were chosen for an investigation into the breakup threshold anomaly [30]. It was found that the real part of the scattering potential decreases at lower energies, while the imaginary potential increases just as in the case of ^{6}Li + ^{208}Pb [30]. This may be the result of strong coupling to breakup channels at near-barrier energies.

In order to more directly evaluate the effect of coupling to breakup channels on fusion cross sections near the barrier, Canto et al. [31,32] have derived a "Universal Fusion Function" (UFF) which they used to compare the systematic behavior of fusion cross sections for stronglybound and weakly-bound stable projectiles, as well as halo nuclei such as ^{6}He. As a first step, the data are compared with predictions based on an optical-model potential that takes into account the target and projectile densities. Canto et al. adopt the S˜ao Paulo folding-model potential [33]. Next, barrier parameters are deduced following the Wong model [34] and a dimensionless fusion function F is defined:

$$F(x) = \frac{2E\sigma_F}{\hbar\omega R_B^2}; \qquad x = \frac{E - V_B}{\hbar\omega}. \tag{1}$$

Here, σ_F is either the experimental cross section or the one extracted from the optical-model potential using incoming-wave boundary conditions. In the case of systems that follow the Wong formula, F(x) reduces to:

$$F_0(x) = \ln[1 + \exp(2\pi x)]. \quad (2)$$

Canto et al. call this the Universal Fusion Function (UFF) since it doesn't depend on the system and can therefore be used to directly compare different systems. Deviations of the reduced experimental data from the UFF are taken as a sign of channel-coupling effects. However, in order to isolate the effect of breakup couplings it's necessary to account for other effects, such as coupling to inelastic channels, by carrying out a coupled-channels calculation and renormalizing the data according to:

$$\bar{F}_{exp}(x) = F_{exp}(x)\frac{F_0(x)}{F_{CC}(x)}. \quad (3)$$

Here, $F_{CC}(x)$ is the fusion function from the coupledchannels calculation. Then, under the assumption that all relevant channels have been included, deviations of the renormalized experimental fusion function from the UFF, $F_0(x)$, arise only from coupling to the breakup channels. Qualitatively, the results of applying this procedure to fusion of ^{6}He with high-Z targets such as ^{209}Bi or ^{238}U gave evidence for suppression of $\approx$ 30% above the barrier and a small enhancement below it [31,32]. In a later article [35], the same group analyzed ^{6}He + ^{64}Zn data [36] and unexpectedly found even greater fusion suppression.

Moro et al. [37] have calculated three-body breakup assuming either transfer to the continuum (TC) or direct breakup (bu). They analyzed data from the ^{6}He + ^{208}Pb reaction at 22MeV [18] and the ^{8}B + ^{58}Ni reaction at 26MeV [38]. The ^{8}B data were equally-well represented in both schemes, but the ^{6}He data were in much-better agreement with the TC calculation, indicating that the α-particles emitted at backward angles arise from 2ntransfer to slightly-unbound states of the target. In this case, the transferred neutrons are highly-correlated in the final state of the target. This calculation assumed a dineutron model for ^{6}He. Ogata et al. [39] have analyzed ^{6}He+^{209}Bi elastic scattering at 19 and 22.5MeV [19] in a four-body CDCC calculation that removes this constraint. They found that the elastic-scattering cross section was much better represented in the four-body approach, but didn't carry out a TC analysis of the α-particle data. An earlier version of this work was reported in ref. [40].

Fusion of ^{6}He

In this section, results from fusion measurements with a ^{6}He projectile that appeared since the review of Keeley et al. [1] are presented. In addition, a selection of some of the many theoretical papers on this subject are discussed.

Recent experimental work on ^{6}He fusion

Scuderi et al. [41] have studied fusion, breakup, and transfer in the ^{6}He + ^{64}Zn reaction at energies near the barrier. The breakup and transfer results, obtained by looking at α-particle production, were consistent with their previous measurements [36]. The fusion cross sections were measured using an activation technique, by detecting delayed x-ray activity following the electron-capture decay of the evaporation residues. Statistical-model calculations showed that electron-capture decay predominates for this beam-target combination and that stable decay products constitute at most 10% of the fusion yield. A stack of four ^{64}Zn targets separated by ^{93}Nb catcher foils was exposed to a ^{6}He beam at two different energies, and the x-ray activity was detected off-line using a Si(Li) detector having an efficiency of 100% for the energies of interest. Five different fusion-evaporation channels were observed and separated by their corresponding x-ray energies and half-lives. Cross sections were determined from the measured number of incident beam and target particles and the known Kα fluorescence probabilities. The beam current was measured as a function of time in order to correct for decay of short-lived activities during the bombardment and transfer to the counting station. The fusion excitation function was initially obtained by summing the contributions from all the reaction channels. However, it was observed that the αn channel leading to ^{65}Zn was very strong compared with a statistical-model calculation. Since this channel could also be reached via the known strong transfer processes, a corrected fusion probability was obtained by replacing its measured cross section by the calculated one. The data were compared with results from the ^{4}He+^{64}Zn system reported in the same experiment. It was concluded that no effect on the fusion cross section at energies above the barrier was evident in the ^{6}He-induced reaction with respect to that induced by ^{4}He. Enhancement of the fusion cross section observed below the barrier was attributed to static effects of the halo structure rather than dynamic coupling effects.

Complete (CF) and incomplete (ICF) fusion reactions of ^{6}He on ^{166}Er were investigated by Fomichev et al. [42] at an energy of 64MeV, well above the Coulomb barrier. The data were compared with results from the ^{6}Li+^{165}Ho and ^{6}Li + ^{166}Er reactions at 58MeV. Charged particles, γ rays, and neutrons were detected. Exit channel identification was obtained via a (γ_1-γ_2-light charged particle) triple coincidence. Relative intensities of γ rays in ^{166}Yb from ^{165}Ho (^{6}Li, 5n) and ^{166}Er (^{6}He, 6n) were compared as a function of the spin of the γ-emitting state, up to $22\hbar$ in the ground-state band and one excited band. The results were compared with the predictions of the EMPIRE code (ref. [43]). From this comparison, the effects of ICF reactions associated with the capture of the α clusters in either ^{6}Li or ^{6}He by the ^{166}Er nuclei were estimated to be less than 10% as compared with the intensities of the ^{166}Yb γ rays and corrections were made for this process. The spin distribution of the compound systems were deduced from the corrected intensities and CF suppression factors of 3.4 for the 6He reaction and 2.4 for the ^{6}Li reactions were calculated, again by comparison with EMPIRE, suggesting a large effect of projectile breakup. A similar analysis was made for γ rays in the ground-state band of ^{164}Er coming from ^{165}Ho (^{6}Li,α3n) and ^{166}Er (^{6}He,α4n), and values for lower and higher orbital angular momentum cutoffs were deduced.

Fusion and nucleon transfer processes in the ^{6}He + ^{197}Au system were studied in the energy range from 40– 120MeV by Skobelev et al. [44]. The stacked-foil activation method was used, with Au foils separated by Al absorbers to degrade the beam energy. A range of Tl, Au, and Hg isotopes were identified off-line using γ detectors. Excitation functions for fusion reactions involving the evaporation of up to ten neutrons from the compound nucleus, as well as reactions with the emission of charged particles and nucleon transfer products, were obtained. It was concluded that, unlike the situation at near-barrier energies, no evidence for fusion inhibition was found. The production cross sections for ^{194,196}Au and $^{192-197}$Hg, however, were shown to be dominated by direct processes over the entire energy range. Isomeric ratios for several isotopes of Tl, Au, and Hg were also determined. Isomeric ratios for ^{196,198}Tl and ^{196,198}Au produced via ^{6}He + ^{197}Au fusion and transfer reactions had already been measured by Kulko et al. [45]. Population probabilities for high- and low-spin states were determined in the range from 7–60MeV and compared with statisticalmodel calculations. They were found to be similar to those from ^{4}He-induced reactions. The isomeric ratios for fusion reactions followed by the evaporation of five and seven neutrons were 1 to 3 orders of magnitude greater than those observed for neutron-transfer products. Finally, the isomeric ratios for ^{196}Au and ^{198}Au produced via transfer processes displayed a different energy dependence and differ in absolute value: in the case of neutron removal from ^{197}Au, the isomeric ratios are higher than for the transfer of a neutron to ^{197}Au

Sub-barrier fusion of 6He with 206Pb was measured by Wolski et al. [46] in the energy range from 14–28MeV using a stacked-foil activation technique and radiochemical analysis via α-particle detection. The elastic scattering at 18MeV was simultaneously measured during the beam exposure period in order to obtain an accurate value for ^{210}Po production in this reaction. This 2n-evaporation channel dominates the fusion cross section at all energies except for the highest one measured. Corrections were made for the naturally-occurring background from this isotope. The deduced fusion cross sections were in disagreement with an earlier measurement [47,48], as illustrated in fig. 5. It was suggested that the disagreemen might be due to the effects of the beam energy spread, which was particularly significant for the earlier data as noted by the large horizontal error bars for this experiment shown in the figure. However, it is not clear that this effect could completely account for the very large differences below 17MeV, and the much smaller discrepancy at 17.5MeV is unexplained. The authors of ref. [47] also used the stacked-foil method with off-line α-particle detection and compared with ^{4}He + ^{208}Pb fusion leading to the same ^{212}Po compound system and the same ^{210}Po 2n-evaporation channel. Good agreement was obtained with data from a previous experiment [50].

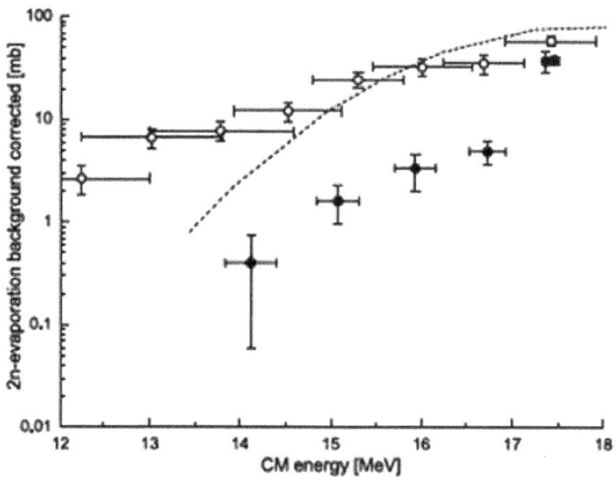

Figure 5: Data for sub-barrier fusion of ^{6}He+^{206}Pb from refs. [46] (solid circles) and [47] (open circles). The dashed curve is the prediction of the sequential fusion model [48,49] for the 2n evaporation channel. Figure reprinted from ref. [46] with kind permission of The European Physical Journal (EPJ).

REACTIONS WITH ^{8}HE

Reaction data for ^{8}He beams at energies near the Coulomb barrier are still very scarce due to the limited availability of this beam. Two early papers, from 2003 and 2004, are reviewed in ref. [1]. More recently, Marqu´ınez-Dur´an et al. [51,52] have studied ^{8}He elastic scattering and reactions on a ^{208}Pb target at energies of 16 and 22MeV. (The Coulomb barrier for this system is at 19MeV). They compared their elastic-scattering results with ^{6}He+^{208}Pb data at the same energies and found that the absorption was greater for 8He (see, e.g. fig. 6). The observed absorption was consistent with the fact that large yields of ^{4}He and ^{6}He fragments were also seen. The data were also compared with a coupled-reaction-channel (CRC) calculation including only 1n-stripping to ^{7}He. The data near the Fresnel peak and at back angles were somewhat overpredicted, probably due to the neglect of other reaction channels.

Transfer/breakup channels in the ^{8}He + ^{197}Au system at E_{lab} = 18.7, 20.1 and 29.4MeV, and the ^{8}He+^{65}Cu system at E_{lab} = 19.9 and 30.6MeV, were studied by Lemasson et al. [53]. Two different experimental methods were used. For the ^{65}Cu target, in-beam measurements of light charged particles, neutrons, and γ-rays were carried out. The Q-value spectrum for ^{6}He in coincidence with the 186- keV γ-ray transition in ^{66}Cu peaked around Q = 0, consistent with expectations for 2n-transfer to states in ^{67}Cu above the neutron-emission threshold. The inclusive Qvalue distribution was nearly identical in shape, suggesting that 2n-transfer processes to low-lying bound states were weak or non-existent. Very large total neutron transfer cross sections ($\sigma_{1n}+\sigma_{2n}$) of 782±78 (759±114)mb at 19.9 (30.6)MeV were deduced from the measurement of characteristic γ-ray transitions, corrected for γ-rays arising from fusion-evaporations reactions.

A stacked-foil radiochemical technique was used with the ^{197}Au target. Gamma rays from ^{198,199}Au were counted off-line, and cross sections for ^{198m}Au were also obtained from γ-γ coincidence data. Surprisingly, a relatively large cross section was obtained for ^{199}Au despite the fact that the optimum Q-value for 2n-transfer would lead to population of states that are well above the neutron-emission threshold (see fig. 7). In the case of the ^{65}Cu target discussed above, survival of ^{67}Cu was not observed. An explanation for this difference was given in terms of mixing of bound and unbound levels in ^{199}Au.

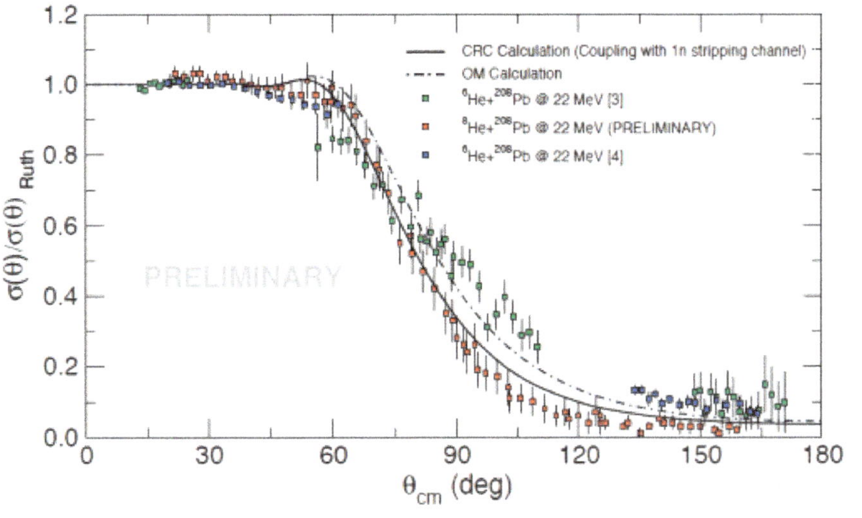

Figure 6: Ratio-to-Rutherford of the elastic cross section for $^{8}He + ^{208}Pb$ and $^{6}He + ^{208}Pb$ at E_{lab} = 22 MeV. The dotdash line is the result of an optical-model calculation using parameters from $^{6}Li + ^{208}Pb$ scattering, while the solid line is the result of a CRC calculation including 1n-stripping. See refs. [51,52] for details.

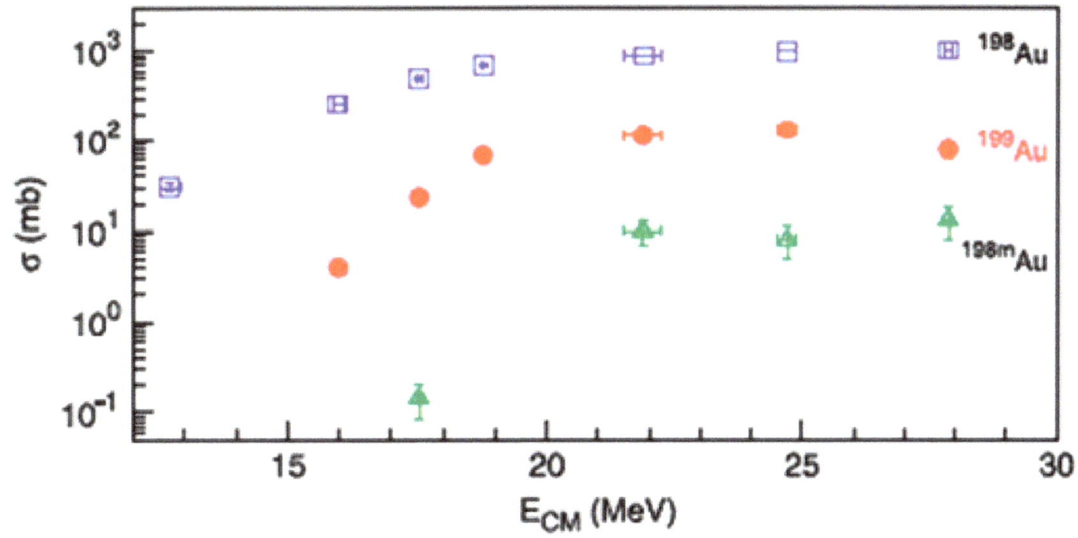

Figure 7: Excitation functions for $^{198,198m,199}Au$ production from ref. [53].

Fusion of ^{8}He with ^{206}Pb was studied by Parkar et al. [54] at energies near the Coulomb barrier, using a radiochemical technique with stacked targets of ^{206}Pb separated by mylar and Al spacers. The yield of the ^{210}Po evaporation residue (half-life of 138 days) was counted off-line after the Pb targets were dissolved and a ^{209}Po tracer was added. The absolute normalization was determined during the beam exposure period by measuring Rutherford scattering at forward angles ($\theta_{lab} < 30°$). However, the results have not yet been published as of the end of 2015.

In addition to the elastic scattering and transfer/breakup data discussed above, Lemasson et al. have also reported on fusion of ^{8}He with ^{197}Au [55] and ^{65}Cu [56] using on-line measurements of characteristic γ radiation. Surprisingly, the ^{197}Au fusion excitation function turned out to be very similar to that measured for ^{6}He incident on the same target [57]. Both display larger cross sections than those for $^{4}He + ^{197}Au$ [58]. However, when scaled according to the procedure described in refs. [31, 32], the sub-barrier fusion was only mildly enhanced and above-barrier fusion was suppressed as might be expected from the large transfer yields reported in ref. [53]. The ^{65}Cu fusion data [56], taken at energies well above the Coulomb barrier, displayed a somewhat different behavior. When compared with ^{6}He data on the same target, the ^{8}He cross sections were considerably greater than those for ^{6}He, as were the neutron transfer cross sections (fig. 8).

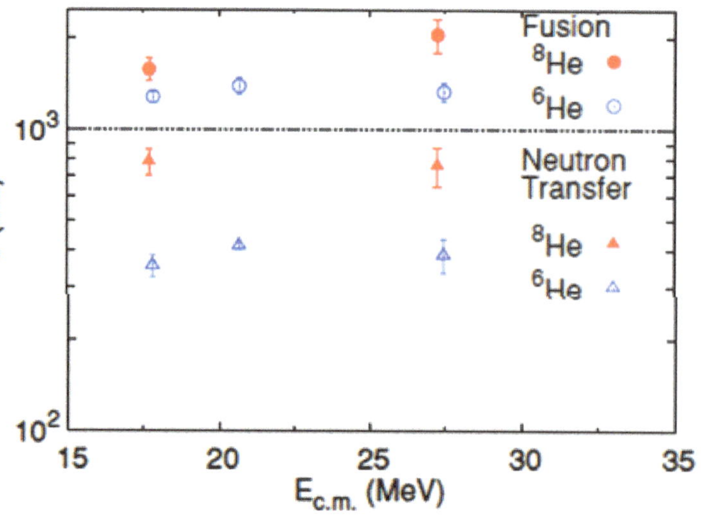

Figure 8: Excitation functions for fusion and transfer in the ^{8}He+ ^{65}Cu system. Note the very large yield for neutron transfer. The comparison ^{6}He + ^{65}Cu cross sections were taken from refs. [59,60]. Figure reprinted with permission from ref. [56]. c 2010, The American Physical Society.

REACTIONS WITH ^{8}LI

Although the short-lived radioactive nucleus ^{8}Li is not considered to be exotic, its low 1n separation energy threshold of 2.03MeV means that it is a weakly-bound nucleus, as are the stable isotopes ^{6}Li and ^{7}Li. The main motivation of the many experiments performed with ^{8}Li beams is its role in primordial nucleosynthesis (see, e.g., ref. [61]). Another important motivation, more related to the scope of this review, is the investigation of the effect of the different cluster configurations between the Li isotopes: ^{8}Li = ^{7}Li + n (BE = 2.033MeV), ^{7}Li = α + t (BE = 2.467MeV), ^{6}Li = α + d (BE = 1.474MeV). The role of neutron stripping in the elastic scattering and fusion reaction has also been studied. For instance, the difference in the neutron separation threshold, Sn, for the ^{8}Li, ^{7}Li and 6Li isotopes (2.03MeV, 7.25MeV and 5.66MeV, respectively) can provide a different interplay between transfer and breakup in coupled-channels treatment of elastic scattering and fusion data.

In-flight beams of ^{8}Li at low energy have been obtained with relatively good intensity at small facilities such as Twinsol at the University of Notre Dame and RIBRAS at Universidade de S˜ao Paulo, and also in bigger facilities such as EXOTIC in the Laboratori Nationali di Legnaro. Some of the works (theoretical and experimental) involving reactions induced by ^{8}Li beams produced in these laboratories are reviewed below.

Elastic scattering of ^{8}Li

^{8}Li + ^{7}Li

An angular distribution for the elastic scattering of ^{8}Li on ^{7}Li has been measured at an incident energy of 11MeV by Howell et al. [62] at both forward and backward angles. The angular distribution was very well described by an optical-model analysis using a Woods-Saxon potential. At forward angles elastic scattering is the most important mechanism, while at backward angles the neutron transfer reaction, ^{7}Li(^{8}Li, ^{7}Li)^{8}Li, dominates. The interference between elastic scattering and transfer at backward angles was used to obtain the asymptotic normalization coefficient (ANC) [63] for the valence neutron in ^{8}Li, and thereby to infer the astrophysical S-factor [64] $S_{17}(0)$ for the isospin symmetric ^{7}Be(p,γ)^{8}B capture reaction.

^{8}Li + ^{9}Be

Angular distributions for ^{8}Li+^{9}Be elastic scattering have been measured at 19.6MeV [65] and at 27MeV [66]. In both measurements the data were well described by ^{7}Li + ^{9}Be optical-model potential parameters at similar ^{7}Li energies. An optical-model analysis with the S˜ao Paulo double-folding potential [33] was also performed for these data, giving good results. Total reaction cross sections were derived from these analyses and compared with the values for other systems with weakly-bound and stable projectiles, applying the two most-used reduction methods [31,32,67,68]. The conclusion is that reactions with non-halo light nuclear projectiles lead to similar total reaction cross sections [65].

^{8}Li + ^{12}C

Two elastic-scattering experiments with ^{8}Li incident on a ^{12}C target, performed in the 1990s, were reviewed in ref. [1]. More recently, an elastic-scattering experiment was performed at an incident energy of 23.9MeV by Barioni et al. [69]. Optical-model analyses with Woods-Saxon and S˜ao Paulo potentials gave a good description of the elastic data. A coupled-channels calculation, including inelastic and breakup channels, was performed and coupling effects were found to be unimportant.

^{8}Li + ^{51}V

Elastic scattering of ^{8}Li + ^{51}V was measured at 18.5MeV [65] and 26.0MeV [9]. Optical-model analyses with Woods-Saxon and S˜ao Paulo potentials were performed and total reaction cross sections were derived from the angular distributions measured at both energies. A comparison of the total reaction cross sections with values from other light weakly- and strongly-bound projectiles on similar mass targets was carried out. Once adequate reductions were applied, a similar behavior as a function of energy was observed (fig. 9).

$^{8}Li + ^{90}Zr$

Very recently, measurement and analysis of two angular distributions for ^{8}Li + ^{90}Zr at 18.5 and 21.5MeV have been reported, in refs. [70] and [71]. The main goal of these measurements was to investigate the influence of the single neutron stripping reaction in the coupling scheme for quasielastic scattering. They found that the dominant contribution to the CRC (Coupled Reaction Channel) calculation was neutron stripping: ^{90}Zr(^{8}Li, ^{7}Li)^{9}1Zr.

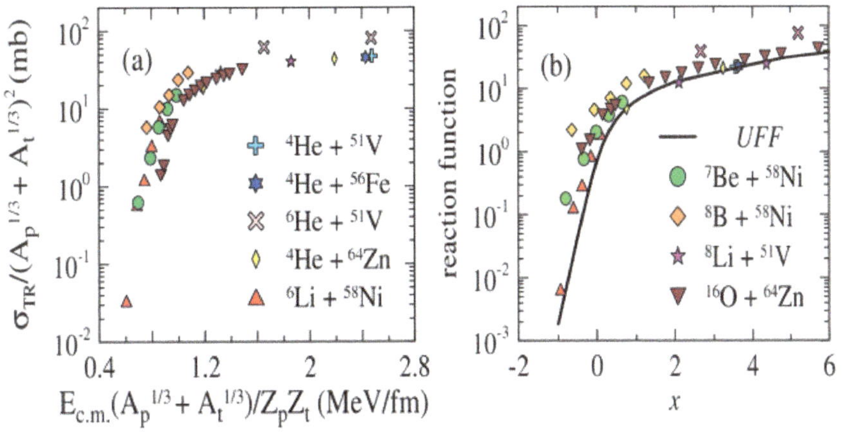

Figure 9: Total reaction cross sections for systems with different projectiles on targets from A = 51 to 64, reduced using two different reduction methods. Figure from ref. [65].

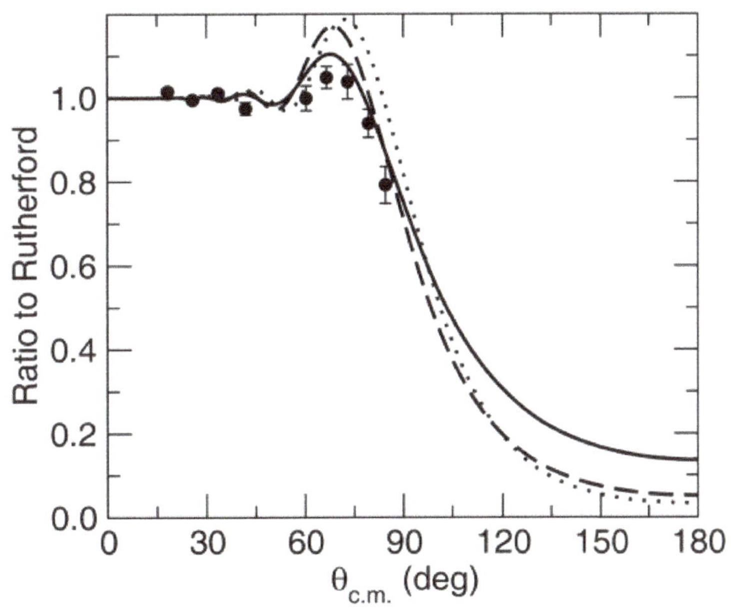

Figure 10: Quasielastic scattering angular distribution for the ^{8}Li+^{90}Zr system at 18.5 MeV compared with no-coupling (dotted curve), CC (dashed curve), and CRC (solid curve) calculations, respectively. Figure reprinted from ref. [71] with kind permission of The European Physical Journal (EPJ).

Figure 10 shows the comparison of this calculation with the experimental angular distribution measured at 18.5MeV. Total reaction cross sections for the two energies were also determined from the analysis and compared with several other similar systems with similar projectile and target mass, applying the reduction suggested in refs. [31,32] but replacing σ_F by σ_R. The results indicate that the total reaction cross sections exhibit rather similar behavior as a function of the reduced energy at higher energy. For lower energy, however, some differences among the systems appear. They attributed the difference between the UFF function and the best fit to the data to a possible role of direct reactions, and determined the ratio of the direct to total reaction cross section for ^{8}Li on several targets. The conclusion was that direct reactions are very important at energies below the Coulomb barrier.

$^{8}Li + ^{208}Pb$

The only existing data on ^{8}Li + ^{208}Pb elastic scattering were already reviewed in ref. [1]. However, the role of the neutron stripping channel in the ^{8}Li + ^{208}Pb elastic scattering has been investigated by Keeley in ref. [72]. The author of this work performed a full CRC calculation for the quasielastic barrier distribution of this system and compared with that for ^{6}He + ^{208}Pb. For ^{8}Li, the effect of the single-neutron transfer coupling was to move the peak of the barrier distribution to lower energies, while the effect in ^{6}He was to split the barrier distribution into two distinct peaks, one at lower energy and one at higher energy. These are actually schematic calculations and the observed structures in the barrier distributions can be washed out in the measurements due to other possible couplings, such as to breakup channels.

Transfer reactions induced by ^{8}Li

Proton and neutron transfer reactions with a ^{8}Li projectile have been measured on ^{9}Be and ^{12}C targets. The ^{9}Be(^{8}Li, ^{9}Be)^{8}Li proton transfer reaction was measured at 27MeV and reported in ref. [66]. In this work, the spectroscopic factor for the $\langle ^9\mathrm{Be}_{gs}|^8\mathrm{Li}_{gs} + \mathrm{p}\rangle$ bound system was obtained from a comparison between the experimental differential cross sections for the transfer reaction and finite-range distorted-wave Born approximation (FRDWBA) calculations, assuming a direct transfer and using the code FRESCO [73,74]. The spectroscopic factor was then used to infer information concerning the ^{8}Li(p,γ)^{9}Be proton capture reaction. The transfer reactions ^{9}Be(^{8}Li, ^{7}Li)^{10}Be and ^{9}Be(^{8}Li, ^{9}Li)^{8}Be, measured in the same experiment, were published in ref. [75]. The $\langle ^9\mathrm{Li}_{gs}|^8\mathrm{Li}_{gs} + \mathrm{n}\rangle$ and $\langle ^8\mathrm{Li}_{gs}|^7\mathrm{Li}_{gs} + \mathrm{n}\rangle$ spectroscopic factors were obtained from a comparison between the experimental differential cross sections and FRDWBA calculations. Cross sections for the direct neutron-capture reactions ^{7}Li(n,γ)^{8}Li and ^{8}Li(n,γ)^{9}Li were then deduced in the framework of a potential model [76] using these spectroscopic factors.

Transfer reactions have also been measured on 12C. The ^{12}C(^{8}Li, ^{7}Li) and ^{12}C(^{8}Li,α) reactions were measured in the early 1990s. The latter was confirmed to be due to two-body transfer rather than a fusion-evaporation mechanism [77]. More recently Barioni et al. [69] took advantage of the higher cross section for the ^{12}C(^{8}Li, ^{9}Be)^{11}B$_{gs}$ proton transfer reaction to obtain the spectroscopic factor for $\langle ^9\mathrm{Be}_{gs}|^8\mathrm{Li}_{gs} + \mathrm{p}\rangle$. The results were compared with shell-model calculations and previous data and good agreement was observed.

Fusion of ^{8}Li

Fusion of ^{8}Li on ^{208}Pb was measured at the University of Notre Dame for energies between 32 and 37.5MeV (centerof-momentum), corresponding to between 3 and 8.5MeV above the Coulomb barrier [78]. The excitation function for complete fusion (CF) of this system (fig. 11) was obtained by detecting delayed α-particle emission from the ^{211,212}At residual nuclei (after 5n and 4n evaporation from the compound nucleus, respectively).

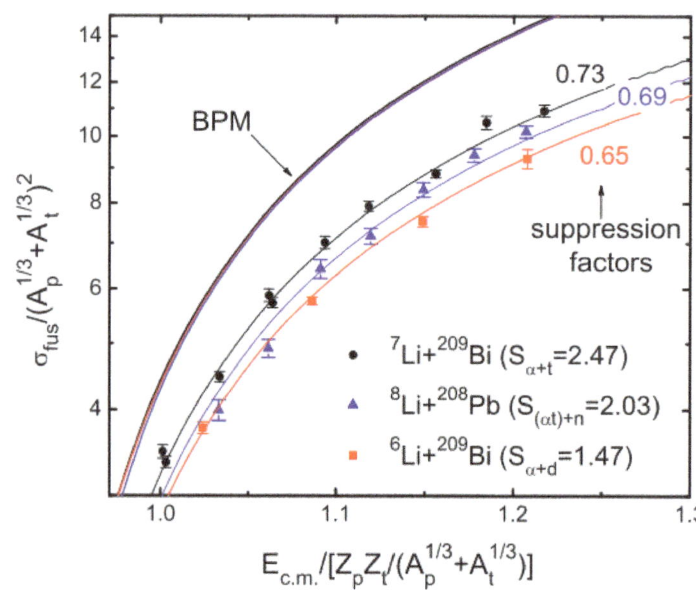

Figure 11: Reduced complete fusion (CF) cross sections for ^{8}Li+ ^{208}Pb. The breakup threshold energies for ^{6}Li, ^{7}Li and ^{8}Li are indicated, in MeV. Figure reprinted with permission from ref. [78]. c 2009, The American Physical Society.

The experimental results were compared with onedimensional barrier-penetration model (BPM) calculations and PACE2 [79] predictions. A suppression of about 30% was observed for energies above the barrier. Using the above comparison, the roles of CF and ICF (incomplete fusion) reactions were disentangled. ICF due to the fusion of the heavier ^{7}Li fragment from the breakup of ^{8}Li was estimated and found to be important for the higher energies. A comparison with fusion data for other Li isotopes was also performed by applying the appropriate scaling factors of $(A_p^{1/3} + A_t^{1/3})^2$ and $Z_pZ_t/(A_p^{1/3} + A_t^{1/3})$ for the cross sections and energies, respectively. Some correlation between the suppression factor used to adjust the BPM calculation to the experimental data for each isotope, ^{6}Li, ^{7}Li and ^{8}Li, and the threshold breakup energies of the projectiles was observed, as shown in fig. 11.

REACTIONS WITH ^{9}LI

Since ^{9}Li is a short-lived radioactive nucleus with oneneutron separation energy Sn = 4.064MeV, it may be considered to be weakly bound. Very few experiments have been carried out with this nucleus as projectile, and most of those were performed in association with the famous halo isotope ^{11}Li. There is, however, some intrinsic interest in elastic-scattering measurements with ^{8,9}Li, related to the relative importance of the one-neutron-transfer and breakup mechanisms. Previous elastic-scattering experiments, reviewed in ref. [1], were performed at relatively high energy beyond the scope of the present review. However, a recent elastic experiment was performed with both ^{9}Li and ^{11}Li beams produced by fragmentation and postaccelerated at the ISAC-II line at the TRIUMF facility (Vancouver, Canada) [80]. Fusion experiments induced by ^{9}Li on ^{70}Zn and on ^{208}Pb have also been performed at the same facility and are reviewed below.

Elastic scattering of ^{9}Li

Angular distributions were measured for ^{9}Li + ^{208}Pb at center-of-mass (c.m.) energies of 23.1 and 28.3MeV [81], corresponding to a little below and a little above the barrier for this system. The lower energy angular distribution showed a typical Rutherford scattering characteristic, as expected. The higher energy angular distribution, however, displayed Coulomb-nuclear interference, with a maximum at $\theta = 75°$ and a smooth decrease beyond that angle. This was interpreted as a typical Fresnel-type pattern, characteristic of the scattering of ordinary stable nuclei near the Coulomb barrier. This result was used to compare with data for the more exotic ^{11}Li projectile as discussed in sect. 6.1 below.

Fusion of ^{9}Li

Fusion measurements induced by ^{9}Li beams have been performed at ISAC-TRIUMF on ^{70}Zn [82] and ^{208}Pb [83] targets. In both experiments, enhancement of the fusion cross section for energies below the barrier and suppression above the barrier was observed. The below-barrier enhancement for ^{9}Li + ^{70}Zn remains a challenge for theoretical calculations. The role of neutron transfer, which is usually the main motivation to investigate reactions with Li isotopes, was explicitly considered by Zagrebaev et al. [84] in a standard coupled-channels calculation. This particular calculation did not describe the data, and the possible role of dineutron transfer was postulated. Later, Balantekin and Kocak [85] reported a coupled-channels calculation with the inclusion of coupling to additional inelastic channels. Unfortunately, this calculation also failed to reproduce the low-energy enhancement in ^{9}Li + ^{70}Zn fusion. The authors then proposed a very sophisticated mechanism in which a molecule with two cores, ^{9}Li and ^{68}Zn, might be formed and share and exchange two neutrons thereby forming ^{11}Li and ^{70}Zn.

The role of the 2n-transfer process was also explicitly taken into account in a quantum diffusion approach by Sargsyan et al. [26]. In contrast to the calculation of Balantekin [85], they considered a 2n-stripping reaction to the target, ^{70}Zn(^{9}Li, ^{7}Li)^{72}Zn, which is expected to be much more favorable (Q_{2n} = 8.6MeV) than the 2n-pickup reaction ^{70}Zn(^{9}Li, ^{11}Li)^{68}Zn (Q_{2n} = −15.4MeV). The agreement with the ^{9}Li + ^{70}Zn fusion data was improved, but the lowest point in the excitation function could still not be reproduced (fig. 12). This data set still deserves further experimental and theoretical investigation.

Data on fusion of ^{9}Li+ ^{208}Pb were reported as a function of energy by Vinodkumar et al. [83]. They detected the decay α particles emitted by the evaporation residues ($^{214-211}$At) using a beam on/beam off method. The fusion cross sections were estimated from statistical model calculations. This data set was compared with the predictions of coupled-channels calculations using the code CCFULL [86]. Only the important inelastic channels were included. The results of this calculation, as well as a one-dimensional barrier-penetration calculation, underestimated the sub-barrier fusion cross sections and overestimated the above-barrier cross sections, indicating that the description of the fusion of ^{9}Li is yet to be understood. On the other hand, the authors also compared the fusion data for ^{9}Li + ^{208}Pb with that for ^{6}Li, ^{7}Li, and ^{8}Li available in the literature. Considering the reduced cross sections $(\sigma_{fus}/\pi R_B^2)$ as a function of reduced energy ($E_{c.m.}/V_B$), they found no obvious difference in behavior for the 9Li projectile. The role of the breakup and transfer was not explicitly considered in any of these calculations, and these could provide further insights.

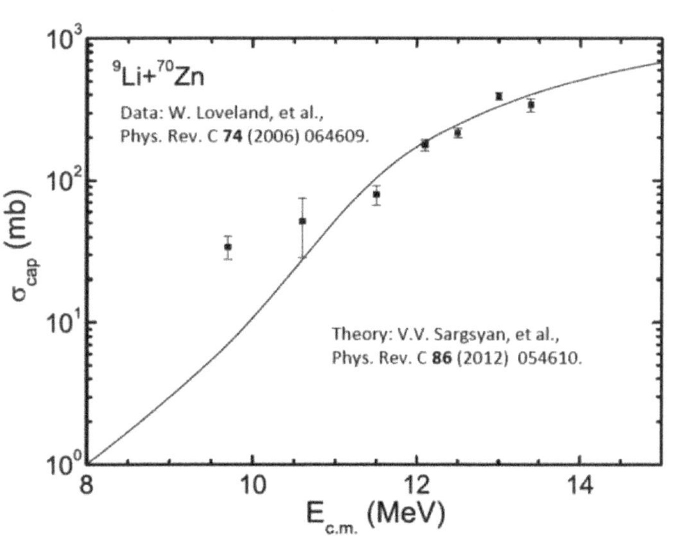

Figure 12: Fusion (capture) cross section for the ^{9}Li + ^{70}Zn system. The data are from ref. [82]. The solid line is the result of a quantum diffusion calculation [26].

REACTIONS WITH ^{11}LI

The ^{11}Li nucleus is the most fascinating case, and the archetype, of a Borromean halo nucleus. Its low binding energy for the di-neutron (S_{2n} = 0.369MeV) generates an extended mass distribution going well beyond its core. Also, due to the loosely bound structure, the di-neutron halo can be easily polarized by the strong electric field of a heavy target such as ^{208}Pb. This polarization would correspond to a distortion of the halo nucleus in which the positively charged core would move away from the target and the halo neutrons would move in the opposite direction. The large E1 transition strength observed in the Coulomb dissociation measurement of ^{11}Li [87] corroborates this possibility. Recently, a reliable beam of ^{11}Li became available at the post-accelerated ISAC-II line of the TRIUMF facility (Vancouver, Canada). The intensity achieved for this beam was about 5×10^3 particles/s, enough for elastic-scattering [81], breakup [88], and fusion [89] measurements on a ^{208}Pb target. These measurements are reviewed below.

Elastic scattering

The strong dynamic polarizability for ^{11}Li in the presence of the electric field of a heavy target can give rise to a long-range Coulomb interaction whose effect would be to reduce the elastic-scattering cross sections even at energies well below the barrier. This possibility was addressed by Cubero et al. in ref. [81], which reported the first measurement of the scattering of ^{11}Li on a ^{208}Pb target at incident energies of 24.3 and 29.8MeV. These are, respectively, below and above the Coulomb barrier ($V_{barrier}$ = 28MeV). In this work, they also reported on the elastic scattering of ^{9}Li + ^{208}Pb with the same setup and at the same c.m. energies of 23.1 and 28.3MeV. This comparison is important to highlight the different characteristics of the elastic scattering of the ^{9}Li and ^{11}Li nuclei, and to disentangle the contribution of the loosely-bound structure of ^{11}Li. Actually, the large polarizability of ^{11}Li was already predicted years earlier by Andr`es and G´omez-Camacho [90]. They also predicted a strong reduction of the elastic cross section for ^{11}Li.

This strong reduction in the elastic cross section was then experimentally corroborated in the paper of Cubero et al. [81]. To investigate this effect in a more quantitative way, they considered the comparison of four-body CDCC calculations with and without coupling to the continuum. The three-body description of the borromean ^{11}Li projectile was based on the idea used to describe the other borromean nucleus, ^{6}He [91]. It is well known that coupling to the continuum allows the evaluation of projectile breakup effects on the elastic scattering. The results of such calculations for the ^{11}Li data can be seen in fig. 13 as the dotted and solid curves. In this figure, a strong cross section reduction can be observed for the 4-body CDCC calculation in the "no-coupling" case, where the extended size of ^{11}Li was taken into account. However, although the cross section decreased faster than the ^{9}Li data beyond 90◦, it still was not sufficient to describe the data. A good description was achieved when coupling to continuum was incorporated into the calculation. The final reduction was then attributed to the strong dipole coupling between the ground and continuum states. They also concluded that inclusion of the low-lying dipole resonance, which is close to the breakup threshold for ^{11}Li, into the calculation improved the description of the data. Keeley et al. [92] simplified these CDCC calculations by employing a two-body di-neutron model for the ^{11}Li nucleus. They claimed that, in principle, this approach would have all the essential physics to investigate the influence of the continuum states on ^{11}Li elastic scattering. In addition to this model applied for ^{11}Li, they also considered dynamic polarization potentials (DPPs) using the trivially equivalent local potential (TELP) derived from the CDCC calculations associated with optical-model analysis. The agreement between the calculations and the data indicated that this methodology could well represent the effects of the large dipole polarizability in describing the ^{11}Li elastic data.

A different approach to analyze the ^{11}Li data on elastic scattering has been reported in refs. [93,94]. In these papers, the effect of ^{11}Li polarizability was investigated using long-range potentials. A dynamic polarization potential (DPP) was used to simulate the Coulomb dipole excitation (CDE) or electric dipole polarization for ^{11}Li, where the charged ^{9}Li core is decelerated by the effect of the Coulomb field of the heavy target while the valence neutrons move in a straight-line direction. The DPP was obtained using a Coulomb dipole strength B(E1) = 1.42 and 1.41 e^2 fm^2, at $E_{c.m.}$ = 23.1 and 28.3MeV, respectively, and a dipole resonance with energy $\epsilon = 0.69$MeV. In the analysis, they incorporated an extra-long-range complex nuclear potential in an attempt to improve the agreement between calculation and the data.

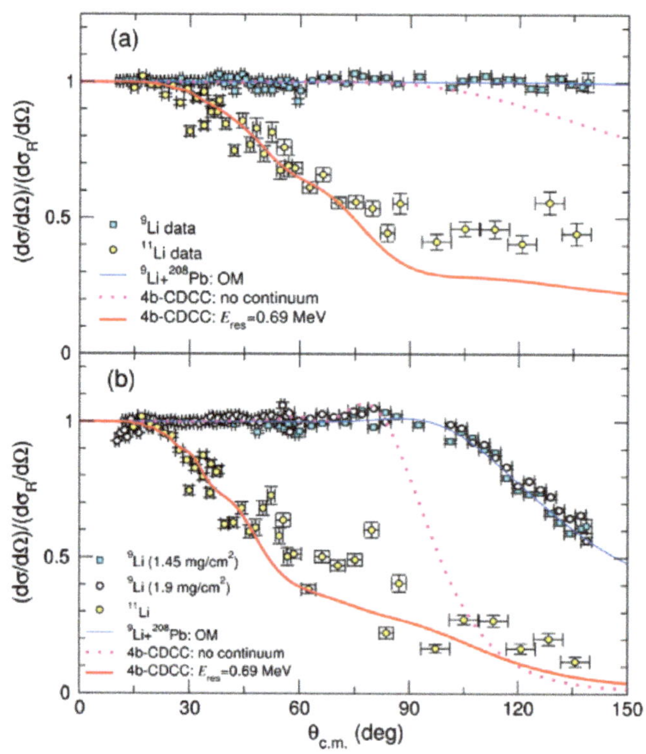

Figure 13: Elastic-scattering cross section (ratio-to-Rutherford) for ^{9}Li and ^{11}Li on a ^{208}Pb target at the same c.m. energies of 23.1 MeV (upper panel) and 28.3 MeV (lower panel). The curves are the results of calculations with a standard opticalmodel potential (thin blue line), 4-body CDCC with no continuum (dotted line), and 4-body CDCC including the continuum (solid red line). Figure reprinted with permission from ref. [81]. c 2012, The American Physical Society.

The results of such calculations can be seen in fig. 14, where eq. (1) stands for optical model with a shortrange complex potential given by the usual Woods-Saxon function, eq. (3) included the Coulomb dipole excitation (CDE) for ^{11}Li and eq. (5) has an additional complex nuclear potential of Woods-Saxon type with large diffuseness parameters to simulate a long-range interaction. In the analysis of ref. [94] the authors used a DPP taking into account the B(E1) strength value from experimental data on 11Li breakup [87]. Due to the spreading of the data it is not possible to disentangle the relative importance of the inclusion of the extra nuclear potential. More recently, another analysis in terms of an OM has been performed by Fern´andez-Garc´ıa et al. [95]. In this work, the authors combined a double-folding potential, the S˜ao Paulo Potential (SPP), and a long-range polarization potential. They used more realistic densities for ^{9}Li and ^{11}Li from proton scattering measurements [96], and the experimental B(E1) distribution from the Coulomb dissociation measurement [87], to obtain both the shortand long-range potentials. Considering only the doublefolding SPP in the form of $U_{opt} \approx V_{SPP} + i(0.78V_{SPP})$, without any long-range potential, the calculation for the ^{9}Li+^{208}Pb system agrees very well with the experimental angular distribution. For the ^{11}Li + ^{208}Pb system it was necessary to include a long-range Coulomb Dipole Potential (CDP). The good agreement between the calculation and data indicates the dominance of the Coulomb dipole polarizability for ^{11}Li. The long-range absorptive tail of the ^{11}Li imaginary CDP extends to a very large radius, suggesting that the Coulomb breakup of ^{11}Li takes place much further from the target. Although there was a good description of the data with these calculations, the role of the halo structure of ^{11}Li is not completely understood. The presence of the strong soft electric dipole (E1) response at low excitation energies for ^{11}Li due to Coulomb excitation is clear. However, the existence of a low-energy dipole resonance in ^{11}Li [97], and what would be its contribution in the B(E1) strength and thus in the elastic scattering, deserves further investigation.

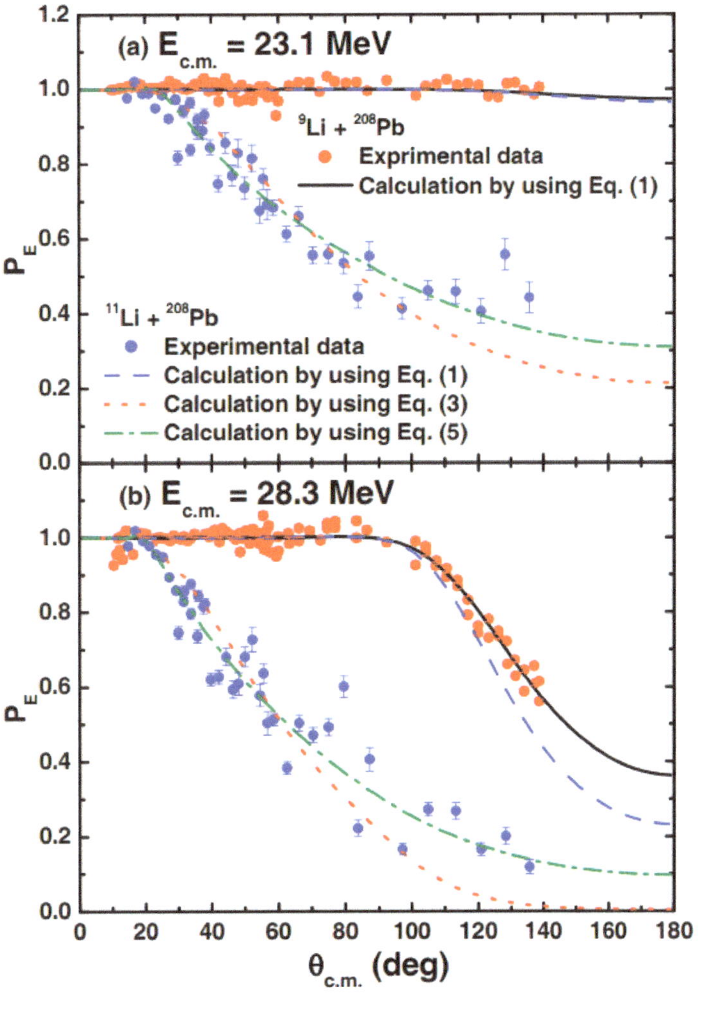

Figure 14: Elastic-scattering cross section ratio to Rutherford for ^{9}Li and ^{11}Li on a ^{208}Pb target. The curves are the results of calculations with a standard Optical-Model (OM) potential for ^{9}Li (black line) or ^{11}Li (dashed blue line), with DPP (dotted red line) and with an extra long-range nuclear potential (dot-dashed green line). Figure reprinted with permission from ref. [93]. c 2014, The American Physical Society.

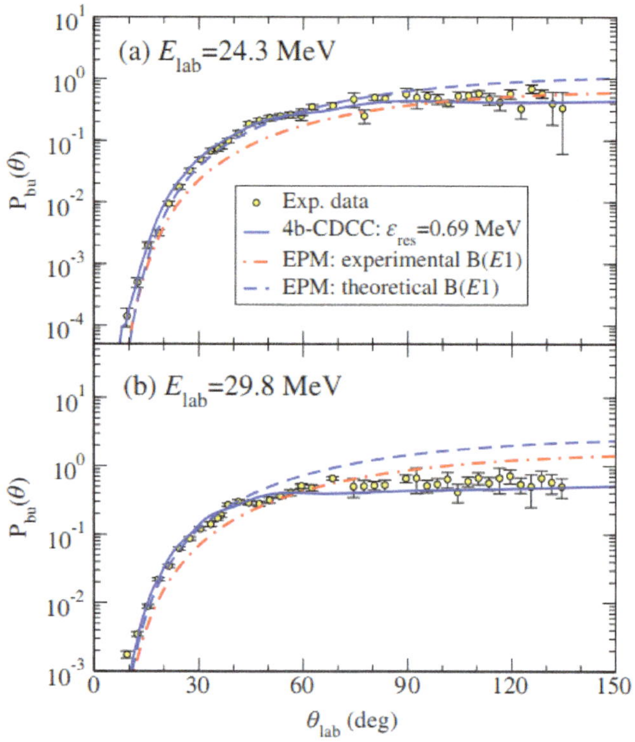

Figure 15: Angular distribution of the ^{11}Li breakup on a ^{208}Pb target at (a) 24.3 MeV and (b) 29.8 MeV. The solid line represents the 4b-CDCC calculation and the dashed and dot-dashed lines are the semi-classical calculation with theoretical and experimental B(E1) distribution, respectively. Figure reprinted with permission from ref. [88]. c 2013, The American Physical Society.

Breakup of ^{11}Li

Breakup reactions play a significant role in the investigation of halo structure in nuclei. From the direct measurement of breakup cross sections, we can infer and better understand the exotic structure caused by the weakly-bound nature of valence neutrons in halo nuclei. Both Coulomb and nuclear breakup are important in the spectroscopy of halo nuclei since the observables are strongly correlated with their ground and continuum structure. Two decades after the first confirmation of halo structure in ^{11}Li, several breakup-reaction measurements of halo nuclei have been performed, and some of them are reviewed here. However, only two experiments involving ^{11}Li as a projectile have been performed, one at high energy in RIKEN, reported in ref. [87], and the other at TRIUMF measured at an energy around the Coulomb barrier and reported in ref. [88]. Both measurements were performed with a ^{208}Pb target, which in principle would probe Coulomb breakup. Although important for the investigation of the low-lying resonances in ^{11}Li, the experiment performed at RIKEN is outside the scope of our review.

The experiment performed at TRIUMF corresponds to an inclusive angular distribution measurement for the breakup of ^{11}Li + ^{208}Pb at 24.3 and 29.8MeV. The setup was composed of 4 telescopes placed around the ^{208}Pb target, covering a large angular range from 10 to 140 degrees. The telescopes consisted of ΔE-E double-sided Si strip detectors (DSSSD). The ^{9}Li fragments from the breakup of ^{11}Li were clearly separated and could be easily identified in a two-dimensional ΔE vs. ΔE + E plot.

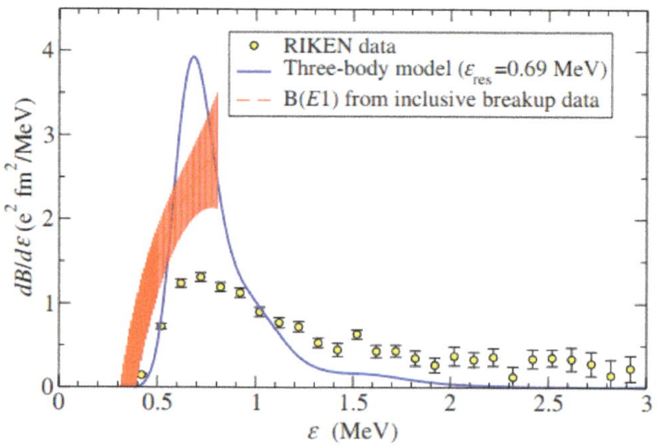

Figure 16: B(E1) distributions of ^{11}Li. The circles represent the experimental data from ref. [87], the solid line is the result of the extraction from the full CDCC calculation including the dipole resonance at $\epsilon = 0.69$ MeV above the ground state and the shaded area is the result of a calculation where the B(E1) distribution was assumed to be a linear function of the excitation energy given by eq. (5) in ref. [88]. Figure reprinted with permission from ref. [88]. c 2013, The American Physical Society.

The experimental angular distributions are shown in fig. 15 together with the results of a 4b-CDCC calculation, using a ^{11}Li model similar to the one proposed for ^{6}He in ref. [91]. As can be seen in the figure, this 4b-CDCC calculation reproduced the data quite well. This good agreement, in particular at forward angles ($\theta < 60°$), indicates the validity of the semiclassical approximation and the dominance of E1 couplings. The figure shows the results of semiclassical Coulomb excitation theory as described in ref. [98], where three-body theoretical (dashed line) and experimental (dot-dashed line) B(E1) distribution were used. As can be seen in the figure, the latter calculation did not reproduce the data. The authors also used the results of the full 4b-CDCC calculation, which better described the data, to extract the B(E1) distribution. It is important to mention that the three-body model used to describe ^{11}Li predicted the existence of a low-lying dipole resonance, but its energy was adjusted to $\epsilon_x \sim 0.69$MeV above the ground state, or 0.32MeV above the 2n breakup threshold. The comparison of the B(E1) distributions obtained from the breakup data from RIKEN [87], from the 4b-CDCC calculation, and from a linear approximation using the inclusive breakup data from TRIUMF, can be seen in fig. 16. The authors claimed that their methodology can be used to obtain the binding energy of other weakly bound nuclei from inclusive breakup angular distributions.

Fusion of ^{11}Li

When it comes to fusion reactions, the relevant question is whether fusion at energies near the Coulomb barrier is enhanced, owing to the large extent of the nuclear matter distribution, or hindered due to the very low threshold of the breakup channel. Since ^{11}Li is a nucleus with more-diffuse density and a high probability of breakup, interest in its fusion reaction was quite high. It was not so long after the measurement of the large interaction cross section for ^{11}Li by Tanihata et al. [99] that theoreticians started making predictions for the fusion excitation function of ^{11}Li + ^{208}Pb. Considering different approaches such as the influence of the soft Coulomb dipole, the breakup channel, and one-dimensional tunneling, Canto [100], Takigawa [101], and Hussein [102,103] used an optical-model approach with a polarization potential to take into account the breakup channel. Dasso [104] used a coupled-channel approach with breakup as one channel. Their predictions are summarized in fig. 17.

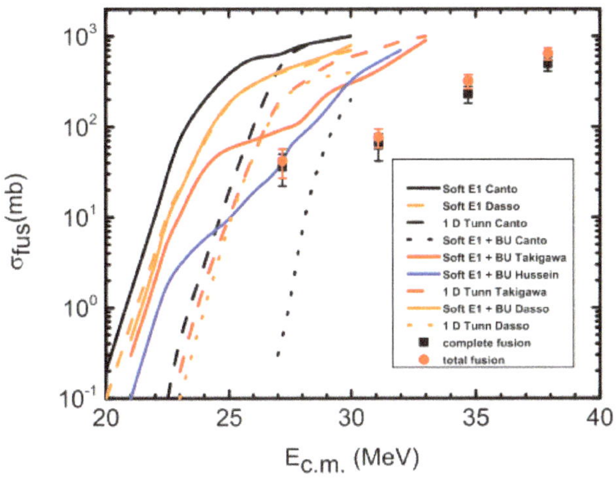

Figure 17: Fusion excitation function for the ^{11}Li+^{208}Pb system. The several lines correspond to predictions from refs. [100–104] and the symbols are the data from ref. [89]. Figure reprinted with permission from ref. [89]. c 2013, The American Physical Society.

These predictions varied by up to one order of magnitude so the need for data was clear. The only fusion measurement induced by a ^{11}Li beam was very recently performed at the ISAC2 facility at TRIUMF on a ^{208}Pb target [89]. In this experiment, the 40MeV post-accelerated ^{11}Li beam, with an intensity of about 10^3 pps, impinged on a stack of ^{208}Pb targets. The α-particle emission yields from fusion-evaporation residues ($^{212-216}$At) were used to obtain the fusion cross section. The beam was pulsed on for 5ns and then data were collected during a 172ns beamoff period. The incident c.m. energy spanned from 27.1 to 37.9MeV, covering a region from below to above the nominal Coulomb barrier. The data are plotted in fig. 17, together with the theoretical predictions.

The main motivation for these studies was related to the role of breakup and other reaction channels in terms of enhancing or reducing the fusion yield. Breakup of the incident ^{11}Li results in incomplete fusion (ICF), where the ^{9}Li fragment fuses with the target. It is very difficult to experimentally separate ICF from complete fusion (CF). However, based on a previous fusion measurement for the ^{9}Li + ^{208}Pb system [83], the authors claimed to have disentangled the contribution of CF from the total fusion (CF+ICF), shown separately in fig. 17. Although it was claimed that the contribution of ICF was quite large, the data were still overestimated by all the previous predictions. The authors also reported a new coupled-channels calculation where they considered the optical-model parameters from the elastic-scattering measurement for ^{9,11}Li with ^{208}Pb [81]. The results of these new calculations are shown in fig. 18, and as one can see, they still overestimate the data for both ^{9,11}Li+^{208}Pb systems at energies above the Coulomb barrier. In ref. [89], a comparison plot of the reduced fusion cross sections, $\sigma_{fus}/\pi R_R^2$, for all the lithium isotopes ^{6,7,8,9,11}Li as a function of $E_{c.m.}$/VB was also given. The differences could be attributed to geometry for the $^{6-9}$Li isotopes, but not for ^{11}Li. It is clearly important to have an improved ^{11}Li beam intensity, and fusion measurement on other targets, to achieve a better understanding of fusion induced by ^{11}Li.

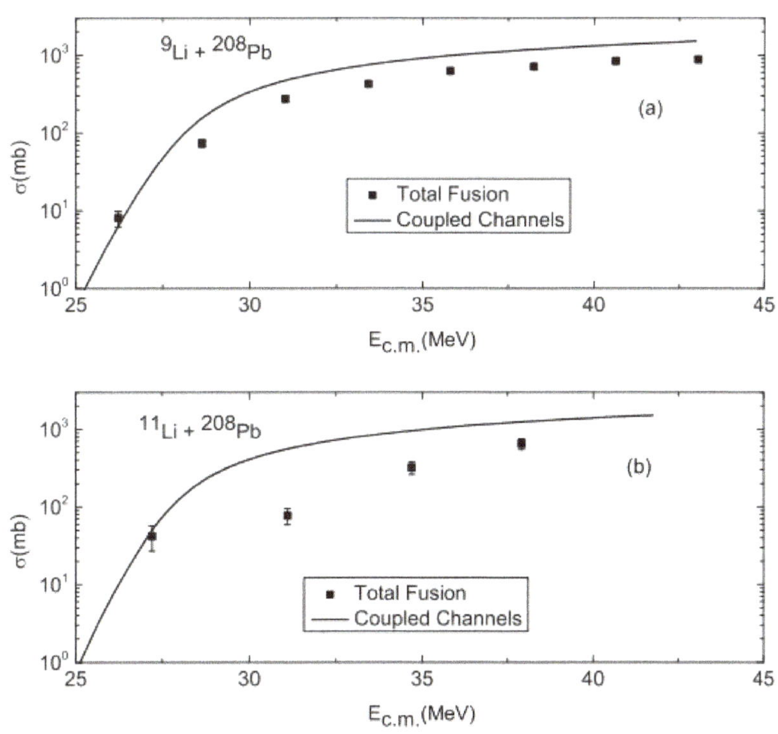

Figure 18: Fusion excitation functions for the ^{9,11}Li+^{208}Pb systems. The solid lines are predictions of the total interaction cross section from coupled-channel calculations. The data are for the total fusion of ^{9,11}Li+^{208}Pb. Figure reprinted with permission from ref. [89]. c 2013, The American Physical Society.

REACTIONS WITH ^{7}BE

^{7}Be lies on the proton-rich side of the line of nuclear stability and it is weakly bound, with a separation energy of 1.59MeV for breakup into ^{3}He and ^{4}He. It is a radioactive nucleus with a half-life of 53.2d and the fact that it is the core for the proton-halo nucleus ^{8}B enhances its own interest. In ref. [1], elastic scattering and reaction data for ^{7}Be on targets of ^{7}Li, ^{10}B, ^{12}C, ^{14}Ni, ^{27}Al, ^{58}Ni, and ^{238}U were reviewed. Data appearing later will be emphasized here.

Elastic scattering of ^{7}Be

The elastic-scattering studies performed so far with a ^{7}Be projectile seem to indicate only small effects of bu couplings, although sizable effects appear for energies above 12MeV per nucleon (sect. 7.1.2). In spite of being a weakly bound nucleus, a behaviour characteristic of the breakup threshold anomaly does not seem to apply to reactions with ^{7}Be (sect. 7.1.6). Recent experiments leading to these conclusions are reviewed below.

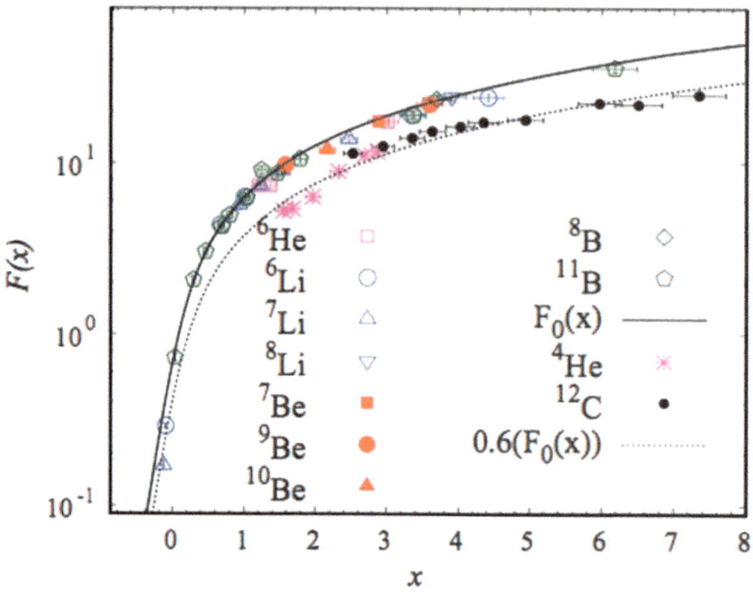

Figure 19: Reduced total reaction cross sections for several light projectiles on ^{12}C. Figure reprinted with permission from ref. [110]. c 2011, The American Physical Society.

$^7Be + {}^9Be$

Quasi elastic scattering angular distributions for this system were reported in refs. [105,106] for energies E_{lab} of 17, 19 and 21MeV in the angular range $24° \leq \theta_{c.m.} \leq 57°$. These, along with similar data for ^{7}Li + ^{9}Be taken from the literature, were analyzed within the CRC scheme in ref. [107]. Inelastic excitations, reorientation of projectile and target nuclei as well as the most relevant transfer channels were included in the CRC calculations. The data for both systems could be well described with very similar energy-dependent OM potentials. The authors concluded that potential scattering and reorientation of projectile and target are the most important mechanisms in both cases, with the former (latter) mechanism dominating at forward (backward) angles.

A quasielastic angular distribution for the same system at E_{lab} = 23.7MeV was reported in ref. [108], covering the angular range between 12.3 and 38.8 degrees in the laboratory frame of reference. An absolute normalization was not obtained from the experiment, so the data were normalized to optical-model predictions. Coupled-channel calculations with and without couplings to the continuum gave very similar results.

$^7Be + {}^{12}C$

Angular distributions for the elastic scattering of ^{7}Be+^{12}C at E_{lab} = 18.8MeV were measured at Notre Dame [109] and S˜ao Paulo [110]. The two data sets, corresponding to different angles that complement each other, gave consistent results and were well described by optical-model calculations using Woods-Saxon potentials. By analyzing both data sets together, a total reaction cross section of 1198mb was deduced in ref. [110]. In order to investigate possible effects of bu couplings, CDCC calculations were performed [110] where the ^{7}Be nucleus was considered as formed by ^{3,4}He clusters. The effects of the couplings were found to be small. Using the "fusion function" type of data reduction described earlier (sect. 2.2.2), but applied now to total reaction cross sections, data for projectiles of ^{4,6}He, ^{6,7,8}Li, ^{7,9,10}Be, ^{8,11}B and ^{12}C on ^{12}C targets were plotted together (fig. 19). All reaction cross section data fell on a trajectory described by the Universal Fusion Function ($F_0(x)$), except those corresponding to the ^{4}He and ^{12}C projectiles which were suppressed by a factor of 0.6. The high proton and neutron separation energies of the latter projectiles could be the reason for the observed suppression. An angular distribution for quasielastic scattering of ^{7}Be + ^{12}C at E_{lab} = 34MeV was reported in ref. [111]. These data were used to obtain the OM potential needed for a DWBA calculation of the ^{12}C(^{7}Be, ^{3}He) reaction, which was the reaction of primary interest in this work. No further analysis of the elastic-scattering data was done.

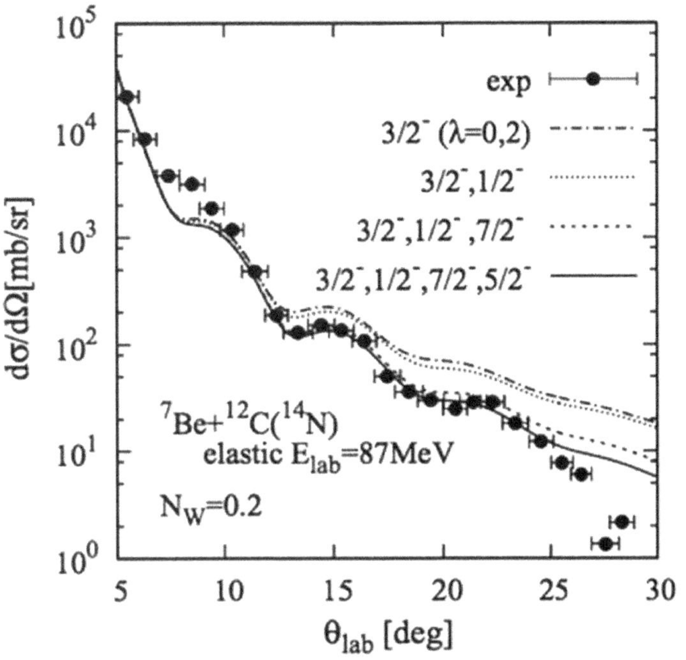

Figure 20: Elastic-scattering angular distribution for $_7$Be on melamine ($C_3N_6H_6$) at 87.7 MeV [112] and respective coupledchannel calculations. Figure reprinted with permission from ref. [113]. c 2010, The American Physical Society.

In ref. [112], data for elastic scattering at 87.7MeV were reported for ^{7}Be on a melamine target (which has a ratio of 0.5 to 1 for ^{12}C to ^{14}N). These data were later theoretically analyzed in ref. [113], where a ^{4}He+^{3}He cluster structure was assumed for ^{7}Be and a coupled-channel calculation was performed including its ground and first bound states $(\frac{3}{2}^-, \frac{1}{2}^-)$ as well as the first two resonant states $(\frac{7}{2}^-, \frac{5}{2}^-)$. As shown in fig. 20, the inelastic couplings have an important effect at backward angles, with the excitation of the $\frac{7}{2}^-$ resonant state making the most significant contribution.

^{7}Be + ^{14}N

The available data for this system refer to the melamine target mentioned in the previous section (reported in ref. [112]) and have been reviewed in ref. [1].

^{7}Be + ^{27}Al

In addition to the data of ref. [114], which were reviewed in ref. [1], elastic-scattering angular distributions for this system have been reported in ref. [115] for E_{lab} = 10.0, 13.8, 15.2, and 15.4MeV. Optical-model fits to the data yielded total reaction cross sections of 171 ± 92, 737±81, 951±70, and 998±74mb, respectively. Although inelastic yields corresponding to excitation of the lowestlying states of projectile and target would in principle also contribute to the data, coupled-channel calculations indicated that such contribution can be neglected. The above cross sections were obtained by using potentials with real and imaginary parts proportional to the S˜ao Paulo potential: $V = N_R V_{SPP}$, $W = N_I V_{SPP}$. A good description of the data was achieved, but normalization factors different from the most usual values (N_R = 1, N_I = 0.78) were generally needed to fit the data. Using two different data reduction procedures, a comparison of reaction cross sections was performed between systems having the same ^{27}Al target and projectiles of ^{6}He, ^{6,7}Li, ^{7,9}Be and ^{16}O, but no definitive conclusion could be drawn. This is actually consistent with part of the conclusions of a recent revision of the most frequently used methods of data reduction [116], where it was warned that these methods in general do not perform well when applied to reaction cross sections and should only be used with some caution.

^{7}Be + ^{51}V

An elastic-scattering angular distribution for the ^{7}Be+^{51}V system at ^{26}MeV was reported in ref. [117]. An OM fit using the S˜ao Paulo potential gave a good description of the data, yielding a total reaction cross section of 1210mb.

^{7}Be + ^{58}Ni

As reviewed in ref. [1], experimental angular distributions for quasielastic scattering of this system (including inelastic excitation of the 0.43MeV state in ^{7}Be) were reported in ref. [118], for Elab energies of 15.1, 17.1, 18.5, 19.9, and 21.4MeV. New measurements were recently reported at E_{lab} = 21.5Mev in ref. [119] where, in addition to the quasielastic-scattering data, angular and energy distributions of the ^{3,4}He residues produced in the reaction also were detected (these will be further discussed in sect. 7.2). The quasielastic data included in principle inelastic yields corresponding to population of the 0.43 (1.45)MeV excited state of the projectile (target), but the estimated contributions were small (negligible), respectively. A comparison with the data of ref. [118] at 21.4MeV shows a remarkably good agreement, as seen in fig. 21. Opticalmodel and CDCC calculations, both with no free parameters, gave satisfactory descriptions of the data. A ^{3}He+^{4}He cluster structure was assumed for ^{7}Be in the latter calculations and the involved interaction potentials were obtained from independent experimental data. The optical-model calculations, represented with the solid curve in fig. 21, yielded a total reaction cross section of 585mb, consistent with the 506 ± 97mb value obtained in ref. [118].

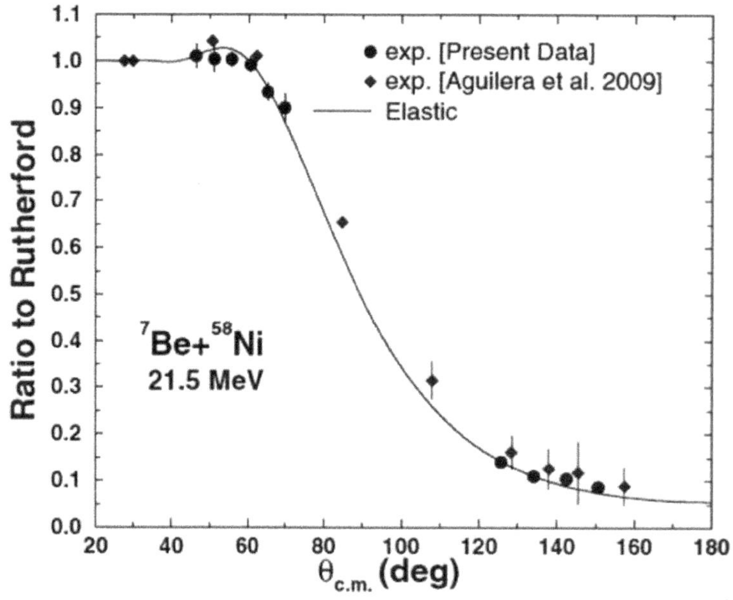

Figure. 21: Quasielastic-scattering angular distribution for ^{7}Be+ ^{58}Ni at 21.5 MeV. Circles are from ref. [119] while diamonds are from [118]. The solid line corresponds to an optical-model calculation using a global potential parametrization for ^{7}Li. Figure reprinted with permission from ref. [119]. c 2015, The American Physical Society.

On the basis of the elastic-scattering data of ref. [118], calculations on the energy dependence of the optical polarization potential were performed in ref. [120]. The experimental uncertainties did not allow to make a definitive conclusion about the presence (or not) of the breakup threshold anomaly (BTA) in the ^{7}Be+^{58}Ni system. However, when the respective fusion data (sect. 7.3) could be also included in the analysis, it was concluded that this system actually shows the more normal threshold anomaly (TA) instead of the BTA [121].

BREAKUP OF ^{7}BE

An unambiguous measurement of non-capture breakup of ^{7}Be into ^{3}He+^{4}He would require coincidence detection of the respective He residues. In the ^{7}Be + ^{58}Ni experiment mentioned in sect. 7.1.6, ref. [119], no such coincidences were observed within the limits of the respective geometrical efficiency, thus providing an indication of a low bu cross section. Upper limits for the experimental bu angular distribution were obtained and, through respective CDCC calculations consistent with such limits, an estimated σ_{bu} = 10.8mb was obtained for this cross section at E_{lab} = 21.5Mev. According to the calculations, bu would make a contribution of 31% (24%) to the ^{3}He (^{4}He) experimental yields corresponding to direct processes, i.e., after subtraction of the evaporation α yield from fusion. The latter yield was obtained by fitting with the statistical code PACE2 the ^{4}He angular distribution measured in the angular range $\theta_{lab} \geq 115$◦, where fusion is likely to be the dominant mechanism. Other direct processes considered in ref. [119] were ^{4}He stripping, which would contribute to ~ 69% of the observed ^{3}He yield, and 1n-pickup (1nstripping) with a 27% (22%) contribution to the measured ^{4}He yield. A remaining ~ 26% ^{4}He yield, which happened to concentrate in the angular region between 50◦ and 70◦, was ascribed to the 3He stripping process.

In ref. [122] (reviewed in ref. [1]), reactions of ^{7}Be with ^{238}U at $E_{c.m.}$ energies of 54.4, 50.5, 48.4, 42.7, and 38.8MeV were measured, stressing fusion-fission events. Light charged-particle yields were observed in coincidence with fission events, but this would exclude most breakup events because the fission signature requires a large energy transfer to the target. However, similar to the conclusions mentioned with respect to ^{7}Be + ^{58}Ni in the preceeding paragraph, these yields could be best explained as the products of ^{3}He and ^{4}He cluster transfers.

^{7}Be breakup events also were searched for in the ^{7}Be + ^{12}C experiment of ref. [111] by generating coincidence spectra for ^{3}He, ^{4}He, and Z = 1 ions. Very few ^{3,4}He coincidences were observed, but no upper limits for the respective bu cross section were estimated. A strong selectivity was observed in the ^{12}C(^{7}Be, ^{3}He) reaction, populating α cluster states in ^{16}O, and the measured angular distributions for ^{3}He were forward peaking. These two features are indicative of a direct α transfer as the main mechanism contributing to the observed ^{3}He yield, consistent with the conclusions of the ^{7}Be + ^{58}Ni experiment mentioned above [119].

Fusion of ^{7}Be

Fusion measurements with ^{7}Be projectiles have been reported for the rather light target ^{27}Al [114], the mediummass target ^{58}Ni [123], and the much heavier target ^{238}U [122]. The ^{27}Al data were obtained by subtracting the measured one-proton-stripping cross sections from the total reaction data deduced from respective quasielasticscattering measurements [114]. It should thus be correct to associate the reported values to the total fusion (TF) cross section or, at worst, to an upper bound for it. Fusion cross sections of 635 ± 76, 858 ± 94 and 922 ± 92mb were obtained for E_{lab} = 17, 19 and 21MeV, respectively. A comparison with fusion data for the mirror nucleus ^{7}Li on the same target yielded similar results.

For the ^{58}Ni target, evaporation proton yields were measured and total fusion cross sections were deduced by using calculated proton multiplicities [123]. Measurements performed at energies $E_{c.m.}$ = 13.9, 15.0, 16.6, 17.4, 17.9, and

19.0MeV gave cross sections of σ_{TF} = 26 ± 3, 61 ± 7, 165 ± 32, 246 ± 78, 292 ± 46, and 395 ± 116mb, respectively. An enhancement with respect to BPM predictions corresponding to a realistic bare potential was observed, even in the region above the barrier.

In the case of the ^{238}U target, fusion-fission was measured [122]. Charged particles were detected with 40 Silicon detectors completely surrounding the target and a fission signature was established through coincidence of products flying in nearly opposite directions. Light charged particles in coincidence with fission also were detected in some events, which discards the possibility of complete fusion (CF) for these events. A complete fusion cross section (σ_{CF}) was thus deduced by subtracting the latter events from the total fission events.

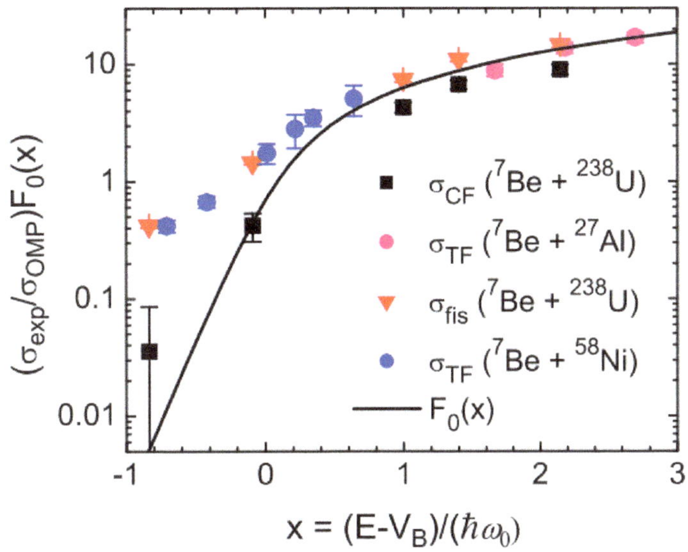

Figure 22: Reduced total fusion cross sections for ^{7}Be + (^{27}Al, ^{58}Ni, ^{238}U), from data published in refs. [114,123,122], respectively. In the case of the ^{238}U target, respective CF data are also shown. The solid curve is the UFF. Adapted with permission from a figure published in ref. [123]. c 2014, The American Physical Society.

Taking the total fission data for the latter system, both CF and ICF would be included in the cross sections for all three systems, as well as possible 3,4He direct transfer to highly excited states. This kind of transfer would produce fission (evaporation protons) in the case of the U (Ni) target, respectively. The three data sets are plotted together in reduced units in fig. 22, which in addition includes the CF data for ^{238}U, for comparison. A good consistency is seen in this figure between the σ_{TF} data (σ_{fis} for ^{238}U) corresponding to the three systems, which could indicate that the dominant mechanism responsible for the observed sub-barrier enhancement is mainly associated with the ^{7}Be projectile, independent of the target. The enhancement factor is still fairly large at the barrier (~ 2.5) and there is also evidence of enhancement for energies slightly above the barrier. For x = 0.35, for instance, the respective factor is ~ 1.5. This corresponds to 1.1 (1.8)MeV above the barrier for the ^{7}Be + ^{58}Ni (^{7}Be + ^{238}U) system, respectively. It was shown in ref. [123] that inelastic excitations of target and projectile can account for part of the observed enhancement, but an additional mechanism would be needed to properly describe the data, especially for energies close to and below the barrier.

REACTIONS WITH ^{10}BE

The ^{10}Be nucleus is tightly-bound, with a neutron separation energy S_n = 6.81MeV and a half-life of 1.51 × 10^6 years. It can therefore be produced as either a primary or secondary beam. Fusion and elastic-scattering experiments with a ^{10}Be beam have been performed for intrinsic interest and also for comparison with data for other isotopes such as the proton-rich ^{7}Be and neutron-rich ^{11}Be projectiles. These works are reviewed below.

Elastic scattering of ^{10}Be

^{10}Be + ^{12}C

The elastic scattering of a low-energy ^{10}Be beam on a ^{12}C target was reported in ref. [110]. The 26MeV secondary ^{10}Be beam was produced by the double solenoid RIBRAS facility at the Universidade de S˜ao Paulo, Brazil [124]. The ^{9}Be(^{11}B, ^{10}Be) reaction was used to produce a beam with an intensity of about 5×10^3 pps. The elastic-scattered particles were identified by three ΔE-E silicon detector telescopes. An analysis with an optical-model potential was performed, considering both volume Woods-Saxon (WS) and double-folding S˜ao Paulo (SP) type potentials, achieving good agreement with the data.

Coupled-channel calculations have also been performed considering the two bound excited states in ^{10}Be (2_1^+ and 0_1^+) . These two excited states correspond to the most probable quadrupole transitions in ^{10}Be. The vibrational model for two-phonon states was adopted. The reduced transition probabilities $B(E2)(2_1^+ \to 0_{gs}^+)$ = 8.0W.u. and $B(E2)(0_1^+ \to 2_1^+)$ = 2.5 W.u. were taken from ref. [125]. These calculations also gave a good description of the angular distribution. It was determined that the coupling to the first excited state (2_1^+) was the most important one in describing the data. A comparison of the reduced total reaction cross section for the ^{7,9,10}Be + ^{12}C systems, also obtained from the doublefolding potential analysis, is shown in fig. 19 (sect. 7.1.2), together with the results for some other light projectiles from the literature. The reduction of the reaction cross section is as given in sect. 2.2.2 above.

As pointed out in ref. [110], this behavior differs from that for weakly-bound projectiles on heavy targets where enhancement of the total reaction cross sections at nearbarrier energies was observed. This can be an indication that dynamic effects have only a small influence on the reactions of these projectiles on the light ^{12}C target.

$^{10}Be + ^{64}Zn$

The elastic scattering of ^{10}Be on ^{64}Zn was reported in ref. [126]. A full description of the experiment, as well as a more detailed analysis, was reported later in ref. [127]. Beams of ^{10}Be and ^{11}Be, produced by spallation of 1.4GeV protons on a Ta target at REX-ISOLDE at CERN [128], were post-accelerated in the REX-Linac to 28.4MeV and 28.7MeV, respectively. The data on ^{11}Be will be discussed in sect. 9 below.

The motivation to perform elastic-scattering measurements with these projectiles was to investigate the in- fluence of their nuclear structure on this process. The measurement with strongly-bound ^{10}Be serves as a base for comparison with the more exotic ^{11}Be isotope. The elastic-scattering angular distribution of strongly-bound light projectiles can show a Coulomb rainbow, when there is little or no absorption, or Fraunhofer/Fresnel diffraction when there is absorption.

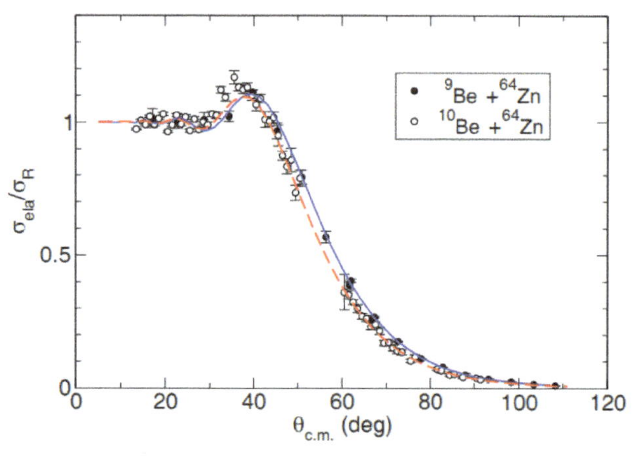

Figure 23: Elastic-scattering angular distribution for $^{9}Be + ^{64}Zn$ (filled symbols) and $^{10}Be + ^{64}Zn$ (open symbols). The curves represent the results of an OM analysis. Figure reprinted with permission from ref. [127]. c 2012, The American Physical Society.

An angular distribution for $^{10}Be + ^{64}Zn$ reported in refs. [126,127] was measured at $E_{c.m.}$ = 24.5MeV. It is shown in fig. 23, together with that measured for the weakly-bound ^{9}Be isotope at ^{29}MeV. The detection setup used in this experiment consisted of an array of 50 × 50mm^2 double sided Si-detector (DSSD) telescopes covering the angular range of 10◦ to 150◦.

An OM analysis was performed with Woods-Saxon–type potentials. Angular distributions for the elastic scattering of both ^{9}Be and ^{10}Be show a Fresnel peak as expected. The most important conclusion from this OM analysis was the observation of a large ambiguity in the depth of the real potential. By continuously varying the radius parameter, equivalent fits could be obtained with different real potential depths, without changing the imaginary potential. Also, all potentials gave equivalent fits for distances larger than about R = 10fm, indicating that the tail for all these potentials is identical. A double-folding potential analysis has also been performed for these systems by Hemalatha [129].

$^{10}Be + ^{19}4Pt$

A Coulomb excitation study of the 2_1^+ state at 3.368MeV in ^{10}Be has been performed via an inelastic-scattering measurement on a ^{194}Pt target [130]. The experiment was performed with a 41MeV secondary ^{10}Be beam, produced at the TRIUMF/ISAC-II radioactive ion beam facility with an intensity of 1×107 ions/s. The beam impinged on a 3.0mg/cm^2 ^{194}Pt target at the center of the TIGRESS γ- ray spectrometer [131]. The 41MeV incident energy is well below the Coulomb barrier for this system. The use of reactions with negligible nuclear contributions is fundamental in such experiments, to avoid Coulomb-nuclear interference. Gamma rays emitted following the de-excitation of states in both beam and target nuclei were detected by eight segmented, highly efficient, Compton-suppressed TIGRESS clover detectors. The γ rays were used to clean the energy spectrum of the scattered 10Be ions.

The quadrupole moment values and γ-ray yields from the de-excitation of the 3.368MeV state were used to obtain the diagonal matrix element of the electricquadrupole tensor in ^{10}Be, which was compared with predictions from several calculations. The experiment showed the feasibility of Coulomb excitation studies of high-lying 2_1^+ states in light nuclei, which is important to improve the understanding of the contributions of 2-nucleon and 3-nucleon potentials to the nuclear spin-orbit interaction, and especially how these contributions affect electricquadrupole matrix elements.

Fusion of ^{10}Be

$^{10}Be + ^{209}Bi$

Fusion reactions induced by $^{10,11}Be$ at energies close to the barrier (E_{lab} = 40 to 50MeV) have been measured at RIKEN, Japan. This experiment was reviewed in [2]. The data were obtained in two different runs and were reported in two papers [132,133]. In the second paper, a new analysis of the first run was presented. The investigation of the fusion dynamics for these nuclei is related to the influence of their cluster structure on the complete and incomplete fusion process.

The data analysis for the $^{10}Be+^{209}Bi$ system has been revised by Hinde and Dasgupta [134]. Their motivation to carry out a new analysis was the fact that the results and conclusions from the previous two papers in refs. [132, 133] were different. The cause of difference was the detection efficiency within the time windows considered in these papers, which used the time of flight of the beams to determine the incident energy on an event by event basis. Depending on the time windows considered, different contributions were obtained from $^{215,216}Fr$ following 4n and 3n evaporation in the case of ^{10}Be projectile. Using the empirical systematics from the cross section for the $^{9}Be + ^{209}Bi$ system [132,135,136], the efficiency could be accounted for. The relative yields for the xn channels from complete fusion, the correct absolute normalization, and the fission contribution to the complete fusion cross section could then be deduced. The results of these corrections, and the newly-obtained cross sections as a function of the incident energy can be seen in fig. 24. The discussion and comparison of the results for the ^{9,10,11}Be+^{209}Bi systems are presented in sect. 9.3 below.

REACTIONS WITH ^{11}BE

The ^{11}Be nucleus has a $J^{\pi} = 1/2^{+}$ ground state with a neutron binding energy of $S_n = 0.501$MeV. It is one of the rare cases of a halo nucleus with a bound excited state. The low binding energy combined with the primarily S-wave relative orbital of the loosely bound neutron produce a one-neutron halo structure for this nucleus, with a compact ^{10}Be(0^{+}) core (not inert). Actually, ^{11}Be is the only presently-known halo nucleus with a strongly deformed core [137]. It has one of the largest known reduced dipole transition probabilities between the first excited state at $E_x = 0.32$MeV and the ground state (B(E1) = 0.116(12) e^2 fm^2 [138]). This observation is corroborated by the recent high precision measurement of the electromagnetic dipole strengths performed at the TRIUMF Isotope Separator and Accelerator (ISAC II) [139]. The unusual features of this halo nucleus strongly affect the interaction with both light and heavy targets at low bombarding energies. Some experiments have recently been performed with a ^{11}Be beam to further investigate the effect of the distinctive features of this nucleus, such as the extended neutron density distribution, on fusion and elastic scattering.

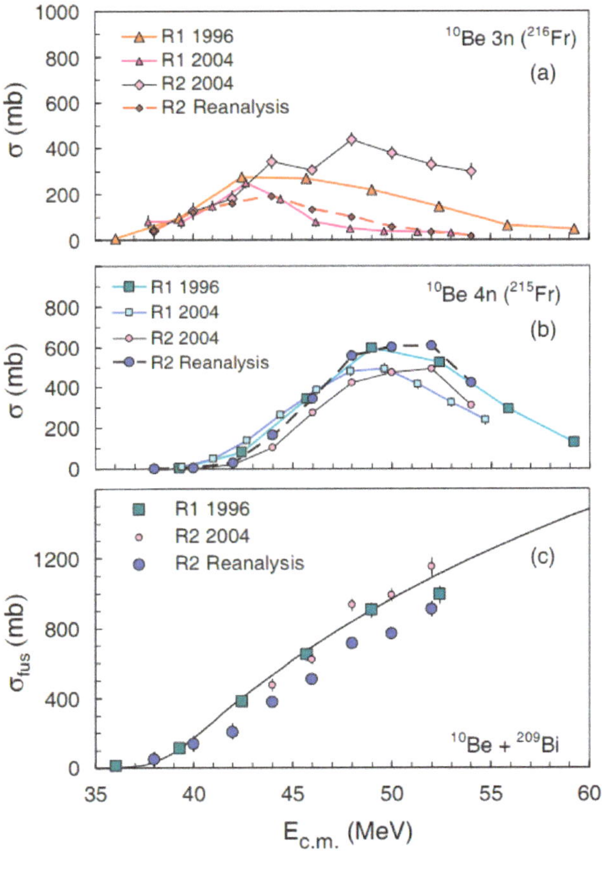

Figure 24: Individual 3n- and 4n-evaporation channels and total fusion cross sections for the ^{10}Be + ^{209}Bi system. The several analyses are indicated. R1 1996 refers to the data and analysis from ref. [132], R1 2004 are the data from ref. [132] with analysis from ref. [133], R2 2004 are the data and analysis from ref. [133] and R2 Reanalysis are the data from ref. [132] and analysis from ref. [134]. Figure reprinted with permission from ref. [134]. c 2010, The American Physical Society.

Elastic scattering of ^{11}Be

The elastic-scattering angular distributions for neutronhalo nuclei such as ^{11}Be from heavy targets at near-barrier energies are strongly affected by dipole polarizability. The combination of the nuclear interaction and the strong electric field make the charged core and the halo neutrons move in opposite directions, which in turn produces strong coupling to dipole modes of the system at relatively low excitation energy. These effects were investigated in the elastic scattering of ^{11}Be.

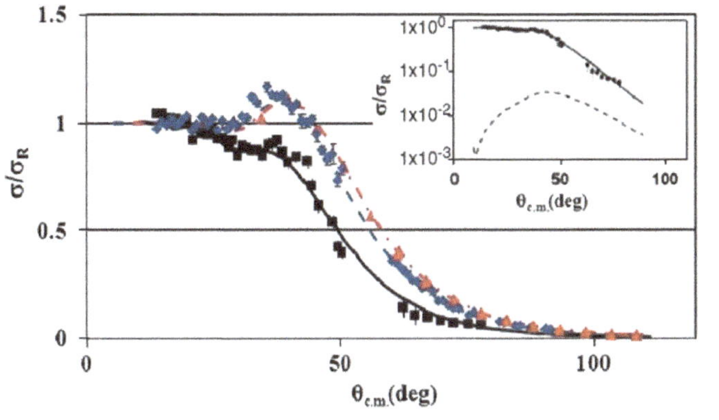

Figure 25: Elastic-scattering angular distributions for ^{9}Be (triangles), ^{10}Be (diamonds), and ^{11}Be (squares) + ^{64}Zn. The data for ^{11}Be is practically elastic (not quasielastic) since the contribution of the inelastic channel (E_x = 0.320 MeV), shown in the inset, is very small. Figure reprinted with permission from ref. [126]. c 2010, The American Physical Society.

^{11}Be + ^{64}Zn

The elastic scattering of a low-energy ^{11}Be beam on a ^{64}Zn target was measured in the same experiment as for the ^{10}Be beam discussed in sect. 8.1.2 above. The angular distribution measured at $_{Ec.m'}$ = 24.5MeV was first reported in ref. [126] and is shown in fig. 25.

Because of the small excitation energy of the first excited state (E_x = 0.320MeV), the elastic cross section data may contain some contribution from inelastic scattering. As can be seen in the figure, the ^{11}Be scattering shows a very different pattern as compared to those for the ^{9}Be and ^{10}Be projectiles. The cross sections at forward angles are strongly reduced in a similar way to what is observed in collisions involving very deformed nuclei, where the damping is due to strong Coulomb excitation of the target [140]. Similar damping of the cross sections at forward angles observed in ^{11}Li and ^{6}He elastic-scattering angular distributions was attributed to a large Coulomb dipole excitation due to the presence of the strong continuum low-lying E1 strength, which is associated with the halo structure of these nuclei. The data shown in fig. 25 were analyzed in terms of the optical model (OM) using Woods-Saxon–type potentials. To describe the unusual pattern of the ^{11}Be angular distribution due to the presence of long-range couplings, a phenomenological surface potential (DPP) having the shape of a WS derivative was added to the imaginary volume potential. The best χ^2 fit was obtained with a large diffuseness, a_{si} = 3.5fm for this DPP. The authors concluded that the suppression of the cross section at forward angles originated from largepartial-wave absorption due to the diffuse halo structure of ^{11}Be. A more detailed analysis of this angular distribution in terms of continuum-discretized coupled-channel (CDCC) calculations was later reported in refs. [127,141], where a full description of the experiment was also given. A comparison of the CDCC results with the experimental data is shown in fig. 26. It's important to mention that the contribution of ^{10}Be core excitation was not taken into account in these calculations.

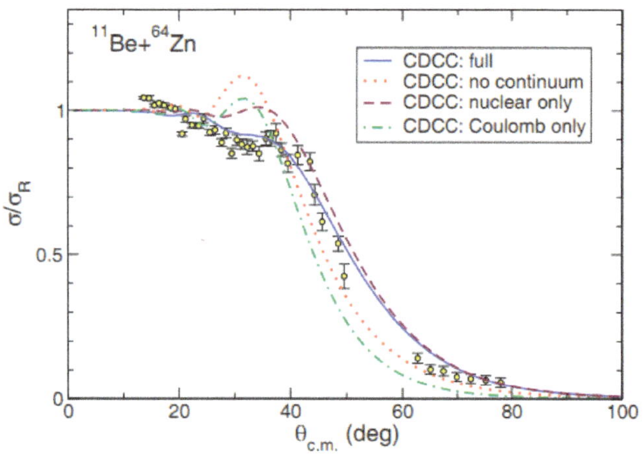

Figure 26: Elastic-scattering angular distribution for the ^{11}Be + ^{64}Zn system at $E_{c.m.}$ = 24.5 MeV. The curves correspond to CDCC calculations as indicated. Figure reprinted with permission from ref. [127]. c 2012, The American Physical Society.

The full CDCC calculation, represented by the solid line in the figure, reproduces the data very well. The importance of the couplings to both nuclear and Coulomb breakup is also clear. As pointed out by Keeley in ref. [142], the largest effect in the damping of the Fresnel peak is due to nuclear breakup coupling. This could be interpreted as an unexpected result and should be further elucidated by more detailed model calculations where ^{10}Be would no longer be treated as an inert core.

A detailed study of the convergence in CDCC calculations for light exotic nuclei, where the importance of spins, binding energy, and number of partial waves were analyzed, was performed by Druet and Descouvemont [143]. They used the ^{11}Be+^{64}Zn elastic-scattering data to show that an accuracy of 10% can be achieved in the CDCC calculations, but for higher accuracy convergence issues must be carefully taken into account.

Another analysis of this set of data was performed in terms of the OM with a bare potential plus a long-range dipole polarization potential (DPP) by Hemalatha [129]. The bare potential was a double-folding potential taking into account a parametrized extended density for ^{11}Be folded with the M3Y effective nucleon-nucleon interaction. The DPP was determined by using the dipole strength distribution from the cluster model and also from experimental data. The results for the calculated differential cross sections show a suppression in the Coulomb-nuclear interference region and reproduce the data quite well.

A more recent analysis of this set of data was performed by So et al. and reported in ref. [144]. Their OM analysis consisted of a short-range potential with a Woods-Saxon shape and parameters from the ^{10}Be+^{64}Zn analysis, plus a long-range dynamic polarization (LRDP) potential consisting of a Coulomb dipole excitation (CDE) potential and a long-range nuclear (LRN) potential. Their conclusion is that both the CDE and LRN potentials are essential to describe the experimental angular distribution. The real parts of both potentials were responsible for the strong absorption by lowering the barrier, and the imaginary parts removed flux from the elastic channel.

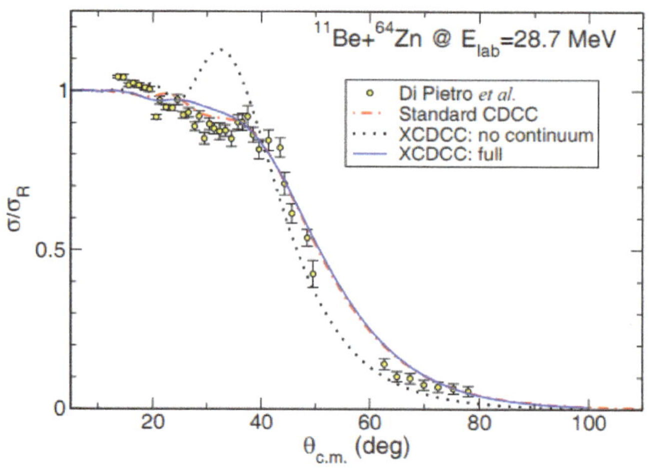

Figure 27: Elastic-scattering angular distribution for the ^{11}Be + ^{64}Zn system at $E_{c.m.}$ = 24.5 MeV. The curves correspond to CDCC calculations as indicated. Figure reprinted with permission from ref. [145]. c 2014, The American Physical Society.

The CDCC calculations described above ignored the possible admixtures of different core states in the wave functions of the ^{11}Be, as well as ^{10}Be core excitation. These admixtures are known to be important, particularly in the case of the well deformed ^{10}Be core. An extended CDCC calculation with core excitation has been performed by de Diego et al. and reported in ref. [145]. The authors improved the XCDCC calculation developed in ref. [146] by using the so-called pseudo-state method, in which the projectile states are approximated by the eigenstates of the Hamiltonian in a truncated basis of square-integrable functions. The results of this calculation for ^{11}Be + ^{64}Zn can be seen in fig. 27. To account for core excitation, the potential was deformed with the same deformation length used in the structure model, i.e., δ2=1.664fm. As can be seen in the figure, the XCDCC calculation is similar to the CDCC result but slightly overpredicts the data near 30°.

^{11}Be + ^{120}Sn

Data on angular distribution for elastic scattering of ^{11}Be + ^{120}Sn were obtained at the REX-ISOLDE facility [147]. The experimental setup was similar to that used to measure elastic scattering and breakup for ^{11}Be+ ^{64}Zn [126,127]. To compensate for the low beam intensity (< 10^5 pps), a thick (3.5mg/cm^2) 120Sn target was used. This limited the energy resolution of their detection system to 350keV. It was therefore not possible to resolve the first excited state in ^{11}Be and the data are actually quasielastic. For angles between 15° and 38°, however, it was possible to identify the ^{10}Be fragments produced by the breakup of ^{11}Be into ^{10}Be+n. The measured angular distribution of quasielastic scattering was analyzed in terms of coupled channel (CC) calculations to study the relevant aspects of the dynamics of the reaction process. The qualitative CC calculation included several excited states in ^{11}Be up to 3.41MeV, taking a vibrational model for the projectile and treating the target as inert. As a result, the couplings were found to be important but insufficient to completely describe the angular distribution.

A CDCC analysis of these data has also been performed [148]. Deformation of the ^{10}Be core was, however, not taken into account. It was found that Coulomb-nuclear interference is very important even at more backward angles where experimental data are unavailable. The effects of neutron transfer were also ignored. The authors suggested that it would be useful to obtain data beyond 40◦ to help disentangle the contributions of breakup and neutron transfer.

^{11}Be + ^{197}Au

Preliminary data on inelastic scattering of ^{11}Be + ^{197}Au at 31.9MeV from the TRIUMF/ISAC-II facility was reported in ref. [149]. The ^{11}Be beam had an intensity of 10^5 pps. The experimental inelastic-scattering probability was compared with semiclassical and coupled-channels calculations. The CDCC calculation overestimated the inelastic-scattering cross section. This can be explained by the fact that pure single-particle configurations of the valence neutron relative to an inert ^{10}Be(0$^+$) core was assumed, where the ground and first excited states correspond to pure $2s_{1/2}$ and $1p_{1/2}$ configurations, respectively. These states could in reality contain significant admixtures of excited states of the ^{10}Be core.

^{11}Be + ^{209}Bi

Data on elastic scattering of ^{11}Be on a ^{209}Bi target were reported in ref. [150]. The experiment was carried out at the RIKEN laboratory in Japan. The ^{11}Be beam was produced in-flight by the fragmentation of ^{13}C ions on a Be target, and selected using the RIKEN Projectile-Fragment Separator (RIPS) [151]. The energy of the ^{11}Be beam (48MeV/u) was strongly degraded to a final energy in the range of E_{lab} = 40–50MeV. Angular distributions of the quasielastic process were obtained at five beam energies between 40 and 48MeV. Since the experimental energy resolution did not allow the separation of inelastic scattering to the first excited state of ^{11}Be (0.320MeV), the differential cross sections are actually quasielastic. The angular distributions were analyzed in terms of the OM with Woods-Saxon potentials. The inelastic contribution (up to 30% of the total yield) was estimated with DWBA and CC calculations. The total reaction cross section for each energy was obtained and scaled by R^2, where R is the sum of the projectile and target radii. The energy was scaled by the Coulomb barrier ($E_{c.m.}$/VC) and the scaled values were compared with those for ^{9}Be+^{209}Bi. At lower energies the reaction cross sections were larger for ^{11}Be, suggesting that direct processes due to the halo structure are more important at energies closer to the barrier.

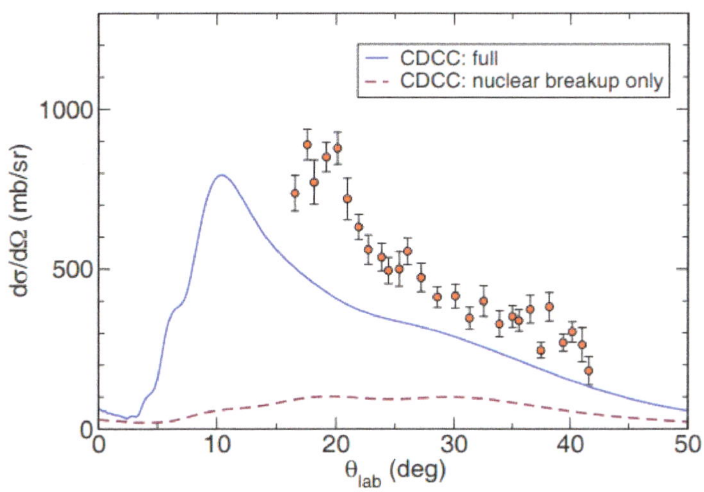

Figure 28: Angular distribution for breakup and/or transfer in the ^{11}Be + ^{64}Zn system at $E_{c.m.}$ 24.5 MeV. The curves correspond to CDCC calculations as indicated. Figure reprinted with permission from ref. [127]. c 2012, The American Physical Society.

Breakup of ^{11}Be

Breakup in the ^{11}Be + ^{64}Zn system

Large numbers of ^{10}Be events from breakup and/or transfer were identified in the elastic scattering of ^{11}Be + ^{64}Zn at $E_{c.m.}$ = 24.5MeV [126].

An angular distribution for these events was obtained, and a total integrated cross section from 0◦ to 60◦ of 1100 ± 150mb was measured. A CDCC calculation was compared with the experimental angular distribution in ref. [127]. The results of this comparison are illustrated in fig. 28, where the solid line represents the full Coulomb + nuclear breakup and the dashed line is the result of nuclear breakup only. This calculation underestimated the data by 30%. This is perhaps to be expected since it corresponds to elastic breakup only. The contribution from the transfer of the undetected valence neutron to the target, as well as the excitation of the target and/or the ^{10}Be core, were not taken into account.

The previously mentioned XCDCC calculations, where admixtures of different core states into the wave functions of ^{11}Be as well as ^{10}Be core excitation were included, is reported in ref. [145]. This method was also used to compute the breakup cross section for the ^{11}Be+^{64}Zn system. The calculation is considered to be only an approximation due to the fact that the XCDCC framework would require an appropriate kinematic transformation that is not yet available. Instead, the ^{10}Be scattering angle was approximated by the ^{11}Be scattering angle. The results of this calculation are shown in fig. 29. As can be seen in the figure, the XCDCC calculation is larger than the CDCC result and improves the agreement with the data. The remaining discrepancy was attributed by the authors to limitations of the ^{11}Be model used in the XCDCC calculations, but also to the contribution of non-elastic breakup events in the data.

Breakup in the ^{11}Be + ^{120}Sn system

Data on the breakup angular distribution for the ^{11}Be + ^{120}Sn system were obtained during the elastic-scattering experiment reported in ref. [147]. However, the results for the experimental breakup cross sections and a CDCC analysis were reported in ref. [148]. The contributions of Coulomb (C) and nuclear (N) breakup were investigated. The calculated breakup cross section with pure nuclear interactions was found to be negligible in comparison with the cross section calculated including Coulomb effects. Coulomb-nuclear interference played an important role in explaining the experimental data, although the calculation could not reproduce the absolute values.

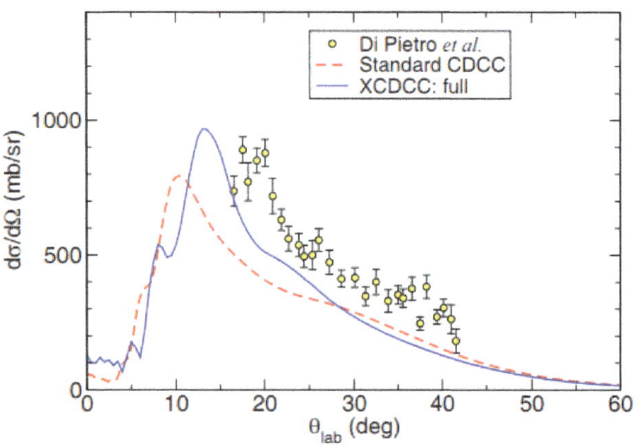

Figure 29: Angular distribution for breakup and/or transfer in the ^{11}Be + ^{64}Zn system at $E_{c.m.}$ 24.5 MeV. The curves correspond to standard CDCC (dashed line) and XCDCC (solid line) calculations as indicated. Figure reprinted with permission from ref. [145]. c 2014, The American Physical Society.

Breakup in the ^{11}Be + ^{197}Au system

Preliminary data on inclusive breakup and/or transfer was obtained in the 31.9MeV inelastic-scattering experiment discussed in sect. 9.1.3 above. An angular distribution measured in the angular range from 13° to 46° was reported in ref. [149]. A large yield of ^{10}Be was observed in this angular range, and the breakup probability was obtained. The data were compared with semiclassical and coupled-channels calculations. The semiclassical calculations, based on the first-order Coulomb theory of Alder and Winther [98] (which gives a breakup probability for a pure dipole excitation) underestimated the data. On the other hand, the CDCC calculation reproduced the experimental breakup cross sections quite well. As already mentioned, the CDCC calculation did not reproduce the inelastic-scattering data, indicating the need for a more sophisticated model to reproduce all the observables. The authors claimed that this was a preliminary analysis, and an extended CDCC calculation incorporating core-excitation effects in the structure of ^{11}Be, as well as in the core-target interaction, was in progress.

Fusion of ^{11}Be

As already mentioned in sect. 8.2 above, fusion induced by ^{10,11}Be at energies close to the barrier has been investigated at RIKEN, and these data were reviewed in ref. [2]. An activation method was used, detecting α particles emitted by the residual nucleus after 4n and 5n evaporation to infer information concerning complete and incomplete fusion. The investigation of the fusion dynamics for these nuclei is related to the influence of their cluster structure on the complete and incomplete fusion processes. In the case of ^{11}Be, a halo nucleus with a large radius, a reduced Coulomb barrier and a larger fusion cross section might be expected. On the other hand, the weak binding energy should increase the breakup probability and consequently may reduce the fusion yield. The data were obtained in two different runs and were reported in two papers [132,133]. In the second paper, a new analysis of the first run was presented. The data obtained with ^{11}Be projectile were also revised by Hinde and Dasgupta [134], who corrected for the time windows considered in the analysis of the 4n and 5n evaporation using the empirical systematics from the cross section for the ^{9}Be + ^{209}Bi system [132,135,136]. Using this set of data, they could correct for the time-window efficiency, calculate the relative yields for xn channels in complete fusion, correct the absolute normalization, and estimate the fission contribution. The results of these corrections, and the newly-obtained cross sections for the ^{11}Be+^{209}Bi system, can be seen in fig. 30.

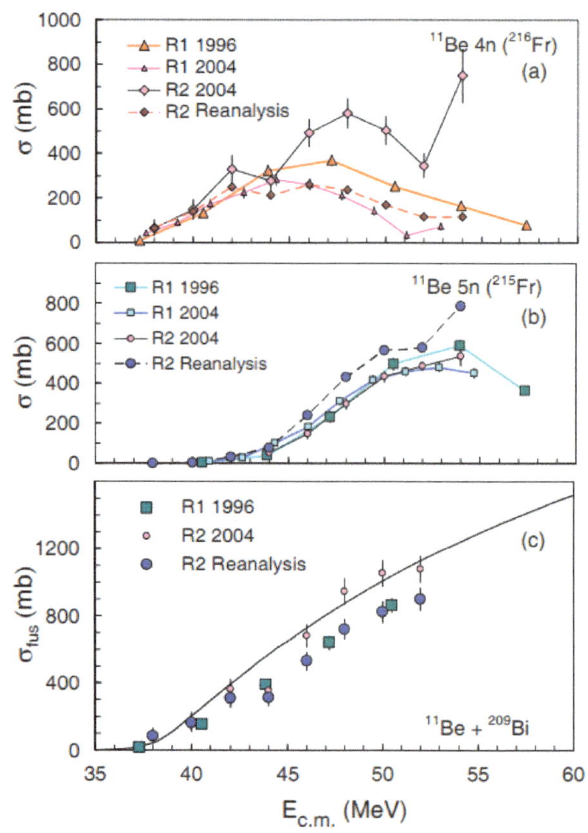

Figure 30: Individual 4n- and 5n-channel evaporation and total fusion cross sections for the ^{11}Be + ^{209}Bi system. R1 1996 are the data and analysis from ref. [132], R1 2004 are the data from ref. [132] with analysis from ref. [133], R2 2004 are the data and analysis from ref. [133] and R2 Reanalysis are the data from ref. [132] but analysis from ref. [134]. Figure reprinted with permission from ref. [134]. c 2010, The American Physical Society.

The final, corrected complete (CF) and total (TF) fusion cross sections for the $^{10,11}Be + ^{209}Bi$ systems were compared with those for the $^{9}Be + ^{209}Bi$ in ref. [134]. The total fusion cross section included the contribution of incomplete fusion. From the ratio of CF to TF, the fraction of incomplete fusion (ICF) was estimated, with surprising results. The fraction of ICF was 32% for ^{9}Be, 15% for 11Be and 25% for ^{10}Be, although it was claimed that these values could be just a lower limit, depending on the identification and separation of Fr and At α decays. It would be worthwhile to repeat this measurement to obtain a better understanding of the role of the neutron-halo and α-cluster structure of $^{9,10,11}Be$ in the fusion process.

REACTIONS WITH ^{8}B

Several pieces of evidence support the presence of protonhalo structure in the ground state of the proton dripline nucleus ^{8}B, and it is now generally agreed that this is the case. The low proton separation energy associated (0.138MeV) should in principle make the respective breakup process an important reaction channel in reactions with ^{8}B. Data with this projectile are still scarce because of the extreme difficulties to get reasonably intense beams, but some low-energy measurements have been performed in the last few years. Including both, experimental and theoretical works involving reactions with ^{8}B, over 40 articles have appeared during this period.

Elastic scattering of ^{8}B

Within the energy range of interest in the present review, elastic scattering of ^{8}B has been measured on targets of ^{12}C and ^{58}Ni. Higher energy data for ^{12}C and ^{208}Pb targets also will be mentioned because they are considered as providing interesting complementary information.

$^{8}B + ^{12}C$

An elastic-scattering angular distribution for $^{8}B + ^{12}C$ has been measured in ref. [109] at an incident energy of 25.8MeV, which is nearly three times the corresponding Coulomb barrier (V_B = 5.14MeV). The data could be well described by an optical-model calculation using the S˜ao Paulo potential (SPP) with conventional normalization, indicating a weak effect of the breakup process on the elastic scattering. This was confirmed by performing CDCC calculations and comparing the full-coupling with the no-coupling results, as shown in fig. 31. Further CDCC studies of this system for near-barrier energies [152] indicated a negligible effect of breakup on the elastic channel at these energies as well. Comparing with previous calculations for a ^{58}Ni target (see sect. 10.1.2), it was concluded that such effect decreases with the charge of the target.

Analyzing data from an experiment performed at the higher energy E_{lab} = 95MeV (ref. [112], reviewed in [1]), it was found in ref. [113] that core excitation has a significant effect on the $^{8}B + ^{12}C$ elastic-scattering channel. In the latter reference, the ^{7}Be core was assumed to have a $^{4}He+^{3}He$ cluster structure and, in addition to the ground state, its three lowest excited states were included in a coupled-channel calculation. The valence proton in ^{8}B, on the other hand, was treated in an adiabatic approximation and, similar to the results of ref. [152], it was shown to have a negligible contribution. However, excitation of the $(7/2)^-$ resonant state of ^{7}Be was found to decrease significantly the elastic-scattering angular distribution for $\theta_{lab} \geq 13°$, thus bringing the predictions into agreement with the respective experimental results (see fig. 20 for the respective effect on ^{7}Be scattering).

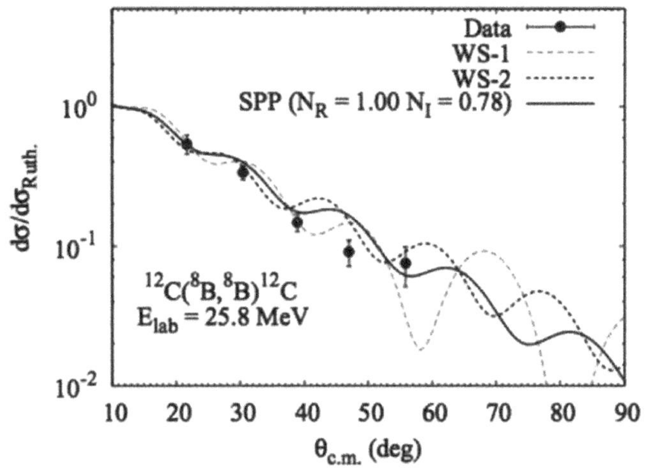

Figure 31: Elastic scattering of ^{8}B on ^{12}C at 25.8 MeV. Figure reprinted with permission from ref. [109]. c 2011, The American Physical Society.

It was also shown in ref. [152] that the continuumcontinuum couplings play an important role on the predicted breakup cross section, a well-known effect in the case of the ^{58}Ni target [153–155].

$^{8}B + ^{58}Ni$

The available elastic-scattering data on this system [118] are reviewed in ref. [1]. It was emphasized that, under a reasonable data reduction, the total reaction cross sections deduced for this system are consistent with those for systems having ^{6}He as projectile. This interesting result was further confirmed in ref. [156], where data for additional systems were also reduced and compared. A figure showing this will be presented later (fig. 41 in sect. 17.1.2). On the theoretical side, the data were analyzed in ref. [154], where the ^{8}B nucleus was described as an inert ^{7}Be core plus one proton and the CDCC method was used to investigate the effects of breakup couplings on the elastic-scattering cross sections. The calculations were in excellent agreement with the experimental data for all five energies measured. Inelastic excitations of the target also were considered and shown to have no important in- fluence. Compared with the no-coupling calculation, the overal effect of bu couplings on the elastic-scattering angular distributions was only modest, but a striking effect of continuum-continuum couplings was noticed.

A comparison with the behavior of ^{6}Li projectiles, which show smaller bu cross sections but quite important bu coupling effects on the respective elastic scattering, has lead Keeley et al. [142,157] to point out that reactions with 8B offer extreme examples where large bu cross sections not necessarily imply a large bu coupling effect. This point has been further discussed in refs. [1,158–160].

In ref. [120], a detailed study of the energy dependence of the optical potential for this system was made. The respective behaviour was found to be more compatible with the Breakup Threshold Anomaly (BTA) than with the more normal Threshold Anomaly. A similar study but including also the respective fusion data (see sect. 10.3) confirmed the BTA behaviour for this system [161].

$^8B + ^{208}Pb$

An elastic-scattering angular distribution for ^{8}B + ^{208}Pb at 170.3MeV was measured in ref. [158]. Even though this energy (nearly three times the Coulomb barrier) is considerably higher than the limit set for the present review, it is mentioned here because the conclusions reached are quite similar as those in the previous lower-enery ^{8}B studies for ^{12}C and ^{58}Ni targets. Arguments were given showing that inelastic scattering makes a negligible contribution to the data and corresponding CDCC calculations suggested a small effect of bu coupling channels on the elastic scattering for this heavier system as well.

Breakup of ^{8}B

The process of ^{8}B breakup in the field of heavy nuclei is of special interest as it can provide information about the ^{7}Be radiative proton capture at low energies, a reaction that plays an important role to understand the solar neutrino emission. Within the energy range covered in the present work, ^{8}B breakup has been measured only with a ^{58}Ni target.

In the first measurement, performed in 2000 [38], a complete angular distribution of the ^{7}Be bu-fragments was measured at 25.8MeV (this has been reviewed in ref. [1]). Later, single angle measurements at $\theta = 45°$ were reported in ref. [162] for the three energies 25.0, 26.9, and 28.4MeV. These measurements yielded results consistent with those at 25.8MeV and showed a slight growth of the bu cross section with energy. CDCC calculations were performed in the same reference, which gave a good description of both, the ^{7}Be angular distribution at 25.8MeV and the single angle bu excitation function. These calculations were later extended to lower energies in ref. [163], where the elasticscattering data for (^{8}B, ^{7}Be) + ^{58}Ni reported in ref. [118] were fitted to determine energy-independent OM potentials and, in addition, DWBA calculations were used to estimate the influence of proton transfer to low-lying states in ^{59}Cu. As a final product, the astrophysical $S_{17}(0)$ factor [64] was found to be 20.8 ± 1.1eV b.

In refs. [155,164,165] the CDCC formalism was used to study the effect of couplings among continuum breakup states on the elastic-scattering and breakup cross sections for the ^{8}B+^{58}Ni system. By calculating the respective polarization potential, the authors predicted that the elastic scattering is enhanced by the couplings while non-capture breakup and fusion processes should be hindered. The later prediction, however, was not confirmed by the fusion measurements that will be described below (sect. 10.3).

The fact that reaction cross sections for different systems can in principle be scaled to each other if the respective charges and masses are not too different, has been used to provide experimental evidence that, in interactions with halo nuclei, the respective cores and haloes decouple from each other. In the case of the ^{8}B+^{58}Ni system, for instance, the interactions of the core (σ_{core}) are scaled from corresponding measurements for ^{7}Be + ^{58}Ni, while the interactions of the halo (σ_{halo}) are accounted for by the bu cross sections for the former system. The fact that for all measured energies the sum $\sigma_{core}+\sigma_{halo}$ coincides with the experimental values obtained for the respective total reaction cross sections verifies the decoupling phenomenon for this proton-halo system [118,166]. Neutron-halo nuclei do also show this phenomenon, as shown in refs. [166, 167] from corresponding data for the ^{6}He + (^{209}Bi, ^{64}Zn) systems.

Fusion of ^{8}B

The methods that have been used to obtain fusion cross sections (σ_{fus}) for ^{8}B consist of measuring a single type of charged evaporated particles and using the results of statistical-model calculations to deduce the respective values of σ_{fus}.

Data were first obtained for the ^{8}B+^{58}Ni system [168], for which σ_{fus} values were reported at ten energies lying in the near- and sub-barrier region. Later, similar data were published for ^{8}B + ^{28}Si [169] at four energies above the respective Coulomb barrier. In ref. [168], evaporation protons were measured while in ref. [169], evaporation alphas rather than protons were measured.

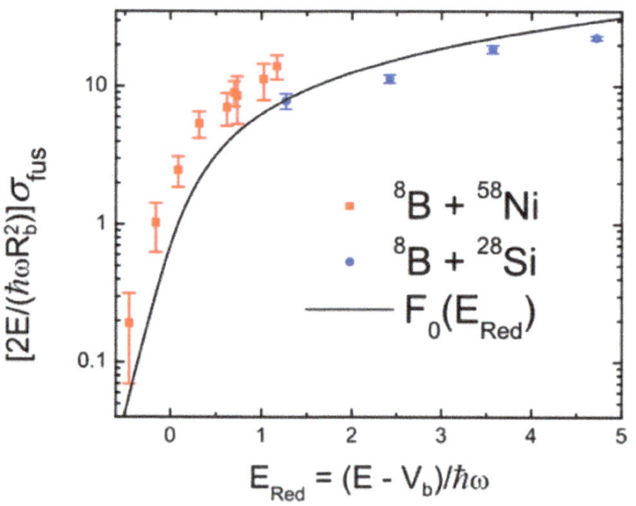

Figure 32: Reduced fusion data for ^{8}B on ^{58}Ni [168] and ^{28}Si [170]. The barrier parameters are V_B = 11.3 MeV, R_B = 8.2 fm, $\hbar\omega$ = 3.38 MeV for ^{28}Si and V_B = 20.8 MeV, R_B = 8.9 fm, $\hbar\omega$ = 4.09 MeV for ^{58}Ni.

The two data sets, plotted in reduced units, are compared to each other in fig. 32. The data for the ^{8}B + ^{58}Ni system show a big fusion enhancement at all energies, even above the Coulomb barrier. This contrasts with what has been observed for neutron-halo systems, whose total fusion cross sections seem to be suppressed for energies above the barrier. On the other hand, the ^{8}B + ^{28}Si system presents a fusion suppression above the barrier, which would indicate a similar behavior with respect to neutron-halo systems, but an apparent inconsistency with the trend observed for the heavier proton-halo system.

It has been suggested [171] that protons coming from breakup (bu) of 8B into ^{7}Be + p may have an important contribution to the data of ref. [168], which could possibly explain the above results. However, it can be shown [172] that this contribution is small, so that the corrected fusion cross sections remain practically unchanged in the region near and above the barrier. This confirms the presence of a fusion enhancement at energies above but not much higher than the barrier for the $^8B + ^{58}Ni$ system. If one considers possible model dependences in the mapping $\sigma_x \rightarrow \sigma_{fus}$, where x stands for either protons [168] or α particles [169], the two data sets are actually consistent with each other [172].

A novel method of data reduction for the purpose of comparing fusion cross sections for different systems was proposed in ref. [173]. Adducing a key role for phase space availability in the compound system, an energy scaling $E_r = (E_{c.m.} + Q)/(V_B + Q)$ is introduced, which depends on the Q value for compound nucleus formation and on the Coulomb barrier height. By applying this scaling to several systems in a $\sigma_{fus}/(A_p^{1/3} + A_t^{1/3})^2$ vs. E_r plot, a similar behavior was found for (^3He, ^{16}O) + 58Ni and $^7Li + ^{59}Co$, with only a slight enhancement around the barrier for the case of $^8B + ^{58}Ni$.

In ref. [174], a simple one-dimensional model was used to analyze the difference between fusion and total reaction cross sections measured for $^8B + ^{58}Ni$ and for other weakly-bound systems, including systems with the neutron-halo projectile ^6He. In this model, all partial waves up to a maximum angular momentum L_f were assumed to fuse, while higher values of L correspond to direct reactions. A simple linear energy dependence was found for L_f in the case of all halo systems analyzed ($^8B+^{58}Ni$, ^6He+[^{209}Bi, ^{64}Zn]), as well as for some weakly bound systems ($^{6,7}Li + ^{59}Co$). These results are consistent with a semiclassical interpretation where the critical distance of approach for fusion to occur grows with decreasing energy.

Within a phenomenological optical-model approach, where both the fusion and the elastic-scattering data were fitted simultaneously, dynamic polarization potentials were derived which describe well the fusion enhancement for the $^8B+^{58}Ni$ system [175]. When these polarization potentials were added to the static Coulomb barrier, the barrier parameters were pushed to larger distances and lower heights than expected for non-halo systems.

In ref. [176], CDCC calculations were carried out including breakup, fusion, and elastic scattering of ^8B on ^{58}Ni. Inelastic excitations in the ^7Be-proton system from the ground state to excited states with orbital angular momenta L = 0–5 and energies up to 8MeV in the continuum were considered. Direct proton transfer (tr) was calculated in the coupled-reaction-channels (CRC) approach and fusion was estimated by assuming $\sigma_{fus} = \sigma_R - \sigma_{bu} + t_r$. The results are presented in reduced form in fig. 33, where a comparison with total reaction data for $^6He + ^{208}Pb$ [17] and $^6He + ^{209}Bi$ [177] also is made. The fusion, total reaction, and bu data for $^8B + ^{58}Ni$ are from refs. [168,118, 38], respectively

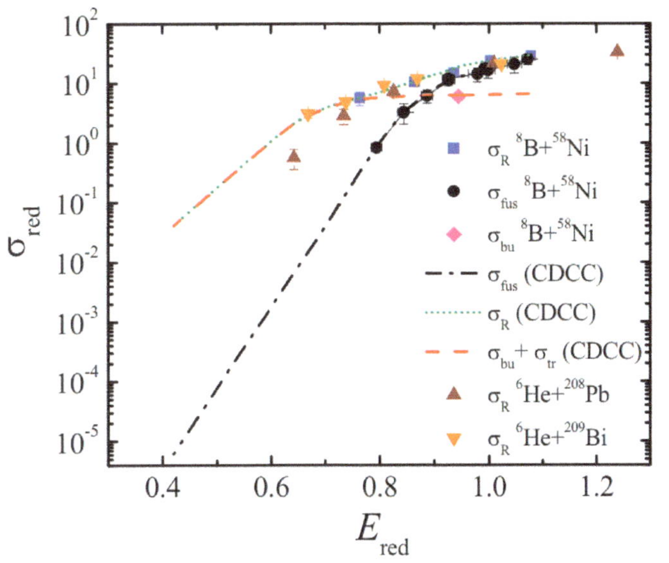

Figure 33: Reduced fusion data for $^8B + ^{58}Ni$ [38,118,168] and respective predictions from ref. [176]. For comparison, total reaction data for $^6He + ^{208}Pb$ [17] and $^6He + ^{209}Bi$ [177] are also included. Figure adapted with permission from ref. [176].

The calculations reproduce well the experimental fusion, breakup, and total reaction cross sections and predict the respective behavior at low incident energies. The predicted cross section for proton stripping accounts for ~ 5% of the total ^7Be emission cross section at $E_{c.m.}$ = 22.5MeV and rapidly decreases for energies below the Coulomb barrier. In this region, breakup is a predominant process and actually saturates the total reaction cross section at the lowest energy measured (20.7MeV), showing a flat behavior down to this energy. Fusion becomes more important above the Coulomb barrier.

Coupled-channel calculations including inelastic excitations of the target were performed in ref. [171] for the $^8B + ^{58}Ni$ system. To describe fusion, an ingoing wave boundary condition was assumed by using an imaginary Woods-Saxon potential internal to the Coulomb barrier (with depth W_0 = 50MeV, radius r_W = 1.06fm, and diffuseness a_W = 0.2fm). The first 2^+ one-phonon state and the triplet of two-phonon states (2^+, 4^+, 0^+) in ^{58}Ni were included. As a conclusion, coupling to inelastic excitations of the target cannot account for the observed fusion enhancement.

Semiclassical arguments indicate that Coulomb polarization might play an important role to produce the enhancement [178]. A time-dependent quantum mechanical treatment that includes both complete and incomplete fusion [179,180] has shown that the fusion probability should be enhanced (suppressed) by the presence of the halo for proton-halo (neutron-halo) nuclei. A gradual process of Coulomb polarization, eventually leading to breakup, was proposed as the mechanism explaining the fusion enhancement. After the breakup, the relevant Coulomb barrier is that between the core and the target, which is lower than the original projectile-target barrier, so the core can more easily penetrate it. A similar conclusion has been reached recently in ref. [181]. However, more theoretical work is needed to achieve a good description of the observed fusion enhancement above the barrier for the proton-halo system $^{8}B + ^{58}Ni$.

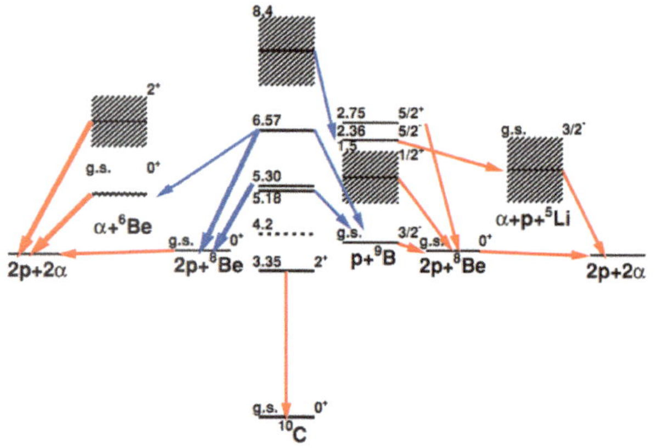

Figure 34: Level scheme and decay paths for 10C. Figure reprinted with permission from ref. [185]. c 2008, The American Physical Society.

REACTIONS WITH ^{10}C

The proton-rich carbon isotope ^{10}C has a half-life of $t_{1/2}$ = 19.3s. Once excited, it can decay by three possible channels: $^{2}p+^{8}Be$, $^{9}B+p$ and $^{6}Be+\alpha$, with threshold energies of 3.820MeV, 4.006MeV and 5.101MeV, respectively (fig. 34). This nucleus is supposed to have a Brunnian (super-borromean) configuration, which would correspond to four interconnected rings (instead of three for a borromean system) associated with the four-body $2\alpha + 2p$ configuration [182,183]. This unusual proposed structure has been investigated by resonance-decay spectroscopy [184–186]. The experiment utilized a 107 MeV ^{10}C beam produced by the p(^{10}B, ^{10}C)n reaction and separated from other reaction products by the MARS spectrometer at the Texas A&M University K500 cyclotron facility [187]. The beam was 99.5% pure and had an intensity of 2×10^{5} pps. The ^{10}C nucleus was excited above 8MeV via interactions with ^{9}Be and ^{12}C targets and all the decay products were detected by an array of four Si-strip ΔE-E telescopes. All the decay channels shown in fig. 34 were observed in coincidence measurements. Proton-proton angular correlations, resonance widths, energy distributions of di-proton and sequential two-proton decays, as well as $2p + 2\alpha$,$p+3\alpha$ and other decay modes from excited resonant states, were studied.

Low-energy radioactive ^{10}C beams have also been used to investigate the unusual inversion of orbitals in the proton-rich nucleus ^{11}N. In general, a one-neutron halo nucleus mainly consists of a valence nucleon in an $s_{1/2}$ orbital coupled to a core. This is indeed the case for ^{11}Be, where the ground state has $J^{\pi} = 1/2^{+}$ in contradiction with the shell-model prediction that the odd neutron should be in a $p_{1/2}$ state leading to $J^{\pi} = 1/2^{-}$. This same inversion was then observed in 11N, which is unbound to proton decay, and could be investigated by resonant elastic scattering with a ^{10}C beam [188,189].

Elastic scattering of ^{10}C

In the last decade only one experiment has been reported on the elastic scattering of a ^{10}C beam on medium- to heavy-mass targets. Although at a little above the energy of the proposed scope of this review, this interesting experiment corresponds to the measurement of angular distributions for quasielastic scattering of $^{10}C+^{nat}Pb$ at 226 and 256MeV (3 times the Coulomb barrier) [190]. The motivation was to search for a Coulomb-rainbow diffraction pattern, which is strongly suppressed in the case of elastic scattering with the neutron-rich nuclei ^{11}Be [126], ^{11}Li [81] and ^{6}He [18], but not for the proton-rich-nuclei ^{8}B [158] and 17F [191]. The experiment was performed with a ^{10}C beam of intensity $\approx 3 \times 10^{4}$ pps, produced at the Heavy Ion Research Facility in Lanzhou, China [192] The setup consisted of two PPACs and time-of-flight (TOF) to identify and reconstruct the incident angles, and two E-ΔE Si-detector telescopes to detect the scattered particles. After a careful OM analysis with Woods-Saxon and doublefolding SP potentials, a clear Coulomb-rainbow pattern was observed in both angular distributions. They also determined the total reaction cross sections and compared with other existing data on ^{6}Li, ^{7}Be, ^{8}B, ^{10}C, ^{10}B and ^{11}C. Using the reduction method proposed in ref. [67] they found that, except for ^{8}B, there were no distinct differences among the weakly-bound and stable projectiles at above barrier energies, indicating that the influence of the additional open channels for these projectiles is small.

Fusion of ^{10}C

A fusion cross section measurement for $^{10}C+^{12}C$ at nearCoulomb-barrier energies was reported in ref. [193]. The motivations for this experiment were i) to investigate the importance of transfer reactions and intrinsic excitations of the target and projectile in the enhancement of fusion cross sections, ii) to compare fusion cross sections of carbon isotopes with theory in preparation for later efforts to investigate the importance of pycnonuclear fusion reactions [194] in the crust of accreting neutron stars, iii) to develop the technique of using a multi-sampling ionization chamber (MUSIC) [195] as an active target for fusion cross section measurements. These experiments were performed at the ATLAS (Argonne Tandem Linac Accelerator System) facility in Argonne National Laboratory, and also included fusion cross section measurements of $^{12,13,14,15}C$. The latter two will be discussed in following sections of this review. The ^{10}C beam was produced with an intensity of about 5×10^{2} pps, which was sufficient to be used in association with the active-target technique. The MUSIC detector was filled with CH_4 gas, and fusion cross sections for $^{10}C + ^{12}C$ were obtained in the range from $E_{c.m.}$ = 9.5–18.5MeV. The averaged cross sections measured between 14 and 17MeV for the carbon isotopes, $^{10,12-15}C$, are shown in fig. 35, where the results of several calculations are also plotted for comparison. As can be observed in the figure, there is an increase in the fusion cross section for higher N − Z that is stronger than the usual R^2 dependence. Also, a coupled-channels calculation with one- and two-phonon excitation and mutual quadrupole and octupole excitations in the projectile and target gives a better prediction as compared with OM (S˜ao Paulo Potential with no free parameters) and TDHF calculations. The OM predictions fail rather badly in this case, possibly indicating some influence of the unusual structure of the ^{10}C nucleus.

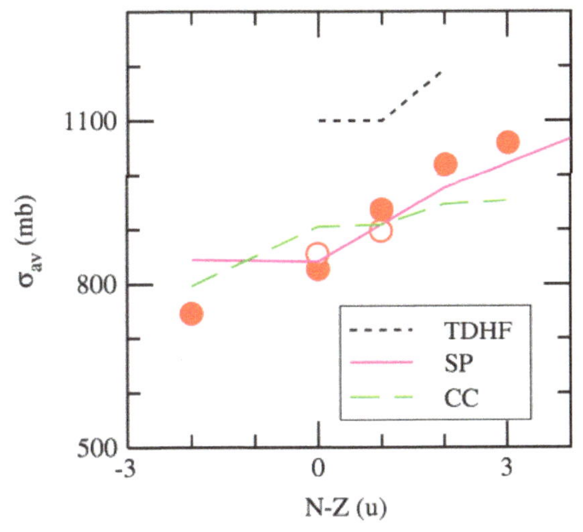

Figure 35: Averaged fusion cross sections for the carbon isotopes $^{10,12-15}C$ on a ^{12}C target. Experimental data (circles) are compared with OM predictions using the S˜ao Paulo (SP) potential (solid line), a coupled-channel calculation (dashed line), and a time-dependent Hartree-Fock (TDHF) prediction (dotted line). The open circles are data taken from ref. [196]. Figure reprinted with permission from ref. [193]. c 2014, The American Physical Society.

REACTIONS WITH ^{11}C

Elastic scattering of ^{11}C

In the past decade, the only experiment performed with a ^{11}C beam was elastic scattering of ^{11}C on ^{nat}Pb, at the Heavy Ion Research Facility in Lanzhou, China [190]. This measurement was reported together with the one for the ^{10}C + ^{nat}Pb system discussed above, and the motivation was the same: to search for a Coulomb-rainbow diffraction pattern in the elastic scattering angular distribution. The ^{11}C beam was 226 and 256MeV with an intensity similar to that for ^{10}C, i.e., 3 × 10^4 pps. The observed angular distributions do not show any suppression of the Coulombrainbow diffraction pattern.

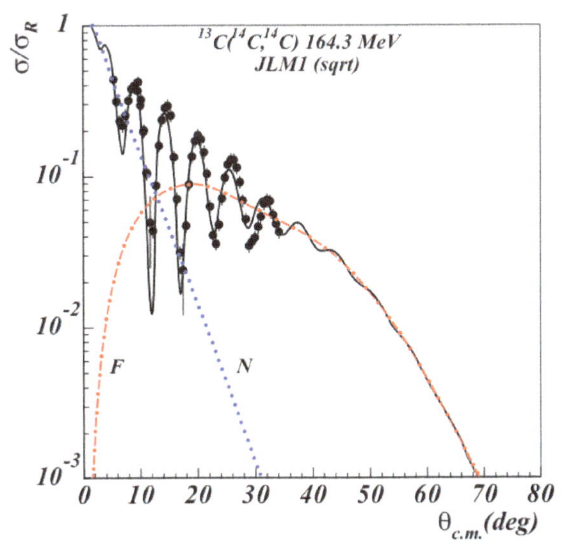

Figure 36: Angular distribution for the elastic-scattering cross section ratio to Rutherford for ^{14}C on ^{13}C. The symbols are the data and the curves are the results of an OM calculation decomposed into near-side (dotted line) and farside (dot-dashed line) contributions, as well as their coherent sum (solid black). Figure reprinted with permission from ref. [201]. c 2014, The American Physical Society.

REACTIONS WITH ^{14}C

Although radioactive, the lifetime of ^{14}C is quite long (5730y). Thus, ^{14}C beams can be produced with similar quality and intensity as stable ^{12}C. Because of this, several experiments for elastic, inelastic, and transfer reactions using ^{14}C beams have been performed, and these were reviewed in ref. [1]. Most of these experiments were performed to investigate the role of one- and two-neutron stripping. Also, some reactions induced by low-energy ^{14}C beam have been used to populate highly-excited levels in some specific neutron-rich nuclei, such as ^{139}Ce populated by the $^{130}Te(^{14}C, 5n)$ [197] reaction, 21F populated by the $^{9}Be(^{14}C, np)$ reaction [198], ^{22}F populated by the $^{9}Be(^{14}C, p)$ reaction [199] and ^{58}Fe populated by the $^{48}Ca(^{14}C, 4n)$ reaction [200]. The final nuclei were studied using γ-ray techniques. The current review is limited to studies of elastic scattering, transfer, breakup and fusion in order to investigate nuclear reaction mechanisms, so these works related to nuclear-structure investigation will not be further discussed.

Elastic scattering of ^{14}C

Elastic scattering of ^{14}C on a ^{13}C target was measured at the Texas A&M Cyclotron Institute (TAMU-CI) [201]. A 164MeV ^{14}C beam obtained from the K500 superconducting cyclotron impinged onto a 104 μg/cm^2 ^{13}C target.

The main purpose of this measurement was to obtain the optical-model parameters for a later DWBA analysis of a transfer reaction measured in the same experiment. The data and the results of the OM analysis with a doublefolding potential are illustrated in fig. 36, where the good quality of the fit can be seen.

Transfer reactions with ^{14}C

Transfer reactions have proven to be a powerful tool in the investigation of nuclear spectroscopy that is important for astrophysical reactions. The indirect method of using transfer reactions to obtain spectroscopic factors (SF) and/or asymptotic normalization coefficients (ANC) was recently applied with ^{14}C beams and these experiments are reviewed below.

^{14}C + $^{6,7}Li$

Two α-particle transfer reactions induced by a ^{14}C beam ($^{6}Li(^{14}C$, d) and $^{7}Li(^{14}C$,t)) were measured at the John D. Fox Superconducting Accelerator Laboratory at Florida State University [202]. Beams of 8.8MeV and 11.5MeV ^{14}C impinged onto ^{6}Li and ^{7}Li targets, respectively. Five silicon ΔE-E telescopes were used to detect the light recoils (deuterons/tritons) produced by the α-transfer reactions. At these sub-Coulomb energies the reaction is assumed to be peripheral. In this case the dependence of the DWBA cross section on the parameters of the opticalmodel potentials is minimized and the ANCs of the 1− (6.198MeV) and 3− (6.404MeV) excited states in ^{18}O could be extracted. These ANCs were then used to infer information concerning the $^{14}C(\alpha,\gamma)$18O capture reaction. The combined final value of the ANC^2 for the state at 6.198MeV in ^{18}O, from both (^{6}Li, d) and (^{7}Li,t) measurements, was 2.8 ± 0.7 (fm^{-1}). The 3^- state is unbound by 177keV but it was treated as a bound state with small binding energy in the analysis. A width Γ_α = (7.8±2.7)×10^{-14} eV was determined from the

^{7}Li(^{14}C,t) reaction. Both of the above parameters were later used to determine the ^{14}C(α,γ)^{18}O reaction rate, improving the uncertainty by five orders of magnitude in the range of interest for core helium flashes in low-mass stars.

$^{14}C + ^{10}Be$

The ^{10}Be(^{14}C, ^{10}Be*)^{14}C and ^{10}Be(^{14}C, ^{12}Be*)^{12}C transfer reactions were used to populate ^{10}Be and ^{12}Be to energies above the α + ^{6}He, α + ^{8}He, and ^{6}He + ^{6}He thresholds [203]. The motivation of this experiment was to investigate the ^{6}He + ^{6}He and ^{4}He + ^{8}He decays of the molecular rotational band in ^{12}Be. An 88.5MeV ^{14}C beam was produced by the superconducting LINAC facility at Florida State University. The He-like decay fragments were detected in coincidence in an array of two (50mm × 50mm) silicon double-sided strip detector (DSSD) telescopes. The excitation energy spectrum for ^{10}Be was reconstructed from the α and ^{6}He relative energy in the ^{10}Be(1^{4}C,α- ^{6}He)^{14}C reaction. Many peaks were observed corresponding to excited states in ^{10}Be. As for the ^{10}Be(^{14}C, ^{6}He-^{6}He)^{12}C and ^{10}Be(^{14}C,α- ^{8}He)^{12}C reactions, no evidence for fragments from the breakup of ^{12}Be* was observed. However, upper limits for the integrated cross sections of 50 and 300nb were assigned, for these transfer reactions, respectively. These values are dependent on the angular distributions used in Monte Carlo simulations of the detection efficiency.

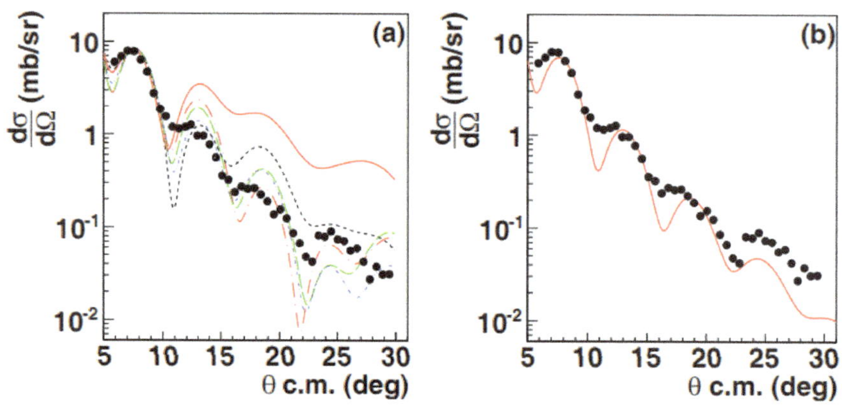

Figure 37: Angular distribution for the ^{13}C(^{14}C, $^{15}C_{gs}$) ^{12}C neutron-transfer reaction to the ground state. The curves are the results of calculations using (a) different WS potentials and (b) a double-folding potential. See ref. [201] for more details. Figure reprinted with permission from ref. [201]. c 2014, The American Physical Society.

$^{14}C + ^{13}C$

The ^{13}C(^{14}C, ^{15}C) transfer reaction was used to obtain spectroscopic factors (SF) for the $\langle ^{15}C_{gs}|^{14}C_{gs}+\mathrm{n}\rangle$ and $\langle ^{15}C^{*}(0.740)|^{14}C_{gs}+\mathrm{n}\rangle$ bound systems [201]. These SFs were then used to infer information concerning the direct, non-resonant 14C(n,γ)$^{15}C_{gs}$ and ^{14}C(n,γ)^{15}C*(0.740) capture reactions with the phenomenological potential method [76] and the code RADCAP [204]. This measurement was performed using a 12MeV/nucleon ^{14}C beam from the K500 superconducting cyclotron at TAMU-CI. The measured angular distribution for the ^{13}C(^{14}C, ^{15}Cgs) transfer reaction can be seen in fig. 37, together with the results of DWBA calculations using different potentials.

The SFs for both the ground and first excited states were obtained by normalizing the calculated differential cross section to the corresponding experimental cross sections at the main stripping peak of the angular distributions. As one can see in this figure, all potentials give good results for the ground state in the region of the main peak (5° to 11°). The beam energy for this experiment was chosen to ensure that the neutron transfer reaction was peripheral. The ANC can thus be extracted and used to normalize the corresponding low-energy capture reaction, which is also a peripheral process governed by the asymptotic tail of the radial overlap function, even for a neutroncapture reaction. In this work, the ANCs for the ground state and first excited state in ^{15}C were evaluated from the SFs obtained from the transfer reaction using a method proposed in ref. [205], according to which the ANCs of the overlap functions are included in the transfer analysis. The obtained values were $ANC^2_{2s1/2} = 2.09 \pm 0.29\ (\mathrm{fm}^{-1})$ and $ANC^2_{1d5/2} = 4.48 \pm 0.58\ (\mathrm{fm}^{-1})$ from ^{13}C(^{14}C, ^{15}C). They also obtained $ANC^2_{2s1/2} = 1.77 \pm 0.21\ (\mathrm{fm}^{-1})$ and $ANC^2_{1d5/2} = 4.08 \pm 0.49\ (\mathrm{fm}^{-1})$ using the d(^{14}C, p)^{15}C reaction. As a final conclusion, the authors claimed that the applied method gave a higher SF value for the first excited state of ^{15}C (E_x = 0.740MeV) which is not consistent with the expected value (less than 1) for a strong single-particle state, and from a mirror-symmetry comparison with the SF for ^{15}F.

Fusion of ^{14}C

Since ^{14}C is the core of the neutron-halo nucleus ^{15}C, fusion measurements induced by this nucleus have also been reported in connection with ^{15}C experiments. These are reviewed in the ^{15}C section below.

$^{14}C + ^{12}C$

Fusion cross sections for ^{14}C + ^{12}C at near-Coulombbarrier energies were reported in ref. [193]. See the section on 10C fusion above.

$^{14}C + ^{208}Pb$

An excitation function for fusion of ^{14}C + ^{208}Pb was obtained in the 1980s [206] at energies close to the Coulomb barrier.

This dataset was recently re-analyzed by Sagaidak et al. [207]. In this new analysis, nuclear potentials which could describe both sub-barrier fusion and heavy-cluster (^{14}C) decay were investigated. A series of nuclear potentials which could be used in barrier-penetration models to obtain the ^{14}C decay probabilities were found.

$^{14}C + ^{232}Th$

Fusion cross section measurements at energies in the vicinity of the Coulomb barrier are reported for the $^{14}C+^{232}Th$ system in ref. [208]. ^{14}C is a spherical system and fusion induced by this nucleus was used as a reference for any unusual characteristics of ^{15}C-induced fusion. ^{14}C was delivered as a primary beam by the ATLAS accelerator at Argonne National Laboratory. The fusion-fission fragments were detected using four $5 \times 5 cm^2$ Si surface barrier detectors, each divided into four quadrants. The experimental fusion-fission cross sections were compared with a coupledchannel calculation. Only the inelastic channels were considered, using deformed Woods-Saxon potentials to describe the static quadrupole and octupole of the ^{232}Th target. Good agreement was obtained. The transfer channel was later considered, in the analysis of Sargsyan et al. [209], without changing the previous conclusions.

REACTIONS WITH ^{15}C

The ^{15}C nucleus is a good candidate for a one-neutron halo system. This nucleus can be described as an $s_{1/2}$ valence neutron coupled to a ^{14}C core with a low separation energy of $S_n = 1.218MeV$. The narrow measured momentum distribution for the one-neutron removal reaction [210] and the observed large interaction radius as compared to $^{14,16}C$ isotopes [211], corroborate this conclusion. The low binding energy of the valence neutron in ^{15}C makes it a good candidate to investigate possible effects of the nucleon-transfer mechanism on elastic scattering and fusion.

Elastic scattering of ^{15}C

$^{15}C + ^{208}Pb$

Although it is potentially very interesting, there are no data related to elastic scattering of ^{15}C on medium to heavy nuclei at energies around the Coulomb barrier. However, coupled-channels calculations for elastic scattering of $^{15}C + ^{208}Pb$ system at 54MeV have been performed by Keeley and Alamanos [212]. They showed that the extended wave function of the valence neutron in ^{15}C leads to a large transfer probability at large radii, larger than for more conventional nuclei. The inelastic channels and (^{15}C, ^{14}C) transfer ($Q = +2.70MeV$) were included, but not (^{15}C, ^{16}C) ($Q = -3.10MeV$) transfer. It was also suggested that breakup couplings might be as important as single-neutron transfer. Data on ^{15}C elastic scattering would be very welcome to check these possibilities.

Fusion of ^{15}C

$^{15}C + ^{12}C$

Fusion cross sections for $^{15}C + ^{12}C$ at near-Coulombbarrier energies were reported in ref. [193]. See the section on ^{10}C fusion above.

$^{15}C + ^{208}Pb$ and $^{15}C + ^{232}Th$

The question of whether fusion with a weakly-bound projectile is enhanced or suppressed has been addressed by Keeley and Alamanos [212], who extended the previously mentioned coupled-channel calculations for elastic scattering to the near-barrier fusion regime for $^{15}C+^{208}Pb$. They found a significant hindrance of the total fusion cross section for energies above the barrier due to single-neutron stripping couplings linked to the dominant $2s_{1/2}$ structure of the ^{15}C ground state.

Fusion data for the $^{15}C + ^{232}Th$ system were reported in ref. [208]. Fission fragments from the decay of the compound nucleus ^{247}Cm were detected in coincidence using four Si surface barrier detectors surrounding a ^{232}Th target. The ^{15}C beam was produced by the in-flight technique with the d(^{14}C, ^{15}C)p reaction. A fusion excitation function from 54 to 70MeV was obtained. The data are shown in fig. 38, together with those for the $^{13,14}C+^{232}Th$ systems and the $^{12}C + ^{232}Th$ system from ref. [213]. The fusion-fission cross section for $^{15}C+^{232}Th$ is enhanced by a factor 5 in comparison with those for the other carbon isotopes at the lowest energies. The results of some coupledchannel calculations using a deformed Woods-Saxon potential with static quadrupole and hexadecapole deformations are also shown. The solid line in this figure corresponds to a coupled-channels calculation for ^{14}C. To describe the data for ^{15}C, the authors assumed a spectator model for the valence neutron, whereby ^{15}C was assumed to have the same excitation-energy spectrum and ion-ion potential as ^{14}C. The result for ^{15}C, therefore, overlaps with that for ^{14}C and is not shown in the figure. The calculation does not reproduce the data at sub-barrier energies, indicating that additional effects are responsible for the strong enhancement. This problem in the calculation was addressed by Sargsyan et al. [209], who investigated the possible influence of neutron transfer in sub-barrier fusion. Their calculation is based in the quantum diffusion approach, which takes into account the fluctuation and dissipation effects in collisions of heavy ions. Low-lying collective modes are included through deformation of the potentials. Transfer reactions were also taken into account by changing the deformations of the interacting nuclei. The agreement between theory and experiment was improved for the $^{12,13,14}C + ^{232}Th$ system but the data for $^{15}C+^{232}Th$ were still underestimated at sub-barrier energies. The authors argued that the role of the breakup channel, which was not included in these calculations, might be much more important than that of transfer. Thus, the $^{15}C + ^{232}Th$ system deserves further investigation.

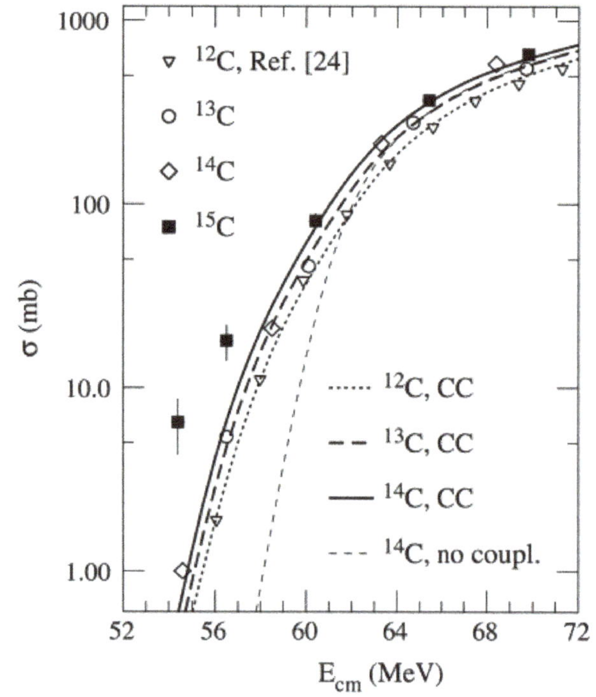

Figure 38: Excitation function of the fusion-fission reactions for $^{12,13,14,15}C+^{232}Th$. The symbols correspond to the data for the isotope indicated. The curves are the result of coupled-channels calculations as indicated. Figure reprinted with permission from ref. [208]. c 2011, The American Physical Society.

REACTIONS WITH ^{12}N

^{12}N decays to ^{12}C by β^+ emission with a half-life of 11ms. It has a low proton separation energy (0.6MeV) and some evidence exists showing that it may have a one-proton halo structure [214,215]. There is no reported experiment with ^{12}N projectiles at energies lower than 10MeV/A, but a measurement taken at an energy slightly above this value is described below.

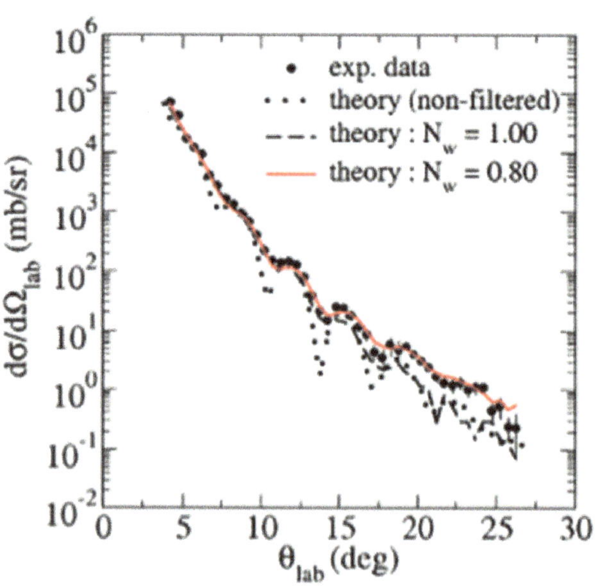

Figure 39: Elastic-scattering angular distribution for ^{12}N on ^{14}N and ^{12}C from a $C_3H_6N_6$ target, at E_{lab} = 137.6 MeV. Figure reprinted with permission from ref. [216]. c 2009, The American Physical Society.

An elastic-scattering angular distribution of ^{12}N on a melamine target, $C_3H_6N_6$, was measured in ref. [216]. The projectile energy at the center of the target was 137.6MeV and the scatterings from ^{12}C and ^{14}N could not be distinguished from each other. The inclusive angular distribution is shown in fig. 39, where the contribution of each target is related to the stoichiometry of carbon and nitrogen in melamine (1:2), as well as to the respective cross sections. The dashed curve corresponds to a calculation using double-folding potentials that had been proved to work well for elastic scattering of radioactive nuclei such as ^7Be, ^8B, ^{11}C, ^{13}N, and ^{17}F. The discrepancy with experiment at large angles indicates that a less absorptive potential is needed to describe the ^{12}N data. Changing the renormalization coefficient for the imaginary part of the potential from 1.0 to 0.8 was sufficient to obtain a good description of the data, as shown with the solid line.

A separate experiment for elastic scattering of ^{12}N from a ^{12}C target was also reported in ref. [216], but the much thicker target prevented observation of the Fraunhofer oscillations in the respective angular distribution. An angular distribution for the one-proton transfer reaction $^{14}N(^{12}N, ^{13}O)^{13}C$ also was reported in ref. [216]. Carrying out appropriate DWBA calculations and using the ANC (Asymptotic Normalization Coefficient) method, the authors were able to deduce the astrophysical S-factor for the $^{12}N(p,\gamma)^{13}O$ proton capture reaction.

REACTIONS WITH ^{13}N

The ^{13}N nucleus is unstable, decaying by β^+ emission with a half-life of 9.97m. It is also weakly bound, with a pro ton separation energy of 1.9MeV. Reactions of this projectile with targets of ^{11}B, $^{12,13}C$ and ^{14}N were reviewed in ref. [1]. Additional measurements with a ^9Be target will be briefly described here.

Evaporation residues from the ^{13}N + ^9Be fusion reaction have been measured in ref. [217] at E_{lab} = 45MeV. The residue distribution was well described by statisticalmodel calculations using the code CASCADE and the extracted fusion cross sections were similar to the ones for the ^{10}B + ^{12}C system at the same energy in the centerof-mass reference frame. Both systems correspond to the same compound nucleus ^{22}Na, so no apparent influence on fusion of the weakly bound nature of the reactants in the former system was found.

REACTIONS WITH ^{17}F

The proton drip line nucleus ^{17}F decays into ^{17}O + β^+ with a half-life of 64.49s. It has a low breakup threshold (0.6MeV) into ^{16}O + p and its first excited state is a proton-halo state [218] which is bound by only 105keV. Some reactions of ^{17}F with targets of ^{12}C, ^{14}N and ^{208}Pb were reviewed in ref. [1]. Additional measurements as well as some theoretical analyses will be described below. It can be anticipated that, given the exotic nature of ^{17}F, a lower reactivity than expected is generally observed.

Elastic scattering of ^{17}F

^{17}F + ^{12}C

An angular distribution for quasielastic scattering of ^{17}F+ ^{12}C at 60MeV was reported in ref. [219]. The experiment was performed at the CRIB facility of the CNS, University of

Tokyo, where a 101MeV ^{16}O primary beam was used to produce about 4 × 10^{5} ^{17}F/s, through the d(^{16}O, ^{17}F)n reaction. Six ΔE-E telescopes were used to detect the scattered particles, covering an angular range of 5°–80°. Energy resolution was not enough to separate inelastic particles populating the 495keV state in ^{17}F, but corresponding CDCC calculations indicated that their contribution is negligible. Optical-model calculations using a potential extrapolated from those fitting ^{16}O + ^{12}C elastic data at several energies produced a reasonable global description of the angular distribution, although not all maxima and minima in the data could be quantitatively reproduced.

The effect of coupling to the first 2^{+} state in ^{12}C also was investigated, but this gave similar quality fits to the data as the single channel calculation. The calculated total reaction cross sections with and without the inclusion of the first 2^{+} state in ^{12}C were 1417.6mb and 1388.5mb, respectively. CDCC calculations where continuum states of ^{16}O+p with relative energies going up to 12MeV were discretized indicated that bu effects on the elastic channel are small.

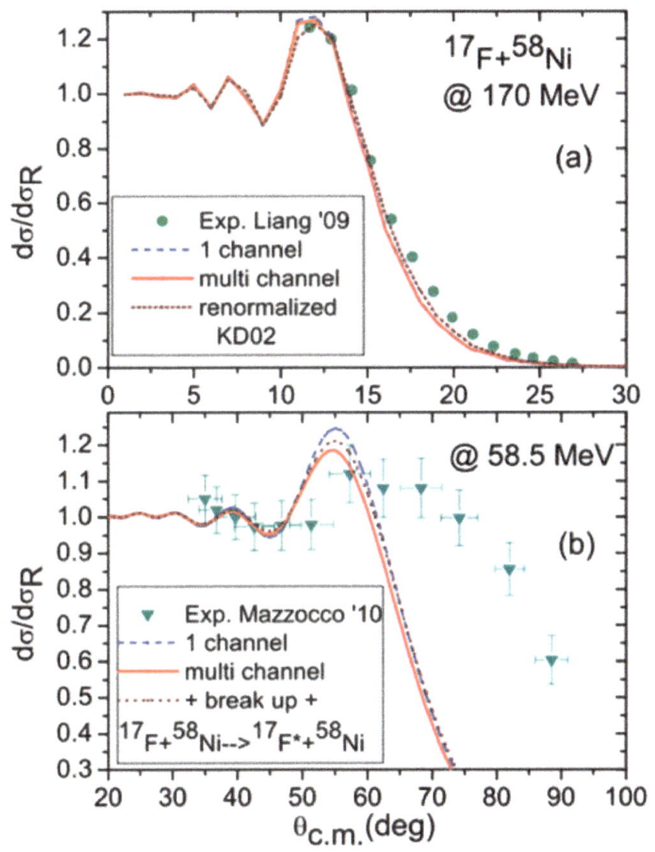

Figure 40: Quasielastic-scattering angular distributions for ^{17}F+ ^{58}Ni at 170 MeV [220] and 58.5 MeV [222], and respective predictions from the MCDCC model [226]. Figure reprinted with permission from ref. [226]. c 2014, The American Physical Society.

^{17}F + ^{58}Ni

An angular distribution for quasielastic scattering of ^{17}F+ ^{58}Ni at 10MeV/A was reported in ref. [220]. A ^{17}F beam of ~ 8 × 10^{6} pps was produced by the ^{16}O(d, n) reaction at the HRIBF facility (Oak Ridge, TN, USA), bombarding a 1.0mg/cm^{2} ^{58}Ni target. Energy resolution prevented separation of several quasielastic processes, including inelastic scattering populating the first excited state of projectile and target and neutron pickup. A silicon surface barrier detector at 10° served to monitor the beam and a three stage telescope using 5×5cm^{2} silicon strip detectors (64 μm, 278 μm, 295 μm) was used to detect the reaction products in the angular range 8°–22°.

The cross sections were first normalized to the 10° monitor assuming Rutherford scattering, but comparison with optical-model predictions using two different potentials (one of them the S˜ao Paulo potential [33]) lead the authors to notice a deficiency in this normalization and introduce a renormalization factor of 1.2. The results are shown in fig. 40(a) with solid circles, along with theoretical predictions that will be described later at the end of the present section. Other reactions measured in ref. [220] will also be discussed later: quasielastic scattering of ^{17}F + ^{208}Pb (sect. 17.1.3), and breakup of ^{17}F on ^{58}Ni and ^{208}Pb (sect. 17.2).

CDCC calculations for this system, assuming a simple p + ^{16}O configuration for the projectile, were reported in ref. [221]. The maximum excitation energy and the number of discrete bins in the continuum were dictated by convergence of the calculated observables. By using three different combinations of core-target and p-target optical potentials, very good agreement with the data was found.

Quasielastic-scattering angular distributions for ^{17}F + ^{58}Ni at two near-barrier energies were reported in ref. [222]. A preliminary account of the same experiment was also given in ref. [223]. The facility EXOTIC at the Laboratori Nazionali di Legnaro, Italy, was used to produce ^{17}F secondary beams through the p(^{17}O, ^{17}F)n reaction, with energies of 54.1 ± 1.1 and 58.5 ± 1.0MeV. The ^{17}F beams (~ 10^{5} pps), with respective purities of 93% and 96%, impinged onto a 1.0mg/cm^{2} ^{58}Ni target, yielding energies at the center of the target of 52.0 and 56.5MeV, respectively. An array of eight ΔE-E silicon telescopes (EXODET), covering ~ 70% of the full solid angle, was used to detect the resulting charged particles. Some inelastic-scattering events, as well as transfer processes and the ^{16}O + p breakup channel, could not be distinguished from the elastic channel. The measured angular distribution corresponding to the highest energy is displayed in fig. 40(b). The meaning of the curves in this figure will be described at the end of the present section.

An optical-model potential of Woods-Saxon shape with $V_0 = 2W_0 = 52.24$MeV, $r_0 = r_i = 1.18$fm, and $a_0 = a_i = 0.63$fm, was used in ref. [222] as a starting point to fit the data. This potential follows a wellknown standard parametrization. The real and imaginary depths were varied and the most relevant direct processes were included in a coupled-channel calculation. Reaction cross sections of 510.5 ± 55.3 (559.7 ± 64.3)mb were obtained for the lower (higher) energy, respectively. The estimated contributions from direct processes, which could not be experimentally separated from the elastic data, were: 67–74mb from inelastic channels, 11–14mb from breakup, and 7–15 mb from p-stripping. It was noted, however, that the later contribution should be considered as a lower estimate because only three final states in ^{59}Cu were considered for the stripped proton, and the actual contribution could in fact be on the order of 50–75mb. A related uncertainty was added by the authors to the error bars reported for the reaction cross sections for the purpose of comparing with results for other systems.

A comparison with other systems was carried out in ref.

[222] with a commonly-used prescription for data reduction. This comparison is extended in fig. 41 to include also all systems considered in ref. [156]. One can see that the reaction cross sections for ^{17}F do not follow the systematics found in ref. [156], whereby the most weakly bound systems (^{6}He, ^{8}B) show the largest reduced cross sections, followed first by less loosely-bound nuclei and finally by well-bound ones. In fig. 41, the reduced cross sections for ^{17}F, which is quite loosely bound (with a bu threshold even lower than those for ^{6}He, ^{6,7}Li and ^{7,9}Be), appear rather close to those of well-bound ^{16}O. One difference with ^{6}He and ^{8}B is that ^{17}F does not have a halo configuration in its ground state, but this would not explain the difference observed with respect to the Li and Be isotopes. The apparently peculiar behaviour of ^{17}F could be at least partially related to an intrinsic limitation of the existing data reduction procedures when comparing total reaction cross sections [116]. This would possibly explain also the fact that different conclusions can be drawn from a comparison of some of the systems of fig. 41, but using an alternate data reduction procedure [224,225].

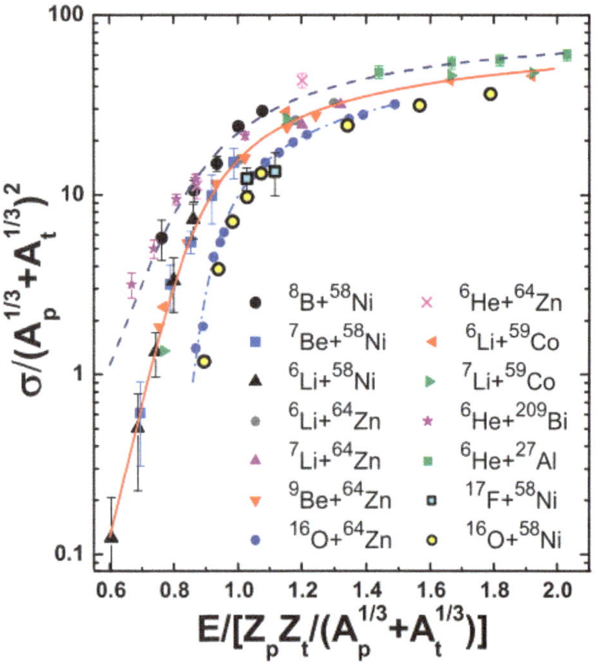

Figure 41: Reduced reaction cross sections for several weakly bound projectiles, compared with the well bound ^{16}O projectile. For the original data sources see refs. [156,222].

Theoretical calculations for the two datasets described above (from refs. [220,222]) were carried out in ref. [226] within a microscopic CDCC model (MCDCC) [227]. This model is based on a nucleon-target optical potential, which is folded with the projectile density to obtain the projectile-target interaction. The ^{17}F nucleus is assumed to have a ^{16}O + p configuration and an effective nucleonnucleon (NN) interaction is used to build the respective internal many-body Hamiltonian, which is in turn used to obtain the projectile density. The target is structureless in this model. A total Hamiltonian for the projectiletarget system was then given, which depends only on the assumed nucleon-target optical potential and the NN interaction, leading to a set of coupled-channel equations involving all projectile states. The CDCC method is then applied to solve these equations, by discretizing the projectile states in the continuum. According to ref. [227], the fact that the nucleon-target optical-model potentials are generally well known should confer a good predictive power on the model. In order to pick a proper nucleon-target optical-model potential, the performance of three different common potentials was carefully analyzed by comparing the results of single-channel calculations with elastic-scattering data for ^{17}F + ^{208}Pb at 170 and 90.4MeV, as well as with data for p and n scattering from the same target at energies near and below 10MeV. The Koning-Delaroche potential [228] was chosen.

The curves in fig. 40 show the MCDCC predictions for the ^{17}F + ^{58}Ni system at the two measured energies. In both cases, the influence of the couplings is small. At 170MeV, the predicted cross sections are in good agreement with the data but at 58.5MeV the description is not satisfactory. Inclusion of inelastic as well as breakup contributions in the calculation does not produce any substantial improvement. The authors of ref. [226] point out that the reasons for the observed discrepancy are not clear. One might notice that there was a rather large energy loss in the target in this particular case (~ 4.3MeV [222]), which could partially explain the discrepancy since apparently the calculations were done assuming a projectile at 58.5MeV.

^{17}F + ^{208}Pb

A quasielastic scattering angular distribution for ^{17}F + ^{208}Pb at 141MeV was reported in ref. [229]. The experiment was performed with the Radioactive Ion Beam Line at Lanzhou, China, using a 70MeV/A ^{20}Ne primary beam on a 3.5mm ^{9}Be primary target. The secondary target was a 2mg/cm^{2} ^{208}Pb foil. Three PPAC's and five Si detectors were used both to locate the beam position and direction and to detect the particles scattered into the angular range 6°–20°. Inelastic scattering populating the first excited state in either ^{17}F or ^{208}Pb could not be separated from the elastic-scattering yield. The data were presented only as a ln(dσ/dθ) vs. θ^2 plot and a semiclassical analysis was attempted. A comparison with data for other systems within the same semiclassical framework was also shown in ref. [230]. The authors claim that a nuclear rainbow is observed for most of the considered systems and they estimate the ratio of the nuclear rainbow angle to the grazing angle. For ^{6}He and ^{17}F projectiles, this ratio is about one half (or less) of that for ^{4}He, ^{6}Li and ^{16,17}O.

A 1.84mg/cm^{2} lead target was also used in the experiment of ref. [220] which was described at the beginning of sect. 17.1.2, so a quasielastic-scattering angular distribution for ^{17}F + ^{208}Pb at 170MeV also was obtained. An angular range of 8°–50° was covered in this case and, in contrast to the case of the ^{58}Ni target, the 10° monitor provided a proper data normalization so no renormalization was necessary. The results are presented in fig. 42 (top panel) along with the predictions of the MCDCC model [226], described in the previous section. Calculations with the S˜ao Paulo potential [33], using default parameters, provided a good description of the data [220].

Figure 42 also includes older data for the same system, not only at 170MeV (top panel) but also at 98MeV (middle panel) and 90.4MeV (bottom panel). References for these older data are given in the figure caption, and they were reviewed in ref. [1]. To get a better description of the 170MeV data within the MCDCC model, a renormalization of the real part of the Koning-Delaroche potential [228] by a factor 0.65 was required [226]. In the case of the data at 98 and 90.4MeV, good agreement was not achieved but the authors claim that similar results are obtained with conventional CDCC calculations. For the case of the 170MeV data in fig. 42, the CDCC approach of ref. [221] that was briefly described in the third paragraph of sect. 17.1.2, was also applied to the ^{17}F + ^{208}Pb system, with very good agreement.

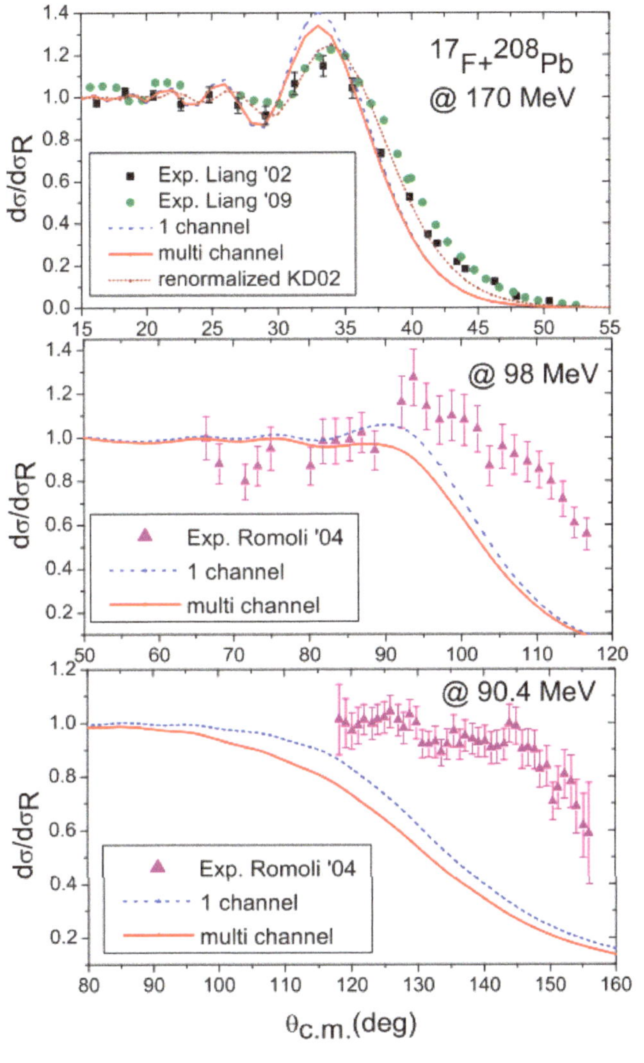

Figure 42: Quasielastic-scattering angular distributions for ^{17}F+ ^{208}Pb at 170 MeV [220,231], 98 MeV [231] and 90.4 MeV [232], and respective predictions from the MCDCC model [226]. Figure reprinted with permission from ref. [226]. c 2014, The American Physical Society.

Quasielastic-scattering data for the ^{17}F+^{208}Pb system including the first excited state of ^{17}F (but not of ^{208}Pb) were reported in ref. [233] at E_{lab} = 86 ± 1MeV. The experiment was performed at the INFN-LNL, Italy, using the EXOTIC facility and the EXODET detection system that were mentioned earlier (sect. 17.1.2) in connection with ref. [222]. The data exhibited a nearly pure Rutherford behaviour so only an upper limit of 5mb was determined for the respective reaction cross section. However, DWBA calculations indicated an inelastic cross section of 35mb to the first excited state of ^{17}F, consistent with a total reaction cross section of 40mb. Breakup data were also reported in this experiment, which will be mentioned in next section.

Breakup of ^{17}F

It was already mentioned in sect. 17.1.1 that the effects of bu couplings on the elastic scattering of ^{17}F + ^{12}C at 60MeV are estimated to be small. The CDCC calculations of ref. [219] yielded 37.5mb for the respective total bu cross section.

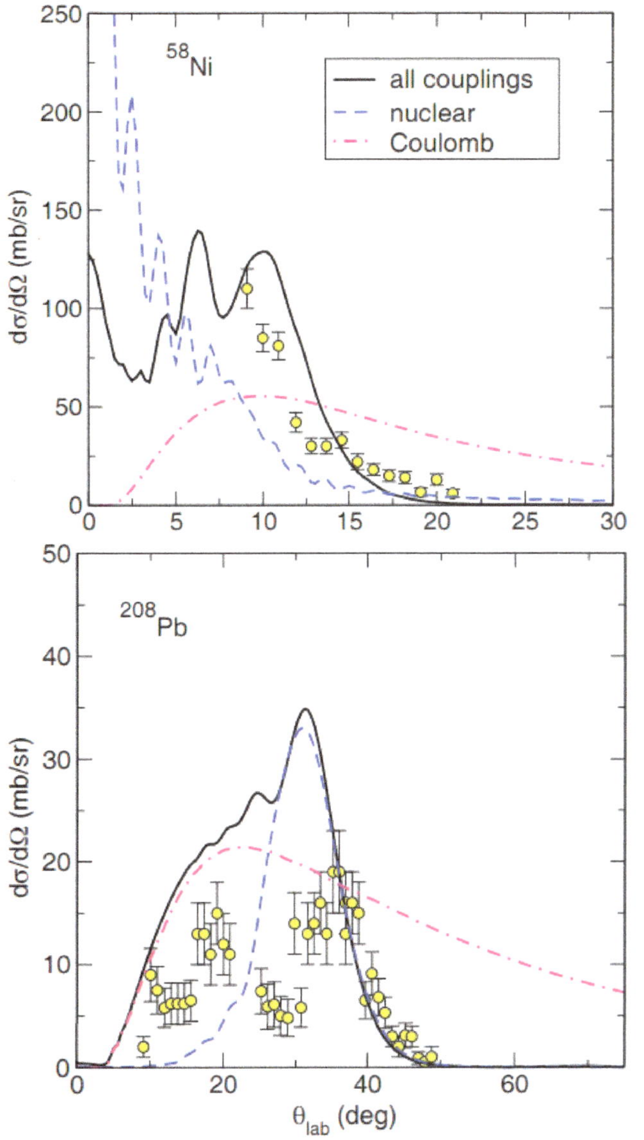

Figure 43: Exclusive breakup angular distributions for ^{17}F + (^{58}Ni, ^{208}Pb) at 170 MeV [220], and respective CDCC calculations [221]. Figure reprinted with permission from ref. [221]. c 2012, The American Physical Society.

As mentioned in sect. 17.1.2, ^{16}O resulting from ^{17}F breakup when bombarding the ^{58}Ni target at 54.1 and 58.5MeV (52.0 and 56.5MeV at the target center, respectively) [222], could not be separated from pure elastic scattering. The CDCC calculations performed in ref. [222], which gave good fits to the quasielastic-scattering data, yielded respective estimations of 11.1 ± 0.7mb and 14.2 ± 0.9mb for the breakup cross sections at the two energies.

Exclusive breakup angular distributions for ^{17}F+(^{58}Ni, ^{208}Pb) at 170MeV were measured at Oak Ridge [220] simultaneously to the respective quasielastic-scattering measurements described above (sects. 17.1.2, 17.1.3). The same experimental arrangement was used. Protons and oxygen ions were detected in coincidence, with efficiencies (estimated from Monte Carlo simulations) of 14%–19% for ^{58}Ni and 18%–40% for ^{208}Pb. The experimental results are shown by the circles in fig. 43. Corresponding bu calculations were performed by assuming an ^{16}O + p configuration for ^{17}F and following the time evolution of the respective wave function along Coulomb trajectories [220]. The agreement with the experiment was good for the ^{58}Ni target but overestimated the forward angle data for ^{17}F+^{208}Pb by a factor of two. The authors mention core excitation, which was not considered in the calculation, as a possible reason for this discrepancy. The predictions of these dynamic calculations look qualitatively similar to the solid curves in fig. 43, which will be described below.

The CDCC calculations of ref. [221], which were briefly described in the third paragraph of sect. 17.1.2, were also used to study the above breakup data for ^{17}F + (^{58}Ni, ^{208}Pb). The respective full calculations, including both nuclear and Coulomb couplings, are presented as solid curves in fig. 43, while the only-Coulomb (onlynuclear) calculations are displayed with dash-dot (dashed) curves, respectively. For ^{58}Ni, nuclear bu is small beyond the grazing angle ($\theta \sim 12°$) and the interference between both couplings is destructive at small angles and constructive around the maximum. The overall description of the data for this target is reasonably good (see the upper part of fig. 43). For ^{208}Pb, the nuclear contribution presents a maximum around the respective grazing angle ($\theta \sim 35°$), and Coulomb bu dominates at small angles but it is strongly suppressed beyond the grazing angle. Additional CDCC calculations in ref. [221] showed that the fact that the proton is charged produces a hindrance of the breakup cross section, consistent with a dynamic polarization effect resulting from a displacement of the proton behind the core, thus screening it from the full Coulomb interaction with the target. The angular distributions of bu protons and oxygen ions were also discussed. While the calculated ^{16}O distribution is fully inside the experimental angular range, a large fraction of the protons would be out of the region covered by the detectors. This could introduce additional uncertainties into the data which, combined with the fact that core excitations were not included in the calculations, may partially explain the discrepancy observed with the bu data for the ^{208}Pb target (see the lower part of fig. 43).

The experiment on ^{17}F + ^{208}Pb that was briefly described at the end of sect. 17.1.3 [233] also reported an exclusive breakup measurement at 86MeV for this system, with a respective cross section of 0.6 ± 0.4mb. Previous data from the same group [232] were also reanalyzed, yielding a breakup (total) reaction cros section of 33 ± 16mb (74mb) for the same system at 90.4MeV. The strong energy dependence of the breakup cross section could not be explained in terms of a simple Coulomb dissociation estimation. A comparison with reaction cross sections induced by other nuclei on heavy targets showed a relatively small reactivity for ^{17}F, indicating that the loosely bound nature of this projectile is not playing an important role in the reaction dynamics.

Fusion of ^{17}F

Only one fusion measurement involving ^{17}F projectiles has been reported, in 1998, corresponding to the ^{17}F + ^{208}Pb system at near-barrier energies [234]. Even though the respective year of publication falls well beyond the general scope of the present review, it is considered that these results provide an important complement to achieve a global view of ^{17}F reactions, so they will be briefly described here.

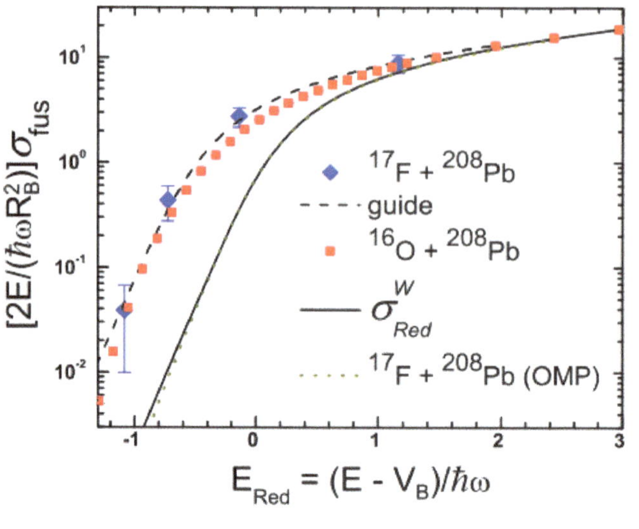

Figure 44: Reduced fusion cross sections for ^{17}F+^{208}Pb [234] and ^{16}O +^{208}Pb [235]. The solid curve is the UFF (sect. 2.2.2) and the dotted curve corresponds to an optical-model calculation using a short range absorptive potential. The dashed curve is to guide the eye. The respective barrier parameters (V_B (MeV), R_B (fm), $\hbar\omega$ (MeV)) are (76, 11.7, 4.6) in the case of ^{16}O and (85.6, 11.6, 4.8) in the case of ^{17}F.

The experiment was performed at the Argonne National Laboratory (USA), where the p(17O, 17F)n reaction was used to produce a ^{17}F beam of (1-2) × 10^5 pps which bombarded a 500 μg/cm^2 ^{208}Pb target. Fission fragments corresponding to four bombarding energies between 87 and 99MeV were measured in coincidence using four 5 × 5cm^2 Si detectors, with a 7.8% total efficiency (determined from a Monte Carlo simulation). For comparison, the ^{19}F + ^{208}Pb fusion-fission reaction at similar energies was also measured. A plot of σ_{fus}/R^2 vs. E/V_C showed very similar results for the fusion with the two projectiles ^{17,19}F, which was somehow unexpected because ^{19}F is well bound. For the lowest energy measured, the reduced cross section for ^{17}F was even lower than that of the heavier isotope. Figure 44 shows a comparison of fusion data for ^{17}F and ^{16}O projectiles with the same target ^{208}Pb, using the data reduction procedure of eq. (1). One can see that the results for both systems behave similarly, so the loosely bound nature of ^{17}F does not seem to produce any appreciable effect on the fusion process. An enhancement is observed with respect to the UFF, even for energies above the barrier. Canto et al. [31,32] analyzed both systems in this context and included the effects of coupling to low-energy inelastic excitations of the target. After renormalizing the experimental cross sections to take account of these effects (see eq. (3)), no enhancement was evident in the remaining values.

REFERENCES1.

1. N. Keeley *et al.*, Prog. Part. Nucl. Phys. 63, 396 (2009)
2. N. Keeley *et al.*, Prog. Part. Nucl. Phys. 59, 579 (2007)
3. L.F. Canto *et al.*, Phys. Rep. 596, 1 (2015)

4. M. Majer *et al.*, Eur. Phys. J. A 43, 153 (2010)
5. E. Muskat *et al.*, Nucl. Phys. A 581, 42 (1995)
6. J. Cook, K.W. Kemper, Phys. Rev. C 31, 1745 (1985)
7. K.C.C. Pires *et al.*, Phys. Rev. C 83, 064603 (2011)
8. K.C.C. Pires *et al.*, EPJ Web of Conferences 66, 03071 (2014)
9. R. Lichtenthäler *et al.*, AIP Conf. Proc. 1139, 76 (2009)
10. V. Morcelle *et al.*, Phys. Lett. B 732, 228 (2014)
11. R. Lichtenthäler *et al.*, AIP Conf. Proc. 1224, 461 (2010)
12. M. Rodríguez-Gallardo *et al.*, Phys. Rev. C 77, 064609 (2008)
13. L. Standyło *et al.*, Phys. Rev. C 87, 064603 (2013)
14. K. Rusek, Eur. Phys. J. A 41, 399 (2009)
15. M. Rodríguez-Gallardo *et al.*, Phys. Rev. C 80, 051601R (2009)
16. V.V. Parkar *et al.*, Acta Phys. Pol. 42, 761 (2011)
17. A.M. Sánchez-Benéz *et al.*, Nucl. Phys. A 803, 30 (2008)
18. L. Acosta *et al.*, Phys. Rev. C 84, 044604 (2011)
19. E.F. Aguilera *et al.*, Phys. Rev. Lett. 84, 5058 (2000)
20. J.P. Bychowski *et al.*, Phys. Lett. B 596, 26 (2004)
21. P. DeYoung *et al.*, Phys. Rev. C 71, 051601 (2005)
22. J.J. Kolata *et al.*, Phys. Rev. C 75, 031302 (2007)
23. P.N. de Faria *et al.*, Phys. Rev. C 82, 034602 (2010)
24. D. Escrig *et al.*, Nucl. Phys. A 792, 2 (2007)
25. J.P. Fernández-García *et al.*, Phys. Lett. B 693, 310 (2010)
26. V.V. Sargsyan *et al.*, Phys. Rev. C 86, 054610 (2012)
27. V.V. Sargsyan *et al.*, Eur. Phys. J. A 45, 45 (2010)
28. P.R.S. Gomes *et al.*, Phys. Rev. C 84, 014615 (2011)
29. Y. Kucuk, Nucl. Phys. A 927, 195 (2009)
30. M.S. Hussein *et al.*, Phys. Rev. C 73, 044610 (2006)
31. L.F. Canto *et al.*, J. Phys. G: Nucl. Part. Phys. 36, 015109 (2009)
32. L.F. Canto *et al.*, Nucl. Phys. A 821, 51 (2009)
33. L.C. Chamon *et al.*, Phys. Rev. C 66, 014610 (2002)
34. C.Y. Wong, Phys. Rev. Lett. 31, 766 (1973)
35. P.R.S. Gomes *et al.*, Nucl. Phys. A 834, 151 (2010)
36. Di Pietro *et al.*, Phys. Rev. C 69, 044613 (2004)
37. A.M. Moro *et al.*, Nucl. Phys. A 787, 463 (2007)
38. V. Guimarães *et al.*, Phys. Rev. Lett. 84, 1862 (2000)
39. K. Ogata *et al.*, Nucl. Phys. A 805, 447 (2008)
40. T. Matsumoto *et al.*, Phys. Rev. C 73, 051602(R) (2006)
41. V. Scuderi *et al.*, Phys. Rev. C 84, 064604 (2011)
42. A.S. Fomichev *et al.*, Phys. Part. Nucl. Lett. 9, 496 (2012)
43. M. Herman *et al.*, Nucl. Data Sheeets 108, 2655 (2007)
44. N.K. Skobelev *et al.*, Phys. Part. Nucl. Lett. 10, 248 (2013)
45. A.A. Kulko *et al.*, J. Phys. G: Nucl. Part. Phys. 34, 2297 (2007)
46. R. Wolski *et al.*, Eur. Phys. J. A 47, 111 (2011)
47. S.M. Lukyanov *et al.*, Phys. Lett. B 670, 321 (2009)
48. Yu.E. Penionzhkevich *et al.*, Phys. Rev. Lett. 96, 162701 (2006)
49. V.I. Zagrebaev, Phys. Rev. C 67, 061601 (2003)
50. A.A. Kulko *et al.*, Phys. At. Nucl. 70, 613 (2007)
51. G. Marquínez-Durán *et al.*, Acta Phys. Pol. B 43, 239 (2012)
52. G. Marquínez-Durán *et al.*, EPJ Web of Conferences 66, 03058 (2014)
53. Lemasson *et al.*, Phys. Lett. B 697, 454 (2011)
54. V.V. Parkar *et al.*, EPJ Web of Conferences 17, 16009 (2011)
55. Lemasson *et al.*, Phys. Rev. Lett. 103, 232701 (2009)
56. Lemasson *et al.*, Phys. Rev. C 82, 044617 (2010)
57. Yu.E. Penionzhkevich *et al.*, Eur. Phys. J. A 31, 185 (2007)
58. M.S. Basunia *et al.*, Phys. Rev. C 75, 015802 (2007)
59. Navin *et al.*, Phys. Rev. C 70, 044601 (2004)
60. Chatterjee *et al.*, Phys. Rev. Lett. 101, 032701 (2008)
61. X. Gu *et al.*, Phys. Lett. B 343, 31 (1995)
62. D. Howell *et al.*, Phys. Rev. C 88, 025804 (2013)
63. H.M. Xu *et al.*, Phys. Rev. Lett. 73, 2027 (1994)
64. I.J. Thompson, F.M. Nunes, Nuclear Reactions for Astrophysics: Principles, Calculations and Applications of Low-Energy Reactions (Cambridge University Press, 2009) p. 5, ISBN 978-0-521-85635-5
65. S. Mukherjee *et al.*, Eur. Phys. J. A 45, 23 (2010)
66. O. Camargo *et al.*, Phys. Rev. C. 78, 034605 (2008)
67. P.R.S. Gomes *et al.*, Phys. Rev. C 71, 017601 (2005)
68. J.M.B. Shorto *et al.*, Phys. Lett. B 678, 77 (2009)
69. Barioni *et al.*, Phys. Rev. C 80, 034617 (2009)
70. Pakou *et al.*, Eur. Phys. J. A 51, 55 (2015)
71. Pakou *et al.*, Eur. Phys. J. A 51, 90 (2015)
72. N. Keeley, Phys. Rev. C 80, 064614 (2009)
73. I.J. Thompson, Comput. Phys. Rep. 7, 167 (1988)
74. I.J. Thompson, www.fresco.org.uk
75. V. Guimarães *et al.*, Phys. Rev. C 75, 054602 (2007)
76. C.A. Bertulani, V. Guimarães, AIP Conf. Proc. 1304, 183 (2010)
77. F.D. Becchetti *et al.*, Phys. Rev. C 71, 054610 (2005)
78. E.F. Aguilera *et al.*, Phys. Rev. C 80, 044605 (2009)
79. Gavron, Phys. Rev. C 21, 230 (1980)
80. G.C. Ball *et al.*, J. Phys. G 38, 024003 (2011)
81. M. Cubero *et al.*, Phys. Rev. Lett. 109, 262701 (2012)
82. W. Loveland *et al.*, Phys. Rev. C 74, 064609 (2006)
83. A.M. Vinodkumar *et al.*, Phys. Rev. C 80, 054609 (2009)
84. V.I. Zagrebaev, V.V. Samarin, W. Greiner, Phys. Rev. C 75, 035809 (2007)
85. A.B. Balantekin, G. Kocak, AIP Conf. Proc. 1072, 289 (2008)
86. K. Hagino, N. Rowley, A.T. Kruppa, Comput. Phys. Commun. 123, 143 (1999)

87. T. Nakamura *et al.*, Phys. Rev. Lett. 96, 252502 (2006)
88. J.P. Fernàndez-García *et al.*, Phys. Rev. Lett. 110, 142701 (2013)
89. A.M. Vinodkumar *et al.*, Phys. Rev. C 87, 044603 (2013)
90. M.V. Andrés, J. Gómez-Camacho, Phys. Rev. Lett. 82, 1387 (1999)
91. M. Rodríguez-Gallardo *et al.*, Phys. Rev. C 80, 051601(R) (2009)
92. N. Keeley, K.W. Kemper, K. Rusek, Phys. Rev. C 88, 017602 (2013)
93. W.Y. So *et al.*, Phys. Rev. C 89, 057601 (2014)
94. W.Y. So *et al.*, Phys. Rev. C 90, 054615 (2014)
95. J.P. Fernández-García *et al.*, Phys. Rev. C 92, 014604 (2015)
96. A.V. Dobrovolsky *et al.*, Nucl. Phys. A 766, 1 (2006)
97. E.C. Pinilla, P. Descouvement, D. Baye, Phys. Rev. C 85, 054610 (2012)
98. K. Alder, A. Winther, Electromagnetic excitations: Theory of Coulomb Excitation with Heavy Ions (North-Holland, Amsterdam, 1975)
99. Tanihata *et al.*, Phys. Rev. Lett. 55, 2676 (1985)
100. L.F. Canto *et al.*, Phys. Rev. C 52, R2848 (1995)
101. N. Takigawa, M. Kuratani, H. Sagawa, Phys. Rev. C 47, R2470 (1993)
102. M.S. Hussein *et al.*, Phys. Rev. C 46, 377 (1992)
103. M.S. Hussein *et al.*, Phys. Rev. C 47, 2398 (1993)
104. C.H. Dasso, A. Vitturi, Phys. Rev. C 50, R12 (1994)
105. S. Verma *et al.*, Eur. Phys. J. ST 150, 75 (2007)
106. S. Verma *et al.*, Eur. Phys. J. A 44, 385 (2010)
107. A.T. Rudchik *et al.*, Eur. Phys. J. A 41, 31 (2009)
108. K.C.C. Pires, Proceedings of the 12th International Conference on Nuclear Reaction Mechanisms, Varenna, Italy, 2009, edited by F. Cerutti, A. Ferrari, Vol. 2 (CERN, 2010) p. 337, http://cdsweb.cern.ch/record/1233497
109. Barioni *et al.*, Phys. Rev. C 84, 014603 (2011)
110. J.C. Zamora *et al.*, Phys. Rev. C 84, 034611 (2011)
111. H. Amro *et al.*, Eur. Phys. J. ST 150, 1 (2007)
112. G. Tabacaru *et al.*, Phys. Rev. C 73, 025808 (2006)
113. K. Horii *et al.*, Phys. Rev. C 81, 061602(R) (2010)
114. K. Kalita *et al.*, Phys. Rev. C 73, 024609 (2006)
115. V. Morcelle *et al.*, Phys. Rev. C 89, 044611 (2014)
116. L.F. Canto *et al.*, Phys. Rev. C 92, 014626 (2015)
117. R. Lichtenthäler *et al.*, Eur. Phys. J. ST 150, 27 (2007)
118. E.F. Aguilera *et al.*, Phys. Rev. C 79, 021601(R) (2009)
119. M. Mazzocco *et al.*, Phys. Rev. C 92, 024615 (2015)
120. Gómez Camacho *et al.*, Nucl. Phys. A 833, 156 (2010)
121. Gómez Camacho, E.F. Aguilera, Phys. Rev. C 90, 064607 (2014)
122. R. Raabe *et al.*, Phys. Rev. C 74, 044606 (2006)
123. E. Martinez-Quiroz *et al.*, Phys. Rev. C 90, 014616 (2014)
124. Lepine-Szily, R. Lichtenthaler, V. Guimarães, Eur. Phys. J. A 50, 128 (2014)
125. D. Tilley *et al.*, Nucl. Phys. A 745, 155 (2004)
126. Di Pietro *et al.*, Phys. Rev. Lett. 105, 022701 (2010)
127. Di Pietro *et al.*, Phys. Rev. C 85, 054607 (2012)
128. D. Voulot *et al.*, Nucl. Instrum. Methods B 266, 4103 (2008)
129. M. Hemalatha, Eur. Phys. J. Web of Conf. 66, 03036 (2014)
130. J.N. Orce *et al.*, Phys. Rev. C 86, 041303 (2012)
131. C.E. Svensson *et al.*, J. Phys. G 31, S1663 (2005)
132. Yoshida *et al.*, Phys. Lett. B 389, 457 (1996)
133. C. Signorini *et al.*, Nucl. Phys. A 735, 329 (2004)
134. D.J. Hinde, M. Dasgupta, Phys. Rev. C 81, 064611 (2010)
135. C. Signorini *et al.*, Eur. Phys. J. A 2, 227 (1998)
136. M. Dasgupta *et al.*, Phys. Rev. C 81, 024608 (2010)
137. H.G. Bohlen *et al.*, Phys. Rev. C 75, 054604 (2007)
138. D.J. Millener *et al.*, Phys. Rev. C 28, 497 (1983)
139. E. Kwan *et al.*, Phys. Lett. B 732, 210 (2014)
140. C.E. Thorn *et al.*, Phys. Rev. Lett. 38, 384 (1977)
141. V. Scuderi *et al.*, Acta Phys. Pol. B 44, 463 (2013)
142. N. Keeley, J. Phys.: Conf. Ser. 381, 012087 (2012)
143. T. Druet, P. Descouvemont, Eur. Phys. J. A 48, 147 (2012)
144. W.Y. So *et al.*, Phys. Rev. C 92, 014627 (2015)
145. R. de Diego *et al.*, Phys. Rev. C 89, 064609 (2014)
146. N.C. Summers, F.M. Nunes, I.J. Thompson, Phys. Rev. C 74, 014606 (2006)
147. L. Acosta *et al.*, Eur. Phys. J. A 52, 461 (2009)
148. K. Rusek *et al.*, Acta Phys. Pol. B 43, 233 (2012)
149. V. Pesudo *et al.*, Acta Phys. Pol. 45, 375 (2014)
150. M. Mazzocco *et al.*, Eur. Phys. J. ST 150, 3740 (2007)
151. T. Kubo *et al.*, Nucl. Instrum. Methods B 70, 309 (1992)
152. B. Paes *et al.*, Nucl. Phys. A 890-891, 1 (2012)
153. F.M. Nunes, I.J. Thomson, Phys. Rev. C 59, 2652 (1999)
154. J. Lubian *et al.*, Phys. Rev. C 79, 064605 (2009)
155. L.F. Canto *et al.*, Phys. Rev. C 80, 047601 (2009)
156. J.J. Kolata, E.F. Aguilera, Phys. Rev. C 79, 027603 (2009)
157. N. Keeley *et al.*, Nucl. Phys. A 834, 792 (2010)
158. Y.Y. Yang *et al.*, Phys. Rev. C 87, 044613 (2013)
159. R.S. Mackintosh, D.Y. Pang, Phys. Rev. C 88, 014608 (2013)
160. R.S. Mackintosh, Phys. Rev. C 88, 054603 (2013)
161. Gómez Camacho *et al.*, Phys. Rev. C 84, 034615 (2011)
162. E.F. Aguilera *et al.*, Phys. At. Nucl. 71, 1163 (2008)
163. T.L. Belyaeva *et al.*, Phys. Rev. C 80, 064617 (2009)
164. J. Lubian *et al.*, Phys. Rev. C 78, 064615 (2008)
165. L. Lubian *et al.*, Nucl. Phys. A 834, 802 (2010)
166. E.F. Aguilera *et al.*, Phys. Rev. C 81, 011604(R) (2010)
167. E.F. Aguilera *et al.*, Phys. Rev. C 83, 021601(R) (2011)
168. E.F. Aguilera *et al.*, Phys. Rev. Lett. 107, 092701 (2011)

169. Pakou *et al.*, Phys. Rev. C 87, 014619 (2013)
170. Pakou *et al.*, Phys. Rev. C 87, 049901(E) (2013)
171. J. Rangel *et al.*, Eur. Phys. J. A 49, 57 (2013)
172. E.F. Aguilera *et al.*, Phys. Rev. C 93, 034613 (2016)
173. R. Wolski, Phys. Rev. C 88, 041603(R) (2013)
174. E.F. Aguilera, J.J. Kolata, Phys. Rev. C 85, 014603 (2012)
175. Gómez Camacho *et al.*, J. Phys. G: Nucl. Part. Phys. 40, 035103 (2013)
176. T.L. Belyaeva *et al.*, EPJ Web of Conferences 66, 03008 (2014)
177. E.F. Aguilera *et al.*, Phys. Rev. C 63, 061603(R) (2001)
178. E.F. Aguilera *et al.*, J. Phys.: Conf. Ser. 492, 012002 (2014)
179. T. Nakatsukasa *et al.*, Prog. Theor. Phys. Suppl. 154, 85 (2004)
180. M. Ito *et al.*, Nucl. Phys. A 787, 267 (2007)
181. R. Kumar *et al.*, Phys. Rev. C 89, 027601 (2014)
182. N. Curtis *et al.*, Phys. Rev. C 77, 021301 (2008)
183. N. Curtis *et al.*, Phys. Rev. C 82, 029907 (2010)
184. R.J. Charity *et al.*, Phys. Rev. C 80, 024306 (2009)
185. K. Mercurio *et al.*, Phys. Rev. C 78, 031602 (2008)
186. R.J. Charity *et al.*, Phys. Rev. C 75, 051304 (2007)
187. R.E. Tribble, R.H. Burch, C.A. Gagliardi, Nucl. Instrum. Methods Phys. Res. A 285, 441 (1989)
188. K. Markenroth *et al.*, Phys. Rev. C 62, 034308 (2000)
189. E. Casarejos *et al.*, Phys. Rev. C 73, 014319 (2006)
190. Y.Y. Yang *et al.*, Phys. Rev. C 90, 014606 (2014)
191. J.F. Liang *et al.*, Phys. Rev. C 65, 051603(R) (2002)
192. Z. Sun *et al.*, Nucl. Instrum. Methods Phys. Res. A 503, 496 (2003)
193. P.F.F. Carnelli *et al.*, Phys. Rev. Lett. 112, 192701 (2014)
194. S. Schramm, S.E. Koonin, Astrophys. J. 365, 296 (1990)
195. P.F.F. Carnelli *et al.*, Nucl. Instrum. Methods Phys. Res. A 799, 197 (2015)
196. D. Kovar *et al.*, Phys. Rev. C 20, 1305 (1979)
197. S. Kaim *et al.*, Phys. Rev. C 91, 024318 (2015)
198. J.M. Von Moss *et al.*, Phys. Rev. C 92, 034301 (2015)
199. S. Lee *et al.*, Phys. Rev. C 76, 034308 (2007)
200. D. Steppenbeck *et al.*, Phys. Rev. C 85, 044316 (2012)
201. M. McCleskey *et al.*, Phys. Rev. C 89, 044605 (2014)
202. E.D. Johnson *et al.*, Phys. Rev. C 80, 045805 (2009)
203. N. Curtis *et al.*, Phys. Rev. C 73, 057301 (2006)
204. C.A. Bertulani, Comput. Phys. Commun. 156, 123 (2003)
205. A.M. Mukhamedzhanov, F.M. Nunes, Phys. Rev. C 72, 017602 (2005)
206. Celler *et al.*, Nucl. Phys. A 432, 421 (1985)
207. R.N. Sagaidak *et al.*, Phys. Rev. C 76, 034605 (2007)
208. M. Alcorta *et al.*, Phys. Rev. Lett. 106, 172701 (2011)
209. V.V. Sargsyan *et al.*, Phys. Rev. C 86, 014602 (2012)
210. E. Sauvan *et al.*, Phys. Lett. B 491, 1 (2000)
211. D.Q. Fang *et al.*, Phys. Rev. C 69, 034613 (2004)
212. N. Keeley, N. Alamanos, Phys. Rev. C 75, 054610 (2007)
213. J.C. Mein *et al.*, Phys. Rev. C 55, R995 (1997)
214. J.-X. Li *et al.*, Chin. Phys. Lett. 27, 032501 (2010)
215. J.-X. Li *et al.*, Chin. Phys. C 34, 452 (2010)
216. Banu *et al.*, Phys. Rev. C 79, 025805 (2009)
217. Di Pietro *et al.*, Phys. At. Nucl. 69, 1366 (2006)
218. R. Morlock *et al.*, Phys. Rev. Lett. 79, 3837 (1997)
219. G.L. Zhang *et al.*, Eur. Phys. J. A 49, 57 (2013)
220. J.F. Liang *et al.*, Phys. Lett. B 681, 22 (2009)
221. Y. Kucuk, A.M. Moro, Phys. Rev. C 86, 034601 (2012)
222. M. Mazzocco *et al.*, Phys. Rev. C 82, 054604 (2010)
223. M. Mazzocco *et al.*, Nucl. Phys. A 834, 488 (2010)
224. M. Mazzocco *et al.*, J. Phys.: Conf. Ser. 312, 082032 (2011)
225. M. Mazzocco *et al.*, EPJ Web of Conferences 17, 13005 (2011)
226. J. Grineviciute, P. Descouvemont, Phys. Rev. C 90, 034616 (2014)
227. P. Descouvemont, M.S. Hussein, Phys. Rev. Lett. 111, 082701 (2013)
228. A.J. Koning, J.P. Delaroche, Nucl. Phys. A 713, 231 (2003)
229. Q. Wang *et al.*, Chin. Phys. Lett. 23, 1731 (2006)
230. Z. Bai *et al.*, Chin. Phys. Lett. 24, 3384 (2007)
231. J.F. Liang *et al.*, Phys. Rev. C 65, 051603 (2002)
232. M. Romoli *et al.*, Phys. Rev. C 69, 064614 (2004)
233. C. Signorini *et al.*, Eur. Phys. J. A 44, 63 (2010)
234. K.E. Rehm *et al.*, Phys. Rev. Lett. 81, 3341 (1998)
235. C.R. Morton *et al.*, Phys. Rev. C 60, 044608 (1999)

CHAPTER 13

PROMPT GAMMA-RAY EMISSION IN NUCLEAR FISSION

S. Oberstedt, R. Billnert, F.-J. Hambsch, M. Lebois, A. Oberstedt and J.N. Wilson

INTRODUCTION

The energy released in nuclear fission is distributed in kinetic and excitation energy of the two fragments. The excitation energy manifests itself in fragment deformation and intrinsic excitation energy, which subsequently is released by the emission of prompt neutrons and γ-rays and, at a later stage, through β-decay towards isotopes in the valley of stability. The first step of de-excitation contributes to the prompt heat released in fission, whereas the second, delayed step is counted as decay heat. Properties of the emitted particles prior to the onset of weak decays are important for nuclear applications as well as for the better understanding of the mechanism of fission-fragment deexcitation. Whilst prompt fission neutron emission is being studied during the last 50 years, see, e.g., refs. [1–6], prompt γ-ray emission was investigated essentially in the early 1970 years and only for a limited number of fissioning isotopes. From measured prompt fission γ-ray spectra (PFGS) average values for the total energy release per fission and γ-multiplicity were obtained. The principal difficulty in prompt fission γ-ray measurements is the wide-spread time distribution, which covers the region from below picoseconds up to several microseconds. The obtained spectral data are, therefore, very sensitive to the particular experiment set-up, the covered time region as well as to the energy range of the emitted γ-rays. An inherent problem of such measurements is the sufficient discrimination of prompt fission neutrons, which may induce the production of γ-rays through inelastic scattering in the detector and the surrounding materials, finally mixing with the signal from prompt fission γ-rays. In the end, those data entered in evaluated nuclear data libraries [7,8] at both thermal and 14.6MeV neutron energies, neglecting a possible dependence on the incoming neutron energy. In addition, those few spectra were scaled to isotopes, which have not yet been measured.

In the following we attempt to summarize and to discuss achievements made in the past without the claim of being complete. However, the data collected during the past and, in particular, their selective entrance into evaluated nuclear data files and nuclear model calculations is the origin of the today's high priority requests on prompt fission γ-ray data [9,10]. Next, we present recent measurement campaigns undertaken in response to the data requests published by the OECD Nuclear Energy Agency (NEA) [11]. We will end with an outlook on open questions and future data needs. Up-coming measurement campaigns and their associated instrument developments will be presented.

HISTORY OF PROMPT FISSION Γ-RAY MEASUREMENTS

The investigation of prompt fission γ-ray emission started in the 1950 years, amongst which the measurement of spectral data from the spontaneous fission of ^{252}Cf using tellurium-doped sodium-iodine scintillation detectors, NaI(Tl) [12]. First correlated measurements of prompt fission γ-ray as a function of the fragment kinetic energy were made by Milton and Fraser [13], and spectra for different fragment mass ratios were published in 1964 [14], employing a germanium detector. In this experiment the Doppler-shift due to the emission from moving fragments were detected putting a

Citation: S. Oberstedt, R. Billnert, F.-J. Hambsch, M. Lebois, A. Oberstedt, and J.N. Wilson, Prompt Gamma-Ray Emission in Nuclear Fission, DOI 10.1140/epja/i2015-15178-8.

limit on the γ emission time relative to the instance of fission. In the 1970 years a systematic study of prompt fission γ-ray (PFG) emission was conducted. There are several studies on spontaneous fission of ^{252}Cf [15] and thermalneutron-induced fission on ^{235}U [15–17], ^{233}U [17,18] and ^{239}Pu [15,18]. The energy range for γ-rays was extended down to 10keV [16] and up to 8MeV. All experiments were carefully designed to match the flight path length with the timing resolution of the NaI γ-ray detectors. Time dependence of PFG emission was investigated up to ^{275}ns after fission [18], showing that the distinction from the neutron-induced γ-ray component becomes increasingly difficult. One reason is the limited pulse-height resolution of sodium-iodine based scintillation detectors, which do not allow suppressing this component by pulseheight analysis. The data from ref. [15] for ^{252}Cf, ^{235}U and 239Pu was used for spectrum evaluation and adapted for neighboring isotopes. Due to the lack of experimental data no change of spectral shape as a function of excitation energy was considered. Very early, attempts were made to determine prompt γ-ray lifetimes and isomer decay of fission fragments [19– 24]. In 1971 Albinsson published decay curves and halflives of γ-emitting states in thermal-neutron-induced fission on ^{235}U. Half-lives down to 7.5ps were measured using a collimator, which was moved along the fission axis [25]. It was concluded that about 70% of the prompt γ-rays is emitted within 60ps. In another experiment, based on the velocity of the fragment and a variable distance between fission source and particle detector, life times in the range of 100ps and 2ns of non Dopplershifted γ-lines could be determined [26]. In a similar experiment shadow cones were used in conjunction with high-resolution X-ray and γ-ray detectors, which allowed the identification of isotopes and associate γ-rays from (nearly) stopped fission fragments [27]. In total 130 transitions in the energy range from 0.01 to 1.5MeV and emission times between 1ns and 3 μs were assigned to specific isotopes.

An interesting spectral measurement up to γ-ray energies of 19MeV was conducted by Dietrich et al. in 1974 [28], which revealed a strong change of the spectral shape at $E_\gamma \approx$ 8MeV. It took about 20 years before another measurement was conducted by van der Ploeg et al. [29,30], which confirmed this observed spectral feature. The angular distribution of prompt fission γ-rays relative to the fission axis started to be investigated in the mid 1960 years [31, 32]. Investigations were resumed in the 1970s which led to the confirmation of the previous results [33,34]. As summarized in the review article from Nifenecker [35] the angular anisotropy amounts to about 15%. Investigations of the angular distribution relative to the fission axis of $2^+ \rightarrow 4^+$ transitions in even-even nuclei ground state bands indicate that the initial fragment spin is oriented perpendicular. When looking on the integral prompt fission spectrum below 1.2MeV the anisotropy still amounts to at least 13% [33, 36]. From that it appears that those γ-rays are essentially from quadrupole transitions (E2) with a possible small admixture of dipole (E1) radiation.

The γ-ray anisotropy as a function of fission-fragment properties was investigated in thermal-neutron-induced fission on 235U only. The anisotropy is constant for total kinetic energies (TKE) between 150 and 170MeV and increases for higher TKE values. An increase of the anisotropy for TKE values smaller than 150MeV cannot be excluded (cf. fig. 7 in ref. [35], p. 127). For average fragment mass-ratios between 1.2 and 1.8 the anisotropy does not change within the uncertainties, still the average anisotropy seems to be higher in heavier fragments with respect to the lighter partner. Up to now, we only considered those experiments where essentially single detector measurements were conducted. Those allow for determining average spectral quantities, as, e.g., the average γ-ray multiplicity and average total energy release per fission. Another type of detector, based on gadolinium-loaded liquid scintillators [37], has been employed providing information about the distribution of the total prompt γ energy per fission. Those instruments are generally used when measuring prompt neutron emission but the time structure of the registered events allows separating between neutrons and prompt γ-rays. Therefore, information also about the total γ-ray energy per fission is accessible [38,39]. With the scintillator tank the dependence of prompt neutron and γ-ray emission on the excitation energy was investigated. In ref. [38] neutron multiplicity, ν_p and average total γ-ray energy, $\overline{E_\gamma}$ were investigated on ^{235}U and ^{239}Pu for the incident neutron energy region between 2 and 50eV, i.e. where the fission cross-section shows a distinct resonance structure.

In the latter fissioning system an anti-correlation between $\overline{E_\gamma}$ and ν_p was observed for 1^+ resonances. This was interpreted as a signature of the (n, γf) reaction, where a γ-ray is emitted prior to fission, leading to an increase of $\overline{E_\gamma}$ and a decrease of vp. No significant correlation was found for 0+ resonances and in the resonance-neutron-induced fission on ^{235}U. In ref. [39] ν_p and the total γ-ray energy, E_γ, were measured on ^{232}Th, ^{235}U and ^{237}Np up to incident neutron energies of 15MeV. A strong linear correlation between both average observables was found for all three fissioning compound nuclei. This correlation was already found in measurements of those quantities as a function of fission fragment mass in spontaneous fission of ^{252}Cf (cf. ref. [35] and references therein). The deduction of the average γ multiplicity and the mean energy per photon was extracted by making an assumption about the multiplicity distribution and the shape of the γ-ray spectrum. Those assumptions of ref. [39] may be subject of discussions as the knowledge of theses quantities improves.

The correlation between prompt neutron multiplicity and average total γ-ray energy was taken up by Valentine [40], who established systematic trends allowing to determine PFGS characteristics for different fissioning isotopes from measured prompt neutron multiplicities.

RENAISSANCE OF PROMPT FISSION Γ-RAY MEASUREMENTS

With the development of advanced Generation-IV nuclear reactors, the need of new PFGS data became imminent. Since four out of six contemplated Generation-IV reactors require a fast neutron spectrum, a wider range of incident neutron energies has to be considered [9, 10]. Modeling of innovative core designs shows that, despite the numerous experiment campaigns reported in the previous section, the uncertainties on the existing PFG data are such that even in the standard thermal power reactors the γ-heating can locally be under-predicted by up to 30%, whereas an accuracy within 7.5% is requested. As a consequence an urgent data request was formulated for the relevant isotopes in the fuel of thermal power reactors, ^{235}U and ^{239}Pu, and included into the high-priority data request list (HPRL) of the OECD/NEA [11]. It requires an accuracy of 7.5% on the total prompt γ-heat production for the whole range of incident neutron energies relevant for thermal and fast power reactors. As a consequence of recent instrumental advancements like the development of new γ-ray detectors [41, 42] as well as digital data acquisition systems, the determination of new and improved PFGS characteristics became possible with high precision. The recent installation of a kinematically focused neutron beam [43] enables the background-free investigation of prompt γ-ray emission in fast-neutron-induced fission on both fissile and non-fissile actinide isotopes.

RECENT ACHIEVEMENTS

During the last five years we have measured PFGS characteristics for the spontaneous fission of ^{252}Cf [44], thermal

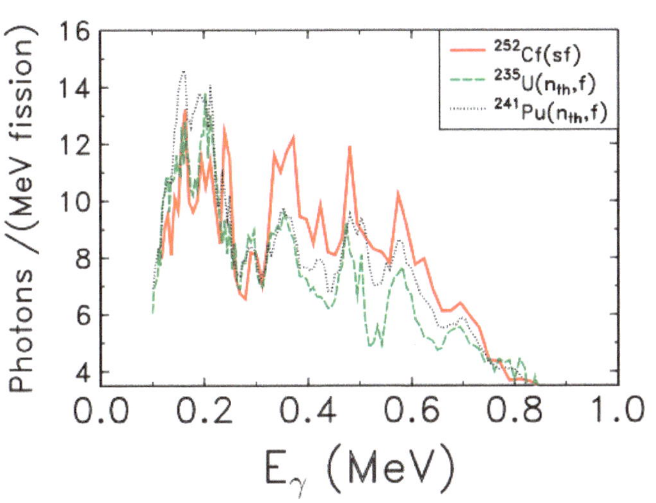

Figure. 1: (Color online) Prompt fission γ-ray spectra from ^{252}Cf(sf) [44], $^{235}U(n_{th}, f)$ [45] and $^{241}Pu(n_{th}, f)$ [46] with focus on the low-energy part of the spectrum showing a distinct structure (for details see text); all spectra were taken with cerium-doped lanthanum-bromide (LaBr3:Ce) detectors of size 5.1 cm × 5.1 cm (diameter × length of the crystal).

neutron-induced fission on ^{235}U [45] and ^{241}Pu [46] as well as fast-neutron-induced fission of $^{235,238}U$ [47,48]. All experiments were performed with different cerium-doped lanthanum-bromide (LaBr3:Ce) and cerium-bromide detectors of size 5.1cm × 5.1cm (diameter × length of the crystal). As fission triggers we used Frisch-grid ionization chambers [49, 50] and poly-crystalline diamond detectors [51]. The response of the γ-detectors was simulated with the Monte-Carlo code PENELOPE2011 [52]. From the emission spectra we determined average values for multiplicity and total energy per fission. As detailed in refs. [44–46] all detectors gave consistent results (within 1σ) with individual uncertainties on the PFGS characteristics smaller than 5%. In all cases the average total γ-ray energy per fission, $\overline{E}_{\gamma,\mathrm{tot}}$, was determined with an accuracy better than 2% (1σ). The quoted uncertainty comprises statistical (≈ 10%) and systematic (≈ 90%) uncertainties. Systematic uncertainties are from simulation of the detector response (≈ 55%), energy calibration (≈ 4%) and fitting the detector response (≈ 31%).

Prompt γ-rays were selected within a time-window of ±3ns. For all isotopes discussed up to here the average total prompt γ-ray energy turns out to be higher by a few percent compared to literature. The observed excess for ^{235}U and ^{241}Pu is 5.2% and 3.2%, respectively, relative to the values found in ENDF/B-VII.1 (see table 1).

Investigation of two adjacent time-bins, ranging from 4–7 and 8–10ns, showed that contributions from isomeric decay may account for about 5% of the prompt photons. However, at this stage of our investigations, this might be regarded as an upper limit, because not all possible components from inelastic prompt neutron scattering, at neutron energies between 6.4 and 12MeV, in the detector are definitely excluded yet in the last time bin. The low-energy part of the emission spectra are depicted in fig. 1. A striking feature is the distinct structure

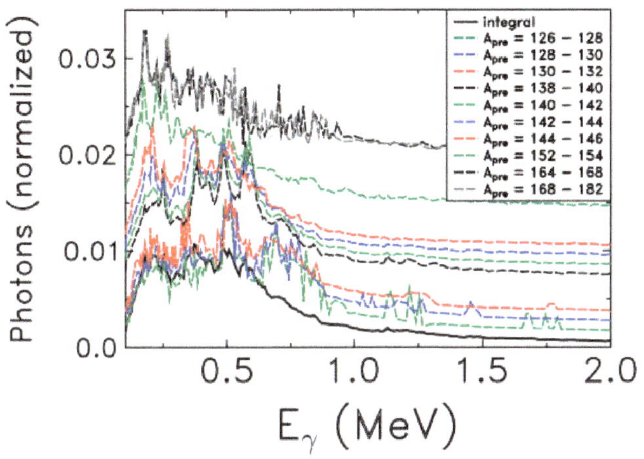

Figure. 2: (Color online) Prompt fission γ-ray spectra for ^{252}Cf (sf) [53] as a function of fragment mass, for nominal mass bins of ΔA = 2, with focus on energies below 2 MeV are shown and compared to the integral spectrum (black full line); each spectrum is normalized and offset for better visibility.

in the spectra of all isotopes, which may be observed at γ-ray energies below 0.8MeV. In a first correlation experiment on spontaneous fission of ^{252}Cf, using a twin Frisch-grid ionization chamber for fission-fragment detection, we started to look into correlations between the observed structure and fission-fragment characteristics, e.g. mass and kinetic energy. In fig. 2 prompt fission γ-ray spectra from the spontaneous fission of ^{252}Cf as a function of the fragment mass, for nominal mass bins of ΔA = 2, are shown with the focus on energies below 2MeV; each spectrum is normalized and offset for better visibility. For further details on this experiment we refer to a forthcoming publication [53]. The strong spectral shape changes with the mass-cut are evident and open up the possibility to benchmark theoretical models on prompt γ-ray emission.

Albeit the energy resolution of lanthanide-halide detectors cannot compete with that of a high-purity germanium (HPGe) detector, the correlation of γ-emission and fragment properties may enable identification and spectroscopy of very neutron-rich isotopes, which can only be produced in nuclear fission. Tagging additionally on isomeric γ-decay in coincidence with prompt γ-rays will enhance the sensitivity of those investigations. First and promising results on this topic will be subject of a forthcoming publication. With the directional fast-neutron beam LICORNE, which became recently operational at the Institut de Physique Nucl´eaire (IPN) in Orsay (France), we started first PFGS studies on ^{235}U and ^{238}U. At LICORNE neutrons are produced in inverse kinematics with 16.5MeV 7Li impinging on a hydrogen-containing material [43] or on a H2-gas cell, which is already being implemented. The generated forward-directed neutrons have an energy between 0.5 and 4MeV with a spectral shape similar to that of a typical fast-reactor spectrum. Details on the experiment will be described elsewhere [48]. As an example for

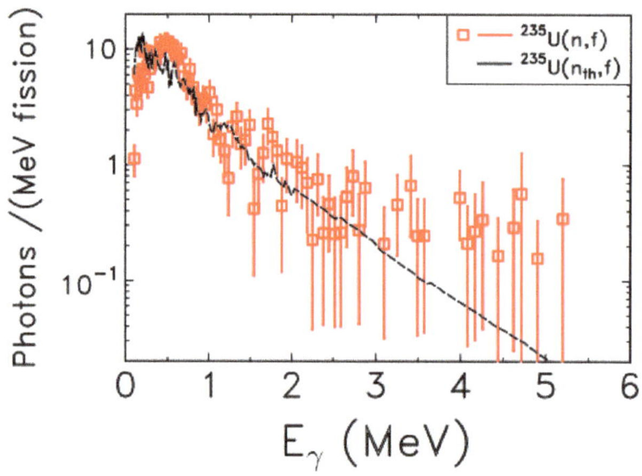

Figure. 3: (Color online) Prompt fission γ-ray spectrum (tentative) from the reaction $^{235}U(n, f)$ at an average incident neutron energy of 1.5 MeV, taken with three LaBr3:Ce detectors of size 5.1 cm × 5.1 cm (diameter × length of the crystal) [48].

Table 1: Average total γ-ray energy per fission, $\overline{E}_{\gamma,tot}$, for the target isotopes discussed in this work [41,45,46,48]; our results are compared to the corresponding data from the ENDF/BVII evaluated data file and ref. [55] together with the observed excess in % (right column). The value marked by (∗) is tentative. The value for $\overline{E}_{\gamma,tot}$ indicated by (†) has been taken up in ref. [56] after the release of ENDF/B-VII.1 and is based on results from ref. [59] and disputed in ref. [57] (see text for more details).

		Our work	Literature	
Target isotope	E_n (MeV)	$\overline{E}_{\gamma,tot}$ (MeV)	$\overline{E}_{\gamma,tot}$ (MeV)	Excess (%)
^{252}Cf	0	6.63 ± 0.08	6.70 [7]	−1.1
^{235}U	thermal	6.92 ± 0.09	6.58 [7]	5.2
	1.5	$7.4 \pm 0.7^{(*)}$	6.58 [7]	12 ± 2
			6.72 ± 0.03 [55]	10 ± 1
^{241}Pu	thermal	6.41 ± 0.06	6.19	3.6
			$7.52^{(\dagger)}$ [56]	−15

the new opportunities open to the nuclear community, a tentative emission spectrum from the reaction $^{235}U(n, f)$ at an average incident neutron energy of 1.5MeV, taken with three LaBr3:Ce detectors similar to those described above, is shown in fig. 3 and compared with the spectrum obtained from thermal-neutron-induced fission. The attenuation visible at low γ-ray energies is corrected for when calculating the spectral characteristics as described in refs. [45,54]. As summarized in table 1, the average total γ-ray energy per fission, $\overline{E}_{\gamma,tot}$ seems to increase with incident neutron energy. However, the presently still large uncertainty does not allow definitive conclusions yet. Next we followed the above-mentioned idea of Valentine [40], who parametrized the PFGS characteristics as a function of average prompt neutron multiplicity, and determined new coefficients based on our new data. Whereas

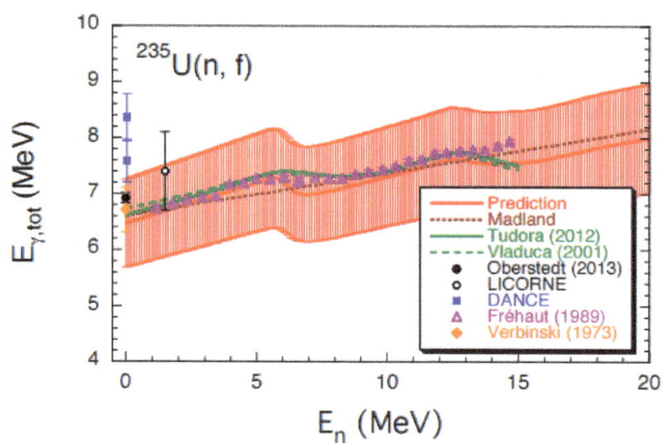

Figure. 4: (Color online) Average total γ-ray energy per fission, $\overline{E}_{\gamma,tot}$ as a function of incident neutron energy, E_n, according to the recipe reported in ref. [57].

in ref. [40] only spontaneous and thermal-neutron-induced fission was taken into account we extended this concept to higher neutron energies [57]. As depicted in fig. 4, a general increase of $\overline{E}_{\gamma,tot}$ as a function of incident neutron energy, E_n, is predicted with distinct slope changes at the thresholds for 2nd and 3rd chance fission. Our tentative experimental value for $\overline{E}_{\gamma,tot}$ from the reaction ^{235}U(n_{th} , f) is in reasonable agreement with the systematic trend.

The DANCE 4π γ-ray spectrometer [58] was recently also employed to measure PFGS data. The spectrometer is based on BaF_2 detectors, which posses fast timing capabilities, but suffer from a much worse pulse-height resolution. The spectrometer allows measuring prompt γ-ray multiplicity distributions and to higher spectral energies. PFGS from the reactions ^{235}U(n, f) and ^{239,241}Pu(n, f) were measured, the results being summarized in ref. [59]. The spectra are integrated over an incident neutron energy range up to 100keV. Details of the experiment are not subject to this paper. However, a general feature of the PFGS is the relatively high low-energy threshold and the negligence of inelastic (prompt fission) neutron scattering in the detector or surrounding structural material.

An interesting attempt was made to investigate the spectral shape as a function of incident neutron energy. With the FIGARO neutron detector array, based on liquid scintillation detectors [60], PFGS were measured for three different incident neutron energy bins, 1–2MeV, 5– 10MeV and 10–20MeV, viz. No spectral change was observed for γ-ray energies up to 4MeV [61]. It is tempting to repeat those measurements with γ-ray detectors having a better pulse-height resolution.

THE MODELING OF PROMPT FISSION Γ-RAY EMISSION

This paper puts a focus on experimental activities. Nevertheless, since there is an important mutual stimulation between theory or modeling and experiment, we would like to mention here briefly ongoing work on the modeling and theoretical description of prompt fission γ-ray emission.

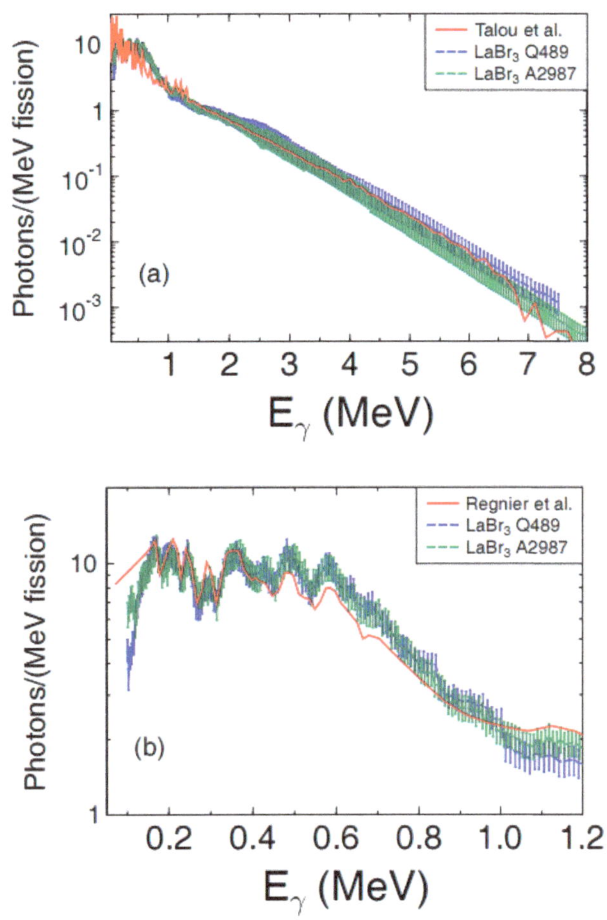

Figure. 5: (Color online) Prompt fission γ-ray spectra from the spontaneous fission of ^{252}Cf for two LaBr3:Ce detectors of different size [44, 54]; (a) Comparison of experimental data with calculations from ref. [62] and (b) with calculations from ref. [70] focussing on γ-ray energies below 1.2 MeV.

Without claiming of being complete we mention work recently performed at the Los Alamos National Laboratory (CGMF [62–65]), the Lawrence Livermore National Laboratory (FREYA [66–68]) and the CEA Cadarache (FIFRELIN [69–72]). Whereas FREYA uses analytical expressions based on Weisskopf's statistical theory to describe neutron and γ-ray emission all other models are based essentially on the Monte Carlo Hauser-Feshbach statistical theory. In the latter type of models competition between neutron and γ-ray emission is incorporated as well.

In fig. 5 simulations of prompt fission γ-ray spectra from the spontaneous fission of ^{252}Cf are shown. In the upper part a simulation performed with the CGMF code is shown, in the lower part the corresponding one obtained with FIFRELIN is shown for γ-ray energies below 1.2MeV. In the latter case the competition between prompt neutron and prompt γ-ray emission is fully accounted for. In this way experimental data, especially when correlated with fission fragment mass and total kinetic energy, may help to reveal distinct ingredients for a proper description of the de-excitation of fission fragments, e.g. the functional dependence of the level density and the γ strength function.

Apart from the models mentioned above, two other approaches are proposed. The phenomenological point-bypoint model [73, 74] is based on available experimental data and systematic behavior of model parameters. While so far no prompt fission γ-ray spectra have been published yet, at least some characteristic parameters are calculated for fast neutrons up to some tens of MeV. As an example some results are shown in fig. 4 for $^{235}U(n, f)$ up to E_n = 15MeV. The GEF code is based on a semiempirical model for a general description of the fission process, which covers most of the properties of fission fragments and the emitted neutrons and photons in a global and consistent way [75–77]. Its parameters are taken from existing experimental data, otherwise they are provided by a model calculation. Among all available output quantities, γ-ray spectra from prompt fission-induced by fast neutrons are given and their characteristics may be deduced. More detailed information about the underlying physics concepts of both approaches are found in the corresponding references.

PROBLEM SOLVED?

In table 1 we compare the average total γ-ray energy per fission, $\overline{E}_{\gamma,tot}$, for the target isotopes discussed with the corresponding values found in the ENDF/B-VII library. The most right column gives the relative excess energy in percent as obtained from the new PFGS data. For all thermal-neutron-induced reactions the new data shows a higher prompt γ-energy release of 3 to 8%. In case of ^{235}U, one of the isotopes listed in the OECD/NEA HPRL, this excess is too small to fully account for the under-prediction of the prompt γ-heat. Since the evaluation for 241Pu is based on that of ^{239}Pu we do not believe that the origin of the problem is caused from this isotope either[1].

Although the heat excess measured for fast-neutroninduced fission is about two times higher than for thermalneutron-induced fission one may not forget that the fission cross-section for fast neutrons is two orders of magnitude lower than for thermal neutrons. We suspect, therefore, that other sources in the prompt heat calculations contribute to the under-prediction. As a first candidate we assume prompt γ-rays from the fast-neutron-induced fission on ^{238}U. This data has thitherto not been measured and present evaluations could not be based on experimental data. However, the low fastneutron-induced fission cross-section might compensate the 30 times higher abundance in the nuclear fuel.

Another source may be γ-induced fission from neutroninduced capture in the construction material around the nuclear fuel. Main constituents of the material are iron and nickel, which both release a considerable fraction of highenergy γ-rays after thermal neutron capture, e.g. $E_\gamma \approx$ 9MeV from the reaction $^{58}Ni(nth,\gamma)$ contributes to about 50%. Of course, photo-fission cross-sections are relatively small reaching not far beyond 10mb above E_γ > 7MeV, but the thermal-neutron capture rate on the construction material is immense inside and around the nuclear core, and this possible contribution is not yet considered in present reactor calculations [78]. Last, but definitely not least, possible isomeric decay has to be investigated more closely. Following consequently the definition of "prompt decay processes", the integral time bin up to the onset of β-decay of fission fragments has to be fully accounted for. Based on the present knowledge about the decay properties of neutron-rich isotopes, decay times up to 1ms might have to be considered. In order to do so, experimental setups have to be modified allowing a more efficient discrimination against ambient background radiation. The contribution of prompt fission neutrons, reaching the detector typically within the first micro-second after fission, must carefully be corrected for. Up to now this contribution has been fully suppressed by means of time-of-flight or corrected for within the first 10ns as described above and in refs. [44–46].

FUTURE Γ-RAY DATA NEEDS

Based on our evaluation of present day's prompt fission γ-ray data, the problem of under-predicting the prompt heat production close to a reactor core is still persisting. In the following we attempt to formulate further data needs and to motivate a future measurement program. The strategy here should be twofold. First, future measurements should produce relevant nuclear data, which may eventually justify re-adjustments of evaluated nuclear data file entries. Second, experiments should aim at investigating more thoroughly correlations between prompt γ-ray and fission-fragment characteristics as well as with prompt neutron properties. Those measurements would help to understand how the excitation energy is shared between the two fission fragments and to improve modeling the de-excitation process in fission fragments, which is a necessary pre-requisite to benchmark nuclear theory and model calculations.

Based on the presently available measurement setup at the Joint Research Centre (JRC) IRMM, attempts should be made to enlarge the data base on PFGS characteristics from, e.g., spontaneous fission of $^{240,242}Pu$ and $^{246,248}Cm$ as input for reducing the uncertainties on systematic trends as shown in fig. 4. Also measurements with thermal beams should be continued. Here, in first instance, prompt fission γ-ray emission in the thermalneutron-induced fission on ^{239}Pu will be measured in 2015 under the CHANDA Trans National Access scheme [79] in order to respond to the remaining entry H.4 on the OECD/NEA HPRL. For those measurements an efficient fission trigger like Frisch-grid ionization chambers [49,50] or ultra-fast diamond detectors [51] should be employed. At JRC-IRMM a new γ-ray spectrometer, called GLADIS (Geel LAthanide halide Detector array for Isomer Studies), was put in operation (see fig. 6). It consists of 8 $CeBr_3$ detectors of size 5.1cm × 5.1cm (diameter × length of the crystal) and 4 high-purity coaxial germanium (HPGe) detectors with 45% relative efficiency. Initially designed for the investigation of shape-isomer population through resonance-neutron capture, it may serve also for a future PFGS measurement program. In a second phase GLADIS will be extended to 15 cerium-bromide detectors and up to 8 HPGe detectors.

Figure. 6: (Color online) Photograph of GLADIS, the new Geel LAthanide halide Detector array for Isomer Studies presently consisting of 4 HPGe detectors (45% rel. eff.) and 8 ceriumbromide detectors of size 5.1 cm × 5.1 cm (diameter × length), taken during commissioning at a neutron flight-path station at JRC-IRMM.

Measurements on fast-neutron-induced fission on ^{235}U and ^{239}Pu are part of the high-priority data request from the OECD/NEA and are possible at the novel LICORNE forward-directed neutron beam. Using 7Li and 11B beams on a hydrogen target allows investigating PFGS characteristics up to an incident neutron energy E_n = 12MeV. When, in the near future, the NFS (Neutrons for Science) [80] facility will be operational, a second neutron beam will be available to the community extending the energy range to E_n = 40MeV. Those data will serve as benchmark for present and future systematics.

At LICORNE PFGS characteristics of non-fissile isotopes relevant for the nuclear fuel cycle can be investigated. Here, of course, emphasis has to be put on the development of compact (multi-layer) fission detectors to provide a sufficiently high fission rate and to employ largearea γ-spectrometers, preferential made of lanthanide-halide (e.g., GLADIS, PARIS [81]) or BaF_2 detectors. While the first detector type provides good energy resolution and fast timing capabilities the latter are available at much larger sizes and, thus higher efficiency, at comparable timing characteristics but much worse energy resolution. Investigations of photon-induced reactions will serve twofold. Firstly, PFGS data from photo-fission of ^{232}Th, $^{234,235,238}U$ and $^{239–242}Pu$ serve as input to establish and test systematic trends and for benchmarking de-excitation models. Those targets can be made available; some of them of spectroscopic quality, to, secondly, allow investigating correlations with fission fragment characteristics. With the S-DALINAC bremsstrahlung facility [82] at the Darmstadt Technical University or the future ELINP [83–85] high-intense mono-energetic photon source at IFINN, Bucharest, excellent facilities for this part of the measurement program are or will soon be available. In order to accomplish a full understanding of the deexcitation mechanism in nuclear fission, prompt neutron- γ-fission fragment correlations should be investigated as well. At the JRC-IRMM an array of 13 neutron detectors, called SCINTIA, was recently made operational. It will be extended with lanthanide-halide detectors. For this an ionization chamber with cos θ- and φ-resolution was developed. In a first experiment prompt fission neutron and γ-ray spectra from resonance-neutron-induced fission on ^{235}U are being measured. Possible contribution from (n,γf) will be investigated. Finally, it is worthwhile to underline that a large fraction of the contemplated measurement program requires considerable beam time at accelerator-driven neutron sources or research reactors together with the associated man power for performing the experiments and the subsequent data analysis. Therefore, we recommend that at least parts of the proposed measurement program will be taken up as high priority nuclear data requests by the OECD/NEA in order to stimulate sufficient support.

One of the authors (RB) is indebted to the European Commission for providing a PhD fellowship at EC JRC-IRMM, during which part of this work was carried out. This work was also supported by the European Commission in the framework of the EFNUDAT program (Agreement No. 31027) as well as the ERINDA program (Agreement No. 269499), which is hereby gratefully acknowledged.

REFERENCES

1. J.C. Browne, F.S. Dietrich, Phys. Rev. C 10, 2545 (1974).
2. G.S. Boikov, V.D. Dmitriev, G.A. Kudyaev, Yu.B. Ostapenko, M.I. Svirin, G.N. Smirenkin, Sov. J. Nucl. Phys. 53, 392 (1991).
3. G.N. Lovchikova, A.M. Trufanov, M.I. Svirin, V.A. Vinogradov, A.V. Polyakov, Phys. At. Nucl. 67, 1246 (2004).
4. S. Lemaire, P. Talou, T. Kawano, M.B. Chadwick, D.G. Madland, Phys. Rev. C 72, 024601 (2005).
5. N. Kornilov, F.-J. Hambsch, I. Fabry, S. Oberstedt, T. Belgya, Z. Kis, L. Szentmiklosi, S. Simakov, Nucl. Sci. Eng. 165, 117 (2010).

6. A. Tudora, F.-J. Hambsch, S. Oberstedt, Nucl. Phys. A 917, 43 (2013).
7. ENDF/B-VII Evaluated Nuclear Data File, http://www.nndc.bnl.gov/exfor/endf00.jsp.
8. JEFF 3.1 Joint European Fusion File, https://www.oecdnea.org/dbdata/nds jefreports/jefreport-22/nea6807-jeff22.pdf.
9. G. Rimpault, A. Courcelle, D. Blanchet, Comment to the HPRL: ID H.3 and H.4.
10. G. Rimpault, in Proceedings of the Workshop on Nuclear Data Needs for Generation IV, April 2005, edited by P. Rullhusen (World Scientific, Singapore, 2006) p. 46.
11. Nuclear Data High Priority Request List of the NEA (Req. ID: H.3, H.4), http://www.nea.fr/html/dbdata/hprl/hprlview.pl?ID=421 and http://www.nea.fr/html/ dbdata/hprl/hprlview.pl?ID=422.
12. A.B. Smith, P.R. Fields, A.M. Friedman, Phys. Rev. 104, 669 (1956).
13. J.C.D. Milton, J.S. Fraser, Phys. Rev. 111, 877 (1958).
14. H.R. Bowman, S.G. Thompson, J.O. Rasmussen, Phys. Rev. Lett. 12, 195 (1964).
15. V.V. Verbinski, H. Weber, R.E. Sund, Phys. Rev. C 7, 1173 (1973).
16. R.W. Peelle, F.C. Maienschein, Phys. Rev. C 3, 373 (1971).
17. F. Pleasonton, R.L. Ferguson, H.W. Schmitt, Phys. Rev. C 6, 1023 (1972).
18. F. Pleasonton, R.L. Ferguson, H.W. Schmitt, Phys. Rev. A 213, 413 (1973).
19. H. Maier-Leibnitz, P. Armbruster, H.J. Specht, Phys. And Chem. of Fission 1965, Vol. II, Proc. of a Symp., Salzburg,
20. 22-26 March 1965, IAEA, Vienna, STI/PUB/101 (1965) 113.
21. F.C. Maienschein, R.W. Peele, W. Zobel, T.A. Love, nd U.N. Int. Conf. on Peaceful Uses of Atomic Energy, Geneva 1958, Vol. 15 (U.N. Geneva, 1958) 366.
22. S.A.E. Johansson, Nucl. Phys. 60, 378 (1964).
23. P. Armbruster, F. Hossfeld, H. Labus, K. Reichelt, Phys. and Chem. of Fission 1973, Vol. II, Proc. of a Symp., Vienna,
24. 28 July-1 August 1969, IAEA, Vienna, STI/PUB/ 234 (1969) 545.
25. G.V. Val'skii, D.M. Kaminker, G.A. Petrov, L.A. Popeko, Sov. At. Energy 18, 279 (1965).
26. J. Higbie, AE-374, AB Atomenergi (1969).
27. H. Albinsson, Phys. Scr. 3, 113 (1971).
28. R.C. Jared, H. Nifenecker, S.G. Thompson, Phys. And Chem. of Fission 1973, Vol. II, Proc. of a Symp., Rochester, New York, 13-17 August 1973, IAEA, Vienna,STI/PUB/347 (1974) 211.
29. R.G. Clark, L.E. Glendenin, W.L. Talbert Jr., Phys. and Chem. of Fission 1973, Vol. II, Proc. of a Symp., Rochester, New York, 13-17 August 1973, IAEA, Vienna, STI/PUB/347 (1974) 221.
30. F.S. Dietrich, J.C. Browne, W.J. O'Connell, M.J. Kay, Phys. Rev. C 10, 795 (1974).
31. H. van der Ploeg, C.R. Laurens, J.C.S. Bacelar, T. van der Berg, V.E. Iacob, J.R. Jongman, A. van der Woude, Phys. Rev. Lett. 68, 3145 (1992) 69, 1148(E) (1992).
32. H. van der Ploeg, J.C.S. Bacelar, A. Buda, C.R. Laurens, A. van der Woude, Phys. Rev. C 52, 1915 (1995).
33. M.M. Hoffman, Phys. Rev. B 133, 714 (1964).
34. O.I. Ivanov, Y.A. Kushner, G.N. Smirenkin, Zh. Ehksp. Teo. Fiz. Pis'ma 6, 898 (1967).
35. P. Armbruster, H. Labus, K. Reichelt, Z. Naturforsch. A 26, 512 (1971).
36. J.B. Wilhelmy, E. Cheifetz, R.C. Jared, S.G. Thompson, H.R. Bowman, J.O. Rasmusson, Phys. Rev. C 5, 2041 (1972).
37. H. Nifenecker, C. Signarbieux, R. Babinet, J. Poitou, Phys. and Chem. of Fission 1973, Vol. II, Proc. of a Symp., Rochester, New York, 13-17 August 1973, IAEA, Vienna, STI/PUB/347 (1974) 117.
38. K. Skarsvag, Phys. Rev. C 22, 638 (1980).
39. J. Frehaut, Nucl. Instrum. Methods 135, 511 (1976).
40. J. Frehaut, D. Shackleton, Phys. and Chem. of Fission 1973, Vol. II, Proc. of a Symp., Rochester, New York, 13- 17 August 1973, IAEA, Vienna, STI/PUB/347 (1974) 201.
41. J. Frehaut, INDC(NDS)-220 (1989) 99.
42. T.E. Valentine, Ann. Nucl. Energy 28, 191 (2001).
43. R. Billnert, S. Oberstedt, E. Andreotti, M. Hult, G. Marissens, A. Oberstedt, Nucl. Instrum. Methods A 647, 94 (2011).
44. A. Oberstedt, S. Oberstedt, R. Billnert, W. Geerts, F.-J. Hambsch, J. Karlsson, Nucl. Instrum. Methods A 668, 14 (2012).
45. M. Lebois, J.N. Wilson, P. Halipr´e, B. Leniau, I. Matea, A. Oberstedt, S. Oberstedt, D. Verney, Nucl. Instrum. Methods A 735, 145 (2014).
46. R. Billnert, F.-J. Hambsch, A. Oberstedt, S. Oberstedt, Phys. Rev. C 87, 024601 (2013).
47. A. Oberstedt, T. Belgya, R. Billnert, R. Borcea, T. Bry´s, W. Geerts, A. G¨o¨ok, F.-J. Hambsch, Z. Kis, T. Martinez, S. Oberstedt, L. Szentmiklosi, K. Tak`acs, M. Vidali, Phys. Rev. C 87, 051602 (2013).
48. S. Oberstedt, R. Billnert, T. Belgya, T. Bry´s, W. Geerts, C. Guerrero, F.-J. Hambsch, Z. Kis, A. Moens, A. Oberstedt, G. Sibbens, L. Szentmiklosi, D. Vanleeuw, M. Vidali, Phys. Rev. C 90, 024618 (2014).
49. M. Lebois, J.N. Wilson, P. Halipr´e, B. Leniau, I. Matea, A. Oberstedt, S. Oberstedt, D. Verney, Phys. Proc. 59, 37 (2014).
50. M. Lebois, J.N. Wilson, P. Halipr´e, B. Leniau, I. Matea, A. Oberstedt, S. Oberstedt, D. Verney, to be submitted.
51. A. G¨o¨ok, F.-J. Hambsch, A. Oberstedt, S. Oberstedt, Nucl. Instrum. Methods A 664, 289 (2012).
52. A. Al-Adili, F.-J. Hambsch, R. Bencardino, S. Pomp, S. Oberstedt, Sh. Zeynalov, Nucl. Instrum. Methods A 671, 103 (2012).
53. S. Oberstedt, R. Borcea, T. Bry´s, Th. Gamboni, W.

Geerts, F.-J. Hambsch, A. Oberstedt, M. Vidali, Nucl. Instrum. Methods A 714, 31 (2013).

54. http://www.oecd-nea.org/tools/abstract/detail/ nea-1525.
55. S. Oberstedt, A. Oberstedt, R. Billnert, to be submitted.
56. A. Oberstedt, R. Billnert, F.-J. Hambsch, S. Oberstedt, Phys. Rev. C, in press.
57. D.G. Madland, Nucl. Phys. A 722, 113 (2006).
58. ENDF/B-VII Evaluated Nuclear Data File, http://www.nndc.bnl.gov/exfor/endf00.jsp (viewed on 2015-06-10).
59. A. Oberstedt, R. Billnert, S. Oberstedt, to be submitted.
60. M. Heil, R. Reifarth, M.M. Fowler, R.C. Haight, F. K¨appeler, R.S. Rundberg, E.H. Seabury, J.L. Ullmann, J.B. Wilhelmy, K. Wisshak, Nucl. Instrum. Methods A 459, 229 (2001).
61. A. Chyzh, C.Y. Wu, E. Kwan, R.A. Henderson, J.M. Gostic, T.A. Bredeweg, A. Couture, R.C. Haight, A.C. HayesSterbenz, M. Jandel, H.Y. Lee, J.M. O'Donnell, J.L. Ullmann, Phys. Rev. C 87, 034620 (2013).
62. R.C. Haight, J.M. O'Donnell, L. Zanini, M. Devlin, D. Rochman, in: J. Kvasil, P. Cejnar, M. Krticka (Editors), Proc. Capture Gamma-Ray Sprectroscopy and Related Topics, Pruhonice, Prague, Czech Republic, September 2-6, 2002, Charles University, Prague, Czech Republic (World Scientific Publishing Co., 2003) 451.
63. E. Kwan, C.Y. Wu, R.C. Haight, H.Y. Lee, T.A. Bredeweg, A. Chyzh, M. Delvin, N. Fotiades, J.M. Gostic, R.A. Henderson, M. Jandel, A. Laptev, R.O. Nelson, J.M. O'Donnell, B.A. Perdue, T.N. Teddeucci, J.L. Ullmann, S.A. Wender, Nucl. Instrum. Methods A 688, 55 (2012).
64. B. Becker, P. Talou, T. Kawano, Y. Danon, I. Stetcu, Phys. Rev. C 87, 014617 (2013).
65. P. Talou, I. Stetcu, T. Kawano, Phys. Proc. 59, 83 (2014).
66. P. Talou, T. Kaweno, I. Stetcu, R. Vogt, J. Randrup, Nucl. Data Sheets 118, 227 (2014).
67. I. Stetcu, P. Talou, T. Kawano, M. Jandel, Phys. Rev. C 90, 024617 (2014).
68. R. Vogt, J. Randrup, Phys. Proc. 47, 3 (2013).
69. R. Vogt, J. Randrup, Nucl. Data Sheets 118, 220 (2014).
70. J.M. Verbeke, J. Randrup, R. Vogt, Compur. Phys. Commun. 191, 178 (2015).
71. D. Regnier, O. Litaize, O. Serot, Phys. Proc. 59, 59 (2012).
72. D. Regnier, Phys. Proc. 47, 47 (2013).
73. O. Litaize, D. Regnier, O. Serot, Phys. Proc. 59, 89 (2014).
74. O. Serot, O. Litaize, D. Regnier, Phys. Proc. 59, 132 (2014).
75. A. Tudora, F.-J. Hambsch, S. Oberstedt, C. Manailescu, Phys. Proc. 31, 43 (2012).
76. A. Tudora, F.-J. Hambsch, S. Oberstedt, G. Giubega, I. Visan, Phys. Proc. 59, 95 (2014).
77. K.-H. Schmidt, OECD/NEA, JEF/DOC 1398 (2012).
78. K.-H. Schmidt, B. Jurado, OECD/NEA, JEF/DOC 1423 (2012).
79. K.-H. Schmidt, B. Jurado, Ch. Amouroux, JEFF-Report 24, Data Bank, Nuclear-Energy Agency, OECD (2015).
80. I. Fabry, Nucl. Eng. Int., May 2014 edition and I. Fabry, private communication (2014).
81. http://www.chanda-nd.eu, and http://win.ciemat.es/chanda/.
82. X. Ledoux, M. A¨ıche, M. Avrigeanu, V. Avrigeanu, L. Audouin et al., Nucl. Data Sheets 119, 353 (2014).
83. http://paris.ijf.edu.pl/.
84. A. G¨o¨ok, M. Chernykh, C. Eckardt, J. Enders, P. von Neumann-Cosel, A. Oberstedt, S. Oberstedt, A. Richter, Nucl. Phys. 851, 1 (2011).
85. The ELI - Nuclear Physics Facility, www.eli-np.ro.
86. D.L. Balabanski, the ELI-NP Science Team, Pramana 83, 713 (2014).
87. P.G. Thirolf, L. Csige, D. Habs, M. Gunther, M. Jentschel, A. Krasznahorkay, D. Filipescu, T. Glodariu, L. Stroh, O. Tesileanu, H. Karwowski, G. Rich, EPJ Web of Conferences 38, 08001 (2012).

SECTION II: STELLAR NUCLEOSYNTHESIS, NEUTRINOS AND DARK MATTER

CHAPTER 14

STELLAR NUCLEOSYNTHESIS NUCLEAR DATA MINING

Boris Pritychenko

INTRODUCTION

In the past 100 years, astronomy, astrophysics and cosmology have evolved from the observational and theoretical fields into more experimental science, when many stellar and planetary processes are recreated in physics laboratories and extensively studied (Boyd et al., 2009). Many astrophysical phenomena have been explained using our understanding of nuclear physics processes, and the whole concept of stellar nucleosynthesis has been introduced. The importance of nuclear reactions as a source of stellar energy was recognized by Arthur Stanley Eddington as early as 1920 (Eddington, 1920). Later, nuclear mechanisms by which hydrogen is fused into helium were proposed by Hans Bethe (Bethe, 1939). However, neither of these contributions explained the origin of elements heavier than helium. Further developments helped to identify the Big Bang, stellar and explosive nucleosynthesis processes that are responsible for the currently-observed variety of elements and isotopes (Burbidge et al., 1957; Cameron, 1957; Hoyle, 1946; Merrill, 1952). Today, nuclear physics is successfully applied to explain the variety of elements and isotope abundances observed in stellar surfaces, the solar system and cosmic rays via network calculations and comparison with observed values. A comprehensive analysis of stellar energy production, metallicity and isotope abundances indicates the crucial role of proton-, neutron- and light ion-induced nuclear reactions and α-, β-decay rates. These subatomic processes govern the observables and predict the star life cycle. Calculations of the transition rates between isotopes in a network strongly rely on theoretical and experimental cross section and decay rate values at stellar temperatures. Consequently, the general availability of nuclear data is of paramount importance in stellar nucleosynthesis research. This chapter will provide a review of theoretical and experimental nuclear reaction and structure data for stellar and explosive nucleosynthesis and modern computation tools and methods. Examples of evaluated and compiled nuclear physics data will be given. Major nuclear databases and their input for nucleosynthesis calculations will be discussed.

NUCLEOSYNTHESIS AND ITS DATA NEEDS

Nucleosynthesis is an important nuclear astrophysics phenomenon that is responsible for presently observed chemical elements and isotope abundances. It started in the early Universe and presently proceeds in the stars. The Big Bang nucleosynthesis is responsible for a relatively high abundance of the lightest primordial elements in the Universe from ^{1}H to ^{7}Li, and it precedes stars formation and stellar nucleosynthesis. The general consistency between theoretically predicted and observed lightest elements abundances serves as a strong evidence for the Big Bang theory (Kolb & Turner, 1988). The currently-known variety of nuclei and element abundances is shown in Fig. 1 and Fig. 2. These Figures indicate a large variety of isotopes in Nature (Anders & Grevesse, 1989) and the strong need for additional nucleosynthesis mechanisms beyond the Big Bang theory.

Citation: Boris Pritychenko, Stellar Nucleosynthesis Nuclear Data Mining, http://arxiv.org/pdf/1110.1076.pdf.

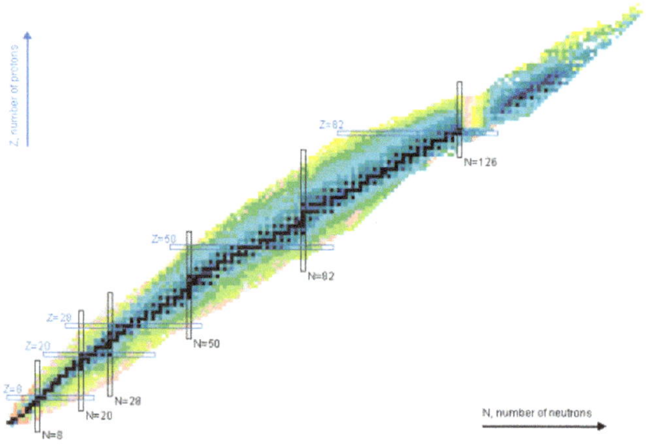

Figure. 1: The chart of nuclides. Stable and long-lived (>10^{15} s) nuclides are shown in black. Courtesy of NuDat Web application (http://www.nndc.bnl.gov/nudat).

These additional mechanisms have been pioneered by Eddington (Eddington, 1920) via introduction of a revolutionary concept of element production in the stars. Present nucleosynthesis models explain medium and heavy element abundances using the stellar nucleosynthesis that consists of burning (explosive) stages of stellar evolution, photo disintegration, and neutron and proton capture processes. The model predictions can be verified through the star metallicity studies and comparison of calculated isotopic/elemental abundances with the observed values, as shown in Fig. 2 . Nowadays, there are many well-established theoretical models of stellar nucleosynthesis (Boyd et al., 2009; Burbidge et al., 1957); however, they still cannot reproduce the observed abundances due to many parameter uncertainties. The flow of the nuclear physics processes in the network calculations is defined by the nuclear masses, reaction, and decay rates, and strongly correlated with the stellar temperature and density. These calculations depend heavily on our understanding of nuclear physics processes in stars, and the availability of high quality nuclear data. The common sources of stellar nucleosynthesis data include KADONIS, NACRE and REACLIB dedicated nuclear astrophysics libraries (Angulo et al., 1999; Cyburt et al., 2010;

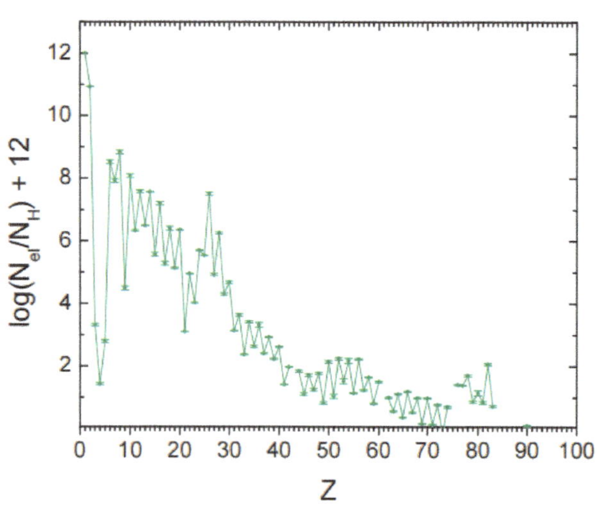

Figure. 2: Solar system elemental abundances; data are taken from (Grevesse & Sauval, 1998)

Dillmann et al., 2006) and general nuclear science and industry databases. Dedicated libraries are optimized for nuclear astrophysics applications, and contain pre-selected data that are often limited to the original scope. In many cases, these sources reflect the present state of nuclear physics, when experimental data are not always available or limited to a single measurement. Such limitation highlights the importance of theoretical calculations that strongly dependent on nuclear models. Another problem arises from the fact that nucleosynthesis processes are strongly affected by the astrophysical site conditions. To broaden the scope of the traditional nuclear astrophysics calculations, we will investigate applicability of nuclear physics databases for stellar nucleosynthesis data mining. These databases were developed for nuclear science, energy production and national security applications and will provide complementary astrophysics model-independent results.

STELLAR NUCLEOSYNTHESIS

This section will briefly consider the light elements and concentrate on the active research subject of production of medium and heavy elements beyond iron via slow and rapid neutron capture and associated data needs. Finally, photo disintegration and proton capture nucleosynthesis processes will be reviewed.

Burning Phases of Stellar Evolution

Fusion reactions are responsible for burning phases of stellar evolution. These reactions produce light, tightly-bound nuclei and release energy. The process of new element creation proceeds before nuclear binding energy reaches maximum value in the Fe-Ni region. Four important cases will be reviewed: pure hydrogen burning, triple alpha process, CNO cycle, and stellar burning. These processes take place in stars with a mass similar to our Sun, as shown in Fig. 3. The data needs for these processes are addressed in the IAEA FENDL (Aldama & Trkov, 2004) and EXFOR Experimental Nuclear Reaction Data(NRDC, 2011) databases.

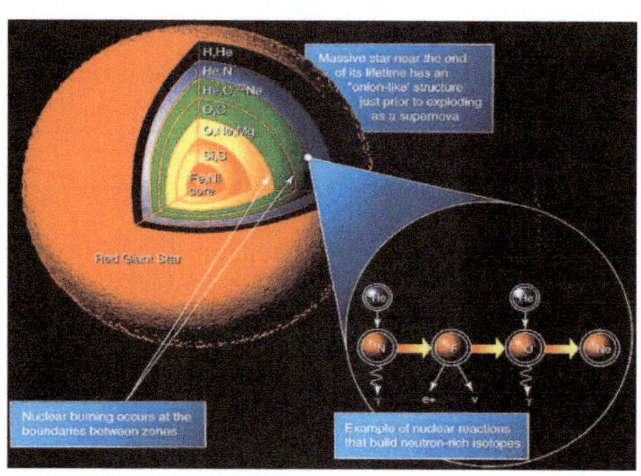

Figure. 3: Cross section of a Red Giant showing nucleosynthesis and elements formed. Courtesy of Wikipedia (http://en.wikipedia.org/wiki/Stellar_nucleosynthesis).

Pure Hydrogen Burning

Hydrogen is the most abundant element in the Universe. The proton-proton chain dominates stellar nucleosynthesis in stars comparable to our Sun

$$p + p \to d + \beta^{+} + \nu + 0.42MeV \tag{1}$$

Further analysis of the pp-process (Burbidge et al., 1957) indicates extremely low cross sections for $E<1$ MeV nuclear projectiles and explains the necessity of large target mass and density for sustainable nuclear fusion reaction

$$\sigma \sim 3E^{4.5} \times 10^{-24} b \tag{2}$$

Next, the deuterium produced in the first stage can fuse with another hydrogen

$$d + p \to^{3} He + \gamma + 5.49MeV \tag{3}$$

Finally, ^{4}He will be produced in the pp I and pp II branches.

Triple Alpha Process

The hydrogen burning in stars leads to the production of the helium core at a star's center. Further helium burning goes through the 3α process

$$\alpha + \alpha \Leftrightarrow {}^{8}Be - 93.7keV$$
$$^{8}Be + \alpha \Leftrightarrow {}^{12}C^{*} + 7.367MeV \tag{4}$$

CNO Cycle

Carbon-Nitrogen-Oxygen cycle leads to the production of elements heavier than carbon and consists of major CNO-I (regeneration of carbon and alpha particle)

$$^{12}C \to^{13} N \to^{13} C \to^{14} N \to^{15} O \to^{15} N \to^{12} C \tag{5}$$

and minor CNO-II branches (production of ^{16}O and proton):

$$^{15}N \to^{16} O \to^{17} F \to^{17} O \to^{14} N \to^{15} O \to^{15} N \tag{6}$$

Advanced Stages of Stellar Burning

Further burning includes carbon, neon, oxygen and silicon burning. Here, we have high-temperature and density burning stages when photonuclear processes produce additional α-particles and create a complex network of reactions with light particles. These reactions produce fusion nuclei up to the Fe-Ni peak. The Fe-Ni region nuclei are the most tightly bound in Nature and fusion reactions stop. Further element production proceeds via neutron and proton captures and photo disintegration.

S-Process

The slow-neutron capture (s-process) is responsible for creation of ~50 % of the elements beyond iron. In this region, neutron capture becomes dominant because of the increasing Coulomb barrier and decreasing binding energies. This s-process takes place in the Red Giants and AGB stars, where neutron temperature (kT) varies from 8 to 90 keV. A steady supply of neutrons is available due to H-burning and He-flash reactions

$$^{13}C(\alpha, n)^{16}O$$
$$^{22}Ne(\alpha, n)^{25}Mg \tag{7}$$

Fig. 2 indicates a high abundance of ^{56}Fe nuclei due to termination of the explosive nucleosynthesis. It is natural to assume that iron acts as a seed for the neutron capture reactions that eventually produce medium and heavy elements. The neutron capture time of s-process takes approximately one year. Consequently, the process path lies along the nuclear valley of stability up to the last long-lived nucleus of 209Bi. In the giant stars, neutron reaction rates define the elemental abundances, and in some cases branching points (created by the competition between neutron capture and β-decay) strongly affect the heavy isotope production rates. Therefore, special attention has to be paid to the branching points at ^{79}Se, ^{134}Cs, ^{147}Pm, ^{151}Sm, ^{154}Eu, ^{170}Yb and ^{185} and the neutron poison (absorption) $^{16,18}O, ^{22}Ne(n,\gamma)$ reactions. The s-process astrophysical site conditions imply the following data needs

- Reaction rates
- neutron-induced
- charged particle
- Half-lives

Convincing proof of s-process existence and its role in Nature could come from the calculation of isotopic abundances and comparison with observed values (Anders & Grevesse, 1989; Grevesse & Sauval, 1998). Present-day s-process nucleosynthesis calculations often are based on the dedicated nuclear astrophysics data tables, such as works of (Bao et al., 2000), and (Rauscher & Thielemann, 2000). These data tables contain quality information on Maxwellian-averaged cross sections $(\langle\sigma_{\gamma}^{Maxw}(kT)\rangle)$ and astrophysical reaction rates (R(T9)). However, it is essential to produce complementary neutron-induced reaction data sets for an independent verification and to expand the boundaries of the existing data tables.

Recent releases of ENDF/B-VII evaluated nuclear reaction libraries (Chadwick et al., 2006) and publication of the Atlas of Neutron Resonances reference book (Mughabghab, 2006) created a unique opportunity of applying these data for non-traditional applications, such as s-process nucleosynthesis (Pritychenko et al., 2010). Many neutron cross sections for astrophysical range of energies, including $^{56}Fe(n,\gamma)$ reaction as shown in Fig. 4, are available in the ENDF (Evaluated Nuclear Data File) and EXFOR libraries. The feasibility study of the evaluated nuclear data for s-process nucleosynthesis will be presented below.

Calculation of Maxwellian-Averaged Cross Sections and Uncertainties

The Maxwellian-averaged cross section can be expressed as

$$\langle\sigma^{Maxw}(kT)\rangle = \frac{2}{\sqrt{\pi}} \frac{(m_2/(m_1+m_2))^2}{(kT)^2} \int_0^{\infty} \sigma(E_n^L) E_n^L e^{-\frac{E_n^L m_2}{kT(m_1+m_2)}} dE_n^L, \tag{8}$$

where k and T are the Boltzmann constant and temperature of the system, respectively, and E is an energy of relative motion of the neutron with respect to the target. Here, E_n^L is a neutron energy in the laboratory system and m_1 and m_2 are masses of a neutron and target nucleus, respectively

The astrophysical reaction rate for network calculations is defined as

$$R(T_9) = N_A\langle\sigma v\rangle = 10^{-24}\sqrt{(2kT/\mu)} N_A \sigma^{Maxw}(kT), \tag{9}$$

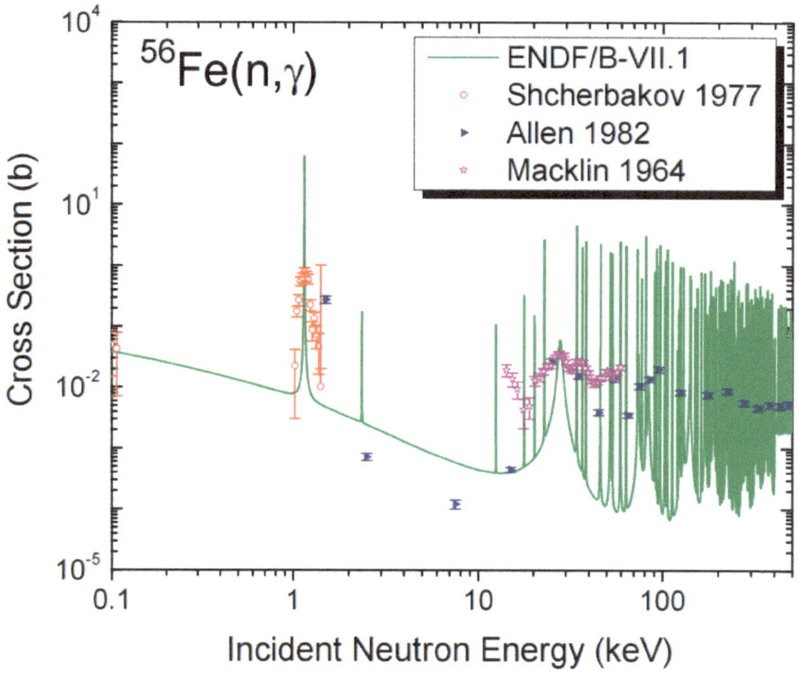

Figure. 4: ENDF/B-VII.1 and EXFOR libraries ^{56}Fe(n,γ) cross sections (Chadwick et al., 2006; NRDC, 2011) for astrophysical range of energies.

where T_9 is temperature expressed in billions of Kelvin, N_A is an Avogadro number. T_9 is related to the kT in MeV units as follows

$$11.6045 \times T_9 = kT \tag{10}$$

It is commonly known that for the equilibrium s-process-only nuclei product of $\langle \sigma_\gamma^{Maxw}(kT) \rangle$ and solar-system abundances ($N_{(A)}$) is preserved (Rolfs & Rodney, 1988)

$$\sigma_A N_{(A)} = \sigma_{A-1} N_{(A-1)} = constant \tag{11}$$

The stellar equilibrium conditions provide an important test for the s-process nucleosynthesis in mass regions between neutron magic numbers N=50,82,126 (Arlandini et al., 1999). To investigate this phenomenon, we will consider ENDF libraries. These data were never adjusted for nuclear astrophysics models and are essentially model-independent. ENDF library is a core nuclear reaction database containing evaluated (recommended) cross sections, spectra, angular distributions, fission product yields, thermal neutron scattering, photo-atomic and other data, with emphasis on neutron-induced reactions. ENDF library evaluations cover all neutron reaction channels within 10^{-5} eV - 20 MeV energy range. In many cases, evaluations contain information on neutron cross section covariances. An example of ENDF/B-VII.1library 56Fe(n,γ) neutron cross section covariances (uncertainties) is shown in Fig. 5.

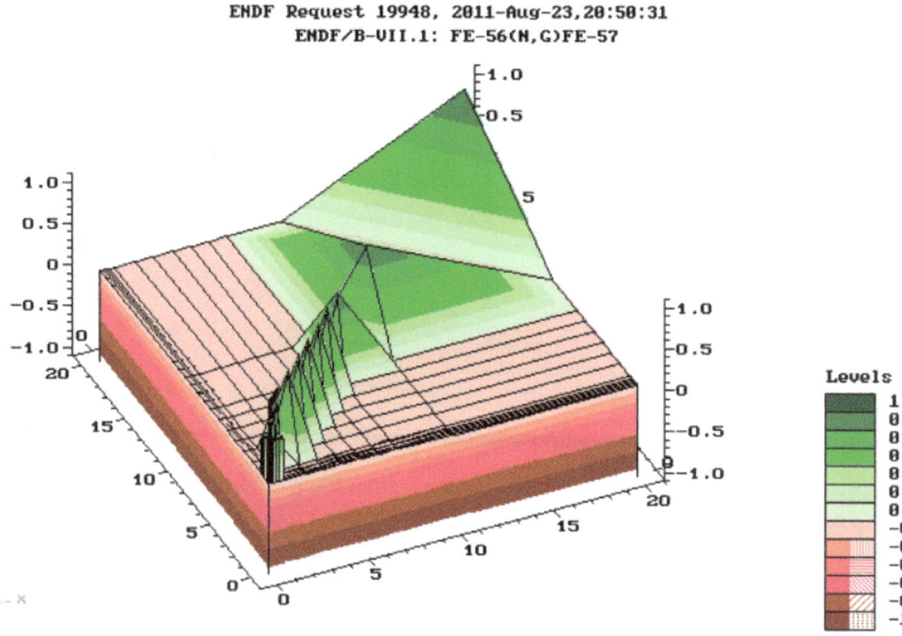

Figure. 5: ENDF/B-VII.1 library ^{56}Fe(n,γ) cross section covariances (Zerkin et al., 2005).

The evaluated neutron libraries are based on theoretical calculations using EMPIRE, TALYS and Atlas collection of nuclear reaction model codes (Herman et al., 2007; Koning et al., 2008; Mughabghab, 2006) that are often adjusted to fit experimental data (NRDC, 2011). The model codes are essential for neutron cross section calculations of short-lived radioactive nuclei where experimental data are not available. The ENDF data files are publicly available from the NNDC (National Nuclear Data Center) Sigma Web Interface: http://www.nndc.bnl.gov/sigma (Pritychenko & Sonzogni, 2008). Previously, Maxwellian-averaged cross sections and astrophysical reaction rates were produced using the Simpson method for the linearized ENDF cross sections (Pritychenko et al., 2010) . A similar effort was completed at the Japanese Atomic Energy Agency (Nakagawa et al., 2005). Fig. 6 shows the NNDC website (Pritychenko et al., 2006) that provides access to nuclear databases and the NucRates Web application (http://www.nndc.bnl.gov/astro). The NucRates application was designed for online calculations of $\langle\sigma^{Maxw}(kT)\rangle$ and $R(T_9)$ using all major evaluated nuclear reaction libraries, astrophysical neutron-induced reactions and kT values ranging from 10^{-5} eV to 20 MeV.

The Simpson method allowed quick calculation of integral values; however, the degree of precision was within ~1%. Precision can be improved with the linearized ENDF files because the cross section value is linearly-dependent on energy within a particular bin (Pritychenko,

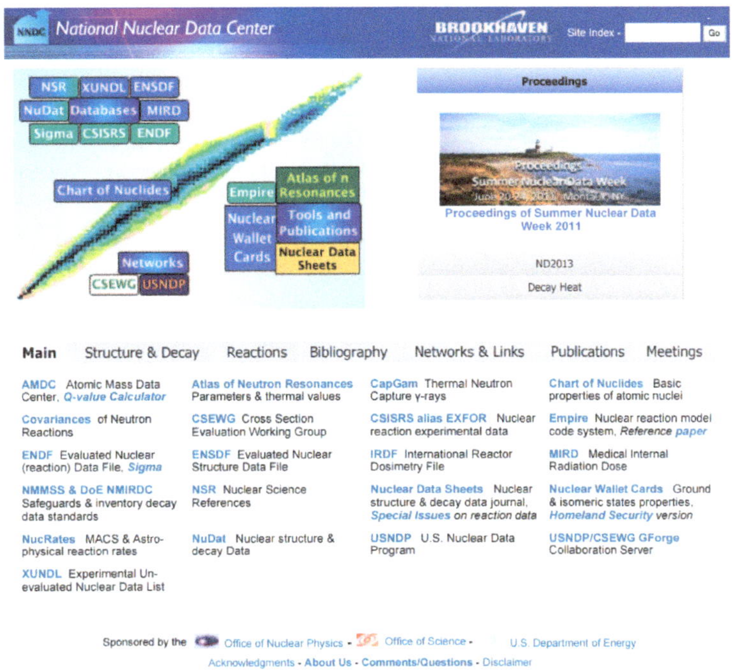

Figure. 6: The NNDC website (http://www.nndc.bnl.gov) provides access to nuclear databases and Maxwellian-averaged cross sections and astrophysical reaction rates online calculations. 2010)

$$\sigma(E) = \sigma(E_1) + (E - E_1)\frac{\sigma(E_2) - \sigma(E_1)}{E_2 - E_1}, \tag{12}$$

where $\sigma(E_1)$, E_1 and $\sigma(E_2)$, E_2 are cross section and energy values for the corresponding energy bin. The last equation is a good approximation of neutron cross section values for a sufficiently dense energy grid. This allowed deduction of $\langle\sigma_\gamma^{Maxw}(kT)\rangle$ definite integrals for separate energy bins using Doppler-broadened cross sections and the Wolfram Mathematica online integrator (Wolram, 2011). Further, summing integrals for all energy bins will produce a precise ENDF value for the Maxwellian-averaged cross section.

The product values of the ENDF/B-VII.1 $\langle\sigma_\gamma^{Maxw}(30keV)\rangle$ times solar abundances (Anders & Grevesse, 1989) are plotted in Fig. 7. They reveal that the ENDF/B-VII.1 library data closely replicate a two-plateau plot (Rolfs & Rodney, 1988). The current result provides a powerful testimony for stellar nucleosynthesis.

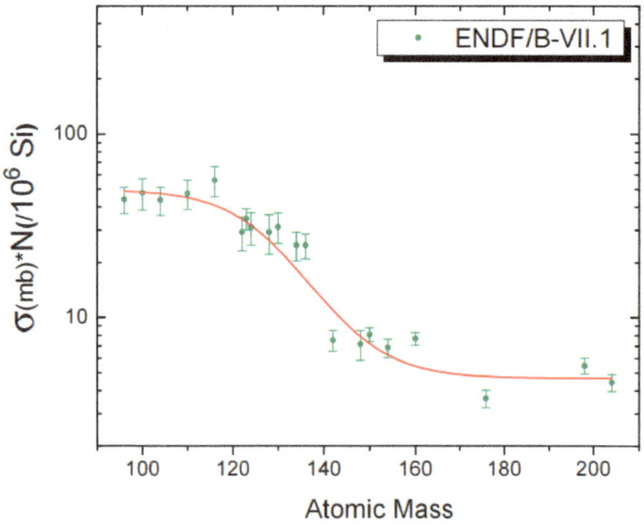

Figure. 7: ENDF/B-VII.1 library product of neutron-capture cross section (at 30 keV in mb) times solar system abundances (relative to Si = 10^6) as a function of atomic mass for nuclei produced only in the s-process.

The predictive power of stellar nucleosynthesis calculations depends heavily on the neutron cross section values and their covariances. To understand the unique isotopic signatures from the presolar grains, ~1% cross section uncertainties are necessary (Käppeler, 2011). Unfortunately, present uncertainties are often much higher, as shown in Fig. 8. The situation gets even more complex after considering s-process branches where β-decay and neutron capture rates are

$$\lambda_n \sim \lambda_\beta \tag{13}$$

Here, the stellar thermal environment may affect β-decay rates and change the process path. Branching is particularly important for unstable isotopes such as ^{134}Cs. This isotope can either decay to ^{134}Ba, if neutron flux is low or capture neutron and produce ^{135}Cs. The ^{134}Cs β-decay lifetime may vary from ~1 y to 30 day over the temperature range of (100-300)×10^6 K, which further complicates the calculations.

Finally, the calculated values of Maxwellian-averaged cross sections at kT=30 keV, for selected s-process nuclei, derived from the JENDL-4.0, ROSFOND 2010 and ENDF/B-VII.1 libraries (Chadwick et al., 2006; Shibata et al., 2011; Zabrodskaya et al., 2007) have been produced and shown in Table 1. The tabulated results are compared with the KADONIS values (Dillmann et al., 2006). Due to a limited number of ENDF covariance files, Low-Fidelity cross section covariances (Little et al., 2008) were used to calculate uncertainties for ENDF/B-VII.1 data.

The complete sets of ENDF s-process nucleosynthesis data sets are available for download from the NucRates Web application http://www.nndc.bnl.gov/astro. These complimentary data

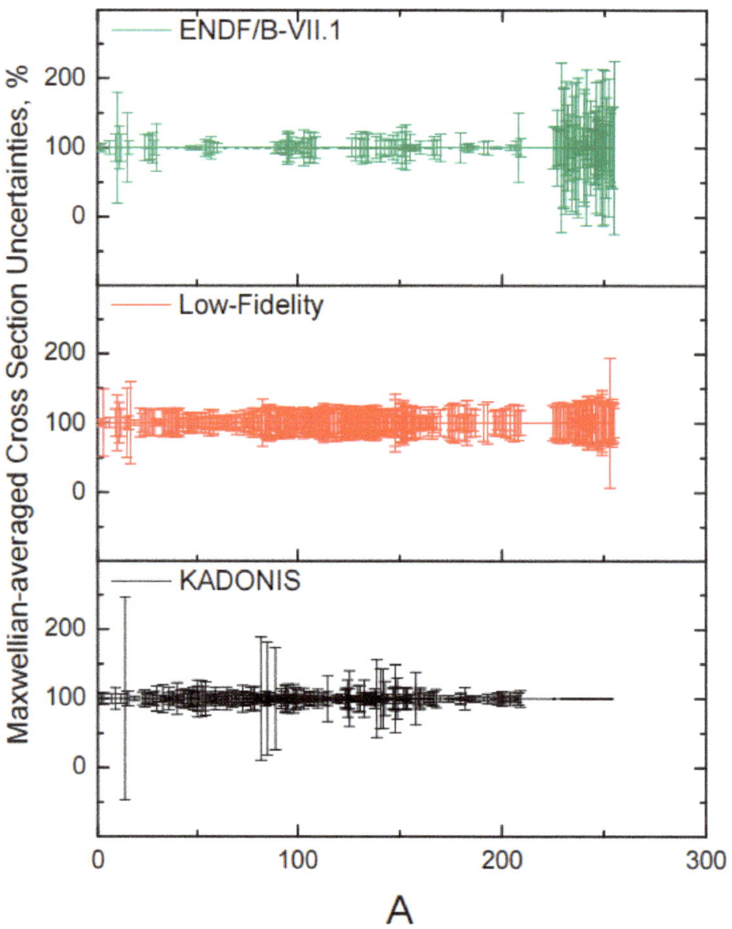

Figure. 8: Maxwellian-averaged cross section uncertainties for ENDF/B-VII.1, Low-Fidelity and KADONIS libraries (Chadwick et al., 2006; Dillmann et al., 2006; Little et al., 2008).

Table 1: Evaluated nuclear reaction and KADONIS libraries Maxwellian-averaged neutron capture cross sections in mb at kT=30 keV for s-process nuclei.

Isotope	JENDL-4.0	ROSFOND 2010	ENDF/B-VII.1	KADONIS
42-Mo- 96	1.052E+2	1.035E+2	1.035E+2±1.700E+1	1.120E+2±8.000E+0
44-Ru-100	2.065E+2	2.062E+2	2.035E+2±3.949E+1	2.060E+2±1.300E+1
46-Pd-104	2.700E+2	2.809E+2	2.809E+2±4.923E+1	2.890E+2±2.900E+1
48-Cd-110	2.260E+2	2.346E+2	2.349E+2±4.263E+1	2.370E+2±2.000E+0
50-Sn-116	9.115E+1	1.002E+2	1.003E+2±1.875E+1	9.160E+1±6.000E-1
52-Te-122	2.644E+2	2.639E+2	2.349E+2±4.882E+1	2.950E+2±3.000E+0
52-Te-123	8.138E+2	8.128E+2	8.063E+2±1.063E+2	8.320E+2±8.000E+0
52-Te-124	1.474E+2	1.473E+2	1.351E+2±2.697E+1	1.550E+2±2.000E+0
54-Xe-128	2.582E+2	2.826E+2	2.826E+2±6.823E+1	2.625E+2±3.700E+0
54-Xe-130	1.333E+2	1.518E+2	1.518E+2±2.835E+1	1.320E+2±2.100E+0
56-Ba-134	2.301E+2	2.270E+2	2.270E+2±4.038E+1	1.760E+2±5.600E+0
56-Ba-136	7.071E+1	7.001E+1	7.001E+1±1.087E+1	6.120E+1±2.000E+0
60-Nd-142	3.557E+1	3.701E+1	3.343E+1±4.251E+1	3.500E+1±7.000E-1
62-Sm-148	2.361E+2	2.444E+2	2.449E+2±4.507E+1	2.410E+2±2.000E+0
62-Sm-150	4.217E+2	4.079E+2	4.227E+2±3.607E+2	4.220E+2±4.000E+0
64-Gd-154	9.926E+2	1.010E+3	9.511E+2±1.096E+2	1.028E+3±1.200E+1
66-Dy-160	8.702E+2	8.293E+2	8.328E+2±6.769E+1	8.900E+2±1.200E+1
72-Hf-176	5.930E+2	4.529E+2	4.531E+2±4.896E+1	6.260E+2±1.100E+1
80-Hg-198	1.612E+2	1.612E+2	1.613E+2±1.635E+1	1.730E+2±1.500E+1
82-Pb-204	8.355E+1	7.242E+1	7.242E+1±7.624E+0	8.100E+1±2.300E+0

sets demonstrate a strong correlation between nuclear astrophysics and nuclear industry data needs, the large nuclear astrophysics potential of ENDF libraries, and a perspective beneficial relationship between both fields.

R-Process

The detailed analysis of stable and long-lived nuclei indicates the large number of isotopes that lie outside of the s-process path peaks near A=138 and 208. In addition, the large gap between s-process nucleus 209Bi and ^{232}Th,^{235},^{238}U effectively terminates the s-process at ^{210}Po. Production of the actinide neutron-rich nuclei cannot be explained by the s-process nucleosynthesis and requires introduction of rapid neutron capture or r-process. In this case, neutron capture timescale has to be less than typical β-decay lifetimes of ~ms for neutron-rich nuclides. It implies neutron fluxes 10^{10}-10^{11} higher than those of the s-process. Such conditions can be found in ν-driven core-collapse supernova and neutron stars. From here, one may conclude that r-process temperature depends on the site and may lie within a (0.5-10) $\times 10^9$ K range. However, it is still not clear where r-process takes place and how it proceeds.

Among many unknowns of the r-process is process path. The path is defined by the nuclear masses and β-decay half-lives. The 2003 & 2011 experimental Atomic Mass Evaluations (Audi et al., 2003; Audi & Meng, 2011) do not cover nuclei far from stability near the r-process expected path. To resolve this problem, theoretical mass calculations based on the FRDM and other models have been performed with 25% uncertainties (Aprahamian, 2011). This calculation helped to identify the list of critical nuclei along the r-process path and demonstrated strong connections between nuclear astrophysics and nuclear structure. Further progress will require mass and half-life measurements of unstable nuclei. The r-process data needs can be summarized as follows

- nuclear masses
- β-decay half-lives
- n-capture rates
- neutrino interaction rates
- fission probabilities
- fission products distribution

Recent r-process estimates (Cowan & Thielemann, 2004) demonstrate sharp abundance peaks for the A=130,195 (N=82,126) nuclei with large N/Z ratios and another broad peak at A=160. Further analysis of r-process abundances shows (Boyd, 2008)

$$dN_{Z,A}/dt = \lambda_{Z-1}N_{Z-1} - \lambda_Z N_Z \quad (14)$$

For neutron closed shells, nuclei most likely will experience β-decay rather than absorb another neutron

$$\lambda_Z N_Z = N_Z/\tau_Z \quad (15)$$

Thus near neutron closed shells, the relationship between abundances and β-decay lifetimes is

$$dN_{Z,A}/dt = N_{Z-1,A-1}/\tau_{Z-1} - N_{Z,A}/\tau_Z \quad (16)$$

Finally, for equilibrium conditions one can deduce r-process analog of the equation 11

$$N_{Z-1}/\tau_{Z-1} = N_Z/\tau_Z \quad (17)$$

From the last formula, one can conclude that closed shell nuclei with the largest half-lives will have the largest abundances. In order to obtain the complete picture, all half-lives and decay modes have to be determined. A list of properties relevant to r-process A~130 nuclei is shown in Table 2. It includes β-decay half-lives and emission probabilities for delayed neutrons. The delayed neutrons provide an additional neutron source for the r-process and may shift the location of the abundance peak. The tabulated data were taken from the Evaluated Nuclear Structure Data File (ENSDF) database (Burrows, 1990) and the relation between nuclear lifetime and half-life is

$$\tau = T_{1/2}/0.693 \tag{17}$$

The regularly updated ENDF and ENSDF database evaluations could provide valuable for r-process data in the actinide region. The extremely neutron-rich superheavy fission nuclei play an important role in element production. Presently, these nuclei can be studied only with theoretical model calculations. These calculations could be calibrated using the existing actinide data for Maxwellian-averaged neutron cross sections, half-lives, spontaneous fission and delayed neutrons probabilities. An example of nuclear reaction and structure data sets for the Z=90-110 region is shown in Table 3. The tabulated values demonstrate an increasingly complex nature of nuclear decay for Z>95 nuclei where spontaneous fission and β-decay play an important role. Spontaneous fission fragments are of interest to the r-process studies. Complimentary information on fission fragments distribution can be obtained from the Sigma Web interface http://www.nndc.bnl.gov/sigma (Pritychenko & Sonzogni, 2008).

Table 2: Properties of neutron-rich nuclides relevant to the A=130 r-process peak. All data are taken from the ENSDF database (http://www.nndc.bnl.gov/ensdf) (Burrows, 1990). The ? symbol was used where data were not available

Isotope	$T_{1/2}$, msec	β^--decay, %	β-n Emission, %
^{127}Ag	109±25	100	?
^{128}Ag	58±5	100	?
^{129}Ag	46±7	100	?
^{130}Ag	≈50	?	?
^{130}Cd	162±7	100	3.5±1.0
^{131}Cd	68±3	100	3.5±1.0
^{132}Cd	97±10	100	60±15
^{131}In	280±30	100	≤2.0±0.3
^{132}In	207±6	100	6.3±0.9
^{133}In	165±3	100	85±10
^{134}In	140±4	100	65
^{135}In	92±10	100	>0
^{130}Sn	223200±4200	100	?
^{131}Sn	56000±5000	100	?
^{132}Sn	39700±800	100	?
^{133}Sn	1460±30	100	0.0294±24
^{134}Sn	1050±11	100	17±13
^{135}Sn	530±20	100	21±3
^{136}Sn	250±30	100	30±5
^{137}Sn	190±60	100	58±15
^{138}Sn	>0.000408	?	?
^{139}Sn	?	?	?
^{140}Sn	?	?	?
^{136}Sb	923±14	100	16.3±3.2
^{137}Sb	450±50	100	49±10
^{138}Sb	≥0.0003	?	?
^{139}Sb	>0.00015	?	?
^{137}Te	2490±50	100	2.99±16
^{138}Te	1400±400	100	6.3±2.1
^{139}Te	>0.00015	?	?

P-Process

A detailed analysis of the Fig. 1 data indicates between 29 and 35 proton-rich nuclei that cannot be produced in the s- or r-processes. A significant fraction of these nuclei originate from the γ-process (Boyd, 2008; Woosley & Howard, 1978). This process could take place in Type-II supernovae at (2-3)×T_9. It begins with (γ,n) reactions that synthesize proton-rich heavy nuclei that are followed by charged-particle emitting reactions. Such process includes an extensive reaction network consisting of approximately 20,000 reactions and 2,000 nuclei. Due to lack of experimental data, p-process network calculations are often based on theoretical model predictions. This situation can be improved via addition of the known experimental reaction cross sections.

Table 3: Nuclear reaction and structure properties of several actinides. Reaction cross sections were calculated from ENDF/B-VII.1 library at kT=400 keV and decay data were taken from the ENSDF database (http://www.nndc.bnl.gov/ensdf). The ? symbol was used where data were not available

Isotope	$\sigma(n,F)$, mb	$T_{1/2}$, y	SF, %	α-decay, %	β-decay, %
^{232}Th	1.672E1	1.40E10	1.1E-9	100	?
^{235}U	1.376E3	7.04E8	7E-9	100	?
^{238}U	7.839E1	4.468E9	5.45E-5	100	?
^{237}Np	9.953E2	2.144E6	≤2E-10	100	?
^{239}Pu	1.885E3	2.411E4	3.1E-10	100	?
^{241}Am	7.767E2	432.6	3.6E-10	100	?
^{250}Cm	3.192E2	8.3E3	74	18	8
^{250}Bk	1.209E3	3.212 h	?	?	100
^{252}Cf	2.133E3	2.645	3.092	96.908	?
^{255}Es	3.108E2	39.8 d	0.0041	8	92
^{255}Fm	2.776E3	20.07 h	2.4E-5	100	?

Table 4: Total number of p-process reaction entries in EXFOR (http://www-nds.iaea.org/exfor) and NSR (http://www.nndc.bnl.gov/nsr) databases as of August 2011.

Reaction	EXFOR	NSR
p,γ	396	2162
p,n	666	2571
p,α	337	1031
α,γ	144	522
α,n	343	1321
α,p	166	848

These reactions have been compiled in the EXFOR database since 70ies, and the database content is shown in Fig. 9. For historic reasons, it is relatively complete for neutron-, proton- and alpha-induced reaction compilations, and has a limited number of compilations for heavy-ion and photonuclear reactions. The IAEA EXFOR Web interface: http://www-nds.iaea.org/exfor (Zerkin et al., 2005) allows user-friendly nuclear astrophysics data search using multiple parameters, such as target, nuclear reaction, cross section, and energy range. The interface is currently used for a p-process nuclei data mining operation at ATOMKI (Szücs et al., 2010). EXFOR is the best source of experimental nuclear reaction data; however, it is not complete, and it takes ~1-2 y before article compilation is completed. To overcome this problem the Nuclear Science References (NSR) database (http://www.nndc.bnl.gov/nsr) (Pritychenko et al., 2011) is recommended. Table 4 shows EXFOR and NSR database content for the ATOMKI project scope of (p,γ), (p,n), (p,α), (α,γ), (α,n) and (α,p) reactions. The tabulated data indicate a factor of 3-5 difference between two databases. This is mostly due to the fact that multiple article can be combined into a single EXFOR entry and gaps in the EXFOR coverage.

Another important tool for p-process nucleosynthesis studies is the nuclear reaction reciprocity theorem. It allows extracting a reaction cross section if an inverse reaction is

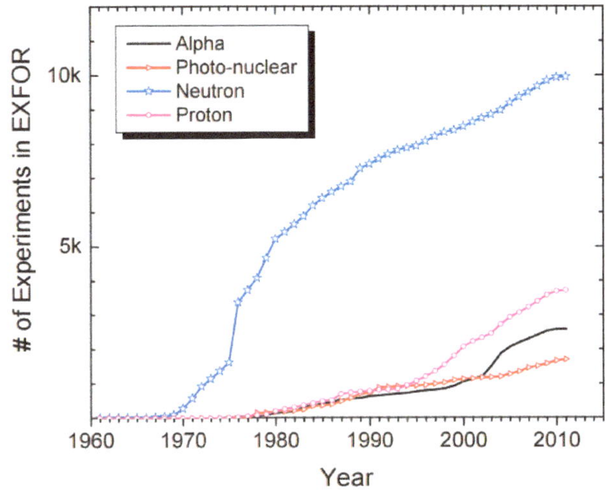

Figure. 9: Time and content evolution of the EXFOR database. Initially database scope was limited to neutron-induced reaction cross sections, later scope expansion included charge particle and photo-nuclear reactions.

known. For 1 + 2 → 3 + 4 and 3 + 4 → 1 + 2 processes the cross section ratio is

$$\frac{\sigma_{34}}{\sigma_{12}} = \frac{m_3 m_4 E_{34}(2J_3+1)(2J_4+1)(1+\delta_{12})}{m_1 m_2 E_{12}(2J_1+1)(2J_1+1)(1+\delta_{34})}, \qquad (19)$$

where E_{12}, E_{34} are kinetic energies in the c.m. system, J is angular momentum, and $\delta_{12}=\delta_{34}=0$.

Rp-Process

The rp-process (rapid proton capture process) consists of consecutive proton captures onto seed nuclei to produce heavier elements. It occurs in a number of astrophysical sites including X-ray bursts, novae, and supernovae. The rp-process requires a high-temperature environment (~10^9 K) to overcome Coulomb barrier for charged particles. This process contributes to observed abundances of light and medium proton-rich nuclei and compliments the p-process. The rp-process data needs include

- nuclear masses
- proton capture rates

CONCLUSION

Finally, a review of stellar nucleosynthesis and its data needs has been presented. Several nuclear astrophysics opportunities and the corresponding computation tools and methods have been identified. Complimentary sets of nuclear data for s-process nucleosynthesis have been produced. These, nuclear astrophysics model-independent data sets are based on the latest evaluated nuclear libraries and low-fidelity covariances data. This analysis indicates that the nucleosynthesis processes and respective abundances are strongly affected by both nuclear reaction cross sections and nuclear structure effects. Several nuclear structure data sets that include β-decay half-lives, spontaneous fission, and delayed neutron emission probabilities have been considered. Present results demonstrate a wide range of uses for nuclear reaction cross sections and structure data in stellar nucleosynthesis. They provide additional benchmarks and build a bridge between nuclear astrophysics and nuclear industry applications. Further work will include extensive data analysis, neutron physics, and network calculations. We are grateful to M. Herman (BNL) for the constant support of this project, to S. Goriely (Universite Libre de Bruxells) and R. Reifarth (Goethe University) for productive discussions, and to V. Unferth (Viterbo University) for a careful reading of the manuscript. This work was sponsored in part by the Office of Nuclear Physics, Office of Science of the U.S. Department of Energy under Contract No. DE-AC02-98CH10886 with Brookhaven Science Associates, LLC.

REFERENCES

1. Aldama, D.L. & Trkov, A. (2004). FENDL-2.1 Update of an evaluated nuclear data library for fusion applications. International Atomic Energy Agency, INDC(NDS)-467, Distribution FE, December 2004.
2. Anders, E. & Grevesse, N. (1989). Abundances of the elements - Meteoritic and solar. Geochimica et Cosmochimica Acta, Vol. 53, January 1989, pp. 197-214.
3. Angulo, C.; Arnould, M.; Rayet, M.; Descouvemont, P.; Baye, D.; Leclercq-Willain, C.; Coc, A.; Barhoumi, S.; Aguer, P.; Rolfs, C.; Kunz, R.; Hammer, J.W.; Mayer, A.; Paradellis, T.; Kossionides, S.; Chronidou, C.; Spyrou, K.; Degl'Innocenti, S.; Fiorentini, G.; Ricci, B.; Zavatarelli, S.; Providencia, C.; Wolters, H.; Soares, J.; Grama, C.; Rahighi, J.;
4. Shotter, A. & Lamehi Rachti, M. (1999). A compilation of charged-particle induced thermonuclear reaction rates. Nuclear Physics A, Vol. 656, No.1, August 1999, pp. 3-183.
5. Aprahamian, A. (2011). R-process mass sensitivities. Proc. XIV International Symposium on Capture γ-ray Spectroscopy, August 28-September 2, 2011, University of Guelph, Guelph, Ontario, Canada.
6. Arlandini, C.; Käppeler, F. ; Wisshak, K.; Gallino, R.; Lugaro, M.; Busso, M. & Straniero, O. (1999). Neutron Capture in Low-Mass Asymptotic Giant Branch Stars: Cross Sections and Abundance Signatures. The Astrophysical Journal, Vol. 525, November 1999, pp. 886-900.
7. Audi, G.; Wapstra, A.H. & Thibault, C. (2003). The AME2003 atomic mass Evaluation (II). Tables, graphs, and references. Nuclear Physics A, Vol. 729, Issue 1, December 2003, pp. 337-676.

8. Audi, G. & Meng, W. (2011). Atomic Mass Evaluation 2011. Private Communication.
9. Bao, Z.Y.; Beer, H.; Käppeler, F.; Voss, F.; Wisshak, K.; Rauscher, T. (2000). Neutron Cross Sections for Nucleosynthesis Studies. Atomic Data and Nuclear Data Tables, Vol. 76, Issue 1, September 2000, pp. 70-154.
10. Bethe, H.A. (1939). Energy Production in Stars. Physical Review, Vol. 55, Issue5, March 1939, pp. 434-456.
11. Boyd, R.N. (2008). An introduction to nuclear astrophysics, The University of Chicago Press, 2008. Boyd, R.N; Bernstein, L. & Brune, C. (2009). Studying Nuclear Astrophysics at NIF. Physics Today, August 14, 2009.
12. Burbidge, E.M.; Burbidge, G.R.; Fowler, W.A. & Hoyle, F. (1957). Synthesis of the Elements in Stars. Review of Modern Physics, Vol. 29, No.4, October 1957, pp. 547-654.
13. Burrows, T.W. (1990). The evaluated nuclear structure data file: Philosophy, content, and uses. Nuclear Instruments and Methods in Physics Research Section A, Vol. 286, Issue 3, January 1990, pp. 595-600.
14. Cameron, A.G.W. (1957). Stellar evolution, nuclear astrophysics, and nucleogenesis. Chalk River Report, CLR-41, Chalk River, Ontario, June 1957, pp. 1-197.
15. Chadwick, M.B.; Obložinský, P.; Herman, M.; Greene, N.M.; McKnight, R.D.; Smith, D.L.; Young, P.G.; MacFarlane, R.E.; Hale, G.M.; Frankle, S.C.; Kahler, A.C.; Kawano, T.; Little, R.C.; Madland, D.G.; Moller, P.; Mosteller, R.D.; Page, P.R.; Talou, P.;
16. Trellue, H.; White, M.C.; Wilson, W.B.; Arcilla, R.; Dunford, C.L.; Mughabghab, S.F.; Pritychenko, B.; Rochman, D.; Sonzogni, A.A.; Lubitz, C.R.; Trumbull, T.H.;
17. Weinman, J.P.; Brown, D.A.; Cullen, D.E., Heinrichs, D.P.; McNabb, D.P.; Derrien, H.; Dunn, M.E.; Larson, N.M.; Leal, L.C.; Carlson, A.D.; Block, R.C.; Briggs, J.B.; Cheng, E.T.; Huria, H.C.; Zerkle, M.L.; Kozier, K.S.; Courcelle, A.; Pronyaev, V. & van der
18. Marck, S.C. (2006). ENDF/B-VII.0: Next Generation Evaluated Nuclear Data Library for Nuclear Science and Technology. Nuclear Data Sheets, Vol. 107, Issue 12, December 2006, pp. 2931-3060.
19. Cowan, J. J. & Thielemann, F.-K. (2004) R-process Nucleosynthesis in Supernovae. Physics Today Vol. 57, October 2004, pp. 47-53.
20. Cyburt, R.H.; Amthor, A.M.; Ferguson, R.; Meisel, Z.; Smith, K.; Warren, S.; Heger, A.;
21. Hoffman, R.D., Rauscher, T.; Sakharuk, A.; Schatz, H.; Thielemann, F.L. & Wiescher, M. (2010). The JINA REACLIB Database: Its Recent Updates and Impact on Type-I X-ray Bursts. The Astrophysical Journal Supplement Series, Vol. 189, July 2010, pp. 240-252.
22. Dillmann, I.; Heil, M.; Käppeler, F.; Plag, R.; Rauscher , T. & Thielemann, F.-K. (2006).
23. KADoNiS - The Karlsruhe Astrophysical Database of Nucleosynthesis in Stars.
24. Proceedings of Capture Gamma-ray Spectroscopy and Related Topics: 12th International Symposium, Vol. 819, pp. 123-127, ISBN: 0-7354-0313-9, Notre Dame, Indiana (USA), 4-9 September 2005, AIP Conference Proceedings.
25. Eddington, A.S. (1920). The Internal Constitution of the Stars. Nature, Vol. 106, September 1920, pp. 14-20.
26. Grevesse, N. & Sauval, A. J. (1998). Standard Solar Composition. Space Science Reviews, v. 85, Issue 1/2, 1998, pp. 161-174.
27. Herman, M.; Capote, R.; Carlson, B.V.; Obložinský, P.; Sin, M.; Trkov, A.; Wienke, H. & Zerkin, V. (2007). EMPIRE: Nuclear Reaction Model Code System for Data Evaluation.
28. Nuclear Data Sheets, Vol. 108, Issue 12, December 2007, pp. 2655-2715.
29. Hoyle, F. (1946). The synthesis of the elements from hydrogen. Monthly Notices of the Royal Astronomical Society Vol. 106, 1946, pp. 343-383.
30. Käppeler, F. (2011). Reaction cross sections for s, r, and p process. Progress in Particle and Nuclear Physics, Vol. 66, Issue 2, April 2011, pp. 390-399.
31. Käppeler, F.; Gallino, R.; Bisterzo, S. & Wako Aoki. (2011). The s-process: Nuclear physics, stellar models, and observations. Review of Modern Physics, Vol. 83, January-March 2011, pp. 157-193.
32. Kolb, E. & Turner, M. (1988). The Early Universe, Addison-Wesley, 1988.
33. Koning, A.J.; Hilaire, S, & Duijvestijn, M.C. (2008). TALYS-1.0. Proceedings of International Conference on Nuclear Data for Science and Technology, Vol. 1, pp. 211-214, ISBN:978-2-7598-0090-2, Nice, France, April 22-27, 2007, EDP Sciences, 2008.
34. Little, R.C.; Kawano, T.; Hale, G.D.; Pigni, M.T.; Herman, M.; Oblozinsky, P.; Williams, M.L.; Dunn, M.E.; Arbanas, G.; Wiarda, D.; McKnight, R.D.; McKamy, J.N. & Felty, J.R. (2008). Low-fidelity Covariance Project. Nuclear Data Sheets, Vol. 109, Issue 12, December 2008, pp. 2828-2833.
35. Merrill, S. P. W. (1952). Spectroscopic Observations of Stars of Class S. The Astrophysical Journal, Vol. 116, February 1952, pp. 21-26.
36. Mughabghab, S.F. (2006). Atlas of Neutron Resonances, Resonance Parameters and Thermal Cross Sections. Z=1-100, Elsevier, 2006.
37. Nakagawa, T.; Chiba, S; Hayakawa, T. & Kajino, T. (2005). Maxwellian-averaged neutron-induced reaction cross sections and astrophysical reaction rates for kT = 1 keV to 1 MeV calculated from microscopic neutron cross section library JENDL-3.3.
38. Atomic Data and Nuclear Data Tables, Vol. 91, Issue 2, November 2005, pp. 77-186.
39. NRDC, International Network of Nuclear Reaction Data Centres. (2011). Compilation of experimental nuclear reaction data (EXFOR/CSISRS), (Available from http://wwwnds.iaea.org/exfor/, http://www.nndc.bnl.gov/exfor/).
40. Pritychenko, B.; Sonzogni, A.A.; Winchell, D.F.; Zerkin, V.V; Arcilla, R.; Burrows, T.W.; Dunford, C.L.; Herman,

M.W.; McLane, V.; Obložinský, P.; Sanborn, Y. & Tuli, J.K.

41. (2006). Nuclear Reaction and Structure Data Services of the National Nuclear Data Center. Annals of Nuclear Energy, Vol. 33, Issue 4, March 2006, pp. 390-399.
42. Pritychenko, B & Sonzogni, A.A. (2008). Sigma: Web Retrieval Interface for Nuclear Reaction Data. Nuclear Data Sheets, Vol. 109, Issue 12, December 2008, pp. 2822 - 2827.
43. Pritychenko, B.; Mughabghab, S.F. & Sonzogni, A.A. (2010). Calculations of Maxwellian-averaged Cross Sections and Astrophysical Reaction Rates Using the ENDF/B-VII.0, JEFF-3.1, JENDL-3.3 and ENDF/B-VI.8 Evaluated Nuclear Reaction Data Libraries. Atomic Data and Nuclear Data Tables, Vol. 96, Issue 6, November 2010, pp. 645-748.
44. Pritychenko, B. (2010). Complete calculation of evaluated Maxwellian-averaged cross sections and their uncertainties for s-process nucleosynthesis. Proceedings of 11th Symposium on Nuclei in the Cosmos, NIC XI, PoS(NIC-XI)197, Heidelberg, Germany, July 19-23, 2010, Proceedings of Science, 2010.
45. Pritychenko, B.; Bˇeták, E.; Kellett, M.A.; Singh, B. & Totans, J. (2011). The Nuclear Science References (NSR) database and Web Retrieval System. Nuclear Instruments and Methods in Physics Research Section A, Vol. 640, Issue 1, 1 June 2011, pp. 213-218.
46. Rauscher T. & Thielemann F. (2000). Astrophysical Reaction Rates from Statistical Model Calculations. Atomic Data and Nuclear Data Tables, Vol.75, Issue 1, January 2000, 1.
47. Rolfs, C.E. & Rodney, W.S. (1988). Cauldrons in the Cosmos, The University of Chicago Press, 1988.
48. Shibata, K.; Iwamoto, O.; Nakagawa, T.; Iwamoto, N; Ichihara, A.; Kunieda, S.; Chiba, S.; Furutaka, K.; Otuka, N.; Ohsawa, T.; Murata, T.; Matsunobu, H.; Zukeran, A.; Kamada, S. & Katakura, J.-i. (2011). JENDL-4.0: A New Library for Nuclear Science and Engineering. Journal of Nuclear Science and Technology, Vol. 48, No. 1, January 2011, pp. 1-30.
49. Szücs, T.; Dillmann, I.; Plag, R. & Fülöp, Zs. (2010). The new p-process database of KADoNiS. Proceedings of 11th Symposium on Nuclei in the Cosmos, NIC XI PoS(NIC-XI)247, Heidelberg, Germany, July 19-23, 2010, Proceedings of Science, 2010.
50. Wolfram Corporation. (2011). Wolfram Mathematica Online Integrator, Available from hhttp://integrals.wolfram.comi.
51. Woosley, S. E. & Howard, W. M. (1978). The p-process in supernovae. The Astrophysical Journal Supplement, Vol. 36, 1978, pp. 285-304.
52. Zabrodskaya, S.V.; Ignatyuk, A.V.; Koscheev, V.N.; Manokhin, V.N.; Nikolaev, M.N. & Pronyaev, V.G. (2007). ROSFOND - Russian National Library of Neutron Data. Voprosy Atomnoj Nauki i Techniki, Seriya: Nuclear Constants Vol. 1-2, 2007, p. 3-21 .
53. Zerkin, V.V.; McLane, V.; Herman, M.W. & Dunford, C.L. (2005). EXFOR-CINDA-ENDF: Migration of Databases to Give Higher-Quality Nuclear Data Services. Proceedings of International Conference on Nuclear Data for Science and Technology, Vol. 769, pp. 586-589, ISBN: 0-7354-0254-X, Santa Fe, New Mexico (USA), 26 September-1 October 2004, AIP Conference Proceedings.

CHAPTER 15

ELECTRON CAPTURE CROSS SECTIONS FOR STELLAR NUCLEOSYNTHESIS

P. G. Giannaka and T. S. Kosmas

INTRODUCTION

Weak interaction processes occurring in the presence of nuclei under stellar conditions play crucial role in the late stages of the evolution of massive stars and in the presupernova stellar collapse [1–6]. As it is known, the core of a massive star, at the end of its hydrostatic burning, is stabilized by electron degeneracy pressure as long as its mass does not exceed an appropriate mass (the Chandrasekhar mass limit, M_{Ch}) [6–10]. When the core mass exceeds M_{Ch}, electron degeneracy pressure cannot longer stabilize the center of the star and the collapse starts. In the early stage of collapse electrons are captured by nuclei in the iron group region [6, 10].

During the presupernova evolution of core collapse supernova, the Fermi energy (or equivalently the chemical potential) of the degenerate electron gas is sufficiently large to overcome the threshold energy E_{thr} (E_{thr} is given by negative Q values of the reactions involved in the interior of the stars) [11] and the nuclear matter in the stellar core is neutronized. This high Fermi energy of the degenerate electron gas leads to enormous e^--capture on nuclei and reduces the electron to baryon ratio Y_e [12, 13]. In this way, the electron pressure is reduced and the energy as well as the entropy drop. One of the important characteristics of the early pre-explosion evolution is the fact that electron capture on nuclei (specifically on nuclei of the pf shell) plays a key role [14, 15]. In the early stage of collapse (for densities lower than a few 10^{10} g cm^{-3}), the electron chemical potential is of the same order of magnitude as the nuclear Q value, and the e^--capture cross sections are sensitive to the details of GT strength distributions in daughter nuclei. For this reason, some authors restrict the calculations only to the GT strength and evaluate e^--capture rates on the basis of the GT transitions (at these densities, electrons are captured mostly on nuclei with mass number $A \leq 60$) [9–12, 15, 16]. Various methods, used for calculating e^--capture on nuclei during the collapse phase, have shown that this process produces neutrinos with rather low energies in contrast to the inelastic neutrino-nucleus reactions occurring in supernova [17–20]. These neutrinos escape the star carrying away energy and entropy from the core which is an effective cooling mechanism of the exploding massive star [21]. For higher densities and temperatures, e^- capture occurs on heavier nuclei $A \geq 65$ [8, 10, 13–15]. As a consequence, the nuclear composition is shifted to more neutron-rich and heavier nuclei (including those with $N > 40$) which dominate the matter composition for densities larger than about 10^{10} g cm^{-3} [1, 15, 21, 22].

The first calculations of stellar electron capture rates for iron group nuclei have been performed by employing the independent particle model (IPM) [2–4]. Recently, similar studies have been addressed by using continuum RPA (CRPA) [25], large scale shell model [26, 27], RPA [11], and so forth [28]. In the present work e^--capture cross sections are obtained within a refined version of the quasiparticle random phase approximation (QRPA) which is reliable for constructing all the accessible final (excited) states of the daughter nuclei in the iron group region of the periodic table [29–38]. For the description of the required correlated nuclear ground states we determine

Citation: P. G. Giannaka and T. S. Kosmas, "Electron Capture Cross Sections for Stellar Nucleosynthesis," Advances in High Energy Physics, vol. 2015, Article ID 398796, 11 pages, 2015. doi:10.1155/2015/398796

single-particle occupation numbers calculated within the BCS theory [29, 31, 32]. Our nuclear method is tested through the reproducibility of experimental muon capture rates relying on detailed calculations of exclusive, partial, and total muon capture rates [23, 24, 39–42]. The agreement with experimental data provided us with high confidence level of our method and we continued with the calculations of electron capture cross sections in supernova conditions (where the densities and temperatures are high) using the pn-QRPA method. In this paper, we performed calculations for ^{66}Zn isotope (it belongs to the iron group nuclei) that plays prominent role in core collapse supernovae stellar nucleosynthesis [18, 19, 43].

Our strategy in this work is, at first, to perform extensive calculations of the transition rates for all the abovementioned nuclear processes, assuming laboratory conditions, and then to translate these rates to the corresponding quantities within stellar environment through the use of an appropriate convolution procedure [9, 14, 15, 21, 26]. To this purpose, we assume that leptons under such conditions follow MaxwellBoltzmann energy distribution [9, 26].

CONSTRUCTION OF NUCLEAR GROUND AND EXCITED STATES

Electrons of energy E_e are captured by nuclei interacting weakly with them via W^- boson exchange as follows:

$$(A, Z) + e^- \longrightarrow (A, Z-1)^* + \nu_e. \tag{1}$$

The outgoing ν_e neutrino carries energy E_ν while the daughter nucleus (A, Z − 1) absorbs a part of the incident electron energy given by the difference between the initial E_i and the final E_f nuclear energies as $E_\nu = E_f - E_i$.

The nuclear calculations for the cross sections of reaction (1) start by writing down the weak interaction Hamiltonian $\widehat{\mathscr{H}}_w$ which is given as a product of the leptonic, j_μ^{lept}, and the hadronic, $\widehat{\mathscr{J}}^\mu$, currents (current-current interaction Hamiltonian) as follows:

$$\widehat{\mathscr{H}}_w = \frac{G}{\sqrt{2}} j_\mu^{\text{lept}} \widehat{\mathscr{J}}^\mu, \tag{2}$$

where $G=G_F \cos\theta_c$ with G_F and θ_c being the well-known weak interaction coupling constant and the Cabibbo angle, respectively [31, 32, 44].

From the nuclear theory point of view, the main task is to calculate the cross sections of reaction (1) which are based on the evaluation of the nuclear transition matrix elements between the initial $|i\rangle$ and a final $|f\rangle$ nuclear states of the form

$$\langle f|\widehat{H_w}|i\rangle = \frac{G}{\sqrt{2}}\ell^\mu \int d^3x e^{-i\mathbf{q}\cdot\mathbf{x}} \langle f|\widehat{\mathscr{J}_\mu}|i\rangle. \tag{3}$$

The quantity $\ell^\mu e^{-i\mathbf{q}\cdot\mathbf{x}}$ stands for the leptonic matrix element written in coordinate space with q being the 3- momentum transfer. For the calculation of these transition matrix elements one may take advantage of the DonnellyWalecka multipole decomposition which leads to a set of eight independent irreducible tensor multipole operators containing polar-vector and axial-vector components [44]

In the present work, in (3) the ground state of the parent nucleus$|i\rangle$ is computed by solving the relevant BCS equations which give us the quasiparticle energies and the amplitudes V and U that determine the probability for each single particle level to be occupied or unoccupied, respectively [31]. Towards this aim, at first, we consider a Coulomb corrected Woods-Saxon potential with a spin orbit part as a mean field for the description of the strong nuclear field [45, 46]. For the latter potential we adopt the parameterization of IOWA group [47]. Then, we use as pairing interaction the monopole part of the Bonn C-D one meson exchange potential. The renormalization of this interaction, to fit in the ^{66}Zn isotope, is achieved through the two pairing parameters $g_{\text{pair}}^{p,n}$, (n) for proton (neutron) pairs, and the values of which are tabulated in Table 1. As it is well known, the pairing parameters, $g_{\text{pair}}^{p,n}$, are determined through the reproduction of the energy gaps, $\Delta_{p,n}^{\text{exp}}$, from neighboring nuclei as follows (3-point formula):

$$\Delta_n^{\text{exp}} = -\frac{1}{4}\left[S_n(A-1,Z) - 2S_n(A,Z) + S_n(A+1,Z)\right],$$

$$\Delta_p^{\text{exp}} = -\frac{1}{4}\Big[S_p(A-1,Z-1) - 2S_p(A,Z) + S_p(A+1,Z+1)\Big], \tag{4}$$

where S_p and S_n are the experimental separation energies for protons and neutrons, respectively, of the target nucleus (A, Z) and the neighboring nuclei (A±1, Z±1) and (A±1, Z). For the readers convenience in Table 2 we show the values of experimental separation energies for the target $^{66}_{30}$Zn and the neighboring nuclei $^{65}_{29}$Cu, $^{67}_{31}$Ga, $^{65}_{30}$Zn, and $^{67}_{30}$Zn.

Subsequently, the excited states $|f\rangle$ of the studied daughter nucleus ^{66}Cu are constructed by solving the pn-QRPA equations [29–38], which are written as follows in matrix form [29]:

$$\begin{pmatrix} \mathscr{A} & \mathscr{B} \\ -\mathscr{B} & -\mathscr{A} \end{pmatrix} \begin{pmatrix} X^\nu \\ Y^\nu \end{pmatrix} = \Omega_{J^\pi}^\nu \begin{pmatrix} X^\nu \\ Y^\nu \end{pmatrix}, \tag{5}$$

$\Omega_{J^\pi}^\nu$ denotes the excitation energy of the QRPA state $|J_\nu^\pi\rangle$ with spin J and parity π. The solution of (5) is an eigenvalue problem which provides the amplitudes for forward and backward scattering X and Y, respectively, and the QRPA excitation energies $\Omega_{J^\pi}^\nu$ [31–34]. In our method the solution of the QRPA equations is carried out separately for each multipole set of states $|J^\pi\rangle$.

For the renormalization of the residual 2-body interaction (Bonn C-D potential), the strength parameters, for the particle-particle (g_{pp}) and particle-hole (g_{ph}) interaction entering the QRPA matrices A and B, are determined (separately for each multipolarity) from the reproducibility of the low-lying experimental energy spectrum of the final nucleus. The values of these parameters in the case of the spectrum of ^{66}Cu are listed in Table 3.

Table 1: Parameters for the renormalization of the interaction of proton pairs, g^p_{pair}, and neutron pairs, g^n_{pair}. They have been fixed in such a way that the corresponding experimental gaps $\Delta^{\exp}_p$ and $\Delta^{\exp}_n$ of ^{66}Zn isotope, are quite accurately reproduced

Nucleus	g^n_{pair}	g^p_{pair}	$\Delta^{\exp}_n$ (MeV)	Δ^{theor}_n (MeV)	$\Delta^{\exp}_p$ (MeV)	Δ^{theor}_p (MeV)
^{66}Zn	1.0059	0.9271	1.7715	1.7716	1.2815	1.2814

Table 2: The experimental separation energies in MeV for protons and neutrons of the target (A, Z) and neighboring $(A \pm 1, Z \pm 1)$ and $(A \pm 1, Z)$ nuclei.

Nucleus	$S_n(A-1,Z)$	$S_n(A,Z)$	$S_n(A+1,Z)$	$S_p(A-1,Z-1)$	$S_p(A,Z)$	$S_p(A+1,Z+1)$
^{66}Zn	7.979	11.059	7.052	7.454	8.924	5.269

At this point, it is worth mentioning that, for measuring the excitation energies of the daughter nucleus ^{66}Cu from the ground state of the initial one ^{66}Zn, a shifting of the entire set of QRPA eigenvalues is necessary. Such a shifting is required whenever in the pn-QRPA a BCS ground state is used, a treatment adopted by other groups previously [9, 40, 48, 49]. The shifting, for the spectrum of the daughter nucleus ^{66}Cu, is done in such a way that the first calculated value of each multipole state of ^{66}Cu (i.e. $1^+_1, 2^+_1, \ldots$, etc.) approaches as close as possible the corresponding lowest experimental multipole excitation. Table 4 shows the shifting applied to our QRPA spectrum for each multipolarity of the parent nucleus ^{66}Zn. We note that a similar treatment is required in pnQRPA calculations performed for double-beta decay studies where the excitations derived for the intermediate odd-odd nucleus (intermediate states) through p-n and n-p reactions from the neighboring nuclei, left or right nuclear isotope, do not match to each other [48, 49]. The resulting low-energy spectrum, after using the parameters of Tables 1 and 3 and the shifting shown in Table 4, agrees well with the experimental one (see Figure 1).

We must also mention that, usually, in nuclear structure calculations we test a nuclear method in two phases: first through the construction of the excitation spectrum as discussed before and second through the calculations of electron scattering cross sections or muon capture rates. Following the above steps, we test the reproducibility of the relevant experimental data for many nuclear models employed in nuclear applications (nuclear structure and nuclear reactions) and in nuclear astrophysics [1, 9].

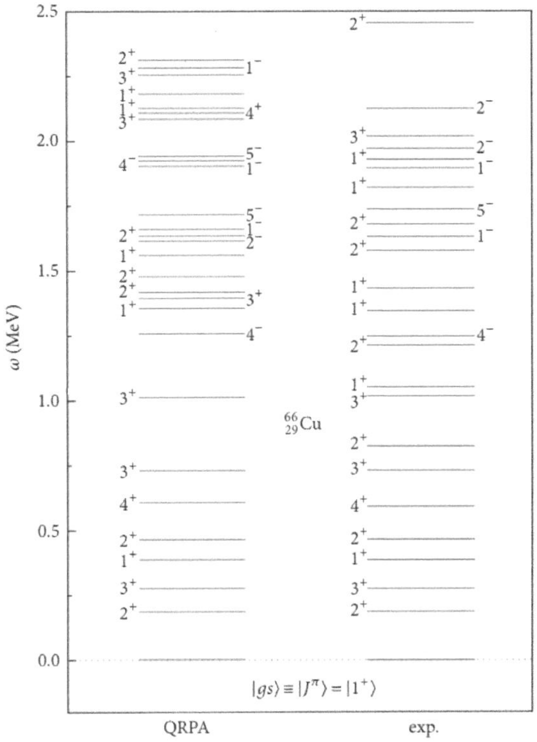

Figure 1: Comparison of the theoretical excitation spectrum (resulting from the solution of the QRPA eigenvalue problem) with the low-lying (up to about 3 MeV) experimental one for ^{66}Cu nucleus. As it can be seen, the agreement is very good below 1 MeV but for higher excitation energies it becomes moderate.

RESULTS AND DISCUSSION

In this work we perform detailed cross section calculations for the electron capture on ^{66}Zn isotope on the basis of the pn-QRPA method. The required nuclear matrix elements between the initial $|J_i\rangle$ and the final $|J_f\rangle$ states are determined by solving the BCS equations for the ground state [29, 31, 32] and the pn-QRPA equations for the excited states [31–34] (see Section 2). For the calculations of the original cross sections, a quenched value of g_A is considered which subsequently modifies all relevant multipole matrix elements [23, 24, 50, 51].

At this point of the present work and in order to increase the confidence level of our method, we perform total muon capture rates calculations [23, 24, 39–42]. The comparison with experimental and other theoretical results is shown in Section 3.1. Afterwards, we study in detail the electron capture process as follows. (i) Initially we consider laboratory conditions; that is, the initial (parent) nucleus is considered in the ground state and no temperature dependence is assumed (see Section 3.2.1). (ii) Second, we consider stellar conditions; that is, the parent nucleus is assumed to be in any initial excited state and due to the e^--capture process it goes to any final excited state of the daughter nucleus. At these conditions it is necessary to take into account the temperature dependence of the cross sections (see Section 3.2.2) [10].

Table 3: Strength parameters for the particle-particle (g_{pp}) and particle-hole (g_{ph}) interaction for various multipolarities (for the rest of multipolarities, $J^{\pi} \leq 5^{\pm}$, the bare 2-body interaction has been used), in the case of the spectrum of ^{66}Cu nucleus

	Positive parity states					Negative parity states		
J^{π}	0^+	1^+	2^+	3^+	4^+	1^-	2^-	3^-	4^-
g_{pp}	0.827	0.547	0.686	0.854	1.300	0.994	0.200	0.486	0.622
g_{ph}	0.336	0.200	1.079	0.235	0.200	1.200	0.200	1.200	1.200

Table 4: The shift (in MeV) applied on the spectrum (seperately of each multipole set of states) of ^{66}Cu isotope, daughter nucleus of the electron capture on ^{66}Zn.

Positive parity states		Negative parity states	
0^+	0.90	0^-	5.00
1^+	2.50	1^-	6.80
2^+	2.55	2^-	3.85
3^+	2.50	3^-	2.60
4^+	1.75	4^-	3.55
5^+	0.55	5^-	3.00

Calculations of Muon Capture Rates for ^{66}Zn

Despite the fact that the muon capture on nuclei does not play a crucial role in stellar-nucleosynthesis, it is, however, important to start our study from this process since the nuclear matrix elements required for an accurate description of the μ-capture are the same for all semileptonic charge-changing weak interaction processes. In addition, the excitation spectrum of the daughter (A, Z − 1) nucleus, as we saw before, is in good agreement with the experimental data. The calculations of the muon capture rates are performed in three steps. In the first step we carry out realistic stateby-state calculations of exclusive ordinary muon capture (OMC) rates in ^{66}Zn isotope for all multipolarities with $J^{\pi} \leq 5^{\pm}$ (higher multipolarities contribute negligibly). The appropriate expression for the exclusive muon capture rates is written as follows:

$$\Lambda_{gs \to J_f^{\pi}} \equiv \Lambda_{J_f^{\pi}} = 2G^2 \langle \Phi_{1s} \rangle^2 R_f q_f^2 \times \left[\left| \langle J_f^{\pi} \| (\widehat{\mathscr{M}}_J - \widehat{\mathscr{L}}_J) \| 0_{gs}^+ \rangle \right|^2 + \left| \langle J_f^{\pi} \| (\widehat{\mathscr{T}}_J^{\text{el}} - \widehat{\mathscr{T}}_J^{\text{magn}}) \| 0_{gs}^+ \rangle \right|^2 \right], \quad (6)$$

where Φ_{1s} represents the muon wave function in the $1s$ muonic orbit. The operators in (6) are refered to as Coulomb $\widehat{\mathscr{M}}_J$, longitudinal $\widehat{\mathscr{L}}_J$, transverse electric $\widehat{\mathscr{T}}_J^{\text{el}}$, and transverse magnetic $\widehat{\mathscr{T}}_J^{\text{mag}}$ multipole operators. The factor R_f in (6) takes into consideration the nuclear recoil which is written as $R_f = (1 + q_f/M_{\text{targ}})^{-1}$, with M_{targ} being the mass of the target (parent) nucleus.

Due to the fact that there are no available data in the literature for exclusive muon capture rates, the test of our method is realized by comparing partial and total muon capture rates with experimental data and other theoretical results [23, 24]. Towards this purpose, our second step includes calculations of the partial μ^--capture rates s for various lowspin multipolarities, $\Lambda_{J^{\pi}}$ (for $J^{\pi} \leq 4^{\pm}$), in the studied nucleus. These partial rates are found by summing over the contributions of all the individual multipole states of the studied multipolarity as follows:

$$\Lambda_{J^{\pi}} = \sum_f \Lambda_{gs \to J_f^{\pi}} = 2G^2 \langle \Phi_{1s} \rangle^2 \left[\sum_f q_f^2 R_f \left| \langle J_f^{\pi} \| (\widehat{\mathscr{M}}_J - \widehat{\mathscr{L}}_J) \| 0_{gs}^+ \rangle \right|^2 + \sum_f q_f^2 R_f \left| \langle J_f^{\pi} \| (\widehat{\mathscr{T}}_J^{\text{el}} - \widehat{\mathscr{T}}_J^{\text{magn}}) \| 0_{gs}^+ \rangle \right|^2 \right] \quad (7)$$

(f runs over all states of the multipolarity $|J^{\pi}\rangle$). We also estimate the percentage (portion) of their contribution into the total μ-capture rate for the most important multipolarities. In Table 5 we tabulate the individual portions of the low-spin multipole transitions ($J^{\pi} = 4^{\pm}$) into the total muon capture rate. As it can be seen, the contribution of the 1^- multipole transitions is the most important multipolarity exhausting about 44% of the total muon capture rate. Such an important contribution was found in ^{16}O and ^{48}Ca isotopes studied in [41].

In the last step of testing our method, we evaluate total muon capture rates for the ^{66}Zn isotope. These rates are obtained by summing over all partial multipole transition rates (up to $J^{\pi} = 4^{\pm}$) as follows:

$$\Lambda_{\text{tot}} = \sum_{J^{\pi}} \Lambda_{J^{\pi}} = \sum_{J^{\pi}} \sum_f \Lambda_{J_f^{\pi}}. \quad (8)$$

For the sake of comparison, the abovementioned μ-capture calculations have been carried out using the quenched value $g_A = 1.135$ [23, 24]. The results are listed in Table 6, where we also include the experimental total rates and the theoretical ones of [23, 24]. Moreover, in Table 6 we show the individual contribution into the total muon capture rate of the polar-vector (Λ^{V}_{tot}), the axial-vector (Λ^{A}_{tot}), and the overlap (Λ^{VA}_{tot}) parts. As it can be seen, our results obtained with the quenched g_A coupling constant are in very good agreement with the experimental total muon capture rates (the deviations from the corresponding experimental rates are smaller than 7%). This agreement provides us with high confidence level for our method.

Table 5: The percentage of each multipolarity into the total muon capture rate evaluated with our pn-QRPA method

Positive parity transitions		Negative parity transitions	
J^π	Portions (%)	J^π	Portions (%)
0^+	8.22	0^-	7.94
1^+	21.29	1^-	44.21
2^+	2.85	2^-	13.32
3^+	1.58	3^-	0.34
4^+	0.01	4^-	0.23

Electron Capture Cross Section.

After acquiring a high confidence level for our nuclear method, we proceed with the main goal of the present study which concerns the calculations of the electron capture cross sections. As mentioned before, this includes original (see Section 3.2.1) and stellar electron capture investigations (Section 3.2.2).

Original Electron Capture Cross Section on ^{66}Zn Isotope

The original cross sections for the electron capture process in the ^{66}Zn isotope are obtained by using the pn-QRPA method considering all the accessible transitions of the final nucleus ^{66}Cu. In the Donnelly-Walecka formalism the expression for the differential cross section in electron capture by nuclei reads [10]

$$\frac{d\sigma_{ec}}{d\Omega} = \frac{G_F^2\cos^2\theta_c}{2\pi}\frac{F(Z,E_e)}{(2J_i+1)} \times \Bigg\{\sum_{J\geq 1}\mathscr{W}(E_e,E_\nu) \times \Bigg\{\left[1-\alpha\cos\Phi+b\sin^2\Phi\right] \times \left[\left|\langle J_f\|\widehat{\mathscr{T}}_J^{\mathrm{mag}}\|J_i\rangle\right|^2+\left|\langle J_f\|\widehat{\mathscr{T}}_J^{\mathrm{el}}\|J_i\rangle\right|^2\right] - \left[\frac{(\varepsilon_i+\varepsilon_f)}{q}(1-\alpha\cos\Phi)-d\right] \times 2\,\mathrm{Re}\langle J_f\|\widehat{\mathscr{T}}_J^{\mathrm{mag}}\|J_i\rangle\langle J_f\|\widehat{\mathscr{T}}_J^{\mathrm{el}}\|J_i\rangle^*\Bigg\} + \sum_{J\geq 0}\mathscr{W}(E_e,E_\nu) \times \Bigg\{(1+\alpha\cos\Phi)\left|\langle J_f\|\widehat{\mathscr{M}}_J\|J_i\rangle\right|^2 + \left(1+\alpha\cos\Phi-2b\sin^2\Phi\right)\left|\langle J_f\|\widehat{\mathscr{L}}_J\|J_i\rangle\right|^2 - \left[\frac{\omega}{q}(1+\alpha\cos\Phi)+d\right]2\,\mathrm{Re}\langle J_f\|\widehat{\mathscr{L}}_J\|J_i\rangle \times \langle J_f\|\widehat{\mathscr{M}}_J\|J_i\rangle^*\Bigg\}\Bigg\}, \tag{9}$$

where $F(Z, E_e)$ is the well-known Fermi function [18]. The factor $W(E_e,E_\nu)=E_\nu^2/(1+E_\nu/M_T)$ accounts for the nuclear recoil [8], MT is the mass of the target nucleus, and parameters α, b, and d are given, for example, in [31]. The nuclear transition matrix elements between the initial state $|J_i\rangle$ and the final state $|J_f\rangle$ correspond to the Coulomb $\widehat{\mathscr{M}}_{JM}$, longitudinal $\widehat{\mathscr{L}}_{JM}$, transverse electric $\widehat{\mathscr{T}}^{\mathrm{el}}_{JM}$, and transverse magnetic $\widehat{\mathscr{T}}^{\mathrm{mag}}_{JM}$ multipole operators

From the energy conservation in the reaction (1), the energy of the outgoing neutrino E_ν is written as follows:

$$E_\nu = E_e - Q + E_i - E_f, \tag{10}$$

which includes the difference between the initial E_i and the final E_f nuclear states. The Q value of the process is determined from the experimental masses of the parent (M_i) and the daughter (M_f) nuclei as $Q=M_f-M_i$ [9].

It is worth mentioning that for low momentum transfer, various authors use the approximation $q\to 0$ for all multipole operators of (9). Then, the transitions of the GamowTeller operator $(GT^+ = \sum_i \tau_i^+\sigma_i)$ provide the dominant contribution to the total cross section [9].

While performing detailed calculations for the original electron capture cross sections on ^{66}Zn isotope we assumed that (i) the initial state of the parent nucleus ^{66}Zn is the ground state $|0^+\rangle$ and (ii) the nuclear system is under laboratory conditions (no temperature dependence of the cross sections is needed). The cross sections as a function of the incident electron energy E_e are calculated with the use of realistic two-body interactions as mentioned before. The obtained total original electron capture cross sections for ^{66}Zn target nucleus are illustrated in Figure 2 where the individual contributions of various multipole channels ($J^\pi \leq 5^\pm$) are also shown.The electron capture cross sections in Figure 2 exhibit a sharp increase by several orders of magnitude within the first few MeV above energy-threshold, and this reflects the GT^+ strength distribution. For electron energy $E_e \geq 10$ MeV the calculated cross sections show a moderate increase. From experimental and astrophysical point of view, the important range of the incident electron energy E_e is up to 30 MeV. At these energies the 1^+ multipolarity has the largest contribution to the total electron capture cross sections [9, 10]. In the present work we have extended the range of E_e up to 50 MeV since at higher energies (around 40 MeV) the contribution of other multipolarities like 1^-, 0^+, and 0^- becomes noticeable and cannot be omitted (see Figure 2).

Table 6: Individual contribution of polar-vector, axial-vector, and overlap parts into the total muon-capture rate. The total muon capture rates, obtained by using the pn-QRPA with the quenched value of $g_A = 1.135$ for the medium-weight nucleus ^{66}Zn, are compared with the available experimental data and with the theoretical rates of [23, 24].

	Total muon capture rates $\Lambda_{tot}(\times 10^6\ s^{-1})$						
	Present pn-QRPA calculations				Experiment	Other theoretical methods	
Nucleus	Λ^V_{tot}	Λ^A_{tot}	Λ^{VA}_{tot}	Λ_{tot}	Λ^{exp}_{tot}	Λ^{theor}_{tot} [23]	Λ^{theor}_{tot} [24]
^{66}Zn	1.651	4.487	−0.204	5.934	5.809	4.976	5.809

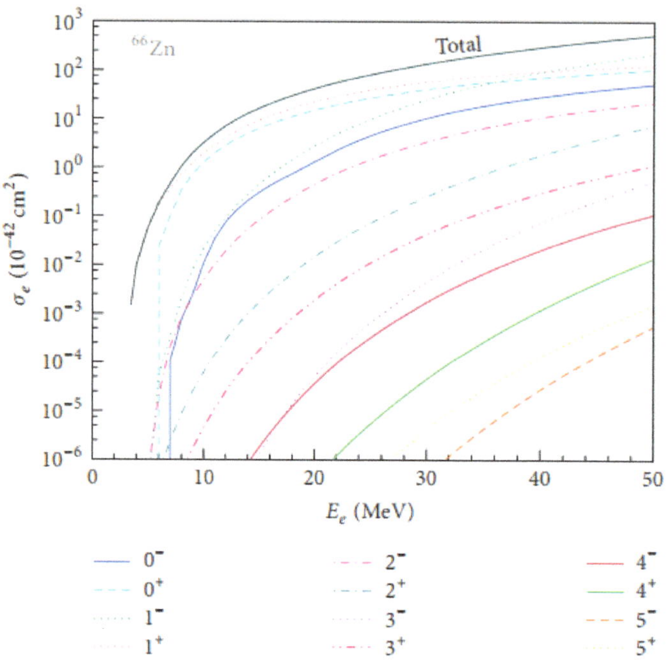

Figure 2: Original total cross sections of electron capture on the ^{66}Zn (parent) nucleus calculated with pn-QRPA method as a function of the incident electron energy E_e. The individual contributions of various multipole channels (for $J^\pi \leq 5^\pm$) are also demonstrated.

From the study of the original electron capture cross sections we conclude that the total cross sections can be well approximated with the Gamow-Teller transitions only in the region of low energies [9–12, 15, 16]. For higher incident electron energies the inclusion of the contributions originating from other multipolarities leads to better agreement [10].

Stellar Electron Capture on ^{66}Zn Isotope.

As it is well known, electron capture process plays a crucial role in late stages of evolution of a massive star, in presupernova and in supernova phases [1–6]. In presupernova collapse, that is, at densities $\rho \leq 10^{10}$ g cm^{-3} and temperatures 300 keV $\leq T \leq$ 800 keV, electrons are captured by nuclei with $A \leq 60$ [9– 12, 15, 16]. During the collapse phase, at higher densities $\rho \geq 10^{10}$g cm^{-3} and temperatures $T \simeq 1.0$ MeV, electron capture process is carried out on heavier and more neutron rich nuclei with $Z < 40$ and $N \geq 40$ [8, 10, 13–15]. In an independent particle picture, the Gamow-Teller transitions (which is the most important in the electron capture cross section calculations) are forbidden for these nuclei [2–4]. However, as it has been demonstrated in several studies, GT transitions in these nuclei are unblocked by finite temperature excitations [21, 22]. At high temperatures, $T \simeq 1.5$ MeV, GT transitions are thermally unblocked as a result of the excitation of neutrons from the pf-shell into the $g9/2$ orbital.

For astrophysical environment, where the finite temperature and the matter density effects cannot be ignored (the initial nucleus is at finite temperature), in general, the initial nuclear state needs to be a weighted sum over an appropriate energy distribution. Then, assuming MaxwellBoltzmann distribution of the initial state $|i\rangle$ in (9) [9, 26], the total e^--capture cross section is given by the expression [10]

$$\sigma(E_e, T) = \frac{G_F^2 \cos^2\theta_c}{2\pi} \sum_i F(Z, E_e) \frac{(2J_i + 1)\, e^{-E_i/(kT)}}{G(Z, A, T)} \times \sum_{f,J} (E_e - Q + E_i - E_f)^2 \frac{|\langle i|\hat{O}_J|f\rangle|^2}{(2J_i + 1)}. \quad (11)$$

The sum over initial states in the latter equation denotes a thermal average of levels, with the corresponding partition function (Z, A, T) [10]. The finite temperature induces the thermal population of excited states in the parent nucleus. In the present work we assume that these excited states in the parent nucleus are all the possible states up to about 2.5 MeV. Calculations involving in addition other states lying at higher energies show that they have no sizeable contribution to the total electron capture cross sections. As mentioned before, for the evaluation of the total electron capture cross sections, the use of a quenched value of g_A is necessary [23, 24, 50, 51]. Since the form factor $F(q^2)$ multiplies the four components of the axial-vector operator (see (A.2)–(A.5)), a quenched value of gA must enter the multipole operators generating the pronounced excitations 0^-, $1^\pm$,..., and so forth. For this reason, in our QRPA calculations we multiplied the free nucleon coupling constant $g_A = 1.262$ by the factor 0.8 [23, 24, 50, 51].

The results coming out of the study of electron capture cross sections under stellar conditions are shown in Figure

3, where the same picture as in the original cross section calculations, but now with larger contribution, is observed. As discussed before, the dominant multipolarity is the 1^+, which contributes by more than 40% to the total cross section. In the region of low energies (up to 30 MeV), the total e −-capture cross section can be described by taking into account only the GT transitions, but at higher incident energies the contributions of other multipolarities become significant and cannot be omitted.

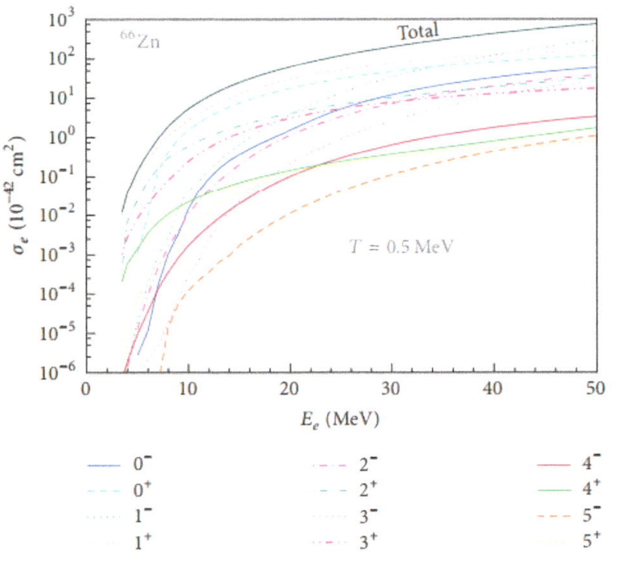

Figure 3: Electron capture cross sections for the ^{66}Zn parent nucleus at high temperature (T = 0.5 MeV) in stellar environment obtained assuming Maxwell-Boltzmann statistics for the incident electrons. The total cross section and the dominant individual multipole channels ($J^\pi \leq 5^\pm$) are demonstrated as functions of the incident electron energy E_e.

The percentage contributions of various multipolarities (with $J^\pi \leq 5^\pm$) into the total e^--capture crosscross section at T = 0.5 MeV and for incident electron energy E_e = 25 MeV are tabulated in Table 7. In addition, in this table we list the values of the individual e^--capture cross cross sections of each multipolarity with $J^\pi \leq 5^\pm$. More specifically, for E_e = 25 MeV the 1^+ multipolarity contributes to about 44%, the 0^+ contributes to about 26%, and the 1^- contributes to about 11%. The contributions coming from other multipolarities are less important (smaller than 5%).

In performing state-by-state calculations for the electron capture cross sections, our code has the possibility to provide separately the contribution of the polar-vector, the axialvector, and the overlap parts induced by the corresponding components of the electron capture operators. In Figure 4 we illustrate the stellar differential cross sections of each individual transition of the polar-vector and axial-vector components.

As mentioned before, our code gives separately the partial e^--capture cross sections of each multipolarity. In order to study the dependence of the differential cross sections on the excitation energy ω throughout the entire pn-QRPA spectrum of the daughter nucleus, a rearrangement of all possible excitations ω in ascending order, with the corresponding cross sections, is required. This was performed by using a special code appropriate for matrices [33]. In the model space chosen for ^{66}Zn isotope, for all multipolarities up to $J^\pi = 5^\pm$, we have a number of 447 final states. The differential electron capture cross sections illustrated in Figure 4 present some characteristic clearly pronounced peaks at various excitation energies ω. These peaks correspond mainly $0^+, 1^+$, and 2^+ transitions. More specifically, in the ^{66}Cu daughter nucleus the maximum peak corresponds to the 0_1^+ QRPA transition at ω = 2.538 MeV and other characteristic peaks correspond to $1_7^+, 1_8^+$, and 1_{10}^+ transitions, located at energies ω = 3.194 MeV, ω = 3.686 MeV, and ω = 6.555 MeV, respectively (see Figure 4). The other less important peaks are also shown in Figure 4.

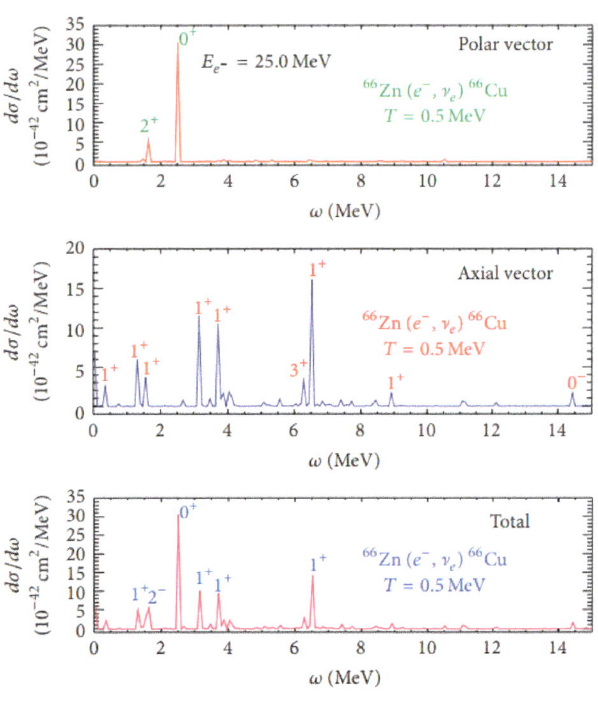

Figure 4: Individual contributions of the polar-vector, (Λ_V), and axial-vector, (Λ_A), components and total electron capture rate as functions of the excitation energy ω (^{66}Zn is the parent nucleus).

Before closing, it should be mentioned that the *e*-capture cross sections presented in this work may be useful in estimating neutrino-spectra arising from *e*-capture on nuclei during supernova phase. The knowledge of]-spectra at every point and time in the core is quite relevant for simulations of the final collapse and explosion phase of a massive star. As it is known [21], in the collapse phase, neutrinos are mainly produced by *e*-capture on nuclei and on free protons. The energy spectra of the emerging neutrinos from both reactions are important ingredients in stellar modelling and stellar simulations [21, 27]. Furthermore, in core collapse simulations one defines the reaction rate of electron capture on nuclei given by

$$R_h = \sum_i Y_i \lambda_i, \tag{12}$$

Table 7: Total e^--capture cross sections (in 10^{-42} MeV^{-1} cm^2) for E_e = 25 MeV. The percentage of each multipolarity into the total e^--capture cross section evaluated with our pn-QRPA code is also tabulated here

	Positive parity transitions			Negative parity transitions	
J^π	$\sigma_e(\times10^{-42}$ cm^2 MeV^{-1})	Portions (%)	J^π	σ_e ($\times10^{-42}$ cm^2 MeV^{-1})	Portions (%)
0^+	31.164	25.96	0^-	5.288	4.41
1^+	52.779	43.98	1^-	13.409	11.14
2^+	6.921	5.77	2^-	3.262	2.72
3^+	5.499	4.58	3^-	0.905	0.75
4^+	0.244	0.20	4^-	0.299	0.25
5^+	0.208	0.17	5^-	0.042	0.04

where the sum runs over all nuclear isotopes present in the astrophysical environment (Y_i denotes the abundance of a given nuclear isotope and λ_i is the calculated electron capture rate for this isotope). The rates of (12) must be known for a wide range of the parameters: T (temperature) and ρ (nuclear density) of the studied star. Thus, for the calculation of the quantity $Y{\cdot}\lambda$ of a specific nuclear isotope one needs to know in addition to the nuclear composition Y the electron capture rates λ calculated as we have shown in our present work. The rates of electron capture on various nuclear isotopes and the corresponding emitted neutrino spectra in the range of the parameters (T, ρ, Y_e) describing the star until reaching equilibrium during the core collapse are comprehensively studied in [21, 26, 27] for a great number of nuclear isotopes by using the large scale shell model.We are currently performing similar calculations for a set of isotopes by employing the present pn-QRPA method [52].

Furthermore, the average neutrino energy, $\langle E_y \rangle$, of the neutrinos emitted by *e*-capture on nuclei can be obtained by dividing the neutrino-energy loss rate (defined by an expression similar to (12) by replacing the rate λ_i with the energy loss rate E_j) with the reaction rate for *e*-capture on nuclei R_h. Assuming, for example, power-law energy distribution for the neutrino spectrum produced by the *e*- capture in supernova phase, the average neutrino-energy $\langle E_v \rangle$ determines a specific supernova-neutrino scenario. In addition, the neutrino emissivity is obtained by multiplying the electron capture rate at nuclear statistical equilibrium with the neutrino-spectra [21, 26, 27]. Finally we note that the rates for the inverse neutrino absorption process are also determined from the electron capture rates obtained as discussed in this section [21].

SUMMARY AND CONCLUSIONS

The electron capture on nuclei plays crucial role during the presupernova and collapse phase (in the late stage e^--capture on free protons is also significant). It becomes increasingly possible as the density in the star's center is enhanced and it is accompanied by an increase of the chemical potential (Fermi energy) of the degenerate electron gas. This process reduces the electron-to-baryon ratio Y_e of the matter composition. In this work, by using our numerical approach based on a refinement of the pn-QRPA that describes reliably all the semileptonic weak interaction processes in nuclei, we studied in detail the electron capture process on ^{66}Zn isotope and calculated original and stellar *e* –-capture cross sections. We tested our nuclear model (the pn-QRPA) through the reproducibility of orbital muon capture rates for this isotope. The agreement with experimental data and other reliable theoretical results of partial and total μ-capture rates and of the percentage contributions of various low-lying excitations is quite good which provides us with high confidence level for the obtained cross sections. Our future plans are to extend the application of this method and make similar calculations for other interesting nuclei [52]. Also this method could be applied to other semileptonic nuclear processes like beta-decay and chargedcurrent neutrino-nucleus processes important in nuclear astrophysics and neutrino nucleosynthesis.

CONFLICT OF INTERESTS

The authors declare that there is no conflict of interests regarding the publication of this paper.

ACKNOWLEDGMENTS

This research has been cofinanced by the European Union (European Social Fund-ESF) and Greek national funds through the Operational Program "Education and Lifelong Learning" of the National Strategic Reference Framework (NSRF)—Research Funding Program: Heracleitus II, investing in knowledge society through the European Social Fund.

REFERENCES

1. K. Langanke and G. Martinez-Pinedo, "Nuclear weakinteraction processes in stars," Reviews of Modern Physics, vol. 75, p. 819, 2003.
2. G. M. Fuller, W. A. Fowler, and M. J. Newman, "Stellar weakinteraction rates for sd-shell nuclei. I. Nuclear matrix element systematics with application to Al-26 and selected nuclei of importance to the supernova problem," Astrophysical Journal Supplement Series, vol. 42, pp. 447–473, 1980.
3. G. M. Fuller, W. A. Fowler, and M. J. Newman, "Stellar weak interaction rates for intermediate-mass nuclei.

II—A = 21 to A = 60," The Astrophysical Journal, vol. 252, pp. 715–740, 1982.

4. G. M. Fuller, W. A. Fowler, and M. J. Newman, "Stellar weak interaction rates for intermediate mass nuclei. III—rate tables for the free nucleons and nuclei with A = 21 to A = 60," Astrophysical Journal Supplement, vol. 48, pp. 279–319, 1982.
5. M. B. Aufderheide, I. Fushiki, S. E. Woosley, and D. H. Hartmann, "Search for important weak interaction nuclei in presupernova evolution," Astrophysical Journal, Supplement Series, vol. 91, no. 1, pp. 389–417, 1994.
6. H. A. Bethe, "Supernova mechanisms," Reviews of Modern Physics, vol. 62, p. 801, 1990.
7. K. Langanke and G. Martinez-Pinedo, "Supernova electron capture rates for 55Co and 56Ni," Physics Letters B: Nuclear, Elementary Particle and High-Energy Physics, vol. 436, no. 1-2, pp. 19–24, 1998.
8. Y. F. Niu, N. Paar, D. Vretenar, and J. Meng, "Stellar electroncapture rates calculated with the finite-temperature relativistic random-phase approximation," Physical Review C, vol. 83, no. 4, Article ID 045807, pp. 045850–10, 2011.
9. D. J. Dean, K. Langanke, L. Chatterjee, P. B. Radha, and M. R. Strayer, "Electron capture on iron group nuclei," Physical Review C, vol. 58, no. 1, pp. 536–544, 1998.
10. N. Paar, G. Colo, E. Khan, and D. Vretenar, "Calculation of stellar electron-capture cross sections on nuclei based on microscopic Skyrme functionals," Physical Review C, vol. 80, Article ID 055801, 2009.
11. J. U. Nabi, M. U. Rahman, and M. Sajjad, "Electron and positron capture rates on 55Co in stellar matter," Brazilian Journal of Physics, vol. 37, no. 4, 2007.
12. J. U. Nabi, "Ground and excited states Gamow-Teller strength distributions of iron isotopes and associated capture rates for core-collapse simulations," Astrophysics and Space Science, vol. 331, no. 2, pp. 537–554, 2011.
13. J. U. Nabi, M. Sajjad, and M. U. Rahman, "Electron capture rates on titanium isotopes in stellar matter," Acta Physica Polonica B, vol. 38, no. 10, pp. 3203–3224, 2007.
14. A. L. Cole, T. S. Anderson, R. G. T. Zegers et al., "GamowTeller strengths and electron-capture rates for pf-shell nuclei of relevance for late stellar evolution," Physical Review C, vol. 86, Article ID 015809, 2012.
15. Q. Zhi, K. Langanke, G. Mart´ınez-Pinedo, F. Nowacki, and K. Sieja, "The 76Se Gamow-Teller strength distribution and its importance for stellar electron capture rates," Nuclear Physics A, vol. 859, no. 1, pp. 172–184, 2011.
16. P. Sarriguren, E. M. de Guerra, and R. Alvarez-Rodr ´ ´ıguez, "Gamow-Teller strength distributions in Fe and Ni stable isotopes," Nuclear Physics A, vol. 716, pp. 230–244, 2003.
17. J. Toivanen, E. Kolbe, K. Langanke, G. Mart´ınez-Pinedo, and P. Vogel, "Supernova neutrino induced reactions on iron isotopes," Nuclear Physics A, vol. 694, no. 1-2, pp. 395–408, 2001.
18. E. Kolbe, K. Langanke, G. Mart´ınez-Pinedo, and P. Vogel, "Neutrino-nucleus reactions and nuclear structure," Journal of Physics G: Nuclear and Particle Physics, vol. 29, no. 11, pp. 2569– 2596, 2003.
19. C. Frohlich, G. Martinez-Pinedo, M. Liebendorfer et al., "Neutrino-induced nucleosynthesis of $A > 64$ nuclei: the $.p$ process," Physical Review Letters, vol. 96, no. 14, Article ID 142502, 4 pages, 2006.
20. A. Juodagalvis, K. Langanke, G. Mart´ınez-Pinedo, W. R. Hix, D. J. Dean, and J. M. Sampaio, "Neutral-current neutrino-nucleus cross sections for A~50–65 nuclei," Nuclear Physics A, vol. 747, no. 1, pp. 87–108, 2005.
21. K. Langanke, G. Martinez-Pinedo, J. M. Sampaio et al., "Electron capture rates on nuclei and implications for stellar core collapse," Physical Review Letters, vol. 90, no. 24, Article ID 241102, 4 pages, 2003.
22. K. Langanke, E. Kolbe, and D. J. Dean, "Unblocking of the Gamow-Teller strength in stellar electron capture on neutronrich germanium isotopes," Physical Review C, vol. 63, no. 3, Article ID 032801, 5 pages, 2001.
23. N. T. Zinner, K. Langanke, and P. Vogel, "Muon capture on nuclei: pandom phase approximation evaluation versus data for $6 \leq Z \leq 94$ nuclei," Physical Review C, vol. 74, Article ID 024326, 2006.
24. T. Marketin and N. Paar, "Relativistic quasiparticle randomphase approximation calculation of total muon capture rates," Physical Review C, vol. 79, no. 5, Article ID 054323, 8 pages, 2009.
25. E. Kolbe, K. Langanke, and P. Vogel, "Comparison of continuum random phase approximation and the elementary particle model for the inclusive muon neutrino reaction on $12C$," Nuclear Physics A, vol. 613, no. 4, pp. 382–396, 1997.
26. K. Langanke and G. Martinez-Pinedo, "Shell-model calculations of stellar weak interaction rates: II. Weak rates for nuclei in the mass range A=45−65 in supernovae environments," Nuclear Physics A, vol. 673, no. 1–4, pp. 481–508, 2000.
27. K. Langanke, G. Martinez-Pinedo, and J. M. Sampaio, "Neutrino spectra from stellar electron capture," Physical Review C, vol. 64, no. 5, Article ID 055801, 7 pages, 2001.
28. R. W. Hix, O. E. B. Messer, A. Mezzacappa et al., "Consequences of nuclear electron capture in core collapse supernovae," Physical Review Letters, vol. 91, Article ID 201102, 2003.
29. P. Ring and P. Schuck, The Nuclear Many-Body Problem, Springer, New York, NY, USA, 1969.
30. T. S. Kosmas and E. Oset, "Charged current neutrino-nucleus reaction cross sections at intermediate energies," Physical Review C: Nuclear Physics, vol. 53, no. 3, pp. 1409–1415, 1996.
31. V. C. Chasioti and T. S. Kosmas, "A unified formalism for the basic nuclear matrix elements in semi-leptonic processes," Nuclear Physics A, vol. 829, no. 3-4, pp. 234–252, 2009.
32. P. G. Giannaka and T. S. Kosmas, "Electron-capture and its role to explosive neutrino-nucleosynthesis," Journal

of Physics: Conference Series, vol. 410, no. 1, Article ID 012124, 2013.

33. V. Tsakstara and T. S. Kosmas, "Low-energy neutral-current neutrino scattering on 128,130Te isotopes," Physical Review C, vol. 83, no. 5, Article ID 054612, 13 pages, 2011.

34. K. G. Balasi, E. Ydrefors, and T. S. Kosmas, "Theoretical study of neutrino scattering off the stable even Mo isotopes at low and intermediate energies," Nuclear Physics A, vol. 868–869, pp. 82– 98, 2011.

35. K. G. Balasi, E. Ydrefors, T. S. Kosmas, and J. Suhonena, "The response of 95,97Mo to supernova neutrinos," Nuclear Physics A, vol. 866, pp. 67–78, 2011.

36. V. Tsakstara and T. S. Kosmas, "Analyzing astrophysical neutrino signals using realistic nuclear structure calculations and the convolution procedure," Physical Review C, vol. 84, no. 6, Article ID 064620, 14 pages, 2011.

37. E. Ydrefors, K. G. Balasi, T. S. Kosmas, and J. Suhonen, "Detailed study of the neutral-current neutrino–nucleus scattering off the stable Mo isotopes," Nuclear Physics A, vol. 896, pp. 1–23, 2012.

38. V. Tsakstara and T. S. Kosmas, "Nuclear responses of 64,66Zn isotopes to supernova neutrinos," Physical Review C, vol. 86, no. 4, Article ID 044618, 10 pages, 2012.

39. T. S. Kosmas, A. Faessler, and J. D. Vergados, "The new limits of the neutrinoless ($\mu-$, $e-$) conversion branching ratio," Journal of Physics G: Nuclear and Particle Physics, vol. 23, no. 6, pp. 693– 703, 1997.

40. R. A. Eramzhyan, V. A. Kuz'min, and T. V. Tetereva, "Calculations of ordinary and radiative muon capture on 58,60,62Ni," Nuclear Physics A, vol. 642, no. 3-4, pp. 428–448, 1998.

41. E. Kolbe, K. Langanke, and P. Vogel, "Muon capture on nuclei with $N>Z$, random phase approximation, and in-medium value of the axial-vector coupling constant," Physical Review C, vol. 62, no. 5, Article ID 055502, 4 pages, 2000.

42. T. S. Kosmas, "Exotic $\mu- \rightarrow e-$ conversion in nuclei: energy moments of the transition strength and average energy of the outgoing," Nuclear Physics A, vol. 683, no. 1, pp. 443–462, 2001.

43. B. S. Meyer, "The r-, s-, and p-processes in nucleosynthesis," Annual Review of Astronomy and Astrophysics, vol. 32, pp. 153– 190, 1994.

44. T. W. Donnelly and R. D. Peccei, "Neutral current effects in nuclei," Physics Reports, vol. 50, no. 1, pp. 1–85, 1979.

45. Y. Tanaka, Y. Oda, F. Petrovich, and R. K. Sheline, "Effect of the spin-orbit potential on the single particle levels in the superheavy region," Physics Letters B, vol. 83, no. 3-4, pp. 279– 283, 1979.

46. E. V. Bugaev, G. S. Bisnovatyi-Kogan, M. A. Rudzsky, and Z. F. Seidov, "The interaction of intermediate energy neutrinos with nuclei," Nuclear Physics A, vol. 324, no. 2-3, pp. 350–364, 1979.

47. J. Vary, Private communication, http://nuclear.physics.iastate .edu/npc.php.

48. M. S. Yousef, V. Rodin, A. Faessler, and F. Simkovic, "Twoneutrino double β decay of deformed nuclei within the quasiparticle random-phase approximation with a realistic interaction," Physical Review C, vol. 79, Article ID 014314, 2009.

49. V. Rodin and A. Faessler, "First application of the continuumQRPA to the description of the double beta decay," Progress in Particle and Nuclear Physics, vol. 57, pp. 226–234, 2006.

50. O. Hausser, M. C. Vetterli, R. W. Fergerson et al., "Nuclear response in the 54Fe($p \rightarrow$, $p \rightarrow$')reaction at 290 MeV," Physical Review C, vol. 43, no. 1, pp. 230–249, 1991.

51. B. H. Wildenthal, "Empirical strengths of spin operators in nuclei," Progress in Particle and Nuclear Physics, vol. 11, pp. 5– 51, 1984.

52. P. G. Giannaka and T. S. Kosmas, In preparation.

53. D. K. Papoulias and T. S. Kosmas, "Nuclear aspects of neutral current non-standard .-nucleus reactions and the role of the exotic $\mu- \rightarrow$ e − transitions experimental limits," Physics Letters B, vol. 728, p. 482, 2014.

54. D. K. Papoulias and T. S. Kosmas, "Exotic lepton flavour violating processes in the presence of nuclei," Journal of Physics: Conference Series, vol. 410, Article ID 012123, 2013.

CHAPTER 16

HIGGS SHIFTS FROM ELECTRON–POSITRON ANNIHILATIONS NEAR NEUTRON STARS

Gary A. Wegner, Roberto Onofrio

INTRODUCTION

The interplay between the mass generation through the Higgs field and gravitation is an active subfield of investigation and the concept of mass is crucial both in quantum field theory and gravitation. Consequently, any common insight may shed light on the possible unification of the two theories. Inflation, although successful in removing potential contradictions to the standard Big Bang model, does not yet have a clear microscopic interpretation and unambiguous observational evidence, and the Higgs field has been conjectured to be responsible for it [1,2]. To these long standing motivations, others have been added since the discovery at the LHC of a scalar particle with the mass and decay branching ratios as expected from the Higgs boson in the minimal Higgs doublet model [3,4]. In the absence of signals for new physics in the current experimental setting, the extrapolation of the standard model to the Planck scale raises an issue of stability of the vacuum for the specific value of the quartic coupling of the Higgs self-interaction which is extracted from its observed mass [5–8]. Various recent contributions point out a solution with no new physics beyond the standard model apart from a non-minimal coupling of the Higgs field to curvature invariants [9–11]. Finally, the tentative interpretation of the BICEP-2 results [12] in terms of gravitational waves produced during inflation, still to be fully scrutinized and compared with Planck results [13], would call for nonminimal coupling of the Higgs field to gravity [14] or to the inflaton [15]. Recently, we discussed upper bounds to the coupling between the Higgs field and a specific curvature invariant, the Kretschmann invariant, based on the analysis of molecular lines of C2 and atomic lines of H, C, Ca, and Mg from the surface of two white dwarfs [16]. This bound, although already competitive with respect to what is achievable in table-top experiments, in principle may be improved by many orders of magnitude by exploiting the strong gravity at the surfaces of neutron stars. The boost in sensitivity is easily estimated as the Kretschmann invariant $K = R_{\mu\nu\rho\sigma} R^{\mu\nu\rho\sigma}$, with $R^{\mu\nu\rho\sigma}$ the Riemann curvature tensor, depends on the sixth power of the radius of the astrophysical object. Thus, for an Earth-radius white dwarf and a 10 km neutron star of equal mass, a gain of order (6000/10)6 4.7 × 1016 is expected if spectral features could be measured with precision comparable to the ones studied in white dwarfs. The crucial issue is to get measurable features from neutron star spectra, and in the following we discuss possibilities to be considered, in the form of reanalysis of already collected data, and future dedicated observations. In Sect. 2 we consider annihilation lines tentatively observed during gammaray bursts with redshifts compatible to that expected from the surface of neutron stars, and discuss the possibility for an anomalous contribution with respect to the redshifts due to lines of nuclear origin. In Sect. 3 this evidence is conservatively considered as a bound on a possible signal due to the Higgs shift, and the related bounds discussed. In Sect. 4 we discuss possible Higgs shifts arising from a comparative analysis of electron–positron and proton–antiproton annihilations, always in proximity of a neutron star, which should result in similar bounds to the one discussed earlier. More general considerations on the possibility of observing simultaneously cyclotron lines from

Citation: Gary A. Wegner, Roberto Onofrio, Higgs shifts from electron–positron annihilations near neutron stars, DOI 10.1140/epjc/s10052-015-3523-5.

electrons and protons are discussed in the conclusions, as well as possible generalizations of the Higgs coupling in theories and models beyond general relativity.

SEARCH FOR ANOMALOUS SHIFTS IN GAMMA-RAY BURSTS

The most likely avenue toward obtaining better bounds to Higgs-Kretschmann couplings from signals in proximity of neutron stars is, to our knowledge, the comparative analysis of electron–positron annihilation lines and narrow lines due to nuclear de-excitations during transients of gammaray bursts. In the following we focus on the event collected by the Goddard germanium Gamma-Ray Burst Spectrometer on board of ISEE-3, GRB 781119, [17,18], as several lines attributed to nuclei like ^{56}Fe, ^{24}Mg, ^{20}Ne, ^{28}Si, ^{14}N were identified. A less prominent peak at 420 keV has also been identified and interpreted as a redshifted e^+e^- annihilation peak. The same event was observed by the Konus gamma-ray observatory of the Leningrad group at the Ioffe Institute [19,20], including the emission feature at 420 keV. The redshift required to justify this line as a e^+e^- annihilation peak is compatible with the gravitational redshift (z 0.2) expected on the surface of a neutron star. More qualitatively, the putative electron–positron line is observed at E_{annih} = (420 ± 20) keV, which is 0.82 ± 0.04 times the value of the electron–positron two-body annihilation line at 511 keV. For the 56Fe line, we have an observed value of EFe = (738 ± 10) keV versus an unshifted energy of 847 k_{eV}, i.e. an energy ratio of 0.87 ± 0.01. Therefore the electron–positron line is more redshifted than the ^{56}Fe line by one standard deviation. Although the discrepancy between the two redshifts is contained within one standard deviation only, by inferring the redshifts of the other identified nuclear lines from Fig. 6 in [18] we notice that they all result in systematically smaller redshifts than the electron–positron annihilation line, at the level of 4.2 standard deviations, as discussed in Table 1.

While it is tempting to interpret this extra shift as due to the Higgs shift, more conservatively it can be used to determine an upper bound on the associated Higgs-Kretschmann coupling, for two reasons. The interpretation of the 420 keV peak as due to electron– positron annihilation is not solid, and needs to be corroborated by more data. There is a general consensus that spectral features in the 300–400 keV region can be interpreted as due to a gravitationally redshifted e^+e^- annihilation line; see [22] for observations of the galactic center and [23] for their

Table 1: Analysis of the line shifts from the ISEE-3 data for the gamma-ray burst event recorded on 19 November 1978 reported in [18]. The origin of each line is in the first column, followed by the transition energy as measured in the laboratory [21], the one observed by ISEE-3, and the related redshift evaluated as $_{zline} = E_{lab}/E_{obs} - 1$. The error bars are evaluated from our analysis of full-width half maxima of the interpolating curves appearing in Fig. 6 of [18], and they are used to obtain weighted average and standard deviation on the eight redshift determinations for the nuclear lines, resulting in $<z_{nucl}>$ = 0.134±0.017, which is smaller by 4.2 standard deviations with respect to the one evaluated using the electron–positron annihilation peak, zannih = 0.217 ± 0.010. The instrumental error for the nuclear transitions reported in [18] is quoted only for the 56Fe line as 10 k_{eV}, corresponding to a relative error of 1.3 %. If we assume the same instrumental error for all remaining nuclear lines, then this last is smaller by one order of magnitude with respect to the statistical relative error on $\langle z_{nucl} \rangle$, and we get $\langle z_{nucl} \rangle$ = 0.143 ± 0.018, which differs by 3.6 standard deviation from zannih, corroborating the former analysis based on our graphical assessment of the error bars in Fig. 6 of [18] Li interpretation. However, the absence of similar signals in later observational campaigns like BATSE strongly constrains the initial interpretation of the observed peaks [24–26]. On the instrumental side, the energy deposition of each photon may not be completely occurring within the detector, and therefore assumptions must be made on the incident spectrum [27–29]. The very existence of the line features has been criticized [30,31], and alternative explanations have been put forward, more specifically as originating from de-excitation of $^7Li^*$ in cosmic rays [32] and from an amplification mechanism through stimulated annihilation radiation [33].

Line	E_{lab} (keV)	E_{obs} (keV)	z_{line}
e^+e^-	511	420 ± 20	0.217 ± 0.010
^{56}Fe	847	738 ± 40	0.148 ± 0.008
^{56}Fe	1238	1076 ± 33	0.151 ± 0.005
^{24}Mg	1369	1164 ± 36	0.176 ± 0.005
^{20}Ne	1634	1444 ± 33	0.132 ± 0.003
^{28}Si	1779	1589 ± 33	0.120 ± 0.002
^{56}Fe	1811	1612 ± 40	0.123 ± 0.003
^{14}N	2313	2011 ± 70	0.150 ± 0.005

Moreover, various environmental factors might create differential shifts between the e^+e^- annihilations and the nuclear lines as there are uncertainties in the models of gamma-ray bursts in neutron stars. Although it is reasonable to assume that the nuclear lines originate from matter on the surface of the neutron star with null Doppler shift, peculiar motions of the e^+e^- plasma clouds may add or subtract a Doppler shift to the gravitational redshift. The broadening and shift due to the finite temperature of the e^+e^- plasma cloud could reduce the observed redshift. The annihilation line broadens proportionally to $T^{1/2}$ for $k_B T << m_e c^2$, and to T for $k_B T >> m_e c^2$, and the peak of the line shifts toward higher energies as $\delta E_{peak}/E_{peak} \simeq 1.25 k_B T/m_e c^2$ [34–38]. This blueshift is going to aggravate the redshift excess we

Table 2: Analysis of the line shifts from emission features in the 70– 470 keV interval from the Konus catalogs [39–41]. The date of the event for each line is in the first column, followed by the observed emission energy, the corresponding redshift if attributed to the e^+e^- annihilation peak, and the M/R parameter expressed in units of $c^2/(2GN)$

Event	E_{obs} (keV)	z_{line}	M/R
18/09/78	380	0.345	0.447
21/09/78	350	0.460	0.531
06/10/78a	420	0.217	0.324
06/10/78b	350	0.460	0.531
23/10/78	280	0.825	0.700
05/03/79	380	0.345	0.447
06/04/79	320	0.597	0.608
02/05/79	470	0.087	0.154
26/05/79	320	0.597	0.608
22/06/79	450	0.136	0.224
28/06/79	410	0.246	0.356
09/11/79	320	0.597	0.608

have discussed. It is also worth remarking that a secondary peak on the right side of the putative e^+e^- annihilation peak is present at an energy of 484 keV in the ISSE-3 data [18]. If this is interpreted as the actual e^+e^- annihilation peak, the corresponding redshift is only $z_{annih} = (0.056 \pm 0.002)$, blueshifted by the average value of the redshift from nuclear lines by about 4.3 standard deviations. Based on the redshift of the nuclear lines, a temperature of the e^+e^- plasma cloud of 20 keV (corresponding to 2.6 × 108 K) should be required to fully justify this blueshift, which seems compatible with the typical surface temperature of neutron stars. A statistical comparison between the distribution of the energy of the putative e^+e^- annihilation peaks in gamma-ray bursts and the mass–radius distribution of neutron stars may also be used both to check the hypotheses attributing these events to neutron stars and to check for systematic deviations. The mass–radius ratio is related to the energy of the observed line and the redshift as

$$\frac{M}{R} = \frac{c^2}{2G_N}\left[1 - \left(\frac{E_{\text{obs}}}{E_{\text{lab}}}\right)^2\right] = \frac{c^2}{2G_N}\left[1 - \frac{1}{(1+z)^2}\right]. \tag{1}$$

The Konus collaboration has published three catalogs of gamma-ray bursts [39–41], and in 25 cases we find evidence for emission peaks at energies in the 100–470 keV range. Within this subsample of events, we apply a lower model-dependent cutoff based on the requirement for causality [42,43]. If the events originate on the surface of a neutron star, this requires $M/R \leq 0.708c^2/(2GN)$ leading to $E_{obs} \geq 276$ keV. The resulting events are reported in Table 2. This leads to a statistical distribution of the annihilation energy peaks, depicted in Fig. 1, which is also compared with the mass-to-radius ratio distribution for a sample of neutron stars obtained from [44–46], considering a range of values for Eobs coming from the allowed intervals for mass and radius (see [44–46] for a discussion of the errors in the determination of mass and radius, and their sensitivity to the equations of state) and summing up the resulting rectangular windows of unit height. The two distributions are at least limited by roughly the same interval, but the Konus distribution seems more peaked at lower energies. The small number of events in the Konus data prevents us to make more quantitative analyses.

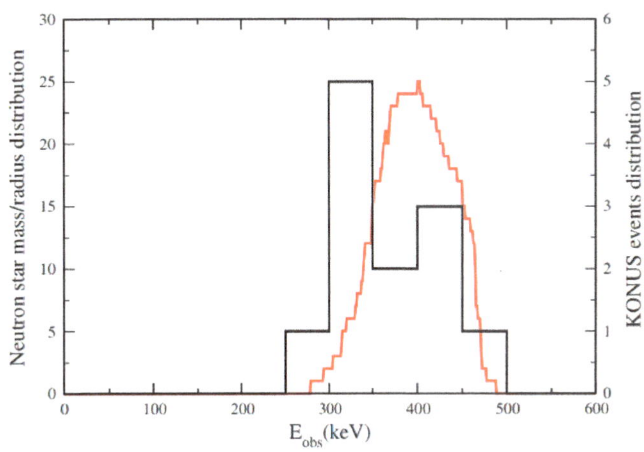

Figure. 1: Statistics of the distribution of the observed energy E_{obs}. Distribution of the redshifts ratio as deduced from the emission lines of the Konus catalog, see Table 2, in the hypotheses they represent e^+e^- annihilation lines (black line, right vertical scale), and from 26 determinations of gravitational redshifts for neutron stars [44–46] (red line, left vertical scale)

BOUNDS ON THE HIGGS-KRETSCHMANN COUPLING

As detailed in former contributions [16,47,48], if the Higgs field φ characterized by quadratic and quartic coefficients μ and λ is coupled to the Kretschmann curvature invariant K via the Lagrangian density term $\xi_K \Lambda_{\rm Pl}^2 \phi^2 K$, with ξ_K their coupling constant and Λ_{Pl} the Planck length, the effective mass parameter of the Higgs field gets an extra-term due to the scalar curvature as $\mu^2 \mapsto \mu^2(1 + \xi_K \Lambda_{\rm Pl}^2 \lambda_\mu^2 K)$, with λ_μ the Compton wavelength of the Higgs field corresponding to its mass of 125 GeV ($\lambda_\mu = 1.6 \times 10^{-18}$ m), and the vacuum expectation value of the Higgs field v depends on space as

$$v = \sqrt{-\frac{\mu^2 + \xi_K \Lambda_{\rm Pl}^2 K}{\lambda}} \simeq v_0 \left(1 + \frac{\xi_K \Lambda_{\rm Pl}^2 K}{2\mu^2}\right), \tag{2}$$

in a weak-curvature limit, i.e. with curvature length scale much larger than λ_μ, a limit well satisfied in all circumstances of astrophysical interest. Then the mass m_e of the electron will be simply changed proportionally to the Higgs vacuum expectation value

$$\delta m_e = \frac{y_e}{\sqrt{2}}(v - v_0) \simeq \frac{y_e \xi_K \Lambda_{\rm Pl}^2 K v_0}{2^{3/2}\mu^2} = \frac{1}{2}\xi_K \Lambda_{\rm Pl}^2 \lambda_\mu^2 K m_e. \tag{3}$$

where y_e is the Yukawa coupling of the electron. Hadrons, having most of their mass arising from QCD vacuum, are instead minimally affected by the curvature invariant. In order to get upper bounds based on the data above, we first check that the Higgs shift expected for the nuclear lines is indeed negligible with respect to the one expected in the e^+e^- annihilation peak. The relationship between the mass of a nucleon and the energy levels of the corresponding nucleus is not trivially available, as there are many phenomenological models based either on a single nucleon approach, in which a nucleon evolves in the mean-field potential created by the remaining nucleons, as in the shell model approach, or on a many-body collective approach as in the droplet model [49]. Considering that strong interactions are mainly responsible for the binding between nucleons, we assume that in a simplified treatment with a harmonic oscillator potential, the energy levels will scale as the inverse of the square root of the involved mass, the mass of the nucleon in the shell models, and the whole mass of the nucleus in collective models. In this case the expected Higgs shift should scale as $\delta\lambda/\lambda \simeq \delta m_n/(2m_n)$, the relevant mass (ranging between the two extreme values of the nucleon mass, for instance the one of the proton m_p, and the nucleus mass m_N). In an infinite square-well model instead the expected scaling should be the inverse of the mass, $\delta\lambda/\lambda \simeq \delta m_n/m_n$, a mere factor 2 larger than in the harmonic potential.

In the case of a single nucleon, relevant for single-particle models such as the shell model, we have, focusing on the proton mass, $m_p \simeq m_{\rm QCD} + (2y_u + y_d)v/\sqrt{2} \simeq 928 + 10(1 + 0.5\xi_K \Lambda_P^2 \lambda_\mu^2 K)$ (with y_u and y_d the Yukawa couplings of the up and down quarks, mQCD the purely gluonic contribution to the proton mass, and all masses and energies expressed in MeV/c^2), which implies $\delta m_p/m_p \simeq 5 \times 10^{-3}\xi_K \Lambda_P^2 \lambda_\mu^2 K$. This has to to be compared with the sensitivity to the Higgs-Kretschmann coupling of the electron, $\delta m_e/m_e \simeq 0.5\xi_K \Lambda_P^2 \lambda_\mu^2 K \simeq 10^2 \delta m_p/m_p$. In order to estimate the mass shift in the case of collective models, we consider the nucleus with the larger number of lines observed as in Table 1, the iron isotope with mass number 56 made of 26 protons and 30 neutrons. Its mass can be written in terms of the Yukawa couplings y_u and y_d of the up and down valence quarks inside protons and neutrons obtaining

$$m_{\rm Fe} = 56 m_{\rm QCD} + \frac{(82y_u + 86y_d)v}{\sqrt{2}} - \Delta m(A, Z), \tag{4}$$

where we have introduced the mass defect $\Delta m(A, Z)$.. In the presence of a curved spacetime with Kretschmann coupling to the Higgs field the mass of a ^{56}Fe nucleus will be

$$m_{\rm Fe}(\xi_K) = m_{\rm Fe}(0)\left(1 + \frac{82y_u + 86y_d}{2\sqrt{2}}\xi_K \Lambda_P^2 \lambda_\mu^2 K\right). \tag{5}$$

Again, the relative mass shift of the nucleus turns out to be smaller than the relative mass shift of the electron by a factor even larger than in the single nucleon case, since $(\delta m_e/m_e)/(\delta m_{\rm Fe}/m_{\rm Fe}) \simeq \sqrt{2} m_{\rm Fe}/(82y_u + 86y_d)v_0 \simeq 87$. Therefore, both the extreme examples of the single nucleon mass and the whole 56Fe mass determining the nuclear spectroscopy show that their contributions are negligible with respect to the mass shift of the electron, the nuclear line thus providing a spectroscopic 'anchor'. This implies that we may attribute the Higgs shift to the electron mass shift alone and, by assuming the Planck length $\Lambda_P = 10^{-35}$ m and a solar mass neutron star with radius 10 km (corresponding to a Kretschmann invariant on the neutron star surface of K = 10^{-16} m^{-4}), the maximum Higgs shift compatible with the observed excess of redshift $\delta z = z_{\rm annih} - \langle z_{\rm nucl}\rangle \geq 1.28 \times 10^{-122}\xi_K$, gives an upper bound on the Higgs-Kretschmann coupling coupling $\xi_K \leq 5.8 \times 10^{120}$ in MKSA units, translated into a value of 7.1 × 10^{35} in natural units. If the line at 484 keV is instead considered responsible for the e^+e^- annihilation, the Higgs shift is negative, $\delta z = z_{\rm annih} - \langle z_{\rm nucl}\rangle = -0.087 \pm 0.018$, corresponding to an upper bound of $\xi_K \geq -6.8 \times 10^{120}$ (i.e. $|\xi_K| \leq 6.8 \times 10^{120}$), always in MKSA units. By assuming $\Lambda_p = 10^{-19}$ m as in models with the Planck scale coinciding with the Fermi scale [50,51], the bounds are correspondingly stronger by a factor $\simeq 10^{32}$.

ELECTRON–POSITRON AND PROTON–ANTI-PROTON ANNIHILATIONS NEAR NEUTRON STARS

An alternative possibility to study Higgs shifts is to compare electron–positron and proton–antiproton annihilation signals from neutron stars, as suggested in [47]. Unlike e^+e^- annihilations, $p\bar{p}$ annihilations produce a continuous photon spectrum since the annihilation produces multiple π^0 mesons in turn decaying into photons. Consequently the photon spectrum has an intrinsically broad peak due to the more than two-body decays, which is further Doppler-broadened by the velocity spread of the involved particles. Due to the flatness of the expected photon spectrum from $p\bar{p}$ annihilation, possible limits on the Kretschmann coupling from this class of events are weaker than the former class of events. The continuous gamma-ray spectrum in the $p\bar{p}$ annihilations has been evaluated in [52] by fitting a Monte Carlo simulation with an analytical function as

$$F(E_\gamma) = N[(E_p - E_\gamma)^{\alpha_1} e^{\beta_1} + (E_p - E_\gamma)^{\alpha_2} e^{\beta_2} + \beta_3 e^{\alpha_3 E_\gamma}], \tag{6}$$

Where $E_p = m_p c^2$ is the proton mass in energy units, and $\alpha_i, \beta_i\ (i = 1 \div 3)$ are fitting parameters available in [52]. The position of the annihilation peak depends upon the assumed proton mass, as shown in Fig. 2, with a best fit yielding $\bar{E}_\gamma = 187.05 - 0.12385 E_p$ (all energies in MeV), with the slope $\delta\bar{E}_\gamma/\delta m_p = 0.12385$ expressing its sensitivity. If the minimum detectable peak shift is then $\delta\bar{E}_\gamma \simeq 0.1$ MeV, the minimum detectable proton mass shift is MeV, the minimum detectable proton mass shift is$\simeq 0.8$ MeV, i.e. $\delta m_p/m_p \simeq 8 \times 10^{-4}$ and the situation is similar to the one in the previous section as the sensitivity of the proton shift to the spacetime curvature is far smaller than the one of the electron. For the electron–positron annihilations, the limitation is due to the intrinsic resolution of the 511 keV peak which depends upon the environmental temperature and energy resolution of the detector, estimated, respectively, to be $\Delta E/E|_{\rm env} = K_B T/E_\gamma \simeq 10^{-5}$, and $\Delta E/E|_{\rm instr} = 1.47 \times 10^{-4}$ [53]. By using the Rayleigh criterion for resolving a shift of the annihilation peak, with a full-width half maximum of 2.37 keV as quoted in [53], and the same values of Λ_{Pl}, λ_μ, mass and radius of the neutron star used above, we get a bound $\xi K = 4.4\times10^{34}$ using natural

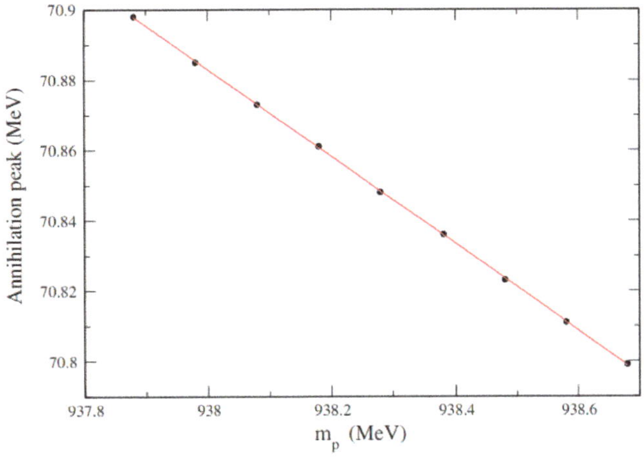

Figure. 2: Sensitivity of the energy distribution peak in $p\bar{p}$ annihilations to the proton mass. The fit yields $\bar{E}_\gamma = 187.05 - 0.12385 m_p$, with energies and the proton mass expressed in MeV

Table 3: Summary of bounds on the Higgs-Kretschmann couplings (in natural units) from various sources, experiments to test the superposition principle of the gravitational force as discussed in [48], the analysis of differential shifts in spectral lines from two white dwarfs [16], the analysis of differential shifts between e^+e^- annihilation line and nuclear lines with two candidates at 420 keV (a) and 484 keV (b) as discussed in Sect. 3, and possible comparisons between e^+e^- and $p\bar{p}$ annihilation lines originating from the surface of neutron stars, as discussed in Sect. 4. The bounds are evaluated for a choice of the Planck length of $\Lambda_{Pl} = 10^{-35}$ m, the bounds for the choice of $\Lambda_{Pl} = 10^{-19}$ m being 10^{32} times stronger

Source	$\xi_K\,(\Lambda_{\rm Pl} = 10^{-35}{\rm m})$
Table-top experiments	2.5×10^{60}
BPM 27606	5×10^{50}
Procyon B	9×10^{50}
e^+e^- vs. nuclei (a)	7.1×10^{35}
e^+e^- vs. nuclei (b)	-8.4×10^{35}
e^+e^- vs. $p\bar{p}$	4.4×10^{34}

units, one order of magnitude stronger than to the one evaluated above for the gamma-ray burst event. Table 3 summarizes our discussion by including upper bounds from laboratory measurements, actual spectroscopy from white dwarfs, and the potential observations from neutron stars reported in this paper.

CONCLUSIONS

In summary, we have discussed neutron stars as potential tools to constrain a specific Higgs-curvature connection. The most promising seems to be a reanalysis of the redshifted signals during GRB events, and analysis of recent data taken with gamma-ray observatories could be targeted looking for this peculiar effect. This suggests the need for a comprehensive reanalysis of gamma-ray bursts in which transient features appear in the energy spectra. There is tension between the various observational parameters involved, as one simultaneously makes three demands: high time resolution to avoid washing out the transient in case of a sampling time too large, high energy resolution to identify with enough precision the location of the lines, and large statistics to avoid the signal being immersed in the background. This also adds motivations to the development of satellite detectors in the 1 keV–10 MeV range with high spectral and temporal resolutions, and large gamma spectrometers on balloons [54–57] in which the shorter observation time could be offset by the larger fiducial detection volume, or the 100-day observation time planned for the Ultra Long Duration Balloon program ([58–60], See for instance [61]). An alternative method could be the simultaneous observation of e^+e^- and $p\bar{p}$ annihilations, and we have shown that bounds are of similar order of magnitude if state of art instrumental resolution can be achieved. We also mention the feasibility of observations of cyclotron lines of electrons and protons or ions in the same region of magnetic fields of neutron stars. So far there have been observations of both lines but in different neutron stars; see [62] for the evidence of a line feature in a soft gamma repeater interpreted as a proton cyclotron resonance, [63] for a feature from a magnetar also interpreted as a proton cyclotron feature, and [64] for a band of electron cyclotron lines from an isolated neutron star. The mismatch of the two cyclotron frequencies by the proton-to-electron mass ratio makes their simultaneous observation on the same neutron star quite dif- ficult. This mismatch is smaller for electron and proton spin- flip resonances, however, qualitative estimates show that their absorption signal is suppressed, with respect to the one due to the cyclotron resonance, by a factor approximately equal to the fine structure constant, ruling out its observability with the current data [65–67]. The work discussed so far on Higgs shifts based on a Kretschmann coupling can also be extended in the analysis of bounds to Higgs-curvature couplings with the Ricci scalar for models beyond general relativity.2 The Jebsen– Birkhoff theorem, i.e. the fact that the Schwarzschild solution is the unique spherically symmetric vacuum solution, does not hold in metric f (R) gravity, and $R \neq 0$ even if T = 0 (see [69] for a dedicated study of possible counterexamples). This means that a putative observed Higgs shift will be differently interpreted in various approaches, as due to the Higgs-Kretschmann coupling in ordinary general relativity, of as a Ricci–Higgs coupling in higher dimensional theories [70,71], f (R)theories [72,73], or in Hoˇrava gravity [74–76]. Further observables will therefore be necessary to disentangle the various theoretical scenarios.

ACKNOWLEDGMENTS

We are grateful to J.M. Lattimer for useful correspondence, and to S. Lenzi , R. M. Millan, and R. Turolla for helpful discussions.

REFERENCES

1. 1. F. Bezrukov, M. Shaposhnikov, Phys. Lett. B 659, 703 (2008)
2. 2. F. Bezrukov, A. Magnin, M. Shaposhnikov, Phys. Lett. B 675, 88 (2009)
3. 3. G. Aad et al., ATLAS Collaboration, Phys. Lett. B 710, 49 (2012)
4. 4. S. Chatrachyan et al., CMS Collaboration, Phys. Lett. B 710, 26 (2012)
5. 5. G. Degrassi, S. Di Vita, J. Elias-Miró J. R. Espinosa, G. F. Giudice, G. Isidori, A. Strumia, J. High Energy Phys. 08, 098 (2012)
6. 6. V. Branchina, E. Messina, Phys. Rev. Lett. 111, 241801 (2013)
7. 7. D. Buttazzo, G. Degrassi, P.P. Giardino, G.F. Giudice, F. Sala, A. Salvio, A. Strumia, J. High Energy Phys. 12, 089 (2013)
8. 8. M. Holthausen, K.S. Lim, M. Lindner, J. High Energy Phys. 12, 037 (2012)
9. 9. F. Wang, Nucl. Phys. B 884, 193 (2014)
10. 10. D.A. Demir, Phys. Lett. B 733, 237 (2014)
11. 11. M. Torabian, arXiv:1410.1744
12. 12. P.A.R. Ade et al., BICEP Collaboration, Phys. Rev. Lett. 112, 241101 (2014)
13. 13. R. Adam et al., Planck Collaboration, arXiv:1409.5738
14. 14. J. Espinosa, G. Giudice, A. Riotto, JCAP 0805, 002 (2008)
15. 15. M. Fairbairn, R. Hogan, Phys. Rev. Lett. 112, 201801 (2014)
16. 16. R. Onofrio, G.A. Wegner, Astrophys. J. 791, 125 (2014)
17. 17. B.J. Teegarden, T.L. Cline, Astrophys. J. 236, L67 (1980)
18. 18. B.J. Teegarden, T.L. Cline, Astrophys. Space Sci. 75, 181 (1981)
19. 19. E.P. Mazets, S.V. Golenetskii, R.L. Aptekar, Yu. A. Guryan, V. N. Ilinskii. Nature 290, 378 (1981)
20. 20. S.V. Golenetskii, E.P. Mazets, R.L. Aptekar, YuA Guryan, V.N. Ilyinskii, Astrophys. Space Sci. 124, 243 (1986)
21. 21. B. Kozlovsky, R.J. Murphy, R. Ramaty, Astrophys. J. 141, 523 (2002)
22. 22. W.M. Johnson III, F.R. Harnden, R.C. Haymes, Astroph. J. Lett. 172, L1 (1972)
23. 23. R. Ramaty, G. Borner, J.M. Cohen, Astrophys. J. 181, 891 (1973)
24. 24. D.M. Palmer et al., Astrophys. J. 433, L77 (1984)
25. 25. K. Hurley, Astrophys. J. Suppl. Ser. 90, 857 (1994)
26. 26. B.J. Teegarden, Astrophys. Space Sci. 231, 137 (1995)
27. 27. E.E. Fenimore, R.W. Klebesadel, J.G. Laros, Adv. Space Res. 3, 207 (1983)
28. 28. B.J. Teegarden, in High Energy Transients in Astrophysics, AIP Conf. Proc. No. 115, ed. by S.E. Woosley (American Inst. Phys., New York, 1984), pp. 352–356
29. 29. T.J. Loredo, R.I. Epstein, Astrophys. J. 336, 896 (1989)
30. 30. A.A. Zdziarski, in Proceedings of 13th Texas Symposium Relativistic Astrophysics, ed. by M. Ulmer (World Scientific, Singapore, 1987), pp. 553–562
31. 31. D.Q. Lamb, in Nuclear Spectroscopy of Astrophysical Sources, ed. by N. Gehrels, G.H. Share (America Inst. Phys., New York, 1988), pp. 265–284
32. 32. G.J. Fishman, R.D. Clayton, Astrophys. J. 178, 337 (1972)
33. 33. R. Ramaty, J.M. McKinley, F.C. Jones, Astrophys. J. 256, 238 (1982)
34. 34. F.A. Aharonyan, A.M. Atoyan, R.A. Sunyaev, The annihilation spectrum of relativistic electron–positron plasma. Yerevan Phys. Inst. 432(39), 80 (1980)
35. 35. A.A. Zdziarski, Acta Astron. 30, 371 (1980)
36. 36. R. Ramaty, P. Mészáros, Astrophys. J. 250, 384 (1981)
37. 37. R. Svensson, Astrophys. J. 258, 335 (1982)
38. 38. F.A. Aharonyan, A.M. Atoyan, R.A. Sunyaev, Astrophys. Space Sci. 93, 279 (1983)
39. 39. E.P. Mazets et al., Astrophys. Space Sci. 80, 3 (1981)
40. 40. E.P. Mazets et al., Astrophys. Space Sci. 80, 85 (1981)

41. 41. E.P. Mazets et al., Astrophys. Space Sci. 80, 119 (1981)
42. 42. N.K. Glendenning, Phys. Rev. D 46, 1274 (1992)
43. 43. J.M. Lattimer, M. Prakash, D. Masak, A. Yahil, Astrophys. J. 355, 241 (1990)
44. 44. A.W. Steiner, J.M. Lattimer, E.F. Brown, Astrophys. J. 722, 22 (2010)

CHAPTER 17

NEUTRINO OSCILLATIONS

S. Bilenky

INTRODUCTION. ON THE HISTORY OF NEUTRINO OSCILLATIONS

Discovery of the neutrino oscillations in the atmospheric Super-Kamiokande [1], solar SNO [2] and reactor KamLAND [3] experiments was a first evidence in favor of a beyond the Standard Model physics in particle physics. Neutrino oscillations were further studied in the long baseline accelerator K2K [4], MINOS [5] and T2K [6] experiments. With the measurement of the small parameter $\sin^2\theta_{13}$ in the accelerator T2K [6], reactor Daya Bay [7], RENO [8] and Double Chooze [9] experiments investigation of neutrino oscillations enters a new era, era of high precision measurements. The 2015 Nobel Prize to T. Kajita and A. McDonald "for the discovery of neutrino oscillations, which shows that neutrinos have mass" is a very important event for the neutrino community which will attract new people and give a great boost to the field.

Idea of neutrino oscillations was first proposed by B. Pontecorvo in 1957–1958 soon after the theory of the two-component neutrino was proposed [10] and confirmed by the Goldhaber et al. experiment [11]. B. Pontecorvo looked in the lepton world for a phenomenon analogous to $K^0 \leftrightarrows \bar{K}^0$ oscillations. In the paper [12] he considered muonium (μ^+e^-) to antimuonium (μ^-e^+) transition. In this paper he mentioned a possibility of the neutrino oscillations. Special paper dedicated to neutrino oscillations was published by B. Pontecorvo in 1958 [13]. At that time only one type of neutrino was known. B. Pontecorvo assumed that in addition to the usual weak interaction there must exist a much weaker interaction which does not conserve the lepton number. Assuming maximum mixing (by the analogy with $K^0 - \bar{K}^0$) he concluded that "...neutrino and antineutrino are particle mixtures, i.e. symmetrical and antisymmetrical combinations of two truly neutral Majorana particles ν_1 and ν_2...":

$$|\bar{\nu}_R\rangle = \frac{1}{\sqrt{2}}(|\nu_1\rangle + |\nu_2\rangle), \quad |\nu_R\rangle = \frac{1}{\sqrt{2}}(|\nu_1\rangle - |\nu_2\rangle). \tag{1}$$

Here $|\bar{\nu}_R\rangle$ is the state of the right-handed antineutrino, $|\nu_R\rangle$ is the state of right-handed neutrino, a particle which does not take part in the weak interaction (later B. Pontecorvo proposed the name sterile for such neutrinos), $|\nu_{1,2}\rangle$ are states of Majorana neutrinos with small masses $m_{1,2}$. As a result of the mixing (1), oscillations $\bar{\nu}_R \leftrightarrows \nu_R$ (sterile) become possible. B. Pontecorvo discussed a possibility to check a hypothesis of neutrino oscillations in the reactor neutrino experiments. In 1958 the only known sources of neutrinos were reactors and the sun. B. Pontecorvo finished the paper [13] with the following remark "...effects of transformation of neutrino into antineutrino and vice versa may be unobservable in the laboratory because of large values of R (oscillation length), but will certainly occur, at least, on an astronomic scale."

In 1962 the idea of neutrino masses and mixing was discussed by Maki, Nakagawa and Sakata [14]. Their proposal was based on the Nagoya model in which nucleons were considered as bound states of a vector boson and neutrino with definite mass. MNS assumed that the fields of the weak neutrinos ν_e and ν_μ are connected with the fields of neutrinos with definite masses ν_1 and ν_2 (they called them true neutrinos) by the orthogonal transformation

Citation: S. Bilenky, Neutrino oscillations 908, July 2016, Pages 2-13, ISSN 0550-3213, http://dx.doi.org/10.1016/j.nuclphysb.2016.01.025.

$$\nu_e = \cos\theta\nu_1 + \sin\theta\nu_2, \quad \nu_\mu = -\sin\theta\nu_1 + \cos\theta\nu_2. \tag{2}$$

The phenomenon of neutrino oscillations was not considered in [14]. However, MNS discussed a possibility of "virtual transmutation" of ν_μ into ν_e. They estimated a time of this transition and discussed how a possible $\nu_\mu \to \nu_e$ transition would influence the interpretation of the results of the Brookhaven experiment [15],[1] which was going on at the time when the MNS paper was written.

In 1967 B. Pontecorvo published the second paper on neutrino oscillations [16]. In this paper he discussed flavor neutrino oscillations $\nu_\mu \leftrightarrows \nu_e$ and also oscillations between flavor and sterile neutrinos ($\nu_{eL} \leftrightarrows \bar{\nu}_{eL}$, etc.). In the paper [16] solar neutrino oscillations were considered. Before the first results of the Davis solar neutrino experiment appeared , B. Pontecorvo pointed out that because of neutrino oscillations the flux of the solar ν_e's could be two times smaller than the expected flux. Thus, he anticipated "the solar neutrino problem".

In the Gribov and Pontecorvo paper [17] it was suggested that only active left-handed neutrinos ν_e and ν_μ and right-handed antineutrinos $\bar{\nu}_e$ and $\bar{\nu}_\mu$ exist in nature (no sterile neutrinos). It was assumed that exist a (miliweak) interaction which does not conserve lepton numbers. After the diagonalization of such an interaction the authors came to the mixing relation

$$\nu_{eL} = \cos\xi\phi_{1L} + \sin\xi\phi_{2L}, \quad \nu_{\mu L} = -\sin\xi\phi_{1L} + \cos\xi\phi_{2L}, \tag{3}$$

where ξ is the mixing angle and φ_1 and φ_2 are fields of the Majorana neutrinos with masses m_1 and m_1. They calculated the probability of ν_e to survive in vacuum. The case of the maximum mixing ($\xi=\pi/4$), analogous to the $K^0 - \bar{K}^0$ case, was considered as the most attractive one. Under this assumption the oscillations of solar neutrinos were discussed.

In the seventies and eighties idea of neutrino masses and oscillations was further developed in Dubna in the papers [18]. In addition to the Gribov–Pontecorvo scheme of the neutrino mixing, based on the Majorana mass term, neutrino mixing based on the Dirac mass term and the most general Dirac and Majorana mass term were considered. Possible reactor, accelerator, solar and atmospheric experiments on the search for neutrino oscillations were discussed. Our general point of view, which we advocated in our papers and in the first review on neutrino oscillations [19] was the following:

1. There are no principles which require that neutrinos are massless particles. It is plausible that neutrinos have small nonzero masses.
2. Neutrino oscillations is an interference phenomenon. Search for neutrino oscillations is the most sensitive method to search for extremely small mass-squared differences.
3. Experiments with neutrinos from different sources are sensitive to different neutrino mass-squared differences. Experiments on the search for neutrino oscillations must be performed with neutrinos from all existing sources.[2]

NEUTRINO MIXING

Neutrino oscillations are based on the mixing of neutrino fields

$$\nu_{lL}(x) = \sum_i U_{li}\nu_{iL}(x). \tag{4}$$

Here U is a unitary mixing matrix and $\nu_i(x)$ is the field of neutrinos (Dirac or Majorana) with mass m_i.

The flavor neutrino fields $\nu_{lL}(x)$ ($l=e,\mu,\tau$) enter into the Standard Model CC and NC interactions are charged leptonic and neutral neutrino currents.

$$\mathcal{L}_I^{CC} = -\frac{g}{\sqrt{2}} j_\alpha^{CC} W^\alpha + \text{h.c.}, \quad \mathcal{L}_I^{NC} = -\frac{g}{2\cos\theta_W} j_\alpha^{NC} Z^\alpha. \tag{5}$$

Here

$$j_\alpha^{CC} = \sum_{l=e,\mu,\tau} \bar{\nu}_{lL}\gamma_\alpha l_L, \quad j_\alpha^{NC} = \sum_{l=e,\mu,\tau} \bar{\nu}_{lL}\gamma_\alpha \nu_{lL} \tag{6}$$

The neutrino mixing takes place if in the total Lagrangian there is a mass term nondiagonal over flavor neutrino fields. In the case of the charged particles (leptons and quarks) only Dirac mass terms are possible. Because the electric charges of neutrinos are equal to zero three different neutrino mass terms are possible (see [20] and [21]).

Dirac Mass Term

$$\mathcal{L}^D = -\sum_{l',l=e,\mu,\tau} \bar{\nu}_{l'L} M_{l',l}^D \nu_{lR} + \text{h.c.}, \tag{7}$$

where M^D is a complex, nondiagonal, 3×3 matrix. After the diagonalization of the matrix M^D we have

$$\nu_{lL}(x) = \sum_{i=1}^{3} U_{li}\,\nu_{iL}(x). \tag{8}$$

Here U is the unitary PNMS mixing matrix and $\nu_i(x)$ is the Dirac field with the mass m_i. The Lagrangian L^D conserves the total lepton number $L=L_e+L_\mu+L_\tau$. Neutrino ν_i and antineutrino $\bar{\nu}_i$ differ by the lepton number: $L(\nu_i) = 1,\ L(\bar{\nu}_i) = -1$.

Majorana Mass Term

$$\mathcal{L}^M = -\frac{1}{2}\sum_{l',l=e,\mu,\tau} \bar{\nu}_{l'L} M_{l'l}^M (\nu_{lL})^c + \text{h.c.}, \tag{9}$$

where M^M is a complex, nondiagonal, symmetrical 3×3 matrix and $(\nu_{lL})^c = C\bar{\nu}_{lL}^T$ is the conjugated field. The mass term (9) violates not only flavor lepton numbers but also the total lepton number L . After the diagonalization of the matrix M^M we have

$$\nu_{lL}(x) = \sum_{i=1}^{3} U_{li}\,\nu_{iL}(x). \tag{10}$$

Here U is a unitary 3×3 mixing matrix and

$$\nu_i(x) = \nu_i^c(x) \tag{11}$$

is the Majorana field with the mass m_i ($\nu_i \equiv \bar{\nu}_i$).

The most general

Dirac and Majorana Mass Term

$$\mathcal{L}^{\mathrm{D+M}} = \mathcal{L}^{\mathrm{M}} + \mathcal{L}^{\mathrm{D}} - \frac{1}{2} \sum_{s',s=s_1,\ldots s_{n_s}} \overline{(\nu_{s'R})^c}\, M^{\mathrm{R}}_{s's}\, \nu_{sR} + \mathrm{h.c.} \quad (12)$$

(M^R is a complex symmetrical matrix) violates lepton numbers and require left-handed and right-handed neutrino fields. After the diagonalization of the mass term L^{D+M} we find

$$\nu_{lL}(x) = \sum_{i=1}^{3+n_s} U_{li}\, \nu_{iL}(x), \quad (\nu_{sR}(x))^c = \sum_{i=1}^{3+n_s} U_{si}\, \nu_{iL}(x). \quad (13)$$

Here U is a unitary $(3+n_s)\times(3+n_s)$ matrix and $\nu_i(x) = \nu_i^c(x)$ is the field of a Majorana lepton with definite mass.

The mixing (13) open different possibilities: the seesaw possibility of the generation of small neutrino masses [22], a possibility of transitions of flavor neutrinos into sterile states, etc.

Let us notice that the Dirac mass term can be generated by the standard Higgs mechanism. The Majorana and the Dirac and Majorana mass terms can be generated only by a beyond the SM mechanisms.

FLAVOR NEUTRINO STATES

There exist different methods of the derivation (of the same) expression for transition probabilities. We will present here a method based on the notion of the coherent flavor neutrino states (see [21])

$$|\nu_l\rangle = \sum_i U^*_{li}\, |\nu_i\rangle, \quad l = e, \mu, \tau. \quad (14)$$

Here $|\nu_i\rangle$ is the state of neutrino (Dirac or Majorana) with mass m_i, momentum $\vec{p}$ and energy $E_i = \sqrt{p^2 + m_i^2} \simeq E + \frac{m_i^2}{2E}\ (E = p)$, and $|\nu_l\rangle$ is the state the flavor neutrino ν_l which is produced together with l^+ in a CC weak decay ($\pi^+ \to \mu^+ + \nu_\mu$, etc.) or produces l^- in a CC neutrino reaction ($\nu_\mu + N \to \mu^- + X$, etc.).

The relation (14) is valid if neutrino mass-squared differences are so small that in weak decays production of neutrinos with different masses cannot be resolved. It follows from the Heisenberg uncertainty relation that this condition is satisfied in neutrino oscillation experiments with neutrino energies many orders of magnitude larger than neutrino masses.

The possibility to resolve small neutrino mass-squared differences is based on the time-energy uncertainty relation (see [24])

$$\Delta E\, \Delta t \gtrsim 1. \quad (15)$$

Here Δt is a time interval during which the state with the energy uncertainty ΔE is significantly changed. In the case of neutrino beams from (15) we find

$$|\Delta m^2_{ki}|\, \frac{L}{2E} \gtrsim 1, \quad (16)$$

where $L \simeq \Delta t$ is the distance between a neutrino source and neutrino detector. For "atmospheric" and "solar" mass-squared differences $\Delta m^2_A \simeq 2.4 \cdot 10^{-3}\ \mathrm{eV}^2$ and $\Delta m^2_S \simeq 7.5 \cdot 10^{-5}\ \mathrm{eV}^2$ the condition (16) is satisfied in the atmospheric Super-Kamiokande [1], long baseline accelerator K2K [4], MINOS [5], T2K [6], reactor KamLAND [3], Daya Bay [7], RENO [8] Double Chooze [9] and other neutrino oscillation experiments.

We will finish this section with a remark about the states of sterile neutrinos which (by definition) do not interact with leptons and quarks via the SM interaction. If in addition to the flavor neutrinos ν_l sterile neutrinos ν_s exist, their states are determined as follows

$$|\nu_s\rangle = \sum_{i=1}^{3+n_s} U^*_{si}\, |\nu_i\rangle, \quad s = s_1, s_1, \ldots \quad (17)$$

where U is a unitary $(3+n_s)\times(3+n_s)$ matrix. The states of active and sterile neutrinos (14) and (17) satisfy the condition

$$\langle\alpha'|\alpha\rangle = \delta_{\alpha'\alpha}, \quad \alpha', \alpha = e, \mu, \tau, s_1, s_1, \ldots s_{n_s}. \quad (18)$$

Neutrino oscillations is a direct consequence of the fact that flavor (and sterile) neutrinos are described by coherent states (14) and (17).

NEUTRINO OSCILLATIONS IN VACUUM

Let us assume that at the initial time $t=0$ a flavor neutrino ν_α was produced. In the general case of flavor and sterile neutrinos at the time t we have

$$|\nu_\alpha\rangle_t = e^{-iH_0 t}\, |\nu_\alpha\rangle = \sum_{i=1}^{3+n_s} |\nu_i\rangle\, e^{-iE_i t}\, U^*_{\alpha i} = \sum_{\alpha'} |\alpha'\rangle (\sum_{i=1}^{3+n_s} U_{\alpha' i}\, e^{-iE_i t}\, U^*_{\alpha i}). \quad (19)$$

Thus, for the $\nu_\alpha \to \nu_{\alpha'}$ transition probability we find

$$P(\nu_\alpha \to \nu_{\alpha'}) = |\sum_{i=1}^{3+n_s} U_{\alpha' i}\, e^{-iE_i t}\, U^*_{\alpha i}|^2. \quad (20)$$

We will present here convenient expression for $\overset{(-)}{\nu}_\alpha \to \overset{(-)}{\nu}_{\alpha'}$ transition probability (see[23]). From (20) we have

$$P(\nu_\alpha \to \nu_{\alpha'}) = |\sum_{i=1}^{3+n_s} U_{\alpha' i}\, e^{-2i\Delta_{pi}}\, U^*_{\alpha i}|^2 = |\delta_{\alpha'\alpha} - 2i \sum_i U_{\alpha' i}\, e^{-i\Delta_{pi}} \sin\Delta_{pi}\, U^*_{\alpha i}|^2 \quad (21)$$

where p is arbitrary, fixed index and

$$\Delta_{pi} = \frac{\Delta m^2_{pi} L}{4E}, \quad \Delta m^2_{pi} = m^2_i - m^2_p. \tag{22}$$

Let us notice that in Eq. (21)

- $i \neq p$,
- we extract the common phase $e^{\frac{-i m^2_p L}{2E}}$,
- we used the unitarity condition $\sum_i U_{\alpha' i} U^*_{\alpha i} = \delta_{\alpha'\alpha}$.

From (21) we find

$$P(\overset{(-)}{\nu}_\alpha \to \overset{(-)}{\nu}_{\alpha'}) = \delta_{\alpha'\alpha} - 4\sum_i |U_{\alpha i}|^2 (\delta_{\alpha'\alpha} - |U_{\alpha' i}|^2)\sin^2\Delta_{pi}$$

$$+8\sum_{i>k}[\mathrm{Re}\,(U_{\alpha' i}U^*_{\alpha i}U^*_{\alpha' k}U_{\alpha k})\cos(\Delta_{pi} - \Delta_{pk})$$

$$\pm\,\mathrm{Im}\,(U_{\alpha' i}U^*_{\alpha i}U^*_{\alpha' k}U_{\alpha k})\sin(\Delta_{pi} - \Delta_{pk})]\sin\Delta_{pi}\sin\Delta_{pk}. \tag{23}$$

Here $+(-)$ sign refer to $\nu_\alpha \to \nu_{\alpha'}$ ($\bar{\nu}_\alpha \to \bar{\nu}_{\alpha'}$) transition.

From our point of view there are some advantages of the expression (23) with respect to the standard expression (for the standard expression see [25]):

- Only independent mass-squared differences enter into this expression.
- The unitarity condition is fully implemented in (23). As a result only independent terms enter into this expression.

We will consider now the most important case of the three-neutrino mixing. Usually neutrino masses are labeled in such a way that $m_2 > m_1$ and solar ("small") mass-squared difference is determined as follows

$$m^2_2 - m^2_1 = \Delta m^2_{12} \equiv \Delta m^2_S. \tag{24}$$

For the neutrino mass spectrum there are two possibilities:

- Normal spectrum (NS): Δm^2_S is the difference between square of masses of the lightest neutrinos. In this case $m_3 > m_2 > m_1$.
- Inverted spectrum (IS): Δm^2_S is the difference between square of masses of the heaviest neutrinos. In this case $m_2 > m_1 > m_3$.

We will determine the atmospheric ("large") neutrino mass squared difference in the following way

$$NS: \Delta m^2_A = \Delta m^2_{23}, \quad IS: \Delta m^2_A = |\Delta m^2_{13}|. \tag{25}$$

Let us notice that there exist different definition of this quantity in the literature

1. The Bari group [26] determines atmospheric mass-squared difference as follows

$$(\Delta m^2_A)' = \frac{1}{2}|\Delta m^2_{13} + \Delta m^2_{23}| = \Delta m^2_A + \frac{1}{2}\Delta m^2_S. \tag{26}$$

2. The NuFit group [27] determines atmospheric mass-squared difference in the following way

$$(\Delta m^2_A)'' = \Delta m^2_{13}\ (NS) = |\Delta m^2_{23}|\ (IS) = \Delta m^2_A + \Delta m^2_S. \tag{27}$$

3. In [28] the parameter Δm^2_{ee} was introduced. It is determined as follows

$$\Delta m^2_{ee} = \cos^2\theta_{12}\Delta m^2_{13} + \sin^2\theta_{12}\Delta m^2_{23}. \tag{28}$$

The parameter Δm^2_{ee} is connected with Δm^2_A and Δm^2_S by the relations

$$\Delta m^2_{ee} = \Delta m^2_A + \cos^2\theta_{12}\Delta m^2_S\ (NS),\ |\Delta m^2_{ee}| = \Delta m^2_A + \sin^2\theta_{12}\Delta m^2_S\ (IS). \tag{29}$$

As it is seen from (25), (26)(27)(29) different definitions of "large" mass-squared difference differ only by a few %. However, neutrino oscillation experiments enter now into precision era when neutrino oscillation parameters will be measured with % accuracy. We believe that the consensus in definition of "large" neutrino mass-squared difference must be found.

For the probability of the transition $\overset{(-)}{\nu}_l \to \overset{(-)}{\nu}_{l'}$ $(l, l' = e, \mu, \tau)$ in the case of normal and inverted mass spectra from (23) we find, correspondingly, the following expressions

$$P^{NS}(\overset{(-)}{\nu}_l \to \overset{(-)}{\nu}_{l'}) = \delta_{l'l} - 4|U_{l3}|^2(\delta_{l'l} - |U_{l'3}|^2)\sin^2\Delta_A$$

$$-\,4|U_{l1}|^2(\delta_{l'l} - |U_{l'1}|^2)\sin^2\Delta_S - 8\,[\mathrm{Re}\,(U_{l'3}U^*_{l3}U^*_{l'1}U_{l1})\cos(\Delta_A + \Delta_S)$$

$$\pm\,\mathrm{Im}\,(U_{l'3}U^*_{l3}U^*_{l'1}U_{l1})\sin(\Delta_A + \Delta_S)]\sin\Delta_A\sin\Delta_S, \tag{30}$$

and

$$P^{IS}(\overset{(-)}{\nu}_l \to \overset{(-)}{\nu}_{l'}) = \delta_{l'l} - 4|U_{l3}|^2(\delta_{l'l} - |U_{l'3}|^2)\sin^2\Delta_A$$

$$-\,4|U_{l2}|^2(\delta_{l'l} - |U_{l'2}|^2)\sin^2\Delta_S - 8\,[\mathrm{Re}\,(U_{l'3}U^*_{l3}U^*_{l'2}U_{l2})\cos(\Delta_A + \Delta_S)$$

$$\mp\,\mathrm{Im}\,(U_{l'3}U^*_{l3}U^*_{l'2}U_{l2})\sin(\Delta_A + \Delta_S)]\sin\Delta_A\sin\Delta_S. \tag{31}$$

The transition probabilities (30) and (31) are the sum of atmospheric, solar and interference terms. Notice that expression (31) can be obtained from (30) by the change $U_{l1} \to U_{l2}$ and $(\pm) \to (\mp)$ in the last term.

The values of the oscillation parameters obtained from global analysis of existing data by the NuFit group [27] are presented in the Table 1.

Table 1: Values of neutrino oscillation parameters obtained in [27] from the global fit of existing data

Parameter	Normal spectrum	Inverted spectrum
$\sin^2\theta_{12}$	$0.304^{+0.013}_{-0.012}$	$0.304^{+0.013}_{-0.012}$
$\sin^2\theta_{23}$	$0.452^{+0.052}_{-0.028}$	$0.579^{+0.025}_{-0.037}$
$\sin^2\theta_{13}$	$0.0218^{+0.0010}_{-0.0010}$	$0.0219^{+0.0011}_{-0.0010}$
δ (in °)	(306^{+39}_{-70})	(254^{+63}_{-62})
Δm^2_S	$(7.50^{+0.19}_{-0.17})\cdot 10^{-5}\ \mathrm{eV}^2$	$(7.50^{+0.19}_{-0.17})\cdot 10^{-5}\ \mathrm{eV}^2$
Δm^2_A	$(2.457^{+0.047}_{-0.047})\cdot 10^{-3}\ \mathrm{eV}^2$	$(2.449^{+0.048}_{-0.047})\cdot 10^{-3}\ \mathrm{eV}^2$

NEUTRINO AND THE STANDARD MODEL

After the discovery of the Higgs boson at LHC the Standard Model acquired the status of the theory of elementary particles in the electroweak range (up to ~300 GeV). The Standard Model is based on the following principles:

- Local gauge symmetry.
- Unification of the weak and electromagnetic interactions.
- Spontaneous breaking of the electroweak symmetry.

It was suggested in [30] that in the framework of these principles nature choose the simplest, most economical possibilities. The Standard Model started with the theory of the two-component neutrino. The two-component, massless, Weil neutrino is thesimplest possibility for the particle with spin 1/2 : only two degrees of freedom. The local $SU_L(2)$ group with the lepton doublets

$$\psi^{lep}_{eL} = \begin{pmatrix} \nu'_{eL} \\ e'_L \end{pmatrix},\ \psi^{lep}_{\mu L} = \begin{pmatrix} \nu'_{\mu L} \\ \mu'_L \end{pmatrix},\ \psi^{lep}_{\tau L} = \begin{pmatrix} \nu'_{\tau L} \\ \tau'_L \end{pmatrix} \tag{32}$$

and corresponding quark doublets is the simplest possibility which allows to include charged leptons and quarks in addition to neutrinos.

In order to unify weak and electromagnetic interactions we need to enlarge the symmetry group: in electromagnetic currents of charged particles enter left-handed and right-handed fields. The simplest enlargement is the $SU_L(2)\times U_Y(1)$ group where $U_Y(1)$ is the group of the weak hypercharge Y determined by the Gell–Mann–Nishijima relation $Q = T_3 + \frac{1}{2}Y$. Neutrinos have no electromagnetic interaction. Unification of the weak and electromagnetic interactions does not require right-handed neutrino fields. The SM interaction of leptons, neutrinos and quarks with gauge vector bosons is the minimal interaction compatible with the local $SU_L(2)\times U_Y(1)$ invariance.

The SM mechanism of the mass generation is the Brout–Englert–Higgs mechanism based on the assumption of the existence of scalar Higgs fields. In order to generate masses of $W^\pm$ and Z^0 bosons we need to have three (Goldstone) degrees of freedom.Minimal possibility is a doublet of complex Higgs fields (four degrees of freedom). With this assumption one scalar, neutral Higgs boson is predicted. This prediction is in a good agreement with existing LHC data.

Masses of $W^\pm$ and Z^0 bosons are given in the SM by the relations

$$m_W = \frac{1}{2} g\, v,\quad m_Z = \frac{1}{2}\sqrt{g^2 + g'^2}\, v = \frac{g}{2\cos\theta_W} v, \tag{33}$$

where $v = (\sqrt{2}G_F)^{-1/2} = 246\ \mathrm{GeV}$ is the parameter which characterizes the scale of the electroweak symmetry breaking. Lepton and quark masses and mixing are due to$SU_L(2)\times U_Y(1)$ invariant Yukawa interactions which generate Dirac mass terms. For the charged leptons we have

$$\mathcal{L}^{lep}_Y = -\sum_l m_l \bar{l}\, l, \tag{34}$$

where $m_l = y_l\ v$ and y_l is the Yukawa constant. Neutrinos in the minimal SM after spontaneous breaking of the electroweak symmetry remain two-component, massless, Weyl particles.

THE WEINBERG MECHANISM OF THE NEUTRINO MASS GENERATION

In the framework of the minimal SM neutrino masses and mixing can be generated only by a beyond the SM mechanism. The most general method which allows to describe effects of a beyond the SM physics is the method of the effective Lagrangian. The effective Lagrangian is a $SU_L(2)\times U_Y(1)$ invariant, dimension five or more local operator built from SM fields. In order to built the effective Lagrangian which generate a neutrino mass term we must use the lepton doublets (32) and the Higgs doublet

$$\phi = \begin{pmatrix} \phi_+ \\ \phi_0 \end{pmatrix} \tag{35}$$

The effective Lagrangian which generates the neutrino mass term has the form [29]

$$\mathcal{L}^{\mathrm{eff}}_I = -\frac{1}{\Lambda}\sum_{l_1,l_2} (\bar{\psi}^{lep}_{l_1 L}\tilde{\phi})\, Y_{l_1 l_2}\, (\tilde{\phi}^T (\psi^{lep}_{l_2 L})^c) + \mathrm{h.c.}, \tag{36}$$

where the parameter Λ characterizes a scale of a beyond the SM physics ($\Lambda \gg v$) and $\tilde{\phi} = i\tau_2\phi^*$ is the conjugated doublet. Let us stress that the Lagrangian (36) does not conserve the total lepton number.[3]

After spontaneous symmetry breaking from (36) we come to the Majorana mass term

$$\mathcal{L}^{M} = -\frac{1}{2}\frac{v^2}{\Lambda}\sum_{l_1,l_2}\bar{\nu}'_{l_1L}\, Y_{l_1l_2}(\nu'_{l_2L})^c + \text{h.c.} = -\frac{1}{2}\sum_{i=1}^{3} m_i\, \bar{\nu}_i\nu_i. \quad (37)$$

Here $\nu_i = \nu_i^c$ is the field of the neutrino Majorana with the mass

$$m_i = \frac{v^2}{\Lambda}y_i = \frac{v}{\Lambda}(y_i v), \quad (38)$$

where y_i is a Yukawa coupling. In (38)$y_i v$ is a "typical" fermion mass in SM. Thus, neutrino masses, generated by the effective Lagrangian (36), are suppressed with respect to "SM masses" by a factor

$$\frac{v}{\Lambda} = \frac{\text{scale of SM}}{\text{scale of a new physics}} \ll 1.$$

The mechanism we have considered is, apparently, the most economical and natural beyond the SM mechanism of the neutrino mass generation. There are two general consequences of this mechanism:

- Neutrinos with definite masses ν_i are Majorana particles.
- The number of neutrinos with definite masses is equal to the number of lepton–quark generations (three). This means that in this scheme there are no transitions of flavor neutrinos into sterile states.

The study of the lepton number violating neutrinoless double β -decay ($0\nu\beta\beta$-decay)

$$(A,Z) \to (A,Z+2) + e^- + e^- \quad (39)$$

of some even–even nuclei is most sensitive way to investigate the Majorana nature of neutrinos with definite masses (see review [31]). The probability of the process (39) is proportional to square of the Majorana neutrino mass

$$m_{\beta\beta} = \sum_i U_{ei}^2 m_i \quad (40)$$

and is very small. It has the following general form

$$\frac{1}{T_{1/2}^{0\nu}} = |m_{\beta\beta}|^2\, |M^{0\nu}|^2\, G^{0\nu}(Q,Z). \quad (41)$$

Here $M^{0\nu}$ is the nuclear matrix element and $G^{0\nu}(Q,Z)$ is known phase factor.

Several experiments on the search for the $0\nu\beta\beta$ of different nuclei are going on and are in preparation. Up to now the process was not observed. From the data of recent experiments EXO-200 [32], KamLAND-Zen [33] and GERDA [34] the following upper bounds were, correspondingly, obtained

$$|m_{\beta\beta}| < (1.9-4.5)\cdot 10^{-1}\,\text{eV},\ (1.4-2.8)\cdot 10^{-1}\,\text{eV},\ (2-4)\cdot 10^{-1}\,\text{eV}. \quad (42)$$

In future experiments on the search for $0\nu\beta\beta$ decay the values $|m_{\beta\beta}| \simeq$ a few$\cdot 10^{-2}$ eV are planned to be reached.

Indications in favor of transitions of flavor neutrinos into sterile states were obtained in the LSND [36] and MiniBooNE short baseline accelerator experiments and in the GALLEX and SAGE calibration experiments and in short baseline reactor experiments which were reanalyzed with a new reactor antineutrino flux (see recent review [35]). Many new short baseline source, reactor and accelerator neutrino experiments on the search for sterile neutrinos with masses $\sim$1 eV are in preparation (see [37]). There is no doubt that in a few years the sterile neutrino anomaly will be resolved.

CONCLUSION

Neutrino masses and mixing, discovered via the observation of neutrino oscillations, is a first particle physics evidence of a new beyond the SM physics. We discuss here briefly first proposals for neutrino oscillations and first steps in the development of the theory of neutrino oscillations. Then we consider basics of neutrino mixing and oscillations, convenient formalism for neutrino oscillations in vacuum and the definition of the atmospheric neutrino mass-squared difference. In the final part of the paper we discuss the important role which play two-component neutrino in the Standard Model and the most economical Weinberg mechanism of the generation of the Majorana neutrino masses.

REFERENCES

1. Y. Fukuda, et al., Super-Kamiokande Collaboration, Phys. Rev. Lett. 82 (1999) 2644, arXiv:hep-ex/9812014.
2. Q.R. Ahmad, et al., SNO Collaboration, Phys. Rev. Lett. 89 (2002) 011301, arXiv:nucl-ex/0204008.
3. K. Eguchi, et al., KamLAND Collaboration, Phys. Rev. Lett. 90 (2003) 021802, arXiv:hep-ex/0212021.
4. C. Mariani, K2K Collaboration, AIP Conf. Proc. 981 (2008) 247.
5. Timmons, arXiv:1511.06178 [hep-ex].
6. K. Abe, et al., T2K Collaboration, Phys. Rev. D 91 (7) (2015) 072010, arXiv:1502.01550 [hep-ex].
7. F.P. An, et al., Daya Bay Collaboration, Phys. Rev. Lett. 115 (11) (2015) 111802, arXiv:1505.03456 [hep-ex].
8. S.B. Kim, Nucl. Part. Phys. Proc. 265–266 (2015) 93, arXiv:1412.2199 [hep-ex].
9. T. Abrahao, et al., Double Chooz Collaboration, arXiv:1510.08937 [hep-ex].
10. L.D. Landau, Nucl. Phys. 3 (1957) 127; T.D. Lee, C.N. Yang, Phys. Rev. 105 (1957) 1671; A. Salam, Nuovo Cimento 5 (1957) 299.
11. M. Goldhaber, L. Grodzins, A.W. Sunyar, Phys. Rev. 109 (1958) 1015.
12. B. Pontecorvo, Sov. Phys. JETP 6 (1957) 429; Zh. Eksp. Teor. Fiz. 33 (1957) 549.
13. B. Pontecorvo, Sov. Phys. JETP 7 (1958) 172; Zh. Eksp. Teor. Fiz. 34 (1957) 247.
14. Z. Maki, M. Nakagawa, S. Sakata, Prog. Theor. Phys. 28 (1962) 870.

15. G. Danby, J.M. Gaillard, K.A. Goulianos, L.M. Lederman, N.B. Mistry, M. Schwartz, J. Steinberger, Phys. Rev. Lett. 9 (1962) 36.
16. B. Pontecorvo, Sov. Phys. JETP 26 (1968) 984; Zh. Eksp. Teor. Fiz. 53 (1967) 1717.
17. V.N. Gribov, B. Pontecorvo, Phys. Lett. B 28 (1969) 493.
18. S.M. Bilenky, B. Pontecorvo, Phys. Lett. B 61 (1976) 248; S.M. Bilenky, B. Pontecorvo, Lett. Nuovo Cimento 17 (1976) 569; S.M. Bilenky, B. Pontecorvo, Yad. Fiz. 24 (1976) 603; S.M. Bilenky, B. Pontecorvo, Sov. J. Nucl. Phys. 24 (1976) 316.
19. S.M. Bilenky, B. Pontecorvo, Phys. Rep. 41 (1978) 225.
20. S.M. Bilenky, S.T. Petcov, Rev. Mod. Phys. 59 (1987) 671.
21. S. Bilenky, Introduction to the Physics of Massive and Mixed Neutrinos, Lecture Notes in Physics, vol. 817, Springer, 2010, pp. 1–255.
22. P. Minkowski, Phys. Lett. B 67 (1977) 421; M. Gell-Mann, P. Ramond, R. Slansky, in: F. van Nieuwenhuizen, D. Freedman (Eds.), Supergravity, North Holland, Amsterdam, 1979, p. 315; T. Yanagida, Proc. of the Workshop on Unified Theory and the Baryon Number of the Universe, KEK, Japan, 1979; S.L. Glashow, NATO ASI Ser., Ser. B: Phys. 59 (1979) 687; R.N. Mohapatra, G. Senjanovic,´ Phys. Rev. D 23 (1981) 165.
23. S. Bilenky, Phys. Part. Nucl. Lett. 12(4), 453, arXiv:1502.06158 [hep-ph].
24. S.M. Bilenky, F. von Feilitzsch, W. Potzel, J. Phys. G 38 (2011) 115002, arXiv:1102.2770 [hep-ph]
25. C. Giunti, C.W. Kim, Fundamentals of Neutrino Physics and Astrophysics, Univ. Press, Oxford, UK, 2007, 710 pp.
26. F. Capozzi, G.L. Fogli, E. Lisi, A. Marrone, D. Montanino, A. Palazzo, J. Phys. Conf. Ser. 598 (1) (2015) 012002.
27. M.C. Gonzalez-Garcia, M. Maltoni, T. Schwetz, J. High Energy Phys. 1411 (2014) 052, arXiv:1409.5439 [hep-ph].
28. H. Nunokawa, S.J. Parke, R. Zukanovich Funchal, Phys. Rev. D 72 (2005) 013009.
29. S. Weinberg, Phys. Rev. Lett. 43 (1979) 1566.
30. S. Bilenky, Phys. Part. Nucl. 46 (4) (2015) 475, arXiv:1501.00232 [hep-ph].
31. S. Bilenky, C. Giunti, Int. J. Mod. Phys. A 30 (04n05) (2015) 1530001, arXiv:1411.4791 [hep-ph].
32. J.B. Albert, et al., EXO-200 Collaboration, Nature 510 (2014) 229, arXiv:1402.6956 [nucl-ex].
33. K. Asakura, et al., KamLAND-Zen Collaboration, AIP Conf. Proc. 1666 (2015), arXiv:1409.0077 [physics.ins-det].
34. T. Wester, AIP Conf. Proc. 1686 (2015) 020026.
35. S. Gariazzo, C. Giunti, M. Laveder, Y.F. Li, E. Zavanin, arXiv:1507.08204 [hep-ph].
36. Aguilar-Arevalo, et al., LSND Collaboration, Phys. Rev. D 64 (2001) 112007, arXiv:hep-ex/0104049.
37. T. Lasserre, Phys. Dark Universe 4 (2014) 81, arXiv:1404.7352 [hep-ex].

CHAPTER 18

STERILE NEUTRINO DARK MATTER AND LOW SCALE LEPTOGENESIS FROM A CHARGED SCALAR

Michele Frigerio and Carlos E. Yaguna

INTRODUCTION

Sterile neutrinos, that is, fermions singlet under the SU(3) × SU(2) × U(1) gauge symmetry, are a very well-motivated extension of the standard model (SM). On the theoretical side, they are a prediction of left-right symmetric theories, they allow to gauge B - L by removing its anomaly and they are necessary in SO(10)grand unification. On the phenomenological side, they provide a non-vanishing mass to the active SM neutrinos, they allow to realize baryogenesis via leptogenesis, and they are also a viable candidate for dark matter, as long as their mass lies in the keV range. In that case, they are naturally long-lived so that, unlike dark matter candidates at the electroweak scale, no additional symmetries are required to stabilize them. Thus, sterile neutrinos can provide a simple solution to the three open problems of the SM: neutrino masses, the baryon asymmetry, and the dark matter.

The minimal model addressing these three issues requires one sterile neutrino N_1 at the keV scale as dark matter candidate [1], and two additional sterile neutrinos $N_{2,3}$ for leptogenesis, which is induced either by N-decays, for sterile neutrino masses above the TeV scale [2], or by N-oscillations, for sterile neutrino masses at the GeV scale [3, 4]. Light active neutrino masses are easily generated, via the seesaw mechanism, provided the sterile neutrino masses are significantly larger than about 1 eV. Thus, one may argue that no new physics is needed above the electroweak scale to explain neutrino masses, baryogenesis, and dark matter, defining the so-called "ν minimal standard model" (νMSM) [5]. Such a model is undoubtedly economical and very predictive, by reason of the small number of degrees of freedom it contains, but it is also strongly constrained. Present experiments, in fact, already exclude the minimal scenario for sterile neutrino dark matter within the νMSM [6]. In the region of mass and mixing angle compatible with current observations, in particular the X-ray bounds [7–9] and the Ly_α forest data [10–12], one cannot produce a large enough population of sterile neutrinos from standard active-sterile oscillations [1]. The only way out within the νMSM is to generate the dark matter neutrinos via resonant active-sterile oscillations triggered by large initial lepton asymmetries [13,14], but that mechanism requires a high degree of fine-tuning [6, 15]. In addition, leptogenesis via $N_{2,3}$-oscillations within the νMSM also requires a significant tuning of parameters, in particular a strong mass degeneracy between N_2 and N_3 [6].

It is important, therefore, to consider alternative ways of realizing leptogenesis and producing sterile neutrinos within extensions of the νMSM. Indeed, most extensions of the SM that are theoretically well-motivated contain new degrees of freedom beside the sterile neutrinos, which may modify significantly the phenomenology of the νMSM. For leptogenesis, the required tuning of parameters can be released by introducing a second Higgs doublet, as recently studied in [16]. Regarding dark matter, several other possibilities have been considered for the production of the sterile neutrinos. They include the production through inflaton decay [17–19], thermal overproduction followed by entropy dilution [20], and the production via the decays of either a neutral scalar in thermal equilibrium [21], or a frozen-in neutral scalar [22].

In this paper we advocate an extension of the SM by three sterile neutrinos, N_i, plus a charged scalar, δ^+, both of which are naturally present in left-right symmetric or unified extensions of the SM [23]. The charged scalar interacts with SM leptons, both doublets and singlets, and with sterile neutrinos, and its mass can be as low as allowed by current collider bounds—

Citation: Michele Frigerio and Carlos E. Yaguna, Sterile neutrino dark matter and low scale leptogenesis from a charged scalar, DOI 10.1140/epjc/s10052-014-3252-1.

about 200 GeV. We will show that the interplay between N_i and δ^+ modifies the dark matter phenomenology, enables new realizations of baryogenesis, and gives rise to novel experimental signatures at colliders. The decays of δ^+, in fact, constitute a new mechanism for the production of sterile neutrino dark matter in the early Universe. An advantage of this mechanism is that the resulting relic density does not depend on the active-sterile mixing angle, allowing to satisfy the strong bounds derived from X-ray observations [7–9]. Besides, the charged scalar induces independent sources of leptogenesis that are effective at scales close to the electroweak scale. If the charged scalar mass were accessible at colliders, the present scenario for dark matter and baryogenesis would be testable in a direct way.

The rest of the paper is organized as follows. In Sect. 2 we introduce the model and fix our notation. Section 3 deals with dark matter production via the decays of δ^+. We obtain, in particular, the dark matter relic density as a function of the parameters of the model. Baryogenesis via leptogenesis is discussed in Sect. 4. We examine two different ways to generate a lepton asymmetry, and outline the region of the parameter space that successfully realizes baryogenesis in each case. The most interesting experimental signatures of this model are discussed in Sect. 5. Section 6 briefly elaborates on the flavour structure of the masses and Yukawa couplings of the model. We present our conclusions in Sect. 7.

THE MODEL

When gauge singlet chiral fermions N_{Ri} are added to the SM, they have in general a Majorana mass term and a Yukawa coupling to the SM lepton doublets,

$$\mathscr{L}_N = \overline{N_{Ri}} i\gamma^\mu \partial_\mu N_{Ri} + \left[-\frac{1}{2}\overline{(N_{Ri})^c}(M_N)_{ij} N_{Rj} - \overline{l_{L\alpha}}(y_\nu)_{\alpha i} N_{Ri} \tilde{H} + h.c. \right]. \tag{1}$$

When a scalar field δ^+, singlet under SU(2) and with electromagnetic charge one, is added to the SM, it has in general, besides its gauge and self-interactions, a quartic coupling to the Higgs doublet and an antisymmetric Yukawa coupling to two lepton doublets,

$$\mathscr{L}_\delta = D_\mu \delta^+ D^\mu \delta^- - M_\delta^2 \delta^+\delta^- - \frac{1}{2}\lambda_\delta(\delta^+\delta^-)^2 - \lambda_{\delta H}\delta^+\delta^- H^\dagger H + \left[-\overline{l_{L\alpha}}(y_L)_{\alpha\beta}(i\sigma_2)(l_{L\beta})^c \delta^- + h.c. \right]. \tag{2}$$

When both the sterile neutrinos N_{Ri} and the charged scalar δ^+ are present, there is an additional Yukawa coupling involving the SM lepton singlets,

$$\mathscr{L}_{\delta N} = -\overline{(e_{R\alpha})^c}(y_R)_{\alpha i} N_{Ri}\delta^+ + h.c. \tag{3}$$

These three terms, Eqs. (1)–(3), plus the well-known SM ones, constitute the Lagrangian considered in this paper. It is the most general Lagrangian consistent with the SU(3) × SU(2) × U(1) gauge symmetry and with the addition of the singlet fermions and the charged scalar. Without loss of generality, we adopt the basis where M_N and the charged lepton mass matrix are real and diagonal. For our following discussion, the parameters of greater relevance are the masses of the singlet fermions (M_i, i = 1, 2, 3) and of the charged scalar (M_δ), and the 3 × 3 Yukawa matrices y_ν and y_R.

In this model active neutrino masses are not affected by the existence of δ^+ and are obtained via the usual seesaw mechanism: $m_\nu \simeq y_\nu M_N^{-1} y_\nu^T v^2$, with v ≃ 174 GeV. In Sect. 3 we will identify the sterile neutrino N_1 with a multi-keV dark matter candidate, with very small mixing angles with active neutrinos, corresponding to tiny neutrino Yukawa couplings, $(y_\nu)_{\alpha 1} \lesssim 3 \times 10^{-13}$. These tiny couplings give rise to a lightest neutrino mass $m_\nu^{\text{lightest}} \lesssim 10^{-6}$ eV. The much larger solar and atmospheric neutrino mass scales are generated instead by the seesaw contribution of the two heavier sterile neutrinos, $N_{2,3}$. The active neutrino mass spectrum is thus hierarchical (either normal or inverted), with one neutrino remaining essentially massless.

The generation of the dark matter energy density and of the baryon asymmetry have been extensively studied in the context of the SM extended with sterile neutrinos only [6]. The SM extension with sterile neutrinos and δ^+ was considered in [23]. It was shown that, taking the sterile neutrinos N_i as light as a few TeVs and heavier than δ^+, one can generate the observed baryon asymmetry via leptogenesis, with no need to enhance resonantly the CP asymmetry.

We explore, instead, the possibility that the lightest singlet fermion, N_1, has a multi-keV mass, $M_1 \ll M_\delta$, and accounts for the observed dark matter density. In this framework, the existence of δ^+ offers an alternative way of producing dark matter that has not been studied before in the literature. Moreover, we will show that this different mass spectrum is still compatible with the leptogenesis mechanism proposed in [23] and, in addition, we will identify a different mechanism for leptogenesis in the presence of δ^+ based on $N_{2,3}$-oscillations.

DARK MATTER PRODUCTION FROM Δ⁺ DECAYS

In this model, the only particle that can play the role of dark matter is the lightest singlet fermion, N_1. It is important, therefore, to determine how it is produced in the early Universe and whether it can account for the observed dark matter density while respecting all other experimental constraints. In this section, after briefly reviewing the standard scenario for sterile neutrino dark matter, we propose a new mechanism for dark matter production via the decays of δ^+.

In the absence of δ^+, our model is described by the seesaw Lagrangian, Eq. (1), and the region of the parameter space compatible with dark matter is the one of the νMSM. It features a singlet fermion (N_1) with a mass at the keV scale, the dark matter particle, and two heavier singlets, $N_{2,3}$. In the νMSM, dark matter is produced at temperatures of order 100 MeV via active-sterile neutrino mixing. The required effective mixing angle, θ_1, is necessarily small and is related to the Yukawa couplings by $\theta_1^2 = \sum_{\alpha=e,\mu,\tau} |(y_\nu)_{\alpha 1}|^2 v^2/M_1^2$. Detailed numerical studies have shown that the observed dark matter density can be reproduced in two different cases [6]. In the non-resonant case (also known as the Dodelson–Widrow mechanism [1]), dark matter neutrinos are thermally and non-resonantly produced with a smooth distribution of momenta. This mechanism yields the minimal amount of dark matter that can be obtained for a given mass and mixing angle. For dark matter masses between 1 keV and 10 keV, the mixing angle required to explain the dark matter density is $\sin^2(2\theta_1) \sim 10^{-8}$- 10^{-9}, as shown e.g. in figure 2 of [6]. These parameters are not consistent with current data [24], which imply $M_1 \lesssim 3$ - 4 keV from the X-ray line bound [7–9] and $M_1 \gtrsim 8$ keV from the Ly_α forest observations [10–12]. The other case in which the dark matter density can be explained, and the only one that is currently viable in the νMSM, is resonant production (also known as the Shi-Fuller mechanism [13]). In it, the dark matter production rate is resonantly amplified by the presence of a lepton chemical potential in the plasma, which enhances the production for particular momenta as they pass through the resonance, giving rise to a non-thermal momentum distribution that is colder than that obtained in the non-resonant case. To explain the observed dark matter density, a lepton asymmetry $|\mu_\alpha| \gtrsim 8 \times 10^{-6}$ at $T \sim 100$ MeV is required [14], where $\mu_\alpha = n_\alpha/s$, s is the entropy density of the Universe, and n_α is the total number density of active leptons of flavour α. In principle, this large lepton asymmetry can be generated within the νMSM via CP-violating oscillations of N_2 and N_3, but only if their mass difference, $M_3 - M_2$, is fine-tuned to the order of 10^{-11} [15, 25].

The existence of δ^+ allows a new production mechanism for sterile neutrinos in the early Universe. Indeed, thanks to the interaction term in Eq. (3), the decays of the charged scalars can produce sterile neutrinos, $\delta^+ \to N_1\ell^+$. This decay is a thermal process that takes place while δ^+ is in equilibrium and it is an example of the so-called freeze-in scenario [26] for dark matter production. The defining feature of freeze-in is that the dark matter interactions are so weak that they never reach thermal equilibrium in the early Universe. As a result, the dark matter abundance is negligible at high temperatures and slowly increases as the Universe cools down. The production ceases when the freeze-in temperature is reached, and the dark matter abundance remains constant from then on.

The N_1 yield, $Y_{N1}(T) = n_{N1}(T)/s(T)$, from δ^+ decays is obtained by solving the following Boltzmann equation [26]

$$sT\frac{dY_{N_1}}{dT} = -\frac{\gamma_{N_1}(T)}{H(T)} \tag{4}$$

where s is the entropy density, H is the expansion rate and $\gamma_{N1}(T)$ is the thermally-averaged production rate. We have that

$$\gamma_{N_1}(T) = \frac{M_\delta^2 T}{2\pi^2}K_1(M_\delta/T) \times \sum_\alpha \left[\Gamma(\delta^- \to N_1\ell_\alpha^-) + \Gamma(\delta^+ \to N_1\ell_\alpha^+)\right] \tag{5}$$

where $K_1(x)$ is the Bessel function of the second kind and the sum runs over the different lepton flavours. Since the lepton and N_1 masses are negligible compared to M_δ, the δ^+ decay rates are calculated as

$$\Gamma(\delta^- \to N_1\ell_\alpha^-) = \Gamma(\delta^+ \to N_1\ell_\alpha^+) = \frac{M_\delta|(y_R)_{\alpha 1}|^2}{16\pi} \tag{6}$$

and their sum over flavours can be conveniently written as

$$\sum_\alpha \left[\Gamma(\delta^- \to N_1\ell_\alpha^-) + \Gamma(\delta^+ \to N_1\ell_\alpha^+)\right] = \frac{M_\delta}{8\pi}\left(y_R^\dagger y_R\right)_{11} = \frac{M_\delta}{8\pi}y_{R1}^2, \tag{7}$$

where we have defined y_{R1} as the combination of couplings that determines the production of N_1 via δ^+decays. Using $s(T) = 2\pi^2 g_s T^3/45$, $H(T) = 1.66\sqrt{g_\rho}T^2/M_{Pl}$, and $x \equiv M_\delta/T$ we can solve Eq. (4) as

$$Y_{N_1}(T_0) = \frac{45}{(1.66)32\pi^5 g_s\sqrt{g_\rho}}\frac{M_{Pl}y_{R1}^2}{M_\delta}\int_{x_{\min}}^{x_{\max}} x^3 K_1(x)\,dx\,, \tag{8}$$

where T_0 is the temperature today and we used $Y(T \gg M_\delta) = 0$ as our initial condition. Integrating this equation from $x_{max} = \infty$ to $x_{min} = 0$ and setting $g_s \approx g_\rho \approx 100$ yields

$$Y_{N_1}(T_0) \approx 1.6 \times 10^{-5}\left(\frac{y_{R1}}{10^{-8}}\right)^2\left(\frac{1\text{ TeV}}{M_\delta}\right). \tag{9}$$

The N_1 relic density is then

$$\Omega_{N_1}h^2 \approx 0.11\left(\frac{M_1}{\text{keV}}\right)\left(\frac{y_{R1}}{5\times 10^{-8}}\right)^2\left(\frac{1\text{ TeV}}{M_\delta}\right). \tag{10}$$

Thus, a keV sterile neutrino can explain the observed dark matter density via δ^+ decays if $M_\delta \sim 1$ TeV and$y_{R1} \sim$ few × 10^{-8}. This analytical result is in very good agreement with the numerical calculations we did for similar models of freeze-in dark matter [27, 28].

A crucial feature of Eq. (10) is that, unlike the production in the νMSM, it does not depend on the active-sterile mixing angle θ_1 that determines the decay rate of the sterile neutrino. Hence, one can take θ_1 small enough to be in agreement with the X-ray bounds without affecting the predicted dark matter density. In other words, in this model one can decouple the dark matter production (determined by the y_R couplings) from the dark matter decay (determined by y_ν via mixing angles), opening new viable regions and alleviating the experimental constraints on sterile neutrino dark matter. In particular, M_1 can be larger than the keV scale, as long as the dark matter sterile neutrino is sufficiently long-lived. Since $\Gamma_{N_1} \propto M_1^5 \sin^2\theta_1$, one cannot exclude, for example, the possibility of vanishing active-sterile mixing, $\theta_1 = 0$, which would allow masses as high as 1 MeV. Above that value, the decay $N_1 \to \nu e^+ e^-$ becomes kinematically available and can proceed through the couplings y_R and y_L, even for $\theta_1 = 0$. In that case, the strong constraints coming from indirect searches of electrons and positrons come into play, but they compete with the smallness of y_{R1}, determined by Eq. (10), and of the Yukawa matrix y_L, that can be very small too.

Regarding structure formation, the sterile neutrinos produced via δ^+ decays are colder than those obtained in the νMSM [29, 30] because the decays take place at a temperature $T \sim M_\delta$ much higher than the QCD scale (~ 150 MeV). In fact, their free-streaming length and phase space density are identical to those of the so-called chilled sterile neutrinos studied in [31]. Consequently, the Lyman-α bounds on the sterile neutrino mass in our scenario are significantly weaker, allowing for a dark matter mass M_1 smaller than in the νMSM [32].

We assume in the following that the decays of the charged scalars are the dominant source of sterile neutrinos, so that Eq. (10) is satisfied. This will be the case in the absence of a large lepton asymmetry and for small mixing angles, $\sin^2 2\theta_1 < 10^{-9}$. Next, we show that this assumption is consistent with baryogenesis, and that it leads to new experimental signatures.

BARYOGENESIS VIA LEPTOGENESIS WITH Δ+

In this section we will discuss leptogenesis in the presence of the charged scalar field δ^+. We will briefly review the two basic mechanism for leptogenesis in the minimal scenario with sterile neutrinos only: N-decays and N-oscillations. We will show that both mechanisms are still operative when one replaces the Yukawa coupling y_ν in Eq. (1) with y_R in Eq. (3), that is, when the role of the SM Higgs doublet H is played by δ^+, and the role of the SM lepton doublet l_L is played by the SM lepton singlet e_R.

The baryon asymmetry of the Universe can be generated from a lepton asymmetry, as long as the latter is present before the electroweak phase transition, which occurs at a temperature $T_{EW} \simeq 150$ GeV. Above this temperature the (B + L)-violating electroweak sphalerons are in thermal equilibrium, thus converting efficiently leptons into baryons. When the SM is extended by sterile neutrinos, one can generate a lepton asymmetry, provided that the set of couplings $(M_N)_{ij}$, $(y_\nu)_{\alpha i}$ and $(y_R)_{\alpha i}$ violates the CP symmetry, and (some of) these couplings are out-of-equilibrium at some epoch before T_{EW}.

Leptogenesis from *N*-decays

The traditional leptogenesis mechanism [2] assumes a Majorana mass matrix M_N for two (or more) sterile neutrinos, with eigenvalues M_i larger than the electroweak scale, and it relies on the out-of-equilibrium decays $N_i \to Hl_{L\alpha}$, at temperatures just below M_i. As N_1 plays the role of keV-scale dark matter candidate, the relevant decays are those of N_2, with $T_{EW} < M_2 < M_3$. Since M_2 violates lepton number, the CP-asymmetry that is generated in the decays is also a lepton asymmetry, or equivalently a (B - L)-asymmetry. After all the N_2 particles decayed, the produced (B - L)-asymmetry remains constant. In this scenario, the couplings that control leptogenesis are the same Yukawa couplings, y_ν, that determine the active neutrino masses. As a consequence, a sufficient amount of baryon asymmetry can be generated only for$M_2 \gtrsim 10^8$ GeV approximately.[1] Thus, no direct test of this scenario can be envisaged.

In the presence of a charged scalar δ^+, an analog source of leptogenesis is provided by the decays $N_i \to \delta^+ e_{R\alpha}$, mediated by the Yukawa coupling y_R, as shown in [23]. Here we adapt the results of that paper to our new context, where the dark matter particle is the keV-scale sterile neutrino N_1. We can safely neglect the presence of N_1 for the computation of the lepton asymmetry because its interaction rates (in particular those violating lepton number) are strongly out-of-equilibrium above T_{EW}, due to the extreme smallness of M_1,$(y_R)_{\alpha 1}$ and $(y_\nu)_{\alpha 1}$. Thus, at least two extra sterile neutrinos $N_{2,3}$ heavier than δ^+ are needed to generate theCP-asymmetry. Taking for simplicity $M_3 \gg M_2 \gg M_\delta$, the CP-asymmetry in N_2-decays is given by

$$\epsilon_{N_2} = \frac{1}{8\pi} \frac{\mathrm{Im}[\sum_\alpha (y_R)_{\alpha 2} (y_R)^*_{\alpha 3}]^2}{\sum_\alpha (y_R)_{\alpha 2} (y_R)^*_{\alpha 2}} \frac{M_2}{M_3}, \tag{11}$$

where we assumed that the charged scalar lepton number is $L(\delta^+) = -2$, since the dominant decay mode is $\delta^+ \to e^+ \bar{\nu}$, through the Yukawa coupling y_L in Eq. (2).[2] To reproduce the observed baryon asymmetry two basic conditions are required [23]: a minimal value for ϵ_{N2} assuming no washout,

$$\frac{|\mathrm{Im}[\sum_\alpha (y_R)_{\alpha 2} (y_R)^*_{\alpha 3}]^2|}{\sum_\alpha (y_R)_{\alpha 2} (y_R)^*_{\alpha 2}} \gtrsim 2 \cdot 10^{-6} \frac{M_3}{M_2}, \tag{12}$$

and N_2-decays out-of-equilibrium at $T = M_2$ to avoid large washout from inverse decays,

$$\sum_\alpha (y_R)_{\alpha 2} (y_R)^*_{\alpha 2} \lesssim 10^{-13} \frac{M_2}{1\ \mathrm{TeV}}. \tag{13}$$

Thus, one can realize leptogenesis for M_2 as small as a few TeVs, as long as $N_{2,3}$ have hierarchical Yukawa couplings: $|(y_R)_{\alpha 2}| \lesssim 3 \times 10^{-7}$ and $|(y_R)_{\alpha 3}| \gtrsim 10^{-3}\sqrt{M_3/M_2}$. It is worth reminding that, when M_2approaches the electroweak scale, a sufficiently large CP-asymmetry implies in general strong (B - L)-washout rates, in particular those mediated by N_3 off-shell, unless special conditions are realized. A neat, model-independent discussion of the lower bound on the leptogenesis scale can be found in [36, 37]. In the present model with $M_2 \sim$ a few TeVs, the washout can be Boltzmann suppressed by raising M_δ sufficiently close to M_2, and taking a sufficiently large coupling y_L in Eq. (2) to avoid washout from the asymmetry stored in δ^+.[3]

Note that the asymmetries generated by N_2-decays through the Yukawa coupling matrix y_ν become negligible in the region $M_2 \ll 10^8$ GeV due to the seesaw relation. The y_R-entries, on the other hand, are not constrained by the light neutrino masses, and leptogenesis can work close to the electroweak scale as described above. This scenario is, therefore, easier to test directly, by the observation of δ^+ (and possibly N_2) at colliders, as discussed in Sect. 5.

Leptogenesis from *N*-oscillations

A complementary mechanism for leptogenesis [3] relies on N-oscillations rather than on N-decays. It also requires two (or more) sterile neutrinos, coupled to the SM through the Yukawa matrix y_ν. In the early Universe, the thermal population of lepton doublets $l_{L\alpha}$ produces, through small, out-of-equilibrium couplings $(y_\nu)_{\alpha i}$, a coherent superposition N_α of the sterile neutrino mass eigenstates N_i. These sterile neutrinos, at temperatures much larger than their masses, coherently oscillate among the different flavours α. Such oscillations conserve lepton number (the N_α conserve their helicity), but violate lepton flavour numbers. If the CP-symmetry is also violated, one generates non-zero flavour asymmetries between the opposite helicities of the N_α. The asymmetry in the flavour α is transferred efficiently to $l_{L\alpha}$, as long as the coupling$(y_\nu)_{\alpha i}$ goes into equilibrium for some i. Since the total lepton asymmetry, that is, the sum over α of the flavour asymmetries vanishes, one needs that some but not all flavours go into equilibrium before T_{EW}, so that a net lepton asymmetry remains stored in the sterile neutrino sector, and an opposite one is available in the SM sector to be transferred to the baryons by electroweak sphalerons. The Yukawa (out-of-)equilibrium condition at T_{EW} reads

$$|(y_\nu)_{\alpha i}| \gtrsim 10^{-7} \quad \text{for some } \alpha \text{ and } i, \quad |(y_\nu)_{\beta i}| \lesssim 10^{-7} \quad \text{for some } \beta \neq \alpha,\ \forall i. \tag{14}$$

A number of additional constraints, on the size of y_ν-entries and on the values of M_i, must be satisfied for this leptogenesis mechanism to work. To facilitate the comparison between the scenarios with and without the charged scalar, we will describe them in some detail. Due to the dark matter constraint, N_1 plays no role in leptogenesis and the asymmetry must be generated by N_2 and N_3 only.[4]

Successful leptogenesis implies a few upper bounds on the y_ν-entries, beside the flavour-dependent one in Eq. (14). First of all, note that the Majorana masses M_i violate lepton number. Therefore, they play the role of washout in this scenario, as they transform opposite helicities of the N_α into one another. The condition to keep the lepton number violation rate out-of-equilibrium down to T_{EW} is approximately

$$|(y_\nu)_{\alpha i}| \lesssim 2 \times 10^{-6} \left(\frac{M_i}{\text{GeV}}\right)^{1/2} \quad \forall\alpha \text{ and } \forall i. \tag{15}$$

which combined with Eq. (14) implies sterile neutrino masses below the electroweak scale, $M_i \lesssim 100$ GeV.[5] In addition, for the case of two sterile neutrinos $N_{2,3}$, analytic and numerical studies [4, 16, 39] show that a sufficient baryon asymmetry requires a strong degeneracy between M_2 and M_3, with $\Delta M/M \lesssim 10^{-5}$. This comes from the interplay of a number of subtle effects: the oscillation time increases as ΔM decreases, and this allows for larger asymmetries because the Yukawa interaction rates are closer to equilibrium at later times; note also that larger Yukawa couplings (for some but not all the flavours) enhance the flavour asymmetries, but also tend to spoil the coherence of the sterile neutrinos, as active-sterile transitions may become faster than the oscillation time. We extrapolate the resulting constraint from figure 7 of [6], that can be written as

$$|(y_\nu)_{\alpha i}| \lesssim 2 \times 10^{-6} \left(\frac{M_i}{\text{GeV}}\right)^{1/2} \quad \forall\alpha \text{ and } \forall i. \tag{16}$$

The masses $M_{2,3}$ are below collider energies, but it is difficult to produce them directly since they have small couplings. Still, the mixing with active neutrinos can be sufficiently large to have an observable effect in various neutrino experiments [40]. The resulting upper bound on active-sterile mixing is also shown in figure 7 of [6]. An experimental proposal to improve significantly the present bound can be found in [41]. In less minimal models, e.g. with three GeV scale sterile neutrinos, there are good detection perspectives in meson decay experiments [42].

On the other hand, several observables other than the baryon asymmetry put lower bounds on the y_ν-entries, that add to the one in Eq. (14). The lower bound on active neutrino masses, $|(m_\nu)_{\alpha\beta}| \gtrsim 0.025$ eV for some α and β, implies a lower bound on the neutrino Yukawa couplings through the seesaw formula,

$$|(y_\nu)_{\alpha i}| \gtrsim 2 \times 10^{-8} \left(\frac{M_i}{\text{GeV}}\right)^{1/2} \quad \text{for some } \alpha \text{ and } i. \tag{17}$$

Indeed, since N_1 gives a negligible contribution to m_ν, one active neutrino is approximately massless, therefore $|(m_\nu)_{\alpha\beta}| \lesssim 0.05$ eV for all α and β. Then, the right-hand side of Eq. (17) provides the natural value of the Yukawa couplings for i = 2, 3: the largest $(y_\nu)_{\alpha i}$ are, the strongest is the cancellation needed among the contributions of N_2 and N_3 to m_ν. These GeV-scale sterile neutrinos should decay (e.g. in 3ν's or νe^+e^-) before T $\simeq$ 1 MeV, not to spoil nucleosynthesis (see [43] for a detailed analysis). We roughly estimate this constraint as

$$|(y_\nu)_{\alpha i}| \gtrsim 3 \cdot 10^{-8} \left(\frac{\text{GeV}}{M_i}\right)^{3/2}, \quad \text{for some } \alpha,\ \forall i. \tag{18}$$

Comparing with Eq. (16), this implies $M_i \gtrsim 0.1$ GeV. All the constraints above confirm that N_1 plays no role in leptogenesis.

Let us show that the same mechanism of leptogenesis through N-oscillations is operative in the presence of the charge scalar δ^+, by replacing the role of y_ν with y_R. Analogously to the previous case, the couplings $(y_R)_{\alpha i}$ must be small to remain out-of-equilibrium while they slowly produce coherent sterile neutrino states. The lepton flavour asymmetries generated by N-oscillations are (partially) transferred to the SM lepton singlets $e_{R\alpha}$. The latter are in equilibrium with $l_{L\alpha}$ through the charged lepton Yukawa couplings (at least for $\alpha = \mu, \tau$), thus electroweak sphalerons transfer the asymmetries to baryons as usual.

One needs that the y_R-entries satisfy the same inequalities as the y_ν-entries in Eqs. (14) and (15), so that some flavour asymmetries are transferred to baryons, while the others remain stored in the sterile sector, and the washout from lepton number violating scattering is small. The only differences amount to (i) order one factors to account for the singlet (doublet) nature of e_R (l_L) in scattering rates; (ii) the range of temperatures where M_δ (M_H) can be neglected: note that the asymmetry is generated at some scale well above T_{EW} through scattering processes mediated by δ^+ (H).

We also expect that, when both y_ν and y_R take values relevant for leptogenesis, there is more freedom to generate large asymmetries, e.g. because of the presence of extra CP-violating phases, and the requirement of a strong degeneracy between M_2 and M_3 could be relaxed. As in the case of y_ν, one cannot raise too much the value of y_R-entries, as the coherence of sterile neutrino oscillations requires the scattering rate with the SM leptons to be out-of-equilibrium at the time of oscillations; this should translate in a bound similar to the one in Eq. (16). A numerical study is needed to establish more precisely the allowed parameter space, and the differences with respect to the νMSM one. The effect of the two sets of Yukawa couplings could be dramatic, as is the effect of a third GeV scale sterile neutrino [42, 44], or of a second Higgs doublet [16].

There is no constraint on y_R coming from the seesaw relation, so one can take couplings smaller than in Eq. (17), slightly enlarging the region of parameters of leptogenesis. Perhaps more importantly, one can take $(y_R)_{\alpha i}$ significantly larger than the right-hand side of Eq. (17), with no need of fine-tuning to keep m_ν small. Note that the coupling y_R does not induce any mixing with active neutrinos. Indeed, direct searches of active-sterile mixing are presently sensitive to y_ν-entries much larger than in Eq. (17): if the seesaw parameters take their natural values, no direct signal of active-sterile mixing is expected.

Finally, the three-body decay rate of N_i through the coupling y_R is proportional to $|(y_R)_{\alpha i}(y_L)_{\beta\gamma}|^2/M_\delta^4$, to be compared with $|(y_\nu)_{\alpha i}|^2/(M_i^2 v^2)$ for a decay through the mixing with active neutrinos. Since the latter is typically much faster, the nucleosynthesis bound applies to y_ν only, in the form of Eq. (18). Then, leptogenesis through y_R could work even for $M_i < 0.1$ GeV, but in this region nucleosynthesis demands large values for the y_ν-entries, that require strong cancellations in the seesaw.

EXPERIMENTAL SIGNATURES FOR N_1 AND Δ^+

Two important differences between our model and the νMSM are the additional region of the parameter space where the dark matter constraint can be satisfied and the presence of the extra charged scalar δ^+. They both give rise to new experimental signatures that may allow to distinguish one model from the other.

Dark matter indirect detection

A multi-keV dark matter neutrino is inherently unstable and decays into three light neutrinos at tree-level ($N_1 \to 3\nu_\alpha$) and radiatively into a light neutrino and a photon ($N_1 \to \nu_\alpha\gamma$). This radiative decay produces an X-ray line at $E_\gamma \simeq M_1/2$ that can be searched for and used to constrain the model or to help determine its parameter space [45]. The presence of the additional scalar δ^+ does not affect the decay modes of the dark matter neutrino, which are still determined by its mass and its mixing with the active neutrinos, just as in theνMSM, but it modifies the regions that are consistent with the dark matter constraint, allowing, in particular, for smaller mixing angles. This fact has important implications, as we show next.

Recently, the detection of an unidentified spectral line at about 3.5 keV has been reported from two independent data sets [46, 47]. Arguments in favour or against the dark matter decay interpretation of the signal can be found in [48–51]. If confirmed, that signal would provide compelling evidence for keV-scale dark matter and, in particular, for dark matter in the form of sterile neutrinos. Within that framework, the signal can be explained if $M_1 \simeq 7$ keV and $\sin^2 2\theta_1 \simeq 5 \times 10^{-11}$. In the minimal scenario for sterile neutrino dark matter, where they are produced non-resonantly, such parameters lead to a relic density way below the range determined by cosmological observations. Thus, an additional source of sterile neutrinos is required. A simple possibility for that new source are the decays of the δ^+ particle, as explained in the previous section. These decays allow to decouple the dark matter production (determined by the y_R couplings) from the dark matter decay (determined by y_ν via mixing angles), opening new viable regions consistent with all bounds. Specifically, the freeze-in production of sterile neutrinos that we have examined in this paper enables to explain the tentative 3.5 keV line and to simultaneously account for the observed relic density, even in the absence of a large, primordial lepton asymmetry. If that line signal turns out to be spurious, X-ray observations will continue to be the main way in which the dark matter sector of this model can be tested in the foreseeable future.

Charged scalar searches at colliders

The singly-charged isosinglet scalar δ^+ could be directly produced at colliders, if its mass were within their energy range. The dominant production channel is the Drell–Yan process $\psi\overline{\psi} \to \gamma/Z \to \delta^+\delta^-$, with ψ= e at LEP and ψ = q at Tevatron and LHC, with a partonic cross-section given e.g. in [52]. The δ^+ decays into one anti-lepton and one anti-neutrino, either through the Yukawa coupling $(y_L)_{\alpha\beta}$, or $(y_R)_{\alpha i}$ when$M_i < M_\delta$.[6]

The standard experimental searches for charged scalars usually assume production and decay modes different from those above. For a singly-charged scalar H^+, the decay channel into lepton plus missing energy has been analyzed, but only for an isodoublet produced from a top-bottom vertex, as in type II two-Higgs doublet models, allowing for an effective background reduction [53, 54]. Indeed, the signal over background ratio is significantly smaller in the case of δ^+; an analysis of the cuts required to maximize the signal has been presented in section 5.2 of [55].

Presently, it appears that the best way to constrain M_δ is to use the experimental searches for supersymmetric particles. Since δ^+ and a right-handed slepton have the same gauge quantum numbers, when the latter is directly pair-produced and decays into a lepton and a light neutralino, it behaves very much as the former, and the same bounds apply to both particles. Recently, that setup was studied in the context of simplified supersymmetric models, for the case of a right-handed s-electron degenerate with a right-handed s-muon, decaying with branching ratio 100 % into electron and muon, respectively, plus missing energy [56, 57]. Assuming a similar efficiency for electron and muon detection and reducing the signal by a factor two, to take into account that the δ^+ production cross-section is the same as for a single right-handed slepton, we can extract a 95 % C.L. lower bound $M_\delta \gtrsim 200$ GeV at ATLAS [56] (see also the auxiliary figures 11a and 12a available online[7]) and $M_\delta \gtrsim 160$ GeV at CMS [57].

Note, however, that these bounds apply as long as the δ^+ branching ratio into taus is negligible. The pair-production of s-taus decaying into taus is presently unconstrained at the LHC (see figure 9 in [58]). As a consequence, if δ^+ decays mostly into taus, its mass can be significantly smaller, say down to the LEP bound ~ 100 GeV. Indeed, in minimal flavour models the δ^+ Yukawa couplings are hierarchical as the SM Yukawa couplings (see Sect. 6), and a large branching ratio into taus is expected.

FLAVOUR STRUCTURE OF THE MODEL

We accomplished our goal of demonstrating that this model can account for dark matter, neutrino masses, and baryogenesis. That is, we have determined, in the previous sections, that there exist regions in the parameter space of the model where the masses of the new particles and their Yukawa couplings are such that these three issues are simultaneously explained. One may wonder, nonetheless, whether those values of the masses and Yukawa couplings have a generic flavour structure, or if they require a special tuning, indicating that some flavour symmetries are operative. Previous works along this line include [59–63], which focused on the generation of the keV scale in neutrino models, and on the flavour structure of the νMSM. In this section, we do not aim to construct complete flavour models, but rather to recap the order of magnitude of the parameters needed phenomenologically, and to suggest some rationale to explain them.

In view of the hierarchical values of masses and couplings that are required for dark matter and leptogenesis, it is useful to describe the required flavour structure in terms of a $U(1)_F$ family symmetry, with different charges assigned to the various fermions. In this framework, the coefficient of each fermion bilinear $\psi_{Li}\psi_{Lj}$ is suppressed by a power $q_i + q_j$ of a small parameter ϵ (we take the fermion $U(1)_F$-charges to be positive). One may assume some underlying flavour dynamics, that generates $\epsilon = \langle\phi\rangle/\Lambda_F \ll 1$, that is, the vev of a spurion field with $q_\phi = -1$, over the cutoff of the flavour theory.

We need $M_1 \ll M_2 \sim M_3$. Note that in the νMSM a strong degeneracy of M_2 and M_3 is required, while in our scenario this constraint is relaxed. The straightforward way to realize this pattern is to take $q_{N1} > q_{N2} = q_{N3}$, so that $M_1/M_2 \sim \epsilon^{2n}$, with $n \equiv q_{N1} - q_{N2}$. For N-oscillation (N-decay) leptogenesis, one needs $\epsilon^{2n} \sim 10$ keV/GeV $= 10^{-5}$ ($\epsilon^{2n} \lesssim 10$ keV/TeV $= 10^{-8}$). This charge assignment implies automatically a hierarchy ϵ^n:1:1 among the three columns of the matrix y_R (as well as of y_ν), in other words, $(y_R)^2_{\alpha i}/(y_R)^2_{\alpha j} \sim M_i/M_j$. Since this relation is not satisfied in certain regions of parameters relevant for dark matter, leptogenesis and neutrino masses, an additional flavour symmetry must be operative. The Yukawa couplings can be further suppressed by introducing parity symmetries $Z_2^{(i)}: N_i \to -N_i$, that allow for M_i but forbid $(y_\nu)_{\alpha i}$ and $(y_R)_{\alpha i}$; these couplings must then be proportional to a small symmetry-breaking parameter ϵ_i. To suppress $(y_\nu)_{\alpha i}$ and not $(y_R)_{\alpha i}$ (or vice versa), one may argue that under the same parity the combination $e_{R\alpha}\delta^+$ (or $l_{L\alpha}H$) is also odd.[8]

Coming to the $U(1)_F$ charges of the three families of lepton doublets and singlets, $q_{L\alpha}$ and $q_{R\alpha}$, first of all they determine the hierarchy of the charged lepton masses, $m_e:m_\mu:m_\tau \sim \epsilon^{q_{Le}+q_{Re}} : \epsilon^{q_{L\mu}+q_{R\mu}} : \epsilon^{q_{L\tau}+q_{R\tau}}$. In turn, the hierarchy among the rows of y_R (y_ν) is determined by the charges $q_{R\alpha}$ ($q_{L\alpha}$). Note that the charges of $l_{L\alpha}$ and $e_{R\alpha}$ are important for charged lepton and active neutrino masses, as well as for leptogenesis from N-oscillations; however they are not very relevant for dark matter production nor for leptogenesis from N-decays.

Let us confront these simple flavour symmetries with the values of the parameters needed for dark matter, neutrino masses and leptogenesis. The freeze-in of the desired amount of N_1 from δ^+ decays requires $\sum_\alpha |(y_R)_{\alpha 1}|^2 \simeq 2.5 \times 10^{-24} M_\delta/M_1$ (see Eq. (10)). The N_1-production from active-sterile oscillations is negligible for $|(y_\nu)_{\alpha 1}| \lesssim 10^{-13}(M_1/\text{keV})$ (barring a large primordial lepton asymmetry). To generate large enough active neutrino masses, Eq. (17) must be satisfied. Clearly N_1 does not contribute significantly to m_ν because of the dark matter constraint, therefore N_2 and N_3 are responsible to generate the atmospheric and solar mass scales. Since $\sqrt{\Delta m^2_{sol}/\Delta m^2_{atm}} \simeq 0.2$, Eq. (17) must hold for both i = 2 and 3, up to a factor of a few. Coming to leptogenesis, the two mechanisms discussed in Sect. 4 correspond to two very different regions of parameters that we discuss in turn.

In the case of leptogenesis from N-decays through the y_ν-interaction, one requires $M_{2,3} >$ TeV, as well as $|(y_R)_{\alpha 3}| > 10^{-3}\sqrt{M_3/M_2}$ for at least one α, and $|(y_R)_{\alpha 2}| < 10^{-4}\sqrt{M_2/(10^8\ \text{GeV})}$ for allα. For definiteness, let us take $M_1 \sim$ keV, $M_\delta \sim$ TeV and $M_{2,3} \sim 100$ TeV (leptogenesis through y_ν is irrelevant at such low scales, barring resonances). Taking $\epsilon = 0.1$ and the $U(1)_F$ charges $q_{N1} = 7$, $q_{N2} = q_{N3} = 2$, $q_{R\tau} = 0 \geq q_{R\mu}, q_{Re}$, $q_{L\tau} = 2 \geq q_{L\mu}, q_{Le}$, we can reproduce the correct size of the M_N, y_R and y_ν entries, except for the second column of y_R, that must be further suppressed by $\epsilon_{R2} \lesssim 10^{-3}$, and the first column of y_ν, to be further suppressed by $\epsilon_{\nu 1} \lesssim 10^{-4}$. As discussed above, ϵ_{R2} ($\epsilon_{\nu 1}$) can be associated to the breaking ofN_2-parity (N_1-parity). Note that here $M_{2,3} \sim \epsilon^4\Lambda_L$, with a lepton number violation scale $\Lambda_L \sim 10^9$ GeV.

In the case of leptogenesis from N-oscillations coupled to the SM through y_R, one requires for i = 2, 3,$M_i < 100$ GeV, $|(y_R)_{\alpha i}|M_i \lesssim 10^{-5}$ GeV, as well as $|(y_R)_{\alpha i}| > 10^{-7}$ for some flavour α and $< 10^{-7}$ for some different flavour β. For definiteness, let us take $M_1 \sim 10$ keV, $M_{2,3} \sim 10$ GeV and $M_\delta \sim$ TeV. Then, one can reproduce the correct size of the M_N and y_R entries taking $\epsilon = 0.1$ and $U(1)_F$ charges $q_{N1} = 8$, $q_{N2} = q_{N3} = 5$,$q_{R\tau} = 1$, $q_{R\mu} = 2$ and $q_{Re} = 3$, and charged lepton masses require $q_{L\tau} = 1 \geq q_{L\mu}, q_{Le}$. Coming to the structure of y_ν, one needs an extra source of suppression for $(y_\nu)_{\alpha 1}$, by a factor $\epsilon_{\nu 1} \lesssim 10^{-3}$, to avoid N_1-overproduction from active-sterile mixing; as before, $\epsilon_{\nu 1}$ can be understood as the breaking parameter of N_1-parity. We note that, for the chosen values of parameters, both y_R and y_ν couplings to $N_{2,3}$ are relevant for leptogenesis. Here lepton number is broken at the scale $\Lambda_L \sim M_{2,3}/\epsilon^{10} \sim 10^{11}$ GeV.

In summary, the regions of parameters where our scenario is effective require a well-defined hierarchy of masses and couplings, that point to specific but relatively natural flavour structures, with no need of extreme tuning of parameters.

CONCLUSIONS

We have shown that, when the SM is extended with three sterile neutrinos N_i and a charged scalar δ^+, baryogenesis via leptogenesis and the production of dark matter can both be mediated by δ^+.

The dark matter candidate in this scenario is the lightest sterile neutrino N_1, with a mass M_1 larger than about a few keVs. Such sterile neutrinos are produced by the decays of δ^+ while it is in thermal equilibrium in the early Universe. We showed that this novel mechanism, which is an example of freeze-in dark matter production, can account for the observed dark matter density. Moreover, since the resulting relic density does not depend on the active-sterile mixing angle θ_1, the constraints from X-ray data are alleviated. The recent hint of a signal at 3.5 keV can be easily explained by N_1-decays, for $\theta_1 \sim 3 \times 10^{-6}$. Alternatively, for a vanishingly small θ_1, the dark matter mass M_1 can be raised to the MeV scale, or beyond.

Leptogenesis can be realized through the coupling y_R between δ^+ and $N_{2,3}$ in two different ways, either viaN-decays or via N-oscillations. Both mechanisms are well-known in the context of the SM plus sterile neutrinos only, as they can both proceed through the neutrino Yukawa coupling y_ν. The latter is constrained by light neutrino masses, restricting significantly the allowed parameter space and the testability of leptogenesis. In contrast, in our scenario y_R is not involved in the neutrino mass generation. As a consequence, the N-decay mechanism is successful for $M_{2,3}$ as small as a few TeVs, with no need of a resonant enhancement. The parameter window for N-oscillation leptogenesis is similar when one employs the coupling matrix y_R instead of y_ν: one needs roughly $M_{2,3} \sim$ GeV and y_R-entries smaller than 10^{-6}. However,y_R introduces additional sources of CP-violation, and moreover it is not constrained neither by light neutrino masses nor by nucleosynthesis. Thus one can largely relieve the fine-tuning of parameters necessary in the case with y_ν only.

Neutrino masses are generated by a low scale seesaw mechanism involving the two heavier singlets, $N_{2,3}$, the N_1-contribution being negligible. In this way one can accommodate the current data on neutrino masses and mixing angles, with the prediction of an almost massless lightest neutrino. At colliders such as the LHC, this scenario can be probed and constrained via the direct production of δ^+. This minimal extension of the SM thus provides a simple and testable scenario to explain neutrino masses, the dark matter, and the baryon asymmetry of the Universe.

ACKNOWLEDGMENTS

MF thanks F. Bezrukov, M. Drewes, T. Hambye, K. Jedamzik, P. Pralavorio and J. Racker for useful discussions. MF is partially supported by the FP7 European ITN project "Invisibles" (PITN- GA-2011-289442-INVISIBLES), and by the OCEVU Labex (ANR-11-LABX-0060) funded by the "Investissements d'Avenir" French government program managed by the ANR. C.Y. is partially supported by the "Helmholtz Alliance for Astroparticle Physics HAP" funded by the Initiative and Networking Fund of the Helmholtz Association.

FOOTNOTES

[1]This lower bound on the leptogenesis scale was pointed out in [33] and its exact value depends on various parameters, in particular the initial N_1 abundance [34]. It can be evaded if one allows for resonance effects, that require a strong degeneracy among the sterile neutrino mass eigenstates [35].

[2]The asymmetry changes by an order one factor if y_L is negligibly small and δ^+ undergoes slower three-body decays [23].

[3]To reduce the washout, one can also delay N_2-decays by taking $(y_R)_{\alpha 2}$ very small, but in this case one may need to produce the initial thermal density of N_2 by some other interaction.

[4]Note that the seesaw lagrangian can violate CP with only two sterile neutrinos, as it contains three physical phases, one combination of them being relevant in oscillations.

[5]We remark that most numerical studies in the literature assume that the lepton number violating rates mediated by M_i are negligible; in view of the size of the Yukawa couplings relevant for leptogenesis, this assumption seems to be justified only for $M_i \lesssim$ GeV. For larger M_i, one should include the lepton number violating rates in the Boltzmann equations. In this regime, it remains conceivable that the CP-asymmetries are large enough to compensate the relatively strong washout. This possibility was recently entertained to make this scenario work even for M_i larger than T_{EW} [38].

[6]In principle these couplings can all be very small, possibly leading to dominant three-body decays into $l^+l^-W^+$ [23]. In this case δ^+ can be sufficiently long-lived to appear as a curved charged track across the whole detector.

[7]https://atlas.web.cern.ch/Atlas/GROUPS/PHYSICS/PAPERS/SUSY-2013-11/.

[8]The other way around, one could also enhance the hierarchy among the M_i relatively to the hierarchy among the columns of y_ν and y_R, by introducing a (family-dependent) lepton number $U(1)_L$, that is conserved in the Yukawa couplings and broken by two units in the Majorana mass term. We will not need such a symmetry in the following.

REFERENCES

1. S. Dodelson, L.M. Widrow, Sterile-neutrinos as dark matter. Phys. Rev. Lett. 72, 17–20 (1994). hep-ph/9303287 [PubMed]
2. Fukugita M, Yanagida T. Baryogenesis without grand unification. Phys. Lett. B. 1986;174:45. doi: 10.1016/0370-2693(86)91126-3. [Cross Ref]
3. K. Akhmedov Evgeny, V.A. Rubakov, A. Yu. Smirnov. Baryogenesis via neutrino oscillations. Phys. Rev. Lett. 81, 1359–1362 (1998). hep-ph/9803255
4. T. Asaka, M. Shaposhnikov, The nuMSM, dark matter and baryon asymmetry of the universe. Phys. Lett. B 620, 17–26 (2005). hep-ph/0505013
5. T. Asaka, S. Blanchet, M. Shaposhnikov, The nuMSM, dark matter and neutrino masses. Phys. Lett. B631, 151–156 (2005). hep-ph/0503065
6. L. Canetti, M. Drewes, T. Frossard, M. Shaposhnikov, Dark matter, baryogenesis and neutrino oscillations from right handed neutrinos. Phys. Rev. D 87, 093006 (2013). arXiv:1208.4607
7. Abazajian K, Koushiappas SM. Constraints on sterile neutrino dark matter. Phys. Rev. D.2006;74:023527. doi: 10.1103/PhysRevD.74.023527. [Cross Ref]
8. A. Boyarsky, D. Iakubovskyi, O. Ruchayskiy, V. Savchenko, Constraints on decaying dark matter from XMM-Newton observations of M31. Mon. Not. R. Astron. Soc. 387, 1361 (2008). arXiv:0709.2301
9. S. Horiuchi, P.J. Humphrey, J. Onorbe, K.N. Abazajian, M. Kaplinghat, et al., Sterile neutrino dark matter bounds from galaxies of the Local Group. Phys. Rev. D 89, 025017 (2014). arXiv:1311.0282
10. A. Boyarsky, J. Lesgourgues, O. Ruchayskiy, M. Viel, Lyman-alpha constraints on warm and on warm-plus-cold dark matter models. JCAP 0905, 012 (2009). arXiv:0812.0010
11. Viel M, Lesgourgues J, Haehnelt MG, Matarrese S, Riotto A. Can sterile neutrinos be ruled out as warm dark matter candidates? Phys. Rev. Lett. 2006;97:071301. doi: 10.1103/PhysRevLett.97.071301. [PubMed][Cross Ref]
12. M. Viel, G.D. Becker, J.S. Bolton, M.G. Haehnelt, Warm dark matter as a solution to the small scale crisis: new constraints from high redshift Lyman-alpha forest data. Phys. Rev. D 88(4), id., 043502 (2013).arXiv:1306.2314
13. Shi X-D, Fuller GM. A new dark matter candidate: nonthermal sterile neutrinos. Phys. Rev. Lett.1999;82:2832–2835. doi: 10.1103/PhysRevLett.82.2832. [Cross Ref]
14. M. Laine, M. Shaposhnikov, Sterile neutrino dark matter as a consequence of nuMSM-induced lepton asymmetry. JCAP 0806, 031 (2008). arXiv:0804.4543

15. M. Shaposhnikov, The nuMSM, leptonic asymmetries, and properties of singlet fermions. JHEP 0808, 008 (2008). arXiv:0804.4542
16. B. Shuve, I. Yavin, Baryogenesis through neutrino oscillations: a unified perspective. Phys. Rev. D 89, 075014 (2014). arXiv:1401.2459
17. M. Shaposhnikov, I. Tkachev, The nuMSM, inflation, and dark matter. Phys. Lett. B 639, 414–417 (2006). hep-ph/0604236
18. F. Bezrukov, D. Gorbunov, Light inflaton Hunter's Guide. JHEP 1005, 010 (2010). arXiv:0912.0390
19. F. Bezrukov, D. Gorbunov, Relic gravity waves and 7 keV dark matter from a GeV scale inflaton (2014). arXiv:1403.4638
20. F. Bezrukov, H. Hettmansperger, M. Lindner, keV sterile neutrino dark matter in gauge extensions of the standard model. Phys. Rev. D 81, 085032 (2010). arXiv:0912.4415
21. A. Kusenko, Sterile neutrinos, dark matter, and the pulsar velocities in models with a Higgs singlet. Phys. Rev. Lett. 97, 241301 (2006). hep-ph/0609081 [PubMed]
22. A. Merle, V. Niro, D. Schmidt, New production mechanism for keV sterile neutrino dark matter by decays of frozen-in scalars. JCAP 1403, 028 (2014). arXiv:1306.3996
23. M. Frigerio, T. Hambye, E. Ma, Right-handed sector leptogenesis. JCAP 0609, 009 (2006). hep-ph/0603123
24. T. Asaka, M. Laine, M. Shaposhnikov, Lightest sterile neutrino abundance within the nuMSM. JHEP0701, 091 (2007). hep-ph/0612182
25. A. Roy, M. Shaposhnikov, Resonant production of the sterile neutrino dark matter and fine-tunings in the [nu] MSM. Phys. Rev. D 82, 056014 (2010). arXiv:1006.4008
26. L.J. Hall, K. Jedamzik, J. March-Russell, S.M. West, Freeze-in production of FIMP dark matter. JHEP1003, 080 (2010). arXiv:0911.1120
27. M. Klasen, C.E. Yaguna, Warm and cold fermionic dark matter via freeze-in. JCAP 1311, 039 (2013). arXiv:1309.2777
28. E. Molinaro, C.E. Yaguna, O. Zapata, FIMP realization of the scotogenic model. JCAP 1407, 015 (2014). arXiv:1405.1259
29. K. Petraki, A. Kusenko, Dark-matter sterile neutrinos in models with a gauge singlet in the Higgs sector. Phys. Rev. D 77, 065014 (2008). arXiv:0711.4646
30. A. Kusenko, Sterile neutrinos: the dark side of the light fermions. Phys. Rept. 481, 1–28 (2009).arXiv:0906.2968
31. K. Petraki, Small-scale structure formation properties of chilled sterile neutrinos as dark matter. Phys. Rev. D 77, 105004 (2008). arXiv:0801.3470
32. Kusenko A, Rosenberg LJ. Working Group Report: non-WIMP dark matter. 2013;1310:8642.
33. Sacha Davidson, Alejandro Ibarra, A Lower bound on the right-handed neutrino mass from leptogenesis. Phys. Lett. B 535, 25–32 (2002). hep-ph/0202239
34. G.F. Giudice, A. Notari, M. Raidal, A. Riotto, A. Strumia, Towards a complete theory of thermal leptogenesis in the SM and MSSM. Nucl. Phys. B 685, 89–149 (2004). hep-ph/0310123
35. Apostolos Pilaftsis, CP violation and baryogenesis due to heavy Majorana neutrinos. Phys. Rev. D 56, 5431–5451 (1997). hep-ph/9707235
36. J. Racker, Mass bounds for baryogenesis from particle decays and the inert doublet model. JCAP 1403, 025 (2014). arXiv:1308.1840
37. J. Racker, N. Rius, Helicitogenesis: WIMPy baryogenesis with sterile neutrinos and other realizations (2014). arXiv:1406.6105
38. B. Garbrecht, More viable parameter space for leptogenesis (2014). arXiv:1401.3278
39. L. Canetti, M. Shaposhnikov, Baryon asymmetry of the universe in the NuMSM. JCAP 1009, 001 (2010). arXiv:1006.0133
40. D. Gorbunov, M. Shaposhnikov, How to find neutral leptons of the ννMSM? JHEP 0710, 015 (2007). arXiv:0705.1729
41. W. Bonivento, A. Boyarsky, H. Dijkstra, U. Egede, M. Ferro-Luzzi et al., Proposal to search for heavy neutral leptons at the SPS (2013). arXiv:1310.1762
42. L. Canetti, M. Drewes, B. Garbrecht, Lab-to-Genesis (2014). arXiv:1404.7114
43. O. Ruchayskiy, A. Ivashko, Restrictions on the lifetime of sterile neutrinos from primordial nucleosynthesis. JCAP 1210, 014 (2012). arXiv:1202.2841
44. M. Drewes, B. Garbrecht, Leptogenesis from a GeV seesaw without mass degeneracy. JHEP 1303, 096 (2013). arXiv:1206.5537
45. Abazajian K, Fuller GM, Tucker WH. Direct detection of warm dark matter in the X-ray. Astrophys. J.2001;562:593–604. doi: 10.1086/323867. [Cross Ref]
46. E. Bulbul, M. Markevitch, A. Foster, R.K. Smith, M. Loewenstein, et al., Detection of an unidentified emission line in the stacked X-ray spectrum of galaxy clusters. Astrophys. J. 789, 13 (2014).arXiv:1402.2301
47. A. Boyarsky, O. Ruchayskiy, D. Iakubovskyi, J. Franse, An unidentified line in X-ray spectra of the Andromeda galaxy and Perseus galaxy cluster (2014). arXiv:1402.4119 [PubMed]
48. A. Boyarsky, J. Franse, D. Iakubovskyi, O. Ruchayskiy, Checking the dark matter origin of 3.53 keV line with the Milky Way center (2014). arXiv:1408.2503 [PubMed]
49. T.E. Jeltema, S. Profumo, Dark matter searches going bananas: the contribution of potassium (and chlorine) to the 3.5 keV line (2014). arXiv:1408.1699
50. D. Malyshev, A. Neronov, D. Eckert, Constraints on 3.55 keV line emission from stacked observations of dwarf spheroidal galaxies (2014). arXiv:1408.3531
51. A. Boyarsky, J. Franse, D. Iakubovskyi, O. Ruchayskiy, Comment on the paper dark matter searches going bananas: the contribution of potassium (and chlorine) to the 3.5 keV line by T. Jeltema and S. Profumo (2014). arXiv:1408.4388
52. M. Muhlleitner, M. Spira, A note on doubly charged Higgs pair production at hadron colliders. Phys. Rev. D 68, 117701 (2003). hep-ph/0305288

53. Serguei Chatrchyan et al., Search for a light charged Higgs boson in top quark decays in pppp collisions at s√=7 TeV. JHEP 1207, 143 (2012). arXiv:1205.5736

54. Search for charged Higgs bosons in the ττ+jets final state with pp collision data recorded at s√=8TeV with the ATLAS experiment. Technical report ATLAS-CONF-2013-090, CERN, Geneva, Aug 2013

55. E. Del Nobile, R. Franceschini, D. Pappadopulo, A. Strumia, Minimal matter at the large hadron collider. Nucl. Phys. B 826, 217–234 (2010). arXiv:0908.1567

56. G. Aad et al., Search for direct production of charginos, neutralinos and sleptons in final states with two leptons and missing transverse momentum in pppp collisions at s√= 8 TeV with the ATLAS detector. JHEP 1405, 071 (2014). arXiv:1403.5294

57. V. Khachatryan et al., Searches for electroweak production of charginos, neutralinos, and sleptons decaying to leptons and W, Z, and Higgs bosons in pp collisions at 8 TeV (2014). arXiv:1405.7570[PMC free article] [PubMed]

58. G. Aad et al., Search for the direct production of charginos, neutralinos and staus in final states with at least two hadronically decaying taus and missing transverse momentum in pppp collisions at s√ = 8 TeV with the ATLAS detector. JHEP 1410, 96 (2014). arXiv:1407.0350

59. A. Merle, V. Niro, Deriving models for keV sterile neutrino dark matter with the Froggatt–Nielsen mechanism. JCAP 1107, 023 (2011). arXiv:1105.5136

60. J. Barry, W. Rodejohann, H. Zhang, Sterile neutrinos for warm dark matter and the reactor anomaly in flavor symmetry models. JCAP 1201, 052 (2012). arXiv:1110.6382

61. M. Shaposhnikov, A possible symmetry of the nuMSM. Nucl. Phys. B 763, 49–59 (2007). hep-ph/0605047

62. M. Lindner, A. Merle, V. Niro, Soft L_e - L_μ - L_τLe−Lμ−Lτ flavour symmetry breaking and sterile neutrino keV dark matter. JCAP 1101, 034 (2011). arXiv:1011.4950

63. A. Merle, keV neutrino model building. Int. J. Mod. Phys. D 22, 1330020 (2013). arXiv:1302.2625

CHAPTER 19

FREE FALL AND GRAVITATIONAL QUANTUM STATES OF ANTIMATTER

G. Dufour, D. B. Cassidy, P. Crivelli, P. Debu, A. Lambrecht, V. V. Nesvizhevsky, S. Reynaud, A. Yu. Voronin, and T. E. Wall

INTRODUCTION

At present, together with the dark matter problem, one of the most tantalizing open questions in physics is the baryon-antibaryon asymmetry; that is, why are we living in a matter-dominated universe? Where did all the antimatter go? Different theoretical and experimental efforts trying to address this question are ongoing, including activities focusing on the gravitational behavior of antimatter [1–5]. No compelling theoretical argument seems to support that a difference between the gravitational behavior of matter and antimatter should be expected [6], although some attempts have been made to show the contrary [7–11]. Moreover observations and experiments have been interpreted as evidence against the existence of "antigravity" type forces [12–15]. However, those could be argued to be model dependent and, therefore, a simple free fall measurement is preferable. This justifies ongoing experimental efforts in that direction. A first attempt in this direction was made recently by the ALPHA collaboration [1] that bounds the ratio of gravitational mass to inertial mass of antihydrogen between −65 and 110.

The idea of directly measuring the gravitational force acting on antiparticles in the Earth's field goes back many decades, from the work of Witteborn and Fairbank, attempting to measure the acceleration of gravity for electrons and, eventually, positrons [16] to the PS200 experiment at CERN in the 1980s which included measurements on antiprotons [17]. Such measurements are extremely difficult because measuring the force of gravity on a charged particle requires a physically unrealistic elimination of stray electromagnetic fields. The obvious solution to this problem is to use neutral antimatter particles. However, at the present time, it is not technically feasible to do so; antineutrons cannot be produced in a controllable manner and antineutrinos are similarly elusive to experimenters. One may instead consider using composite systems that are electrically neutral, in which case it is only necessary to contend with dipole moments. Only a few systems that are composed of or contain some fraction of antimatter are available for scientific study. These are antihydrogen, muonium, and positronium, which have all been suggested as possible candidates for gravity measurements [18].

Selecting between different experimental methods, one should aim at precision experiments as they are much more strongly motivated theoretically. Among these, the method of quantum gravitational spectroscopy stands out by its remarkable statistical sensitivity and its cleanness from a systematic point of view.

Gravitational quantum states are solutions of the Schrodinger equation in a gravity field above a surface. They are characterized by the following energy (E_n) and spatial scale (H_n):

$$E_n = \varepsilon_0 \lambda_n, \quad \varepsilon_0 = \sqrt[3]{\frac{\hbar^2 M^2 g^2}{2m}}, \tag{1}$$

$$H_n = \frac{E_n}{Mg}, \tag{2}$$

$$\mathrm{Ai}\left(-\lambda_n\right) = 0,$$
$$\lambda_n \approx \{2.34, 4.09, 5.52, 6.79, 7.94, 9.02, 10.04, \ldots\}. \tag{3}$$

Here, M is the gravitational mass of the particle, m is its inertial mass (we distinguish between M and m in view of discussing EP tests), g is the gravitational field intensity near the Earth's surface, $\bar{g} = Mg/m$ is the acceleration of the

Citation: G. Dufour, D. B. Cassidy, P. Crivelli, et al., "Free Fall and Gravitational Quantum States of Antimatter," Advances in High Energy Physics, vol. 2015, Article ID 379642, 16 pages, 2015. doi:10.1155/2015/379642.

particle in that field, and Ai(x) is the Airy function [19, 20]. For neutrons and antihydrogen atoms, the height of the lowest gravitational level is 13.7 μm. For positronium, whose mass is approximately 1000 times smaller, it extends over 1.3 mm.The frequency of transitions between first and second quantum states equals 254 Hz for neutrons and antihydrogen and 26 Hz for positronium. The corresponding characteristic times needed to form quantum states are 0.5 ms and 5 ms, respectively.

Quantum gravitational states were observed for the first time with neutrons by measuring their transmission through a slit made of a mirror and an absorber in the GRANIT experiment [21]. If the distance between the mirror and the absorber (which is a rough surface used as a scatterer to mix the velocity components) is much higher than the turning point for the corresponding gravitational quantum state, the neutrons pass through the slit without significant losses. As the slit size decreases, the absorber starts approaching the size of the neutron wave function and the probability of neutron loss increases. If the slit size is smaller than the characteristic size of the neutron wave function in the lowest quantum state, the slit is not transparent for neutrons as was demonstrated experimentally.

Here, we analyze, in detail, several experiments which will study the free fall of antiatoms and argue that the observation of gravitational quantum states of antimatter is feasible. In Section 2, we describe the forthcoming $\overline{H}$ experiment GBAR. We explain in Section 3 the quantum reflection mechanism which allows the formation of gravitational quantum states of $\overline{H}$ above material surfaces. In Section 4, we show how the filtering scheme of the GRANIT experiment could be implemented in GBAR and, in Section 5, we describe a possible spectroscopy of gravitational quantum states of $\overline{H}$. Section 6 reviews the status of positronium free fall experiment at UCL and Section 7 explores the possibility of observing gravitational quantum states of positronium.

THE GBAR EXPERIMENT

GBAR is an experiment in preparation at CERN. Its goal is to measure the gravitational acceleration ($\overline{g} = Mg/m$) imparted to freely falling antihydrogen atoms, in order to perform a direct experimental test of the Weak Equivalence Principle with antimatter. The objective is to reach a relative precision on $\overline{g}$ of 1% in a first stage, with the perspective to reach a much higher precision using quantum gravitational states in a second stage, as is described in Section 5.

The principle of the experiment is described in detail in [22] and is briefly recalled here. It is based on an idea proposed in [23]. Antihydrogen ions $\overline{H}^+$ are produced, trapped, and sympathetically cooled to around $10\,\mu$K. The excess positron is detached by a laser pulse, which gives the start signal for the free fall of the ultracold antihydrogen atom $\overline{H}$. The $\overline{H}$ subsequent annihilation on a plate is detected and provides the information to measure $\overline{g}$. The choice of producing $\overline{H}^+$ ions to get ultracold antihydrogen atom is the specificity of the GBAR experiment. It is very costly in statistics but makes the cooling to μK temperatures a realistic aim.

We report, in this section, on three recent progresses in the preparation of the experiment: estimations of the $\overline{H}^+$ production cross-sections, accumulation of positrons, and cooling of the $\overline{H}^+$ ions.

Production Cross-Sections of $\overline{H}^+$ Ions

The $\overline{H}$ production proceeds in two steps: $\overline{p} + \text{Ps} \rightarrow \overline{H} + e^-$ (1) followed by $\overline{H} + \text{Ps} \rightarrow \overline{H}^+ + e^-$ (2).The Ps symbol stands for positronium. The cross-sections of these reactions are not well known and are very low. The matter counterpart of the first one has been measured. It is around 10^{-15} cm^2 (10^9 barn) for tens of keV protons [24]. The second one is estimated to be around 10^{-16} cm^2 (10^8 barn) [25].

New calculations of these reactions have been performed in which the first excitations levels for the Ps (up to n=3) and the $\overline{H}$ (up to n=5) have been considered. The results suggest that the production of $\overline{H}^+$ can be efficiently enhanced by using either a fraction of Ps(2p) and a 2 keV antiproton beam or a fraction Ps(3d) and antiprotons with kinetic energy below 1 keV [26]. The product of the cross-sections of reactions

(1) and (2) reaches values around 10^{-29} cm^4 (10^{19} barn^2) for an optimized fraction of excited Ps. Simulations are underway to estimate the effective gain with a realistic experimental setup.

This shows that very low energy antiprotons are needed. The extremely low energy antiproton (ELENA) ring which is in construction at CERN and which will complement the antiproton decelerator (AD) will provide 75 ns rms bunches of 5 × 10^6 100 keV antiprotons every 100 s. Those have to be further decelerated and cooled to match the GBAR requirements. The decelerator is under construction at CSNSM (Centre de Sciences Nucleaires et de Sciences de la Mati ´ ere) ` in Orsay, France.

Positron Accumulation

In addition to a high flux of low energy antiprotons, the production of $\overline{H}^+$ via reactions (1) and (2) requires to form a dense cloud of positronium. It has been shown that Ps can be efficiently produced by dumping few keV positrons on mesoporous silica films. Yields of 30 to 40% depending on the incident positron energy (few keV) have been measured [27, 28].The accumulation of a very large number of positrons, around 2 × 10^{10}, between two ejections of antiprotons from ELENA is thus necessary to produce a dense enough positronium cloud.

A demonstration facility for the production and accumulation of positrons is currently running at CEA/Saclay. It consists of a low energy electron linear accelerator (LINAC), a high field Penning-Malmberg trap from the Atomic Physics Laboratory in RIKEN, Japan, and a dedicated beam line for further studies of positron-positronium conversion and for applications in material science. In addition, a laser system is now being built at LKB (Laboratoire Kastler-Brossel) in Paris to test the excitation of the positronium which will be formed downstream of the trap.

The LINAC produces a 4.3 MeV electron bunched beam. The bunch length is 200 μs, and the LINAC runs at 200 Hz, producing a mean current of 120 μA. Electrons are sent onto a tungsten mesh moderator. A flux of typically 3×10^6 slow (few 100 eV) positrons per second is driven towards the Penning trap through a vacuum tube equipped with solenoid coils producing a 80 mT field. They are accelerated to around 1 keV to enter the high magnetic field (5 T) region. They reach the Penning trap which is made of 23 cylindrical electrodes, surrounded by 4 additional long electrodes to control the admission and the trapping of incident particles. Positrons make a round trip in less than 100 ns. In order to trap them, it is necessary to compress the 200 μs bunch. This is done by applying a varying voltage (20 to 150 V) when extracting the slow positron from the moderator. In this way, it is possible to close the entrance of the trap before the bunch escapes. With this method, one can trap one single bunch.

In order to accumulate a large number of bunches, positrons have to be slowed down and stored in a dedicated potential well formed by a subset of electrodes of the trap (see Figure 1) before the next bunch arrives. Positrons are cooled by passing through a preloaded electron plasma in another dedicated potential well. This method has been set up and demonstrated in [29] with a continuous positron beam issued from a ^{22}Na source. With such a beam, it is not possible to close the entrance gate, and positrons must be slowed down in one step. This was done with a remoderator downstream of the trap. An efficiency of 1% was obtained. With a bunched positron beam, the remoderator is not necessary, and a much higher efficiency is expected.

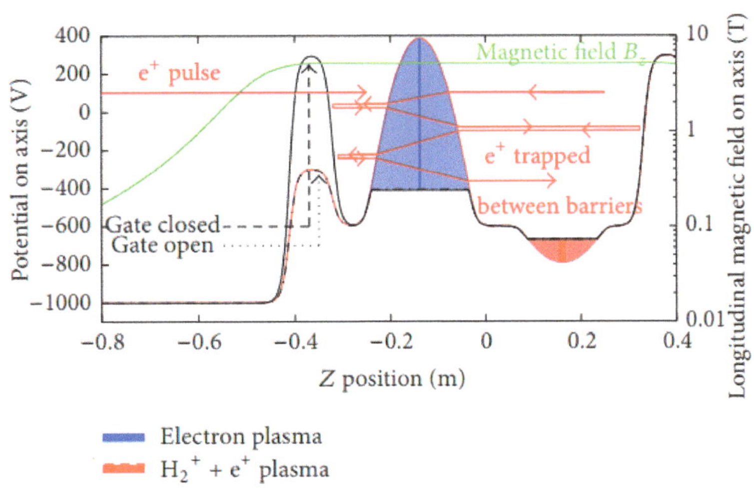

Figure 1: The positron trapping mechanism (from [30]). The horizontal axis is the position along the trap axis. The left vertical axis is the voltage seen by particles. The magnetic field strength is shown by the green curve, with the scale on the right vertical axis. In blue, the electron potential well filled with electrons is drawn, reducing the apparent voltage shown by black curve with the value on the left vertical axis. The positron potential is shown well in red. When a positron bunch arrives, the entrance electrode voltage is low (dashed dotted line). It is then increased and positrons go back and forth (it is depicted by red arrows) between this gate and the downstream part of the trap. They pass many times through the electron plasma and are eventually slowed down and fall into their well. The presence of residual H_2^+ helps the final catching of positrons. In order to reach the desired number of trapped positrons, the positron well has to be enlarged and deepened during the accumulation process from around 200 V to around 1 kV. Positrons may also be trapped in the potential well formed between the entrance electrode and the electron well, but its depth remains constant and the induced loss is small.

The cooling time fixes the maximal LINAC frequency and depends on the density of the electron plasma. With 10^{17} e$^-$/m^3, simulations show that the cooling time is around 3 ms.

The principle of this accumulation scheme has now been successfully demonstrated at Saclay. The details of the experimental setup used during accumulation are given in [30]. The result of a successful set of accumulation trials is shown in Figure 2. Given the characteristics of the demonstrator facility at Saclay, a realistic objective is now to accumulate around 10^8 positrons in the trap within 2 minutes.

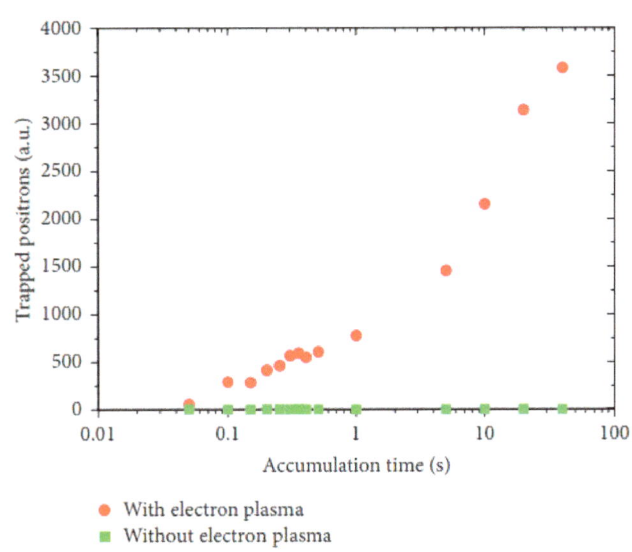

Figure 2: The accumulation of positrons (from [30]).

Cooling of the $\overline{H}^+$ Ions

Recent progresses have been made for the design of the cooling of $\overline{H}^+$ ions. The cooling proceeds in two steps: Doppler cooling at the mK level and Raman side band cooling to reach 10 μK.

In the first step, ions are captured in a linear Paul trap inside which Be^+ ions are preloaded and laser cooled. In the original scheme of GBAR, it was assumed that the $\overline{H}^+$ ions would be cooled by Coulomb interactions with the Be^+ ions (sympathetic cooling). Simulations show that this process is very slow. This is due to the large mass ratio between the ions. As a consequence, one cannot reach the mK level in a short enough time to avoid the destruction of the H+ : the laser cooling of Be^+ induces the photodetachment of the excess positron in a fraction of a second. However, the simulations show also that the addition of a third species of ions of intermediate mass, namely, HD^+ ions, makes the process efficient enough [31]. Starting with 1800 Be^+ and 200 HD^+ ions, cooling times of ms are achievable

In the second step, to reach the 10 μK level necessary for the free fall experiment, a $Be^+/\overline{H}^+$ ion pair must be transferred to a precision trap to undergo a ground state Raman side band cooling. Calculations show that one may achieve the desired cooling in less than a second. This is shown in [31] and references therein. This method will be tested with matter ions (Ca^+/Be^+, $H_2{}^+/Be^+$) before being implemented for the GBAR experiment. Traps are being mounted at LKB in Paris and at Mainz University.

Since the uncertainty on the measurement of $\overline{g}$ is fully dominated by the initial velocity dispersion due to both the vertical velocity after cooling and the recoil due to the positron photodetachment, the implementation of a vertical velocity selector will allow a drastic gain in the statistics needed to reach the 1% precision on $\overline{g}$ as is described in Section 4.

QUANTUM REFLECTION OF ANTIHYDROGEN ON MATERIAL SURFACES

In the ultracold regime, the interaction of antihydrogen ($\overline{H}$) atoms with a surface is governed by the phenomenon of quantum reflection. Although the atoms are strongly attracted to the surface, the atomic wave function can be partly reflected on the steep atom-surface potential, leading to a nonzero probability of classically forbidden reflection. This effect is relevant to experiments such as GBAR where ultracold $\overline{H}$ atoms are detected by annihilation on a plate (see Section 2).

A single atom placed in vacuum near a material surface experiences an attractive Casimir-Polder (CP) force [32, 33]. This force is a manifestation of the electromagnetic quantum fluctuations which are coupled to the atomic dipole. Quantum reflection occurs if an atom impinges with low velocity on such a rapidly varying potential [34]. We will give a more explicit condition later on.

In this section, we first describe how the CP potential is calculated for realistic experimental conditions. We then go on to compute the scattering amplitudes of an atom on this potential. We show that quantum reflection can be understood as a deviation from the semiclassical approximation. Finally, we describe materials from which quantum reflection is enhanced and above which gravitationally bound states of $\overline{H}$ could be observed.

Calculation of the Casimir-Polder Potential

We use the scattering approach to Casimir forces [35] to give a realistic estimation of the atom-surface interaction energy. In this approach, the interacting objects are described by reflection matrices for the electromagnetic field. Reflection on a plane is described by Fresnel coefficients, while reflection on the atom is treated in the dipolar approximation and depends on the dynamic polarizability [36]. This allows an evaluation of the CP potential for any material when its optical properties are known. Those used here are detailed in [37]. Note that since the typical length scale for quantum reflection (~100 nm) is below the thermal wavelength at 300 K (~1 μm), we carried out all calculations at null temperature.

The CP potential for $\overline{H}$ at a distance x of a perfectly conducting plane and thick silicon and silica slabs are presented in Figure 3. For a perfectly conducting mirror in the long-distance regime, we recover the historic result of Casimir and Polder [32, 33]:

$$V(x) \underset{x\gg\lambda}{\approx} V^*(x) = -\frac{3\hbar c}{8\pi x^4}\frac{\alpha(0)}{4\pi\epsilon_0}, \tag{4}$$

where $\alpha(0)$ is the static polarizability of the atom.

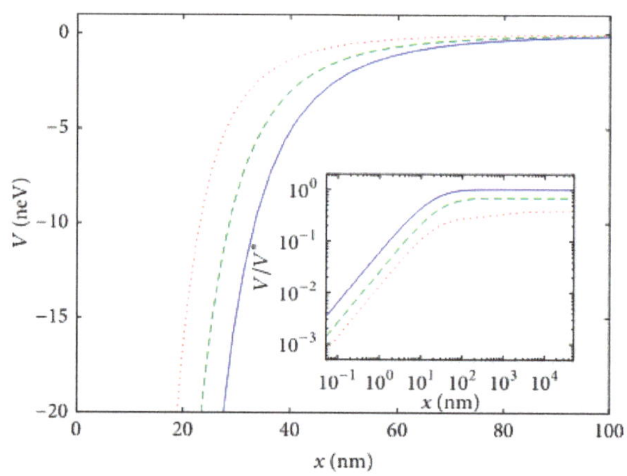

Figure 3: Casimir-Polder (CP) potential for antihydrogen in the vicinity of a material bulk; from top to bottom: perfect conductor (full line), silicon (dashed line), and silica (dotted line); inset: ratio to the retarded potential for a perfectly conducting mirror; see text.

For real mirrors, the potential is reduced but it shows the same power law dependence in the van der Waals (short distance) and retarded (long-distance) regimes:

$$V(x) \underset{x\ll\lambda}{\approx} -\frac{C_3}{x^3}, \qquad V(x) \underset{x\gg\lambda}{\approx} -\frac{C_4}{x^4}, \tag{5}$$

where λ is a typical wavelength associated with the optical response of atom and plane.

Scattering on the Casimir-Polder Potential

We now solve the Schrodinger equation for an atom of energy $E>0$ scattering on the CP potential (x):

$$\frac{d^2}{dx^2}\psi(x) + \frac{p(x)^2}{\hbar^2}\psi(x) = 0, \tag{6}$$

with $p(x) = \sqrt{2m(E - V(x))}$ being the classical momentum. We write the exact wave function as a sum of counterpropagating WKB waves whose coefficients are allowed to vary:

$$\psi(x) = \frac{c_{in}(x)}{\sqrt{p(x)}}\exp\left(-\frac{i}{\hbar}\int^{x} p(x')\,dx'\right) + \frac{c_{out}(x)}{\sqrt{p(x)}}\exp\left(\frac{i}{\hbar}\int^{x} p(x')\,dx'\right). \tag{7}$$

Upon insertion in the Schrodinger equation, we obtain ¨ coupled first-order equations for the coefficients $c_{in}(x)$, $c$out(x) [38]. The annihilation of $\overline{\mathrm{H}}$ on the material surface translates as a fully absorbing boundary condition on the surface: $c_{out}(x = 0) = 0$. This is in contrast with matter atoms, for which more complicated surface physics is involved in the boundary condition. We emphasize that $\overline{\mathrm{H}}$-surface interaction at distances below few atomic units differs significantly from H-surface interaction. It is remarkable that, due to complete annihilation of $\overline{\mathrm{H}}$ in the bulk of the surface, the details of such short-range physics are not important [39]. Indeed, any information about antiatom-wall interaction is encrypted in the reflected wave, which, as we will show in the following, is produced at asymptotically large distances.

Close to the surface, the energy becomes negligible compared with the potential, which takes the van der Waals form and $c_{in}(x)$, $c_{out}(x)$ can be solved analytically [37]. The equations are then integrated numerically until $c_{in}(x)$, $c_{out}(x)$ become constants.

The reflection probability $|r|^2 = \lim_{x\to\infty}|c_{out}(x)/c_{in}(x)|^2$ is plotted against the energy E in Figure 4 for various semiinfinite media. Note that the quantum reflection probability is larger for materials with a weaker CP interaction, such as silica.

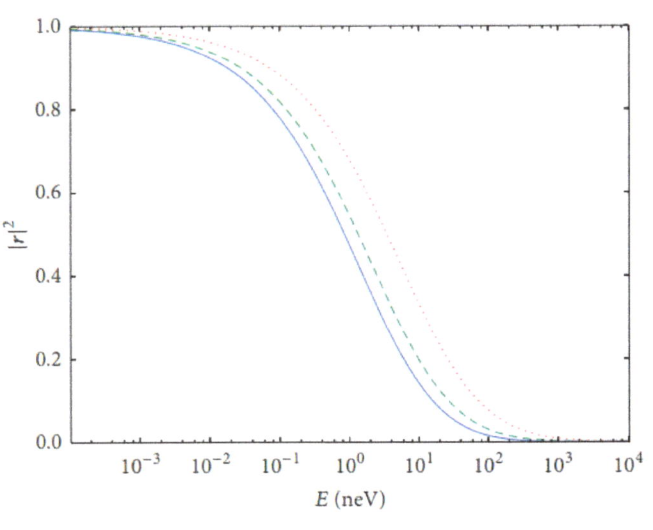

Figure 4: Quantum reflection probability $|r|^2$ as a function of the energy for antihydrogen atoms on bulk mirrors; from bottom to top: perfect conductor (full line), silicon (dashed line), and silica (dotted line).

The Badlands Function

To understand this surprising result, we look more closely at what distinguishes the exact solution of the Schrodinger equation from the reflectionless ¨ WKB approximation. If c_{in}, c_{out} are no longer allowed to vary, one can show that the wave function (7) obeys a modified Schrodinger equation where ¨ $(x)^2$ is replaced by $\tilde{p}(x)^2 = p(x)^2(1 + Q(x))$ [38]. (x) is known as the badlands function since the WKB approximation is not valid in regions where it is nonnegligible:

$$Q(x) = \frac{\hbar^2}{2p(x)^2}\left(\frac{p''(x)}{p(x)} - \frac{3}{2}\frac{p'(x)^2}{p(x)^2}\right). \tag{8}$$

For the CP potential, the badlands function exhibits a peak in the region where $|(x)| = E$ but goes to zero both as $x\to\infty$ (where the potential cancels) and as $x\to 0$ (where the classical momentum diverges).

As the energy is decreased, the semiclassical approximation breaks down and the badlands function's peak becomes larger. But, for a given energy, the peak is larger and closer to the surface when the potential is weak, as shown in Figure 5. The difference between exact and WKB solutions is larger in weaker CP potentials, leading to enhanced quantum reflection.

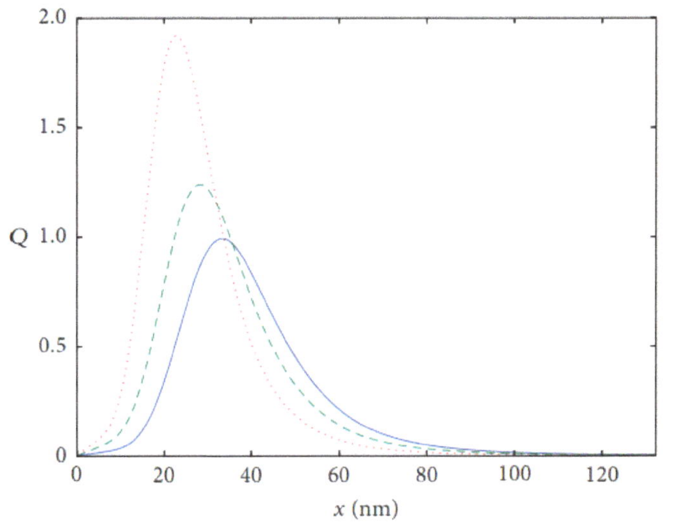

Figure 5: The badlands function (x) for an antihydrogen atom with energy $E = 10$ neV; from bottom to top: perfect conductor (full line), silicon (dashed line), and silica (dotted line).

Enhancing Quantum Reflection

Quantum reflection first appears as a bias in the context of the GBAR experiment, since it tends to exclude low energy atoms from the statistics. However, this phenomenon opens perspectives for the storage and guiding of antimatter with material walls. With this in mind, we consider materials which couple weakly to the electromagnetic field and are therefore good mirrors for atoms, as we have seen in the previous paragraph.

A simple strategy is to remove matter from the reflective medium, by using thin slabs or porous materials for example. Our versatile approach allowed us to compute the CP interaction near thin slabs, an undoped graphene sheet [37] and nanoporous materials [40]. The latter consist of a solid matrix which forms an array of nanometric pores. Aerogels, which are obtained by supercritically drying a silica gel, are a well-known example. We also consider porous silicon and powders of diamond nanoparticles formed by explosive shock.

From a distance larger than the typical pore size, such materials can be modeled as homogeneous effective media with properties averaged between that of vacuum and that of the solid matrix. In consequence, their effective dielectric constant is extremely low, as a result of which quantum reflection is exceptionally efficient. In Table 1, we show the lifetime of an antihydrogen atom in the first gravitationally bound state above a surface (see Section 5 for more details).

Table 1: The lifetime of the first gravitationally bound state of antihydrogen above various surfaces.

Surface (porosity)	Lifetime (s)
Perfect conductor	0.11
Bulk silicon	0.14
Bulk silica	0.22
Nanodiamond powder (95%)	0.89
Porous silicon (95%)	0.94
Silica aerogel (98%)	4.6

Note that this approach does not take into account the possible presence of stray charges on the surface, a question that would have to be addressed to observe the predicted reflection probabilities. Moreover, the effective medium approximation is applicable only for low atom velocities, such that the atom is reflected far enough from the surface. With these caveats, nanoporous materials are an outstanding candidate for the manipulation and study of antihydrogen and its gravitationally bound states over lifetimes of a few seconds.

SHAPING OF VERTICAL VELOCITY COMPONENTS OF ANTIHYDROGEN ATOMS FOR GBAR

The main source of uncertainty on the determination of $\overline{g}$ in the GBAR experiment is the width of the vertical velocity distribution of the atom at the beginning of the free fall. This spread in velocities is due to the quantum uncertainty on the momentum of $\overline{H}^+$ in the ground state of the harmonic Paul trap and to the additional recoil associated with the photodetachment of the extra positron (see Section 2).

In this section, we give an estimation of the uncertainty on the arrival time associated with the initial vertical velocity spread and show how it can be reduced by filtering out the fastest atoms. Since slow antihydrogen atoms bounce on material surfaces thanks to quantum reflection (see previous section), the filtering scheme used in GRANIT with ultracold neutrons [21] can also be applied in GBAR.

Width of the Arrival Time Distribution

We consider a wave packet falling in a linear gravitational potential and want to determine the arrival time distribution on a fixed horizontal plane, supposing there is no reflection from that (ideal) detector. In this case, classical and quantum calculations give identical results, as can be seen by noticing that the Wigner quasidistribution function obeys the classical equations of motion if the potential is at most quadratic. Therefore, in a linear potential, a given initial phase-space distribution simply propagates along the classical trajectories.

For a wave packet initially centered at a height H above the detector, with zero mean velocity and uncorrelated vertical position and velocity distributions of widths Δz and ΔV, respectively, the spread of the arrival time distribution is

$$\frac{\Delta t}{t_H} = \sqrt{\left(\frac{\Delta z}{2H}\right)^2 + \left(\frac{\Delta v}{\sqrt{2\overline{g}H}}\right)^2}, \tag{9}$$

with $t_H = \sqrt{2H/\overline{g}}$ being the classical free fall time. This translates as a statistical uncertainty $\Delta\overline{g}/\overline{g} = 2\Delta t/\sqrt{N}t_H$ on the determination of $\overline{g}$ after N independent measurements.

If the particle is initially in the ground state of a harmonic trap, the distribution is Gaussian and saturates the Heisenberg inequality: $\Delta z \Delta V = \hbar/2m$. Then, the time uncertainty is minimal for

$$\Delta v = \Delta v_{\text{opt}} = \sqrt{\frac{\hbar}{2m}\sqrt{\frac{\overline{g}}{2H}}}. \tag{10}$$

For $H = 30$ cm and $\overline{g}=g$, this evaluates $\Delta v_{\text{opt}} \approx 3.6 \times 10^{-4}$ m/s, and the relative uncertainty on the arrival time is 2×10^{-4}. However, the current expected value for GBAR is three orders of magnitude larger $\Delta V_0 \approx 0.5$ m/s, which leads to a relative uncertainty of 0.2.

The uncertainty in GBAR is largely dominated by the vertical velocity dispersion. If the initial velocity dispersion can be reduced from ΔV_0 to ΔV by filtering out the hottest atoms, the single-shot precision and the number of atoms are both reduced by a factor $\Delta V/\Delta V_0$. Despite the loss in statistics, this results in a net reduction of the statistical uncertainty on $\overline{g}$.

Shaping of the Vertical Velocity Distribution

Our proposal [41] to realize this filtering is to let the atoms pass through a horizontal slit between two disks. The bottom disk has a smooth top surface on which atoms reflect with high probability whereas the top disk has a rough bottom surface which effectively acts as an absorber for the atoms (see Figure 6). Antihydrogen is initially trapped in the center of the two disks (openings are made in the center of the disks to allow operation of the trap). If its vertical velocity is high enough to reach the rough surface, it is reflected nonspecularly and remains inside the device until it annihilates with high probability. On the contrary, if it cannot reach the top disk, the atom will exit the device with high probability after bouncing on the bottom mirror a few times. It then falls freely to a detector located at a height H below. Since the horizontal velocity is conserved, the knowledge of the total time between photodetachment and annihilation and of the total horizontal L distance traveled allows one to correct for the time spent inside the device before the free fall.

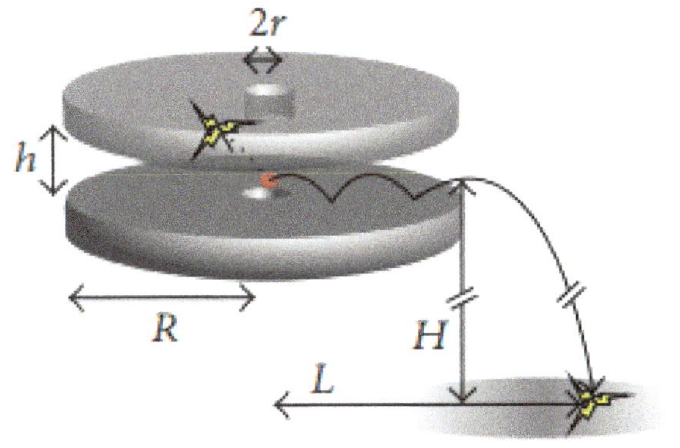

Figure 6: Scheme of the proposed device to reduce the vertical velocity spread of the falling wave packet (see text).

If h is the height of the slit, the velocity spread at the output is $\Delta v \approx \sqrt{2gh}$ and the proportion of atoms that exit the device is $N/N_0 \approx \Delta v/\sqrt{2\pi}\Delta v_0$. Using the shaping device therefore reduces the statistical uncertainty on $\bar{g}$ by a factor which scales $h^{1/4}$.

Classically, going to ever smaller slit heights leads to arbitrarily good precision. For example, for h=1 mm, $\Delta V \approx 0.14$ m/s, $N/N_0 \approx 5\%$, and the accuracy is improved by a factor 2, whereas, for $h = 50$ μm, $\Delta V \approx 0.03$ m/s, $N/N_0 \approx 1\%$, and the accuracy improved by a factor 4.

There are, however, two limits: the number of repetitions of the experiment must be large enough that at least some atoms make it through the filter, but, more fundamentally, the wave function of the atom must fit inside the slit. Indeed for slit sizes below 50 μm the discrete spectrum of states bound by gravity must be taken into account. For a slit size of 20 μm, only the ground state can travel through the guide; below that, the transmission drops to zero.

This fact has been used to demonstrate the existence of gravitationally bound states for neutrons [21]. The next section explores the possibilities of similar experiments with antihydrogen to further increase the precision of equivalence principle tests on antimatter.

RESONANCE SPECTROSCOPY OF GRAVITATIONAL STATES OF ANTIHYDROGEN NEAR MATERIAL SURFACE

In this section, we will study a motion of an $\bar{\mathrm{H}}$ atom, localized in a gravitational state near a horizontal plane mirror. The existence of such states though counterintuitive is explained by the phenomenon of quantum reflection of ultracold (anti-) atoms from a steep attractive Casimir-Polder atomsurface potential. Such states have similar properties with those discovered for neutrons [21, 42–45].

To account for the interaction of $\bar{\mathrm{H}}$ with a material wall, the gravitational quantum states (1) receive a complex energy shift $\varepsilon_0\Delta$, with $\Delta \simeq -i0.005$ for a perfectly conducting wall [46]. All states, therefore, acquire equal shift and width, which is a function of a material surface substance $\Gamma = 2|\Delta|\varepsilon_0$. This width corresponds to the lifetime of 0.1 s in case of a perfectly conducting surface and is twice longer for silica [37, 39, 40], for instance. The equal shifts of all gravitational states energy levels mean that already small modification of gravitational states due to antiatom-surface interaction is canceled out in the transition frequencies. This makes resonance spectroscopy of gravitational states particularly interesting for measuring gravitational properties of $\bar{\mathrm{H}}$.

The interest to study gravitational quasistationary states of $\bar{\mathrm{H}}$ is due to their comparatively long life-time on one hand and easy identification of certain state because of it is mesoscopic spatial scale. This opens an interesting perspective to apply potentially very precise resonance spectroscopy method to establish the gravitational properties of antiatoms. These methods are based on inducing an observation of resonance transitions between gravitational states. One possible approach is to use an alternating inhomogeneous magnetic field for such a purpose.

The interaction of a magnetic field with a ground state H atom moving through the field [47–49] is dominated by the interaction of an average magnetic moment of the atom [20] in a given hyperfine state with the magnetic field. We are going to focus on an alternating magnetic field with a gradient in the vertical direction. This condition is needed for coupling the field and the center of mass (c.m.) $\bar{\mathrm{H}}$ motion in the gravitational field of the Earth. It allows one to induce resonant transitions between quantum gravitational states of $\bar{\mathrm{H}}$ [46].

We will consider the magnetic field in the following form:

$$\vec{B}(z,x,t) = B_0\vec{e}_z + \beta\cos(\omega t)\left(z\vec{e}_z - x\vec{e}_x\right). \tag{11}$$

Here, B_0 is the amplitude of a constant, vertically aligned, component of magnetic field, β is the value of magnetic field gradient, z is a distance measured in the vertical direction, and x is a distance measured in the horizontal direction, parallel to the surface of a mirror. A time-varying magnetic field (11) is accompanied with an electric field ($[\vec{\nabla}\vec{E}] = -(1/c)\partial\vec{B}/\partial t$). However, for the velocities of ultracold atoms, corresponding interaction terms are small and thus will be omitted.

An inhomogeneous magnetic field couples the spin and the spatial degrees of freedom. A $\bar{\mathrm{H}}$ wave function is described in this case using a four-component column (in a nonrelativistic treatise) in the spin space, each component being a function of the c.m. coordinate $\vec{R}$, relative $\bar{p} - \bar{e}$ coordinate $\vec{\rho}$, and time t. The corresponding Schrodinger equation is

$$i\hbar\frac{\partial\Phi_\alpha\left(\vec{R},\vec{\rho},t\right)}{\partial t} = \sum_{\alpha'}\left[-\frac{\hbar^2}{2m}\Delta_R + Mgz + V_{\mathrm{CP}}(z) + \widehat{H}_{\mathrm{in}} + \widehat{H}_m\right]_{\alpha,\alpha'}\Phi_{\alpha'}\left(\vec{R},\vec{\rho},t\right). \tag{12}$$

A subscript α in this equation indicates one of four spin states of the $\bar{p} - \bar{e}$ system. The meaning of the interaction terms is the following. $V_{\mathrm{CP}}(z)$ is an atom-mirror interaction potential, which turns into the Casimir-Polder potential at an asymptotic atom-mirror distance (see [39, 50] and references therein). $\widehat{H}_{\mathrm{in}}$ is the Hamiltonian of the internal motion, which includes the hyperfine interaction:

$$\widehat{H}_{\text{in}} = -\frac{\hbar^2}{2\mu}\Delta_\rho - \frac{e^2}{\rho} + \frac{\alpha_{\text{HF}}}{2}\left(\widehat{F}^2 - \frac{3}{2}\right). \tag{13}$$

Here, $\mu = m_1 m_2/m$, m_1 is the antiproton mass, m_2 is the positron mass, $m = m_1 + m_2$, α_{HF} is the hyperfine constant, and $\widehat{F}$ is the operator of the total spin of the antiproton and the positron. We will treat only $\overline{\text{H}}$ atoms in a 1S-state (below, we will show that the excitation of other states in the studied process is improbable). The term $(\alpha_{\text{HF}}/2)(\widehat{F}^2 - 3/2)$ is a model operator, which effectively accounts for the hyperfine interaction and reproduces the hyperfine energy splitting correctly. The term $\widehat{H}_m$ describes the field-magnetic moment interaction:

$$\widehat{H}_m = -2\vec{B}(z,x,t)\left(\mu_{\bar{e}}\widehat{s}_{\bar{e}} \times \widehat{I}_{\bar{p}} + \mu_{\bar{p}}\widehat{s}_{\bar{p}} \times \widehat{I}_{\bar{e}}\right). \tag{14}$$

Here, $\mu_{\bar{e}}$ and $\mu_{\bar{p}}$ are magnetic moments of the positron and the antiproton, respectively, $\widehat{s}_{\bar{e}}, \widehat{s}_{\bar{p}}$ is a spin operator, acting on spin variables of positron (antiproton), and $\widehat{I}_{\bar{e}}, \widehat{I}_{\bar{p}}$ is a corresponding identity operator. As far as the field $\vec{B}(z,x,t)$ changes in space and in time, this term couples the spin and the c.m. motion.

We will assume that in typical conditions of a spectroscopy experiment the $\overline{\text{H}}$ velocity component V parallel to the mirror surface (directed along x-axis) is of the order of a few m/s and is much larger than a typical vertical velocity in lowest gravitational states (which is of the order of cm/s). We will treat the motion in a frame moving with the velocity V of the $\overline{\text{H}}$ atom along the mirror surface. Thus, we are going to consider the x-component motion as a classical motion with a given velocity V, and we will substitute a x-dependence by a t-dependence. We will also assume that $B_0 \gg \beta L$, where $L \sim 30$ cm is a typical size of an experimental installation of interest. This condition is needed for "freezing" the magnetic moment of an atom along the vertical direction; it provides the maximum transition probability.

We will be interested in the weak field case, such that the Zeeman splitting is much smaller than the hyperfine level spacing $\mu_B B_0 \ll \alpha_{\text{HF}}$. The hierarchy of all mentioned above interaction terms could be formulated as follows:

$$\frac{m_2 e^2}{\hbar^2} \gg \alpha_{\text{HF}} \gg \mu_{\bar{e}}\left|B_0\right| \gg E_n, \tag{15}$$

and, thus, it justifies the use of the adiabatic expansion for solving (12); it is based on the fact that an internal state of an $\overline{\text{H}}$ atom follows adiabatically the spatial and temporal variations of an external magnetic field. Neglecting nonadiabatic couplings, an equation system for the amplitude (t) of a gravitational state $\psi_n(z)$ has the following form:

$$i\hbar\frac{dC_n(t)}{dt} = \sum_k C_k(t)\,V_{n,k}(t)\exp\left(-i\omega_{nk}t\right). \tag{16}$$

The transition frequency $\omega_{nk} = (E_k - E_n)/\hbar$ is determined by the gravitational energy level spacing. This fact is used in the proposed approach to access the gravitational level spacing by means of scanning the applied field frequency, as will be explained in the following.

Within this formalism, the role of the coupling potential $V(z, t)$ is played by the energy of an atom in a fixed hyperfine state thought of as a function of (slowly varying) distance z and time t. Consider

$$V_{n,k}(t) = \int_0^\infty \psi_n(z)\,\psi_k(z)\,E(t,z)\,dz. \tag{17}$$

Here, (z) is the gravitational state wave function, which is known in terms of the Airy function [46].

The energy (z, t) is the eigenvalue of the internal and magnetic interactions $\widehat{H}_{\text{in}} + \widehat{H}_m$, where the c.m. coordinate $\vec{R}$ and time t are treated as slow-changing parameters. Corresponding expressions for the eigenenergies of a 1S manifold are

$$E_{a,c} = E_{1s} - \frac{\alpha_{\text{HF}}}{4} \mp \frac{1}{2}\sqrt{\alpha_{\text{HF}}^2 + \left|\left(\mu_B - \mu_{\bar{p}}\right)B(z,t)\right|^2},$$

$$E_{b,d} = E_{1s} + \frac{\alpha_{\text{HF}}}{4} \mp \frac{1}{2}\left|\left(\mu_B + \mu_{\bar{p}}\right)B(z,t)\right|. \tag{18}$$

Subscripts a, b, c, d are standard notations for hyperfine states of a 1S manifold in a magnetic field. The presence of a constant field B_0 produces the Zeeman splitting between states b and d. As far as the energy of states b, d depends on magnetic field linearly, while, for states a, c, it depends on magnetic field quadratically, only transition between b, d states takes place in case of a weak field. In the following, we will consider only transitions between gravitational states in a 1(b, d) manifold.

A qualitative behavior of the transition probability is given in the Rabi formula, which can be deduced by means of neglecting the high frequency terms compared to the resonance couplings of only two states, initial i and final f, in case the field frequency ω is close to the transition frequency $\omega_{if} = (E_f - E_i)/\hbar$:

$$P = \frac{1}{2}\frac{\left(V_{if}\right)^2}{\left(V_{if}\right)^2 + \hbar^2\left(\omega - \omega_{if}\right)^2} \times \sin^2\left(\frac{\sqrt{\left(V_{if}\right)^2 + \hbar^2\left(\omega - \omega_{if}\right)^2}}{2\hbar}t\right)\exp(-\Gamma t). \tag{19}$$

The factor 1/2 appears in front of the right-hand side of the above expression due to the fact that only two (b, d) of four hyperfine states participate in the magnetically induced transitions.

It is important that the transition frequencies ω_{if} do not depend on the antiatom-surface interaction up to the second order in the splitting Δ. This is a consequence of the already mentioned fact that all energies of gravitational states acquire equal shift due to the interaction with a material surface.

A resonant spectroscopy of $\overline{\text{H}}$ gravitational states could consist of observing $\overline{\text{H}}$ atoms localized in the gravitational field above a material surface at a certain height as a function of the applied magnetic field frequency. A "flow-through type" experiment, analogous to the one discussed for the spectroscopy of neutron gravitational states [51], includes three main steps. A sketch of a principle scheme of an experiment proposed in [41] is shown in Figure 7 (see also Section 4).

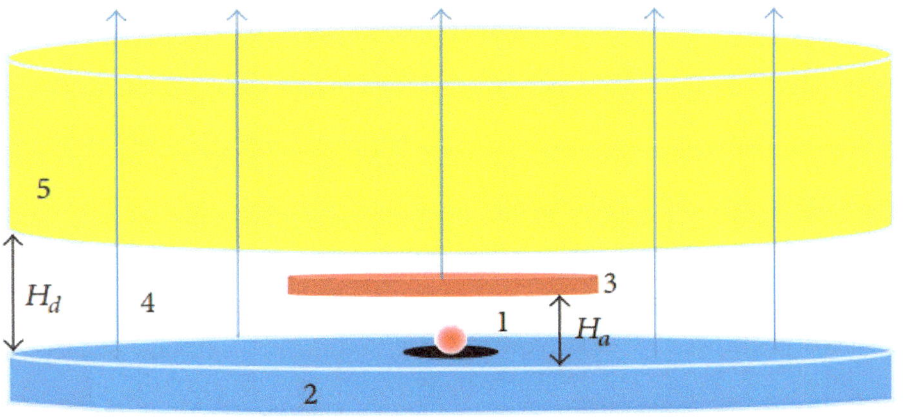

Figure 7: A sketch of the principle scheme of an experiment on magnetically induced resonant transitions between $\overline{\mathrm{H}}$ gravitational states. 1: a source of ultracold antihydrogen, 2: a mirror, 3: an absorber, 4: a magnetic field, and 5: a detector.

First, an atom of $\overline{\mathrm{H}}$ is shaped in a ground gravitational state. This is achieved by means of passing H through a slit, formed by a mirror and an absorber, which is placed above the mirror at a given height H_a. The mirror and the absorber form a waveguide with a state-dependent transmission [45]. The choice of $H_a = H_1 \simeq 13.6\ \mu$m implies that only $\overline{\mathrm{H}}$ atoms in the ground gravitational state pass through the slit. Second, H atoms are affected by an alternating magnetic field (11) while they are moving parallel to the mirror. An excited gravitational state is resonantly populated. Third, the number of $\overline{\mathrm{H}}$ atoms in an excited state is measured by means of counting the annihilation events in a detector, which is placed at a height H_d above the mirror. The value of H_d is chosen to be larger than the spatial size of the gravitational ground state and smaller than the spatial size of the final state (2), $H_1 \ll H_d < Hf$, so that the ground state atoms pass through, while atoms in the excited state are detected.

We present a simulation of the number of detected annihilation events as a function of the field frequency in Figure 8 for the transition from the ground to the 6th excited state, based on a numerical solution of the equation system (16). The corresponding resonance transition frequency is ω = 972.46 Hz. The value of the field gradient, optimized to obtain the maximum probability of 1→6 transition during the time of flight $t_{fl} = \tau = 0.1$ s turned to be equal to β = 27.2 Gs/m; the corresponding guiding field value, which guarantees the adiabaticity of the magnetic moment motion, is B_0 = 30 Gs.

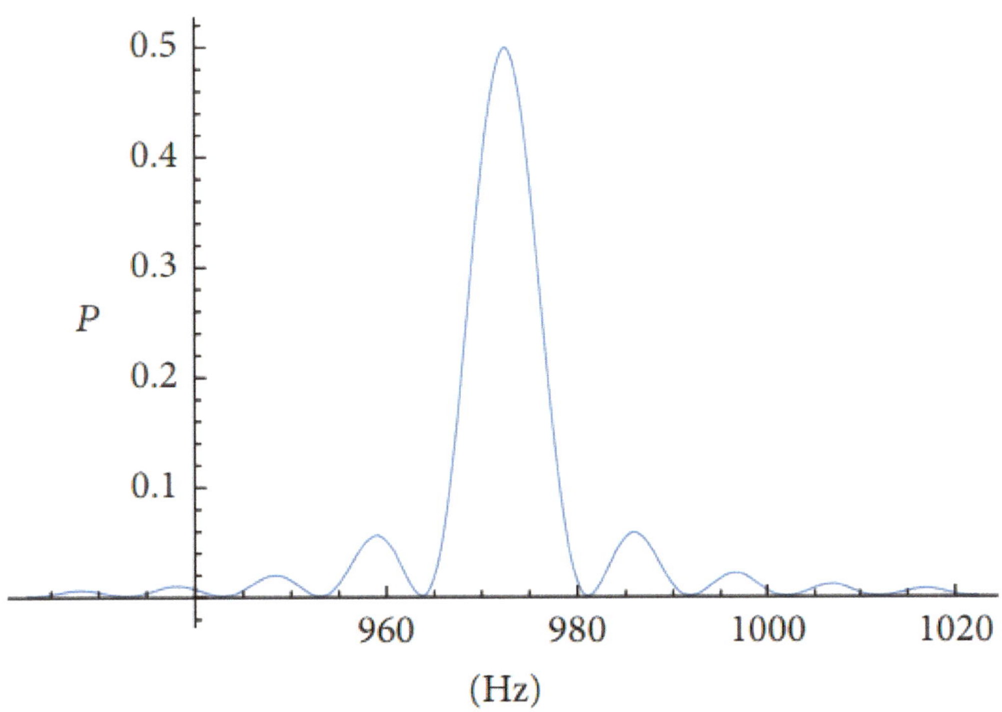

Figure 8: The transition probability as a function of the magnetic field frequency for the transition from the ground state to 6th gravitational state.

It follows from (1) that the $\overline{\mathrm{H}}$ gravitational mass could be deduced from the measured transition frequency ω_{nk} as follows:

$$M = \sqrt{\frac{2m\hbar\omega_{nk}^3}{g^2(\lambda_k - \lambda_n)^3}}. \quad (20)$$

Let us mention that g in the above formula means the gravitational field intensity near the Earth's surface, a value which characterizes properties of the field and is assumed to be known with a high precision. At the same time, all the information about gravitational properties of $\overline{\mathrm{H}}$ is included in the gravitational mass M. Equality of the gravitational mass M and the inertial mass m, imposed by the equivalence principle, results in the following expression:

$$M = \frac{2\hbar\omega_{nk}^3}{g^2(\lambda_k - \lambda_n)^3}. \quad (21)$$

Estimation of the accuracy of the above expression requires accounting for different effects, including dynamical Stark shift of the resonance line, nonadiabatic corrections to the transition probability, and interaction of alternating magnetic field with a mirror. The detailed study of different systematic effects is under way. Assuming that the spectral line width is determined by the lifetime $\tau \approx 0.1$ s of gravitational states, we estimate that the gravitational mass M can be deduced with the relative accuracy $\epsilon_M \sim 10^{-3}$ for 100 annihilation events for the transition to the 6th state.

GRAVITATIONAL FREE FALL OF COLD POSITRONIUM

Antihydrogen, muonium, and positronium are the possible candidates for gravity measurements on antimatter, with various pros and cons. Antihydrogen and muonium [4] are extremely difficult to produce, requiring large facilities (i.e., PSI, CERN), whereas positronium is relatively easy to produce in smaller university laboratories. However, Ps has an inconvenient propensity to self-annihilate; the triplet ground state vacuum lifetime of only 142 ns would seem to preclude using this system for a free fall measurement. As has been pointed out by various authors, in particular Mills, Jr., [52], this is not the case, since one need only excite Ps atoms into long-lived Rydberg states to prevent self-annihilation. Indeed, for any Ps state with $n>1$ the radiative lifetime is always less than the annihilation lifetime. The only excited state for which this is not true is the metastable 2s state. That is to say, for excited states, the overlap of the positron and electron wave functions is sufficiently low that annihilation can be considered to be negligible (see Figure 9).

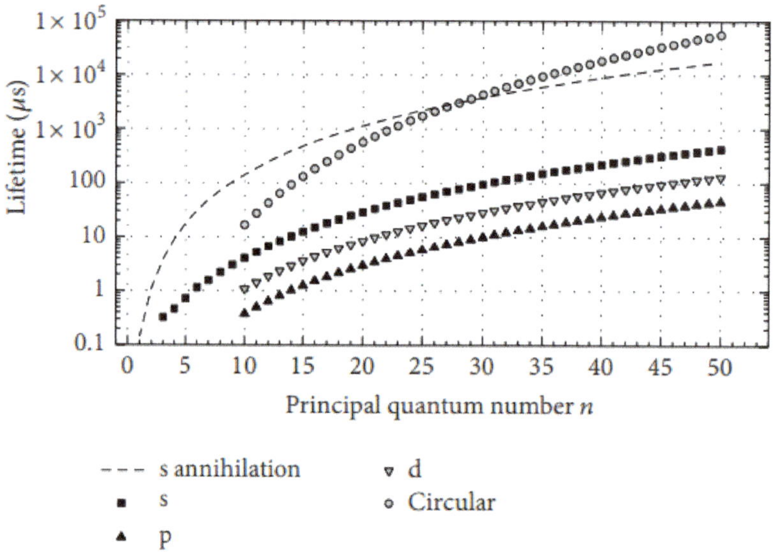

Figure 9: Radiative lifetimes of various Ps states as a function of the principal quantum number n. The lifetimes were calculated by summing the Einstein A coefficients of all electric-dipole-allowed decay channels from each Rydberg state. For each A coefficient the appropriate radial integrals were determined using analytic expressions for the radial wave functions in a pure Coulomb potential [53].The dashed line is the annihilation lifetime of ns states. After [5].

The radiative lifetimes of excited Ps states, shown in Figure 9, are almost twice those of the corresponding states in hydrogen. For practical reasons, the smallest Ps beam deflections one can expect to observe will be 10s of micrometers or more. Therefore, if Ps falls with the usual gravitational acceleration, it would be necessary to produce states with lifetimes of the order of a few ms to observe such deflections. As is evident from Figure 9, achieving such long radiative lifetimes requires exciting either low Rydberg levels (i.e., or) to extremely high principal quantum numbers, or going to lower states (perhaps around or so) and then transferring the atoms to circular states, or if not true circular states, at least states with higher angular momentum. For a discussion of the properties of circular states and methods for producing them, see [54].

Aside from the creation of sufficiently long-lived Rydberg levels, conducting a Ps free fall experiment will require solving many other problems. In order to accomplish an experiment of the type first outlined by Mills and Leventhal [55], it will be necessary to produce a small (10–50 micron) "point" source of slow positronium in a cryogenic environment. The resulting long-lived Rydberg atoms will then have to be formed into a beam, perhaps by electrostatic manipulation (focusing and deceleration) via their electric dipole moments [56], and finally detected with good spatial resolution (as a function of flight time) in order to observe a deflection due to gravity. A schematic view of such an experiment is presented in Figure 10. Possible methods to accomplish some of these tasks are considered elsewhere [5, 55].

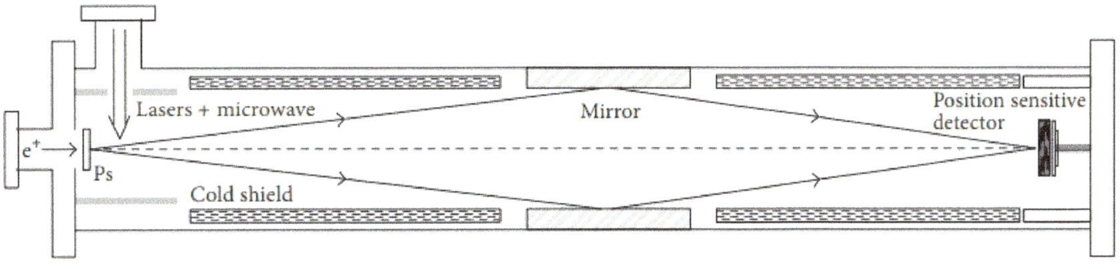

Figure 10: A schematic representation of a Mills-Leventhal type of Ps free fall experiment. A real experiment will undoubtedly be significantly different from this illustration, which is intended only to highlight some of the different steps involved. Of distinct practical concern will be the need to keep the apparatus at low temperatures to mitigate effects of black body radiation, as well as minimizing the Ps speed, which will determine the length of the flight path and hence the experiment.

The production of Ps Rydberg states with principal quantum numbers around 30 can be accomplished using a two-step process (1s → 2p → nd) and has already been experimentally demonstrated using broadband (~100 GHz) lasers to accommodate the large Doppler-broadened width of the transitions [57]. However, this methodology is not well suited to the requirement that these atoms are subsequently transferred to higher angular momentum states, and, in order to achieve the required state selectivity, it may be necessary to use a different excitation mechanism; that is, a Dopplerfree two-photon transition from the ground state directly to a well-defined Rydberg Stark state [58].

As is well known, Rydberg atoms exhibit exaggerated properties [54] (see Table 2). In the present case, this is critical, since we seek to produce Ps atoms with very long lifetimes, and also take advantage of the large electric dipole moments of Rydberg atoms to create and control an atomic beam. However, insofar as we are compelled to make use of electrically neutral systems to measure the weak gravitational force acting on antimatter particles without extraneous electromagnetic fields dominating their motion, excitation to states with very large dipole moments brings us back to the original problem of extraneous field effects. The situation is considerably less dire when dealing with electric dipoles and, to a much lesser extent, magnetic dipoles, since, in this case, only field gradients give rise to forces; nevertheless, in an experiment designed to probe the weak force of gravity with Rydberg atoms, forces due to stray fields must be taken into account. Magnetic dipoles are less important in this regard. Ordinarily, atoms in weak magnetic fields will experience linear Zeeman shifts, but, in the case of ortho-Ps, there is no permanent magnetic dipole moment because the electron and positron contributions cancel exactly. Thus, only diamagnetic interactions or induced dipole moments have to be taken into account, which we expect to be negligible for realistic experimental conditions.

Table 2: The n-dependence of several properties of Rydberg atoms, with examples shown for the 30 d state of Ps, H, and He. The state separation is calculated for 30 d→31 d. The orbital radius is defined here as the expectation value $\langle r \rangle = (1/2)(3n_{\mathrm{eff}}^2 - l(l+1))$, where n_{eff} includes the relevant quantum defect. The electric dipole momentto-mass ratios are calculated for the outermost state of the $n = 30$, $m = 2$ Stark manifold. The radiative lifetime n-dependence applies only to low l states: for circular states, the scaling is closer to n^5 (see Figure 9).

	n-Scaling	Ps	H	He
Binding energy (meV)	n^{-2}	−7.56	−15.11	−15.12
State separation (meV)	n^{-3}	0.48	0.96	0.96
Orbital radius (a_0)	n^2	2694	1347	1347
Radiative lifetime (μs)	n^3	28.4	14.2	12.2
Dipole moment/mass (ea_0/amu)	n^2	2.2×10^6	1206	304

When an atom is placed in an external electric field of strength F and direction z, the field mixes the atom's angular momentum states. To first-order, the state $|n, l, m\rangle$ is mixed with states of adjacent l but the same n and m [59]. The resulting Stark states repel each other, causing them to spread out as the electric field strength is increased, as shown in Figure 11. Following the example of hydrogen, where the first-order Stark shift is analytically calculable, the Schrodinger equation for an atom in an electric field ¨ can be written in cylindrical coordinates, where the relevant quantum numbers are n, m and the parabolic quantum numbers are n_1 and n_2, which together satisfy the condition $n=n_1 + n_2 + |m| + 1$.

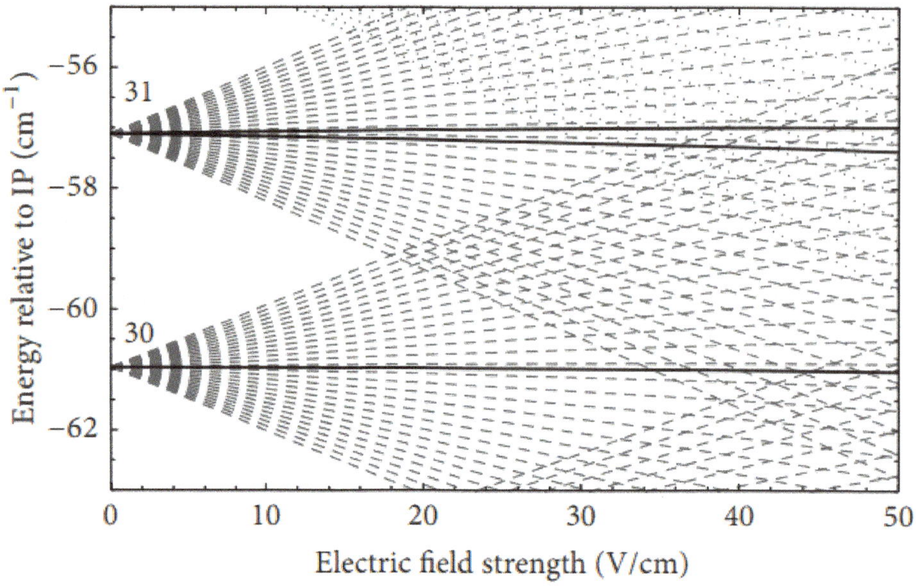

Figure 11: Stark states of $n = 30$ and 31 states of Ps, with m=2 (grey dashed) and $m = 29$ (black). In the $n = 30$ level, the $m = 29$ state is a circular state and experiences no first-order Stark shift and only a very weak second-order shift, as explained in the text.

The Stark states in a given $|n, m\rangle$ manifold are described by the index $k=n_1 - n_2$, where k has values in the range from $k_{min} = -(n - |m| - 1)$ up to $k_{max} = n - |m| - 1$ (with $\Delta k = 2$). The first-order Stark shift in Ps is given by $E_S = -\mu \times F$, where the electric dipole moment has magnitude $|\mu| = (3/2)|k|a_{Ps}$ (where the Ps Bohr radius a_{Ps} is (almost) twice that of hydrogen, i.e., $2a_0$). For a high-n Rydberg state with low angular momentum, for example, the 30d state with $m = 2$, the value of the electric dipole moment can be very large. The Stark state with $k_{max} = 27$ has an electric dipole moment of $2430ea_0$. This large electric dipole moment arises because within this n-state there are many degenerate angular momentum states with the same value of m that are coupled by the electric field. While this is extremely useful for atomic control [56, 60, 61], it presents a significant problem for gravity measurements, since the electric field gradient experienced by a Ps atom in this state which would result in a force equal to that of "normal" gravity ($\sim 2 \times 10^{-29}$ N) is only $\sim 10^{-3}$ V/m^2 . Although this is by no means insignificant, it does compare favorably with the $\sim 5 \times 10^{-11}$ V/m electric field that would apply a g-like force to a bare positron (or electron).

States with the maximum absolute values for the orbital and magnetic quantum numbers for a given n, the so-called circular states, experience no first-order Stark shift. For these states, $m = |n| - 1$, meaning that there is only one Stark state associated with this value of m, which has k=0, and, thus, to first-order, no electric dipole moment (see Figure 11). The explanation for this is that, within a given n-manifold, the circular states have unique values of m, and thus are not coupled to any other degenerate angular momentum states. In the classical limit, these states correspond to circular orbits, in which the average z-position of the electron is zero, resulting in no electric dipole moment, unlike the lower angular momentum Rydberg states where, in the classical limit, the electronic orbit is highly anisotropic, with the electron having a large average displacement from the atomic core. Although there is no atomic core or nucleus in the case of Ps, the wave function is nevertheless hydrogenic, and the same arguments apply. The circular states do experience a second-order Stark shift, from coupling of adjacent n-states; however, this shift is extremely weak. It should, therefore, be possible to produce Ps states with high angular momentum and minimize the effects of stray fields while simultaneously extending the lifetimes to useful levels. However, any manipulation techniques that rely on large dipole moments will obviously have to be performed after the optical excitation to the relevant n states, but before transferring the atoms to states with high angular momentum.

Performing a gravity measurement on any system containing antimatter is clearly very challenging, and many of the potential obstacles are currently being investigated. The ability to produce controllable beams of Ps atoms may also open the door to other types of experiments, such as interferometry [62, 63], which could provide an alternative route to an antimatter gravity measurement.

CAN WE OBSERVE GRAVITATIONAL QUANTUM STATES OF POSITRONIUM?

Positronium is about 1000 times lighter than a neutron or antihydrogen. Therefore, the expected height of the gravitational quantum state is 100 larger corresponding to a macroscopic size of $H_1 = 1.3$ mm while the energy is 10 times smaller, $E_1 = 0.13$ peV (see (1)-(2)). The observation time to resolve a quantum gravitational state can be estimated using the Heisenberg uncertainty principle to be of the order of $\hbar/E_1 \simeq 4.5$ ms. This value is much larger than the longlived triplet positronium lifetime in the ground state which is 142 ns (the Ps singlet state only lives 125 ps and thus, in the following, we will only consider the triplet state and refer to it as Ps). Hence, as for the case of a measurement of the gravitational free fall of Ps described in the previous section, the Ps lifetime must be increased by excitation to a higher level. A possible scheme to observe the Ps gravitational quantum states could employ the flow-through technique used for the first observation of this effect with neutrons (see Figure 12). Greater details of the proposed experimental setup and technique are described in a dedicated contribution to this workshop [64]. Here, we describe the main idea.

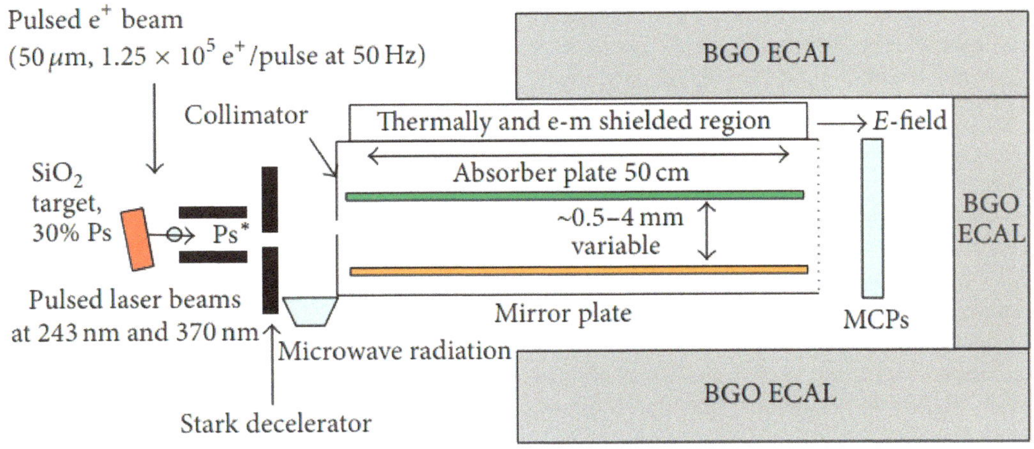

Figure 12: Possible scheme for the observation of the gravitational quantum states of positronium.

Positronium is formed by implanting keV positrons from a remoderated pulsed slow positron beam in a positronpositronium converter. To observe the quantum mechanica behavior of Ps in the gravitational field, its vertical velocity should be of the same order of the gravitational energy levels and thus $V_y < 0.15$ m/s. Furthermore, to resolve the quantum state, the Ps atom has to interact long enough with the slit and, therefore, it has to be laser excited to a Rydberg state with $n > 30$ and maximum l quantum number (see previous section). To keep a reasonable size of the experimental setup (i.e., a slit size of the order of 0.5 m) and minimize the number of detectors, the velocities in the horizontal plane should be smaller than $V_{x,} < 100$ m/s. Similar to neutrons a collimator could be used to select the velocity components V_x, V_y of the positronium distribution. However, since no reliable thermal cold source of positronium exists, the velocity component perpendicular to the surface V_z has to be lowered by some other means. Relying on the fact that atoms in Rydberg states have a large dipole moment, Stark deceleration can be used for this purpose. This method has been demonstrated for different atomic species (including hydrogen) [61] and molecules [65]. Atoms in Rydberg states have large dipole moments; thus, electric field gradients can be used to manipulate them. The acceleration/deceleration a imparted to the Rydberg atoms is given by

$$a = 76\nabla F \frac{1}{m} nk, \tag{22}$$

where ∇F is the gradient of the electric field in V cm^{-2}, m is the mass of the decelerated particles in atomic units, and n and k are the Stark state quantum numbers. H atoms in $n = 25$ and an initial velocity of 700 m/s can be brought at rest in 3 mm [61].

As Ps is 1000 times lighter, decelerations exceeding 10^9 m/s could be realized and therefore the vertical velocity of Ps emitted from thin silica films with initial velocities of the order of 10^5 m/s [66, 67] could be reduced to below 100 m/s. Since one is interested only in decelerating the distribution that is almost perpendicular to the surface of the Ps target, one can expect for those atoms an efficiency close to 100%. This is confirmed by preliminary simulations. The collimator will be placed after the deceleration stage and the microwave region where circularly polarized radiation will spin up the Ps to the maximum l so that kicks to the momentum imparted to the atoms in the vertical direction during these processes will be accounted for.

The fraction of atoms with $Vy < 0.15$, Vx, $Vz < 100$ m/s is estimated to be of the order of 2 × 10−9. After the collimator, the Ps will fly through the slit made of a mirror and the absorber. If the distance between them is smaller than the first expected gravitational state (i.e < 1mm), this will not be transparent and, therefore, no signal will be detected above the expected background in the detectors. If the width of the slit is increased to a value lying between the first and the second gravitational state (i.e.<2mm), the Ps wave function can propagate and a signal is expected to be detected via field-ionization and subsequent detection with MCPs. This quantum jump would provide the unambiguous indication of the observation of a quantum gravitational state of positronium.

As a mirror for Ps, it was proposed to exploit a gradient of magnetic field created using wires arranged parallel to each other with a constant current to create a uniform gradient of the magnetic field. Only the Ps triplet atoms with m=0 have a nonzero net magnetic moment. For the$m = \pm 1$, the electron and the positron magnetic moments cancel and therefore those are insensitive to the magnetic field. Therefore, only one third of the initial population will be reflected. To equate the $E_y = 0.1$ peV a field of few mG at the wire surface will be sufficient.

Because of the large spacial size of gravitational quantum states and the very large characteristic length of the mirror needed to form the gravitational states that is much larger than a characteristic interwire distance, we expect that the very weak magnetic gradient will not perturb the gravitational states. The strict theoretical analysis of this clearly mathematically defined problem is ongoing. A matter mirror could also be considered. Due to the large spacial size of the gravitational quantum state, the surface potential is expected to be very sharp and therefore results in efficient quantum reflection (see Section 3). In both cases (magnetic or material mirror), we expect to have effectively (quasiclassically) only a few collisions with the surface. Nevertheless, the transitions rates due to quenching and ionization caused by the electric or magnetic fields have to be calculated. The absorber as for the neutrons is a rough surface on which the impinging Ps will mix its velocity components and therefore be lost.

With such a scheme assuming a monoenergetic slow positron beam flux of 9×10^8 e^+/s (this being the highest intensity reported so far reached at the FMR II NEMOPUC source in Munich [68]), an event rate of 0.8 events/day with a background 0.05 events/day might be achievable with a realistic extrapolation of current technologies. Possible losses due to spurious effects like stray electric or magnetic fields or black body radiation seem to be negligible but as for the case of a free gravity fall further calculations and preliminary experiments should be done to confirm this assumption and that all the required efficiencies (e.g., Ps excitation in the $n = 33$, $l = 32$ state) can be attained.

Note that the expected height of the gravitational state is related to the gravitational mass M by (2). This means that, for an uncertainty in the determination of H_1 of δH_1, one can get an accuracy in the determination of M at the level of $\delta M/M = 3\delta H_1/H_1\sqrt{N}$ where N is the number of detected signals. Assuming an uncertainty of $\delta H_1 = 0.1$ mm which is mainly determined by the finite source size, the value of $\delta M/M$ can be determined to be at a level of 3% in three months. This precision is comparable to the one that is aimed for by the antihydrogen experiments at CERN [69–71]. Therefore, observation of Ps gravitational quantum states offers a complementary approach to test the effect of gravity on a pure leptonic system. Most of the techniques required for such an experiment are under development for the ongoing free gravity fall experiment of Ps (see Section 6) and Rydberg Ps deceleration experiments are being considered at ETH Zurich where Professor B. Brown's (Marquette University) buffer gas trap is being commissioned. The advantage of using gravitational quantum states is that unpredicted perturbations of the Ps atoms will not result in a systematic effect for the experiment but will only affect the signal rate. Therefore, as for the case of antihydrogen, this approach seems promising to provide a much higher accuracy than a free fall experiment.

CONCLUSION

In this contribution, we have reported the progress of ongoing experiments to measure gravitational free fall of antimatter. The GBAR experiment will produce antihydrogen atoms in the ultracold regime where quantum reflection from surfaces takes place. Quantum reflection will allow the observation of gravitational quantum states of antimatter which promise to lead to a very sensitive probe of the effect of gravity on antiatoms (2 orders of magnitude improvement compared to the free fall experiments).

The techniques developed in experiments designed to produce a cold beam of Ps for a free fall measurement will also eventually find application in creating ultracold Ps atoms, as required for observing gravitational quantum states. They will also enable a wide variety of other experimental areas, such as precision spectroscopy.

Antimatter atoms in gravitational quantum states also provide a unique opportunity to constrain experimentally extra short-range forces between the mirror and the antiatom with about the same sensitivity as we do for normal matter [72].

ACKNOWLEDGMENTS

The authors wish to thank the GRANIT collaboration and the GBAR collaboration (http://gbar.in2p3.fr/) for providing excellent possibilities for discussions and exchange. In addition to the involved institutes and laboratories, preparatory work for GBAR has been funded by the "Conseil Général de l'Essonne," by the "Agence Nationale de la Recherche" in France and by the LABEX P2IO. P. Crivelli acknowledges the support by the Swiss National Science Foundation (Grant PZ00P2_132059) and ETH Zurich (Grant ETH-47-12-1). D. B. Cassidy and T. E. Wall gratefully acknowledge funding from the EPSRC (EP/K028774/1), the Leverhulme Trust (RPG-2013-055), and the ERC (CIG 630119). This work reviews contributions made at the GRANIT 2014 workshop on prospects for the observation of the free fall and gravitational quantum states of antimatter.

REFERENCES

1. The ALPHA Collaboration and A. E. Charman, "Description and first application of a new technique to measure the gravitational mass of antihydrogen," Nature Communications, vol. 4, article 1785, 2013.
2. AEGIS Collaboration and M. G. Giammarchi, "AEGIS at CERN: measuring antihydrogen fall," Few-Body Systems, vol. 54, no. 5-6, pp. 779–782, 2013.
3. P. Pérez and Y. Sacquin, "The GBAR experiment: gravitational behaviour of antihydrogen at rest,"Classical and Quantum Gravity, vol. 29, no. 18, Article ID 184008, 2012.
4. K. Kirch and K. S. Khaw, "Testing antimatter gravity with muonium," International Journal of Modern Physics: Conference Series, vol. 30, Article ID 1460258, 8 pages, 2014.
5. D. B. Cassidy and S. D. Hogan, "Atom control and gravity measurements using Rydberg positronium,"International Journal of Modern Physics: Conference Series, vol. 30, Article ID 1460259, 9 pages, 2014.
6. M. M. Nieto and T. Goldman, "The arguments against "antigravity" and the gravitational acceleration of antimatter," Physics Report, vol. 205, no. 5, pp. 221–281, 1991. ·
7. J. Scherk, "Antigravity: a crazy idea?" Physics Letters: B, vol. 88, no. 3-4, pp. 265–267, 1979.
8. G. Chardin and J.-M. Rax, "CP violation. A matter of (anti)gravity?" Physics Letters B, vol. 282, no. 1-2, pp. 256–262, 1992.
9. G. Chardin, "CP violation and antigravity (revisited)," Nuclear Physics, Section A, vol. 558, pp. 477–495, 1993.
10. G. Chardin, "Motivations for antigravity in general relativity," Hyperfine Interactions, vol. 109, no. 1–4, pp. 83–94, 1997.

11. V. A. Kostelecky and J. D. Tasson, "Matter-gravity couplings and Lorentz violation," Physical Review D, vol. 83, no. 1, Article ID 016013, 2011.
12. M. L. Good, "K20 and the equivalence principle," Physical Review, vol. 121, pp. 311–313, 1961.
13. S. Pakvasa, W. A. Simmons, and T. J. Weiler, "Test of equivalence principle for neutrinos and antineutrinos," Physical Review D, vol. 39, no. 6, pp. 1761–1763, 1989.
14. A. Apostolakis, E. Aslanides, G. Backenstoss et al., "Tests of the equivalence principle with neutral kaons," Physics Letters B: Nuclear, Elementary Particle and High-Energy Physics, vol. 452, no. 3-4, pp. 425–433, 1999.
15. G. Gabrielse, A. Khabbaz, D. S. Hall, C. Heimann, H. Kalinowsky, and W. Jhe, "Precision mass spectroscopy of the antiproton and proton using simultaneously trapped particles," Physical Review Letters, vol. 82, no. 16, pp. 3198–3201, 1999.
16. F. C. Witteborn and W. M. Fairbank, "Experimental comparison of the gravitational force on freely falling electrons and metallic electrons," Physical Review Letters, vol. 19, no. 18, pp. 1049–1052, 1967.
17. M. H. Holzscheiter, "Constraints on gravitational properties of antimatter from cyclotron-frequency measurements," International Journal of Modern Physics: Conference Series, vol. 30, Article ID 1460260, 9 pages, 2014.
18. Antimatter and Gravity, International Journal of Modern Physics: Conference Series, 2014.
19. M. Abramowitz and I. A. Stegun, Handbook of Mathematical Functions with Formulas, Graphs, and Mathematical Tables, Wiley, New York, ny, usa, 1972.
20. L. D. Landau, E. M. Lifshitz, J. B. Sykes, and J. S. Bell, Quantum Mechanics: Non-Relativistic Theory, Royaume-Uni, Oxford, UK, 1965.
21. V. V. Nesvizhevsky, H. G. Börner, A. K. Petukhov et al., "Quantum states of neutrons in the Earth's gravitational field," Nature, vol. 415, no. 6869, pp. 297–299, 2002.
22. G. Chardin, P. Grandemange, D. Lunney, et al., "Proposal to measure the gravitational behaviour of antihydrogen at rest," Tech. Rep. CERN-SPSC-2011-029. SPSC-P-342, CERN, Geneva, Switzerland, 2011.
23. J. Walz and T. W. Hänsch, "A proposal to measure antimatter gravity using ultracold antihydrogen atoms," General Relativity and Gravitation, vol. 36, no. 3, pp. 561–570, 2004. · ·
24. J. P. Merrison, H. Bluhme, J. Chevallier et al., "Hydrogen formation by proton impact on positronium,"Physical Review Letters, vol. 78, no. 14, pp. 2728–2731, 1997. ·
25. H. R. J. Walters and C. Starrett, "Positron and positronium scattering," Physica Status Solidi (C), vol. 4, no. 10, pp. 3429–3436, 2007.
26. P. Comini and P.-A. Hervieux, "H-+ ion production from collisions between antiprotons and excited positronium: cross sections calculations in the framework of the GBAR experiment," New Journal of Physics, vol. 15, no. 9, Article ID 095022, 2013.
27. L. Liszkay, C. Corbel, P. Pérez et al., "Positronium reemission yield from mesostructured silica films,"Applied Physics Letters, vol. 92, no. 6, Article ID 063114, 2008. ·
28. D. B. Cassidy, T. H. Hisakado, H. W. K. Tom, and A. P. Mills, "Photoemission of positronium from Si,"Physical Review Letters, vol. 107, no. 3, Article ID 033401, 2011.
29. N. Oshima, T. M. Kojima, M. Niigaki, A. Mohri, K. Komaki, and Y. Yamazaki, "New scheme for positron accumulation in ultrahigh vacuum," Physical Review Letters, vol. 93, no. 19, Article ID 195001, 2004.
30. P. Grandemange, Piegeage et accumulation de positons issus d'un faisceau pulse produit par un accelerateur pour l'etude de l'interaction gravitationnelle de l'antimatiere [Ph.D. thesis], Universite Paris Sud, Paris, France, 2013.
31. L. Hilico, J.-P. Karr, A. Douillet, P. Indelicato, S. Wolf, and F. S. Kaler, "Preparing single ultra-cold antihydrogen atoms for free-fall in GBAR," International Journal of Modern Physics: Conference Series, vol. 30, Article ID 1460269, 11 pages, 2014.
32. H. B. G. Casimir and D. Polder, "Influence of retardation on the London-van der Waals forces," Nature, vol. 158, no. 4022, pp. 787–788, 1946.
33. H. B. G. Casimir, "On the attraction between two perfectly conducting plates," Proceedings of the Koninklijke Nederlandse Akademie van Wetenschappen, vol. 51, p. 793, 1948.
34. H. Friedrich, G. Jacoby, and C. G. Meister, "Quantum reflection by Casimir-van der Waals potentials tails," Physical Review A, vol. 65, no. 3 B, Article ID 032902, 2002.
35. A. Lambrecht, P. A. Maia Neto, and S. Reynaud, "The Casimir effect within scattering theory," New Journal of Physics, vol. 8, article no. 243, 2006.
36. R. Messina, D. A. R. Dalvit, P. A. M. Neto, A. Lambrecht, and S. Reynaud, "Dispersive interactions between atoms and nonplanar surfaces," Physical Review A, vol. 80, no. 2, Article ID 022119, 2009.
37. G. Dufour, A. Gerardin, R. Guérout et al., "Quantum reflection of antihydrogen from the Casimir potential above matter slabs," Physical Review A, vol. 87, no. 1, Article ID 012901, 2013. ·
38. M. V. Berry and K. E. Mount, "Semiclassical approximations in wave mechanics," Reports on Progress in Physics, vol. 35, no. 1, article no. 306, pp. 315–397, 1972. ·
39. A. Y. Voronin, P. Froelich, and B. Zygelman, "Interaction of ultracold antihydrogen with a conducting wall," Physical Review A, vol. 72, no. 6, Article ID 062903, 2005.
40. G. Dufour, R. Guerout, A. Lambrecht, V. V. Nesvizhevsky, S. Reynaud, and A. Y. Voronin, "Quantum reflection of antihydrogen from nanoporous media," Physical Review A, vol. 87, no. 2, Article ID 022506, 2013.
41. G. Dufour, P. Debu, A. Lambrecht, V. V. Nesvizhevsky, S. Reynaud, and A. Y. Voronin, "Shaping the distribution of vertical velocities of antihydrogen in GBAR," European Physical Journal C, vol. 74, no. 1, article 2731, 2014.

42. V. V. Nesvizhevsky, H. Börner, A. M. Gagarski et al., "Search for quantum states of the neutron in a gravitational field: gravitational levels," Nuclear Instruments and Methods in Physics Research, Section A: Accelerators, Spectrometers, Detectors and Associated Equipment, vol. 440, no. 3, pp. 754–759, 2000.

43. V. V. Nesvizhevsky, H. G. Börner, A. M. Gagarski et al., "Measurement of quantum states of neutrons in the Earth's gravitational field," Physical Review D, vol. 68, no. 10, Article ID 102002, 2003.

44. V. V. Nesvizhevsky, A. K. Petukhov, H. G. Börner et al., "Study of the neutron quantum states in the gravity field," European Physical Journal C, vol. 40, no. 4, pp. 479–491, 2005.

45. A. Y. Voronin, H. Abele, S. Baeßler et al., "Quantum motion of a neutron in a waveguide in the gravitational field," Physical Review D, vol. 73, no. 4, Article ID 044029, 2006.

46. A. Y. Voronin, P. Froelich, and V. V. Nesvizhevsky, "Gravitational quantum states of Antihydrogen,"Physical Review A: Atomic, Molecular, and Optical Physics, vol. 83, no. 3, Article ID 032903, 2011.

47. W. E. Lamb, "Fine structure of the hydrogen atom. III," Physical Review, vol. 85, no. 2, pp. 259–276, 1952.

48. L. P. Gor'kov and I. E. Dzyaloshinskii, "Contribution to the theory of the Mott exciton in a strong magnetic field," Soviet Physics—JETP, vol. 26, pp. 449–453, 1968.

49. Y. E. Lozovik and S. Y. Volkov, "Hydrogen atom moving across a magnetic field," Physical Review A: Atomic, Molecular, and Optical Physics, vol. 70, no. 2, Article ID 023410, 2004.

50. A. Y. Voronin and P. Froelich, "Quantum reflection of ultracold antihydrogen from a solid surface,"Journal of Physics B: Atomic, Molecular and Optical Physics, vol. 38, no. 18, pp. L301–L308, 2005.

51. M. Kreuz, V. V. Nesvizhevsky, P. Schmidt-Wellenburg et al., "A method to measure the resonance transitions between the gravitationally bound quantum states of neutrons in the GRANIT spectrometer,"Nuclear Instruments and Methods in Physics Research A: Accelerators, Spectrometers, Detectors and Associated Equipment, vol. 611, no. 2-3, pp. 326–330, 2009. ·

52. A. P. Mills Jr., "Positron moderation and remoderation techniques for producing cold positron and positronium sources," Hyperfine Interactions, vol. 44, no. 1–4, pp. 105–123, 1989. ·

53. H. A. Bethe and E. E. Salpeter, Quantum. Mechanics of One- And Two-Electron Atoms, Dover, New York, NY, USA, 1957.

54. T. F. Gallagher, Rydberg Atoms, Cambridge University Press, Cambridge, UK, 1994.

55. A. P. Mills Jr. and M. Leventhal, "Can we measure the gravitational free fall of cold Rydberg state positronium?" Nuclear Instruments and Methods in Physics Research, Section B: Beam Interactions with Materials and Atoms, vol. 192, no. 1-2, pp. 102–106, 2002. ·

56. C. Seiler, S. D. Hogan, H. Schmutz, J. A. Agner, and F. Merkt, "Collisional and radiative processes in adiabatic deceleration, deflection, and off-axis trapping of a Rydberg atom beam," Physical Review Letters, vol. 106, no. 7, Article ID 073003, 2011.

57. A. C. L. Jones, T. H. Hisakado, H. J. Goldman, H. W. K. Tom, A. P. Mills Jr., and D. B. Cassidy, "Doppler-corrected Balmer spectroscopy of Rydberg positronium," Physical Review A, vol. 90, no. 1, Article ID 012503, 2014.

58. T. E. Wall, D. B. Cassidy, and S. D. Hogan, This has not yet been demonstrated for positronium; the feasibility of doing so is considered, to be published.

59. T. F. Gallagher, "Rydberg atoms," Reports on Progress in Physics, vol. 51, no. 2, pp. 143–188, 1988.

60. E. Vliegen and F. Merkt, "Normal-incidence electrostatic rydberg atom mirror," Physical Review Letters, vol. 97, no. 3, Article ID 033002, 2006.

61. S. D. Hogan and F. Merkt, "Demonstration of three-dimensional electrostatic trapping of state-selected rydberg atoms," Physical Review Letters, vol. 100, no. 4, Article ID 043001, 2008.

62. T. J. Phillips, "Antimatter gravity studies with interferometry," Hyperfine Interactions, vol. 109, no. 1–4, pp. 357–365, 1997.

63. M. K. Oberthaler, "Anti-matter wave interferometry with positronium," Nuclear Instruments and Methods in Physics Research, Section B: Beam Interactions with Materials and Atoms, vol. 192, no. 1-2, pp. 129–134, 2002.

64. P. Crivelli, V. V. Nesvizhevsky, and A. Y. Voronin, "Can we observe the gravitational quantum states of Positronium?" Advances in High Energy Physics. In press.

65. S. D. Hogan, C. Seiler, and F. Merkt, "Rydberg-state-enabled deceleration and trapping of cold molecules," Physical Review Letters, vol. 103, no. 12, Article ID 123001, 2009.

66. D. B. Cassidy, P. Crivelli, T. H. Hisakado et al., "Positronium cooling in porous silica measured via Doppler spectroscopy," Physical Review A: Atomic, Molecular, and Optical Physics, vol. 81, no. 1, Article ID 012715, 2010.

67. P. Crivelli, U. Gendotti, A. Rubbia, L. Liszkay, P. Pérez, and C. Corbel, "Measurement of the orthopositronium confinement energy in mesoporous thin films," Physical Review A—Atomic, Molecular, and Optical Physics, vol. 81, no. 5, Article ID 052703, 2010. ·

68. C. Hugenschmidt, B. Löwe, J. Mayer et al., "Unprecedented intensity of a low-energy positron beam,"Nuclear Instruments and Methods in Physics Research A: Accelerators, Spectrometers, Detectors and Associated Equipment, vol. 593, no. 3, pp. 616–618, 2008. ·

69. A. Kellerbauer, M. Amoretti, A. S. Belov et al., "Proposed antimatter gravity measurement with an antihydrogen beam," Nuclear Instruments and Methods in Physics Research B: Beam Interactions with Materials and Atoms, vol. 266, no. 3, pp. 351–356, 2008. ·

70. D. P. van der Werf, "The GBAR experiment," International Journal of Modern Physics: Conference Series, vol. 30, Article ID 1460263, 2014.

71. A. I. Zhmoginov, A. E. Charman, R. Shalloo, J. Fajans, and J. S. Wurtele, "Nonlinear dynamics of anti-hydrogen in magnetostatic traps: implications for gravitational measurements," Classical and Quantum Gravity, vol. 30, no. 20, Article ID 205014, 2013.

72. I. Antoniadis, S. Baessler, M. Büchner et al., "Short-range fundamental forces," Comptes Rendus Physique, vol. 12, no. 8, pp. 755–778, 2011.

CHAPTER 20

QUANTUM HAPLODYNAMICS, DARK MATTER, AND DARK ENERGY

Harald Fritzsch and Joan Solà

INTRODUCTION

The standard model (SM) of strong and electroweak interactions may not be the final theory of the universe. The dark matter (DM) and dark energy (DE) are fundamental problems awaiting for an explanation [1–5]. In this paper we consider the possible impact of the chiral gauge theory QHD (quantum haplodynamics [6, 7]) on these problems. QHD is a theory of bound states for all the SM particles and the dark matter particles.

If the QCD and QHD couplings take definite values around the Planck scale and this scale is permitted to slowly change with time, the QCD and QHD couplings should change also with time. Hints that the electromagnetic fine structure constant α_{em} might change with the cosmic evolution are reported in the literature [8, 9]. If α_{em} changes in time, we expect that all fundamental coupling constants change in time, including the gravity constant [10–14]. In particular, let us note that the particle masses could also change in time, such as the proton mass. Different experiments have explored this possibility; see, for example, [15–17] and the reviews [18, 19]. Since the gravity constant G_N determines the Planck mass $M_P = G_N^{-1/2}$, one expects that M_P depends also on time and slowly evolves with the cosmic expansion.

The present framework obviously implies a link between gravity and particle physics. We will come back to the cosmological implications after describing the essentials of QHD.

Citation: Harald Fritzsch and Joan Solà, "Quantum Haplodynamics, Dark Matter, and Dark Energy," Advances in High Energy Physics, vol. 2014, Article ID 361587, 6 pages, 2014. doi:10.1155/2014/361587.

QHD

In QHD all of the SM particles (except the photon and the gluons) are bound states of the fundamental constituents called haplons, h, and their antiparticles. A first model of this type was introduced in 1981 (see [20–25]). Here we extend it and assume that the QHD chiral gauge group is the unitary left-right group $SU(2)_L \times SU(2)_R$, which we will denote by $SU(2)_h$ for short. All species of haplons h are $S(2)$ doublets; hence each one has two internal states h_i represented by the $SU(2)_h$ quantum number $i = 1, 2$. Rotations among these states are performed by the exchange of two sets of massless $S(2)_h$ gauge bosons $(X^r_{L,R})_\mu$ $(r = 1, 2, 3)$ for each chirality.

There are six haplon flavors, two of them are electrically charged chiral spinors ($\chi = \alpha, \beta$) and four are charged scalars

Table 1: Quantum numbers of the six haplons: spin (s), electric charge Q (in units of $|e|$), and corresponding representations of $S(3)_c$ and $SU(2)_h$.

	s	Q	$SU(3)_c$	$SU(2)_h$
α	1/2	+1/2	1	2
β	1/2	−1/2	1	2
ℓ	0	+1/2	1	2
c_k	0	−1/6	3	2

S. One scalar (ℓ) has electric charge (+1/2) and carries leptonic flavor. The other three scalars have charge (−1/6) and carry color: $c_k = R, G, B$ ("red, green, and blue"). In Table 1 we indicate the relevant quantum numbers.

The complete QHD gauge group is $S(3)_c \times SU(2)_h \times U(1)_{em}$, with the coupling constants (g_s, g_h, e). There is a gauge coupling $g_h = g_h^L, g_h^R$ h for each chiral factor in $S(2)_h$. The QHD part of the interaction Lagrangian involving a generic chiral haplon χ (χ_L or χ_R) and a scalar haplon S reads

$$\mathscr{L}_{\text{int}} = \overline{\chi}^i \gamma_\mu i \mathscr{D}^\mu_{ij} \chi^j + \left(D^\mu_{ij} S^j\right)^* \left(D^{ij}_\mu S_j\right) + \cdots, \quad (1)$$

where $\mathscr{D}^\mu_{ij} = \delta_{ij}\partial^\mu + i g^{L,R}_h \, (X^r_{L,R})^\mu (\sigma_r/2)_{ij}$ is the $SU(2)_h$ covariant derivative and σ_r are the Pauli matrices. The other covariant derivative is coincident with the previous one if S is any scalar haplon.

For colored scalar haplons c_k, however, we have additional terms where the covariant derivative is $D^\mu_{ij} = \delta_{ij}\partial^\mu + i g_s A^\mu_a (\lambda^a/2)_{ij}$ and involves the gluons A^μ_a and Gell-Mann matrices λ. Only the scalar haplons c_k interact with the gluons, since these are the only colored constituents. Besides the fine structure constant $\alpha_{\text{em}} = e^2/4\pi$ there is the strong coupling $\alpha_s = g_s^2/4\pi$ and its $S(2)_h$ counterpart $\alpha_h = g_h^2/4\pi$, which is also strong at energies $\mu \lesssim \Lambda_h$ but asymptotically free well above it. Here Λ_h denotes generically any of the two confining scales Λ^L_h and Λ^R_h associated with the chiral factors of $S(2)_h$.

From the various haplon flavors the bound states of QHD can be constructed. Only for energies μ well above Λ_h these states break down into the fundamental haplons. The weak gauge bosons are s-wave bound states of left-handed haplons α and β and their antiparticles: $W^+ = \overline{\beta}\alpha$, $W^- = \overline{\alpha}\beta$, and $W^3 = (\overline{\alpha}\alpha - \overline{\beta}\beta)/\sqrt{2}$.

The neutral weak boson mixes with the photon (similar to the mixing between the photon and the neutral ρ-meson). One obtains the physical Z-boson with a mass slightly heavier than the W-boson:

$$\frac{M_Z^2 - M_W^2}{M_W^2} = \frac{m^2}{1 - m^2}, \quad (2)$$

where the mixing parameter $m = eF_W/M_W$ is related to the W decay constant F_W [6, 7]. The confinement scale Λ^L_h for $S(2)_L$ defines the Fermi scale $G_F^{-1/2} \sim 0.3\,\text{TeV}$ and the size of the weak gauge bosons of the SM.

Owing to the $S(2)_L \times SU(2)_R$ chiral structure of QHD, besides the three observed weak bosons there are also vector bosons that are coupled to the right-handed leptons and quarks. We assume that the right-handed confining scale Λ^R_h is higher so that the masses of the new vector bosons lie well above 1 TeV. They might be observed in the new experiments at the LHC. The observed scalar resonance is a p-wave excitation of the Z. Here we will not discuss these aspects of QHD in detail [6, 7].

The leptons and quarks are themselves bound states. They are composed of a chiral haplon (α or β) and a scalar haplon: ℓ for leptons and c_k for quarks. The electron and its neutrino have the structure $\nu = (\alpha\overline{\ell})$ and $e^- = (\beta\overline{\ell})$, which is consistent with the quantum numbers of Table 1. Similarly, the up and down quarks (with c_k color) are given by $u = (\alpha\overline{c}_k)$ and $d = (\beta\overline{c}_k)$.

In QHD the first generation of leptons and quarks describes the ground states of the fermion-scalar bound states; the second and third generation must be dynamical excitations. Likewise the c-quark and t-quark families are excitations of the first quark generation. Compared to the QHD mass scale the masses of the observed leptons and quarks are essentially zero.

The outcome is an effective theory equivalent to the electroweak SM in good approximation. However, new matter content is predicted. In particular, the simplest neutral bound state of the four scalars with haplon number $\mathscr{H} = 4$ is a stable color singlet spinless boson: $D = (lRGB)$. It is stable due to haplon number conservation, which is similar to the conservation of baryon number.

A NEW DARK MATTER CANDIDATE

The mass of the D-boson is expected to be in the region of a few TeV. It can be produced together with its antiparticle by the LHC accelerator, and it can be observed by the large missing energy. We interpret it as the particle providing the DM in the universe. The properties of this DM particle are similar to a "weakly interacting massive particle" (WIMP) but it can be much more elusive concerning the interactions with nuclei.

After the Big Bang the universe is filled not only with a gas of quarks and antiquarks but also with a relativistic gas of D-bosons and the corresponding antiparticles, which annihilate into other particles. Due to the CP violation there is an asymmetry in the number of D-bosons and anti-D- bosons. After freeze-out, at temperatures roughly $\sim 1/20$ of the D-boson mass, the D-particles abandon the equilibrium. A relic density remains today, a gas of D-bosons, forming the dark matter. If we use the average density of matter in our galaxy, we find that there should be $\mathcal{O}(100)$ D-particles per cubic meter.

In contrast to more conventional WIMPs, the QHD candidate for DM can escape more easily the recent, highly restrictive bounds obtained from scattering of DM particles off nuclei [26]. It is not difficult to estimate the cross-section for the D-boson off a nucleon $\mathscr{N}$ (of mass $m_{\mathscr{N}}$). It should be of order

$$\sigma_{D\mathscr{N}} \sim f_D^2 \frac{\alpha_h^2}{\Lambda_h^4} m_{\mathscr{N}}^2 \sim f_D^2 \alpha_h^2 G_F^2 m_{\mathscr{N}}^2, \quad (3)$$

where $G_F^{-1/2} \sim \Lambda_h \sim 300\,\text{GeV}$ according to our definition of Fermi's scale in QHD. Here f_D is the dimensionless form factor of the D-meson, which describes the confinement of the haplons by the $S(2)_h$ strong gauge force. All QHD bound states have a form factor, which is of order one only for gauge boson mediated interactions, which are described by the exchange of weak bosons $(M_W^2 \lesssim \Lambda_h^2)$. For a deeply bound state as D, however, we rather expect $f_D \sim \Lambda_h^2/B_D^2 \ll 1$, where B_D is the characteristic binding energy scale.

In QCD the proton mass is given by a few times the value of the confining scale Λ_{QCD}, with only a tiny contribution from the quark masses. Similarly, in QHD the masses of the bound states are proportional to Λ_h, although here the spectroscopy is richer and the proportionality factor is bigger for the more deeply bound states. Whereas for a weak boson such factor is of order one, in the case of the DM candidate $D = (lRGB)$ it can be much larger. This way we can explain the small cross-section of the DM particles with ordinary matter. For $B_D > \mathcal{O}(10)\,\text{TeV}$ the scattering cross-section of D-bosons off nucleons, (3), can be approximately reduced to the level of

$\sim 10^{-45}$ cm^2 and become roughly compatible with the current bounds [26]. Let us note that, for energy scales $\mu > 10$ TeV, we have $\alpha_h(\mu) = \mathcal{O}(0.1)$ owing to the asymptotic freedom of the haplon interaction constant (cf. next section). One cannot be more precise at this point because at present we cannot work out the details of the form factor f_D, and therefore (3) is only indicative of the kind of result that could be obtained. The correct order of magnitude could be reached if the bound state associated with DM in this model is very tightly bound and massive. For the exceptional fourhaplon state $D = (lRGB)$ this situation should be regarded, in principle, as possible. We believe that as a possibility it is worthwhile to take into account, still more considering that the present bounds from direct searches put very stringent constraints on virtually every DM candidate, including the more familiar ones from more conventional extensions of the SM.

UNIFICATION AT THE PLANCK SCALE

The QHD, QCD, and QED couplings might unify at the Planck scale. We can check it at one-loop level, starting from their low-energy values and using the renormalization group equations (RGEs) to compute the running of these parameters. For $S(N)$ groups (N>1) one has [27]

$$\frac{d\alpha_i}{d\ln\mu} = -\frac{1}{2\pi}\left(\frac{11}{3}N - \frac{2}{3}n_f - \frac{1}{6}n_s\right)\alpha_i^2 \equiv -\frac{1}{2\pi}b_N\alpha_i^2. \tag{4}$$

Here we have $\alpha_i = \alpha_h$, α_s (n_f and ns are the number of fermion flavors and scalars). For the (1) coupling α_{em} we have a formula similar to (4), but in this case

$$b_1 = -N_h\left(\frac{4}{3}\sum Q_f^2 + \frac{1}{3}\sum Q_s^2\right). \tag{5}$$

Here $N_h = 2$ for $S(2)_h$. The electric charges Q are defined in Table 1. For energies below Λ_h we have to replace N_h in $b1$ with $N_c = 3$ (or 1) and use the electric charges of the quarks (leptons) rather than those of the haplons.

Table 2: The QED, QCD, and QHD fine structure constants $\alpha_i = g_i^2/4\pi$ at the Z-pole scale $\mu_0 = M_Z$, at an intermediate high energy scale μ_1 = 2 TeV (around the haplon continuum threshold), and at the Planck energy $M_P \sim 1.2 \times 10^{19}$ GeV, for the $S(2)_h$ chiral gauge group of QHD.

	μ_0 M_Z	μ_1 2 TeV	M_P 10^{19} GeV
α_{em}	0.007816	0.008092	0.008727
α_s	0.1184	0.08187	0.01370
α_h	—	0.62	0.030

For the fine structure constant α_{em} we extrapolate its value from low energies to the Planck scale $MP \simeq 1.22 \times 10^{19}$ GeV. At the mass of the Z-boson we have $\alpha_{em}^{-1}(M_Z) = 127.94 \pm 0.014$. From the mass scale of the Z-boson, $\mu = M_Z$, until a scale well above Λ_h, say $\mu \sim 2$ TeV, we use the RGE, taking into account the charges of the three charged leptons and of the five quarks, not including the top quark: $\alpha_{em}^{-1}(2\,\text{TeV}) = 123.57$. From ~ 2 TeV up to the Planck mass we take into account the electric charges of the two spin-1/2 haplons and of the four scalar haplons (cf. Table 1). The result is

$$\alpha_{em}^{-1}(M_P) = 114.58, \tag{6}$$

equivalently, $\alpha_{em}(M_P) = 0.008727$ (cf. Table 2).

A similar procedure can be followed to compute the QCD coupling constant at various energies. The accurate measurement of this constant at the Z-pole yields $\alpha(M_Z) = 0.1184 \pm 0.0007$. At the Fermi scale $\Lambda_h \sim 0.3$ GeV we find $\alpha(\Lambda_h) = 0.1010$. Well above 1 TeV up to the Planck scale the renormalization proceeds via haplon pairs (as indicated in Table 2).

For the $S(2)_h$ group, we focus here on the left-handed sector and assume $\Lambda_h \simeq 0.3$ TeV. We find, for example, α_h(2 TeV) = 0.62, and eventually at the Planck energy $\alpha_h(M_P) \simeq 0.030$. From Table 2 we see that the three couplings approach each other at the Planck scale. The details of the unification will depend on the particular GUT group and can be affected by Clebsch-Gordan coefficients of $\mathcal{O}(1)$.

We note that $SU(3) \times SU(2)_L \times SU(2)_R \times U(1)$ is a natural breakdown step for GUT groups such as, for example, $SO(10)$. In our case we do not have spontaneous symmetry breaking (SSB); the breaking is always meant to be dynamical. The complete QHD group can thus be naturally linked to the GUT framework without generating unconfined vacuum energy.

If the three couplings come close at the Planck scale, interesting consequences can be derived in connection to the time variation of the fundamental constants, of which hints in the literature appear quite often [8, 9,15–19]. Exact unification is not essential; we only require that the three couplings take fixed values at or around M_P.

TIME EVOLUTION OF FUNDAMENTAL "CONSTANTS"

A cosmic time change of Newton's constant G_N (and hence of M_P) is conceivable in the same way as one admits a possible time change of α_{em} [8, 9, 18, 19]. If the QED, QCD, and QHD coupling constants emerge at the Planck epoch, their primeval values should be very close and not be time dependent. Since the Planck energy changes in time, there must be time evolution of the gauge couplings at lower energies, say around the confining scale of the weak bosons, $\Lambda_h \sim 300$ GeV. At the same time the masses of all the particles (including of course the baryons and the D-bosons) are forced to slowly evolve with the cosmic expansion since their binding energies are functions of the coupling strengths.

We can estimate the time change of G_N in QHD. We use the approximate time variation of α_{em} suggested in a typical measurement where the current value of the QED coupling is compared with that of a quasar some 12 billion years ago [8, 9]: $\Delta\alpha_{em}/\alpha_{em} = (-0.54 \pm 0.12) \times 10^{-5}$.

From the RGEs and setting $\mu=M_P$ we can obtain the time variation (indicated by a dot) of the Planck scale. Since $b_1 = -14/9$ in this case, we find

$$\frac{\dot{M}_P}{M_P} = -\frac{\dot{\alpha}_{em}(M_Z)}{\alpha_{em}(M_Z)}\left[\ln\frac{M_P}{M_Z} + \frac{9\pi}{7\alpha_{em}(M_P)}\right]. \tag{7}$$

It follows that $\Delta M_P/M_P \simeq 0.0027$ or $\Delta G/G \simeq -0.0054$

The time variation of the non-Abelian gauge couplings α_i (i.e., α_s and α_h) at an arbitrary scale μ below M_P is also determined:

$$\frac{\dot{\alpha}_i(\mu)}{\alpha_i(\mu)} = \frac{\dot{M}_P}{M_P}\left[-\ln\frac{M_P}{\mu} + \frac{2\pi}{b_N\alpha_i(M_P)}\right]^{-1}, \tag{8}$$

where b_N was defined in (4).

Since $\dot{M}_P/M_P$ is fixed from (7), the above equation enables us to compute the cosmic time variation of the QCD and QHD couplings within the last 12 billion years at any desired energy well above Λ_h, for example, at $\mu_1 = 2$ TeV (cf. Table 2):

$$\frac{\Delta\alpha_s}{\alpha_s} \simeq 1.1 \times 10^{-4}, \qquad \frac{\Delta\alpha_h}{\alpha_h} \simeq 6.3 \times 10^{-4}. \tag{9}$$

Using the definition of the corresponding confining scales Λ_i (namely, Λ_{QCD}, Λ_h) we can check from the above formulas that their cosmic time evolution [10] is renormalization group invariant and is directly tied to the cosmic evolution of M_P itself:

$$\frac{\dot{\Lambda}_i}{\Lambda_i} = \frac{\dot{\alpha}_i(\mu)}{\alpha_i(\mu)}\frac{2\pi}{b\alpha_i(\mu)} = \frac{\dot{M}_P}{M_P}. \tag{10}$$

Numerically, $\Delta\Lambda_i/\Lambda_i \simeq 3 \times 10^{-3}$ for the indicated period.

QHD AND DARK ENERGY

The unification of the QCD and QHD couplings can have nontrivial significance for the combined framework of particle physics and general relativity (GR). It suggests cosmic evolution of all the masses in the universe, both of the nuclei and of the DM particles.This can be perfectly compatible with GR.

In order to preserve the Bianchi identity that is satisfied by the Einstein tensor of the gravitational field equations $(\nabla^\mu G_{\mu\nu} = 0)$, the time evolution of the masses can be compensated for by the time variation of one or more fundamental gravitational parameters, typically the gravitational constant G_N, or the cosmological constant Λ, or both [10–14].

With the help of the Friedmann-Lemaˆıtre-RobertsonWalker (FLRW) metric one finds that the most general local conservation law preserving the Bianchi identity, corresponding to an isotropic and homogeneous dust matter fluid with density ρ_m, reads

$$\frac{G_N'}{G_N}(\rho_m + \rho_\Lambda) + \rho_m' + \rho_\Lambda' + \frac{3}{a}\rho_m = 0, \tag{11}$$

where a is the scale factor and the primes denote derivatives of the various quantities with respect to it. ρ_Λ is the vacuum energy density.

From (11) we can understand how a theory of bound states at low energies can lead to general evolution of the vacuum energy density ρ_Λ and Newton's coupling G_N in combination with the particle masses.

Let us note that although many of the studies that motivated the possibility of having variable fundamental constants of Nature were mainly focused on possible time variation of the fine structure constant from QSO absorption spectra [8, 9], subsequent investigations admitted the possibility of a time variation of the particle masses, for example, the proton mass [15–17]. The predicted time evolution of the particle masses in QHD can be parameterized as $\rho_m \sim a^{-3(1-\nu)}$ [10], where the presence of $|\nu| \ll 1$ denotes a very small departure from the standard conservation law $\sim a^{-3}$. Such a departure is not viewed here as a loss or an excess in the number of particles in a comoving volume (beyond the normal dilution law), but rather as a change in the value of their masses. Models with anomalous matter conservation laws of the above type have been carefully confronted with the precise cosmological data on distant supernovae, baryonic acoustic oscillations, structure formation, and CMB anisotropies, and one finds the upper bound $|\nu| \lesssim \mathcal{O}(10^{-3})$ [28–30].

The anomalous matter conservation law $\rho_m \sim a^{-3(1-\nu)}$ implies via (11) that a dynamical response will be generated from the parameters of the gravitational sector, G_N and ρ_Λ. This is how the dark energy can emerge: it is related to the change of the vacuum energy density triggered by the time evolution of all the masses in the universe. While we cannot predict its value, we suggest that it is time dependent, which should be regarded as natural for the vacuum energy density of an expanding universe.

The time variation of the proton mass (and in general of all masses) within the aforementioned parameterization is approximately given as follows:

$$\left|\frac{\dot{m}_p}{m_p}\right| \simeq 3|\nu| H. \tag{12}$$

Notice that the index ν need not be universal. Here for simplicity we mention only the case of the proton [10]. The corresponding change of the vacuum energy density reads

$$\left|\frac{\dot{\rho}_\Lambda}{\rho_\Lambda}\right| \simeq -3|\nu|\frac{\Omega_m^0}{\Omega_\Lambda^0} H, \tag{13}$$

where Ω_m^0, Ω_Λ^0 are the current cosmological parameters associated with matter and vacuum energy. Similarly, if the gravitational constant can change with time, we expect [10]

$$\left|\frac{\dot{G}}{G}\right| \lesssim |\nu| H. \tag{14}$$

Using the current value of the Hubble parameter as a reference, $H_0 = 1.0227h \times 10^{-10}$ yr^{-1}, where $h \simeq 0.70$, and the mentioned limit $|\nu| \lesssim \mathcal{O}(10^{-3})$, we find that the time variations of the above parameters are at most of order $\lesssim 10^{-13}$ yr−1. These changes are of course only indicative since ν could be smaller, but at that level they are approximately within the present bounds. Such bounds are obtained from many sources and usually with large errors [18, 19]. Let us mention that a new generation of high precision lab experiments could reach the level $\lesssim 10^{-14}$ yr^{-1} [31]. In that case they might be sensitive to the possible time variations, if they are there.

One cannot exclude a priori that various sorts of effects are involved at the same time. In such case the results can be more difficult to interpret and may require the use of more than one observable. In general this situation might enforce some reinterpretation of the previous observations, in which more emphasis is given to the variation of one constant with respect to another. For example, it is interesting to note that in the context of conventional GUTs one finds that if the Λ_{QCD} scale (and hence the proton mass) changes with time, such change would be significantly larger (by more than one order of magnitude) than the possible time variation of the fine structure constant—see [10, 32, 33]. At the end of the day the most important feature is that a positive observational effect (irrespective of the various sources of time variation concurring in it) would signal the new qualitative fact that the constants of Nature may be varying.

Finally, we should also mention that the models (11) where the variable vacuum energy and gravitational constant can be linked to the variation of the particle masses are compatible with the primordial nucleosynthesis bounds on the chemical species. The important constraint to be preserved here is that the vacuum energy density remains sufficiently small as compared to the radiation density at the time of nucleosynthesis. At the same time, potential variations of the gravitational constant should also be moderate enough to avoid a significant change in the expansion rate at that time. Once more one can show that these situations are under control provided the limit $|\nu| \lesssim \mathcal{O}(10^{-3})$ is fulfilled [30, 34, 35]. Under these circumstances the abundances of primordial chemical elements remain essentially the same.

CONCLUSIONS

The framework we have outlined here proposes a new formulation of GUT's involving gravity ab initio (in contrast to more conventional formulations). It also proposes a new candidate for the dark matter that is not in conflict with the recent, highly restrictive bounds for the scattering of DM particles off nuclei. The neutral and stable D-boson is the particle of the dark matter.

QHD is not based on the conventional SSB mechanism and it does not lead to a large contribution to the cosmological term. The DE appears here as the tiny (but observable) dynamical change of the vacuum energy density of the expanding background and hence is a part of the generic response of GR to the cosmic time variation of the masses of all the stable baryons and dark matter particles in the universe.

These ideas can be tested by future astrophysical and laboratory tests in quantum optics, which are expected to detect potential proton mass variations of order $\lesssim 10^{-14}$yr^{-1} [31], hence at the level of the expected running of the cosmological parameters.

ACKNOWLEDGMENTS

Harald Fritzsch thanks the BKC excellence programme and the Department of. ECM, University of Barcelona, for hospitality and support. Joan Solà is supported in part by MICINN, DEC, and CPAN. The authors are both grateful to the Institute for Advanced Study at the Nanyang Technological University in Singapore for hospitality and support.

REFERENCES

1. S. Weinberg, "The cosmological constant problem," Reviews of Modern Physics, vol. 61, no. 1, pp. 1–23, 1989.
2. V. Sahni and A. Starobinsky, "The case for a positive cosmological Lambda-term," International Journal of Modern Physics A, vol. 9, pp. 373–444, 2000.
3. T. Padmanabhan, "Cosmological constant—the weight of the vacuum," Physics Reports, vol. 380, no. 5-6, pp. 235–320, 2003.
4. P. J. E. Peebles and B. Ratra, "The cosmological constant and dark energy," Reviews of Modern Physics, vol. 75, no. 2, pp. 559–606, 2003.
5. E. J. Copeland, M. Sami, and S. Tsujikawa, "Dynamics of dark energy," International Journal of Modern Physics D, vol. 15, article 1753, no. 11, 2006.
6. H. Fritzsch, "The size of the weak bosons," Physics Letters B, vol. 712, pp. 231–232, 2012.
7. H. Fritzsch, "Composite weak bosons, leptons and quarks," Modern Physics Letters A, vol. 26, no. 31, pp. 2305–2312, 2011.
8. M. T. Murphy, J. K. Webb, and V. V. Flambaum, "Further evidence for a variable fine-structure constant from Keck/HIRES QSO absorption spectra," Monthly Notices of the Royal Astronomical Society, vol. 345, no. 2, pp. 609–638, 2003.
9. H. Rahmani, N. Maheshwari, and R. Srianand, "Constraining the variation in fine-structure constant using SDSS DR8 QSO spectra," http://xxx.tau.ac.il/abs/1312.5324.
10. H. Fritzsch and J. Solà, "Matter non-conservation in the universe and dynamical dark energy," Classical and Quantum Gravity, vol. 29, no. 21, Article ID 215002, 25 pages, 2012.
11. J. Solà, "Cosmological constant and vacuum energy: old and new ideas," Journal of Physics: Conference Series, vol. 453, Article ID 012015, 2013.
12. J. Solà, "Cosmologies with a time dependent vacuum," Journal of Physics: Conference Series, vol. 283, no. 1, Article ID 012033, 2011.
13. J. Solà, "Dark energy: a quantum fossil from the inflationary Universe?" Journal of Physics A:

Mathematical and Theoretical, vol. 41, no. 16, Article ID 164066, 2008.

14. J. Solà, "Vacuum energy and cosmological evolution," AIP Conference Proceedings, vol. 1606, pp. 19–37, 2014.

15. E. Reinhold, R. Buning, U. Hollenstein, A. Ivanchik, P. Petitjean, and W. Ubachs, "Indication of a cosmological variation of the proton-electron mass ratio based on laboratory measurement and reanalysis of H2 spectra," Physical Review Letters, vol. 96, no. 15, Article ID 151101, 2006.

16. W. Ubachs and E. Reinhold, "Highly accurate H2 lyman and werner band laboratory measurements and an improved constraint on a cosmological variation of the proton-to-eiectron mass ratio," Physical Review Letters, vol. 92, no. 10, 2004.

17. A. Ivanchik, P. Petitjean, D. Varshalovich et al., "A new constraint on the time dependence of the proton-to-electron mass ratio: analysis of the Q 0347-383 and Q 0405-443 spectra," Astronomy and Astrophysics, vol. 440, no. 1, pp. 45–52, 2005.

18. J.-P. Uzan, "Varying constants, gravitation and cosmology," Living Reviews in Relativity, vol. 14, p. 2, 2011.

19. T. Chiba, "The constancy of the constants of nature: updates," Progress of Theoretical Physics, vol. 126, p. 993, 2011.

20. H. Fritzsch and G. Mandelbaum, "Weak interactions as manifestations of the substructure of leptons and quarks," Physics Letters B, vol. 102, no. 5, pp. 319–322, 1981.

21. H. Fritzsch and G. Mandelbaum, "The substructure of the weak bosons and the weak mixing angle,"Physics Letters B, vol. 109, no. 3, pp. 224–226, 1982.

22. R. Barbieri, R. N. Mohapatra, and A. Masiero, "Compositeness and a left-right symmetric electroweak model without broken gauge interactions," Physics Letters B, vol. 105, no. 5, pp. 369–374, 1981.

23. H. Fritzsch, D. Schildknecht, and R. Kogerler, "Weak current algebra, composite W bosons and universality of the weak interactions," Physics Letters B, vol. 114, pp. 157–161, 1982. ·

24. L. F. Abbott and E. Farhi, "Are the weak interactions strong?" Physics Letters B, vol. 101, no. 1-2, pp. 69–72, 1981.

25. T. Kugo, S. Uehara, and T. Yanagida, "Weak bosons as composite gauge fields of hidden symmetries,"Physics Letters B, vol. 147, no. 4-5, pp. 321–324, 1984.

26. LUX Collaboration, D. S. Akerib, H. M. Araujo, et al., "First results from the LUX dark matter experiment at the Sanford underground research facility," Physical Review Letters, vol. 112, Article ID 091303, 2014.

27. T. P. Cheng, E. Eichten, and L. Li, "Higgs phenomena in asymptotically free gauge theories," Physical Review D, vol. 9, no. 8, pp. 2259–2273, 1974.

28. S. Basilakos, M. Plionis, and J. Solà, "Hubble expansion and structure formation in time varying vacuum models," Physical Review D, vol. 80, no. 8, Article ID 083511, 2009.

29. J. Grande, J. Solà, S. Basilakos, and M. Plionis, "Hubble expansion and structure formation in the "running FLRW model" of the cosmic evolution," Journal of Cosmology and Astroparticle Physics, vol. 1108, 2011.

30. S. Basilakos, D. Polarski, and J. Solà, "Generalizing the running vacuum energy model and comparing with the entropic-force models," Physical Review D, vol. 86, Article ID 043010, 2012. ·

31. T. W. Hänsch, Private communication.

32. X. Calmet and H. Fritzsch, "Symmetry breaking and time variation of gauge couplings," Physics Letters B, vol. 540, no. 3-4, pp. 173–178, 2002.

33. X. Calmet and H. Fritzsch, "A time variation of proton-electron mass ratio and grand unification,"Europhysics Letters, vol. 76, no. 6, pp. 1064–1068, 2006.

34. J. Grande, J. Solà, J. C. Fabris, and I. L. Shapiro, "Cosmic perturbations with running G and Lambda,"Classical and Quantum Gravity, vol. 27, Article ID 105004, 2010.

35. I. L. Shapiro, J. Solà, and H. Štefančić, "Running G and Λ at low energies from physics at MX: possible cosmological and astrophysical implications," Journal of Cosmology and Astroparticle Physics, vol. 2005, article 012, 2005.

INDEX